Oceanic Hotspots

Springer-Verlag Berlin Heidelberg GmbH

Roger Hekinian · Peter Stoffers · Jean-Louis Cheminée † (Eds.)

Oceanic Hotspots

Intraplate Submarine Magmatism and Tectonism

With 217 Figures and 34 Tables

Springer

Editors

Dr. Roger Hekinian
Keryunan, 29290 Saint Renan, France
E-mail: hekinian@wanadoo.fr

Dr. Jean-Louis Cheminée †

Prof. Peter Stoffers
Institut für Geowissenschaften,
Universität Kiel
Olshausenstr. 40, 24098 Kiel, Germany
E-mail: pst@gpi.uni-kiel.de

Cover Montage – Background: A three-dimensional map of the Pitcairn hotspot seafloor's volcanic landscape located in the South Pacific near 25°30' S–129°30' W using multibeam data processed by E. Le Drezen and A. Le Bot (IFREMER and GENAVIR). Overlay photographs (courtesy of IFREMER): An active hydrothermal chimney at 1457 m depth on top of the Teahitia Volcano (Society hotspot) and the submersible *Nautile*.

ISBN 978-3-642-62290-8

Library of Congress Cataloging-in-Publication Data data

Oceanic hotspots : intraplate submarine magmatism and tectonism / Roger Hekinian, Peter Stoffers, Jean-Louis Cheminée (eds.).
   p. cm.
   Includes bibliographical references and index.
   ISBN 978-3-642-62290-8   ISBN 978-3-642-18782-7 (eBook)
   DOI 10.1007/978-3-642-18782-7
   1. Magmatism–Pacific Ocean. 2. Submarine geology. 3. Geology–Pacific Ocean. 4. Seamounts–Pacific Ocean. I. Hékinian, R. (Roger), 1935- II. Stoffers, P. (Peter) III. Cheminée, Jean-Louis.
QE461.O28 2004
551.46'814--dc22

Bibliographic information published by Die Deutsche Bibliothek
Die Deutsche Bibliothek lists this publication in the Deutsche Nationalbibliografie;
detailed bibliographic data is available in the Internet at http://dnb.ddb.de

This work is subject to copyright. All rights are reserved, whether the whole or part of the material is concerned, specifically the rights of translation, reprinting, reuse of illustrations, recitation, broadcasting, reproduction on microfilms or in any other way, and storage in data banks. Duplication of this publication or parts thereof is permitted only under the provisions of the German Copyright Law of September 9, 1965, in its current version, and permission for use must always be obtained from Springer-Verlag. Violations are liable for prosecution under the German Copyright Law.

springeronline.com
© Springer-Verlag Berlin Heidelberg 2004
Originally published by Springer-Verlag Berlin Heidelberg New York in 2004
Softcover reprint of the hardcover 1st edition 2004

The use of general descriptive names, registered names, trademarks, etc. in this publication does not imply, even in the absence of a specific statement, that such names are exempt from the relevant protective laws and regulations and therefore free for general use.

Cover Design: Erich Kirchner
Dataconversion: Büro Stasch (*stasch@stasch.com*) · Bayreuth

Printed on acid-free paper – 32/3141 LT – 5 4 3 2 1 0

We were deeply sorry to learn that Jean-Louis Cheminée, one of the co-editors of this volume, died after a long illness on October 15, 2003, before our volume had gone to press. In addition to being a geochemist, volcanologist and an excellent field geologist, Jean-Louis was also a true friend of the editors of this volume and probably of many of you who will read this book.

Jean-Louis devoted his career to the enjoyment of studying subaerial and submarine volcanoes and was one of the main coordinators of the French project dedicated to the study of Intraplate Volcanism. He was a friendly, communicative person, and was always available to talk about volcanoes. We have lost a good friend and colleague with whom we shared and enjoyed numerous sea-going expeditions. We wish to salute his spirit of adventure and dedication to helping the scientific community enhance its knowledge. We will keep him in our thoughts and when a volcano trembles in the future, we will remember hearing him say, "The Earth is still alive!"

This book is dedicated to the spirit of international scientific cooperation and collaboration and to an ideal future world of freedom, justice, tolerance, equality and fraternity, where humankind's ethnic and ideological differences will be respected and considered as a source of intellectual wealth and well-being.

Preface

During the past thirty years, studies on intraplate and ridge-centered volcanism have become an important aspect of planetary geology and have helped to elucidate the mechanisms of mantle convection and crust-lithospheric construction. This work has stimulated and improved multiple lines of scientific research and led to many seagoing expeditions in uncharted oceanic basins where there was very little previous information.

The fourteen papers presented in this volume were contributed by some of the outstanding members of a pioneering oceanographic community whose work has enhanced our understanding of the geophysical, morpho-structural, geochemical, hydrothermal and associated phenomena related to oceanic hotspots.

All the authors and editors are particularly indebted to the support from government agencies in France, Germany, the USA, Canada and England as well as several research institutions in these countries. It is thanks to their funding and logistic support that we have been able to carry out extensive exploration in such remote areas of the world's oceans.

The data and samples obtained at sea are also the result of a valuable collaborative effort among the captains, officers, crewmembers and the scientific teams on board the oceanographic vessels. In addition, the engineers and pilots from "GENAVIR" (Groupement pour la Gestion des Navires Océanographiques) who worked on board the submersibles *Cyana* and *Nautile* were extremely helpful in gathering the in situ observations and samples that were the basis of some of the research presented in this volume. Without all this precious collaboration in obtaining the samples and data used by the authors of this volume, our work would not have been nearly as successful.

The editors especially acknowledge the support received from the three research centers that were mainly responsible for seventeen years of Franco-German collaboration related to the program on Intraplate Volcanism. Dr. R. Hekinian has recently retired from the Institut Français de Recherche et Exploitation de la Mer (IFREMER) in Brest, France, Prof. P. Stoffers is Research Director at the Geoscience Institute of Kiel University in Kiel, Germany, and Dr. J. L. Cheminée † was Directeur de Recherche at the Centre Nationale de Recherche Scientifique (CNRS), Observatoires Volcanologiques, Jussieu, in Paris, France.

The volume "Oceanic Hotspots" has also been made possible thanks to the Alexander von Humboldt Fellowship awarded to one of the editors (Hekinian) and to the many years of work, research and publishing by all the authors and editors who have contributed to this book. In addition, we are thankful to Drs. R. Batiza,

I. Campbell, J. Casey, J.-L. Charlou, C. Deplus, P. England, J. Francheteau, A. Hirn, J. Natland, E. Okal, J. Pearce, and M. Regelous for their comments and criticisms of the various chapters of this volume.

Finally, the three editors sincerely thank Virginia Hekinian for her editorial assistance, and the staff of Springer-Verlag Publishing for their collaboration. R. Hekinian is also grateful to the Department of Marine Geosciences at IFREMER Brest and to the University of Kiel for the generous help and technical support received during the preparation of this volume.

October 2003, Saint Renan, France

Contents

Introduction 1
References 7

1 Sea-Floor Topography and Morphology of the Superswell Region 9
1.1 Introduction 9
1.2 Data Sources and Methods 12
1.3 Sea-floor Morphology in French Polynesia 15
    1.3.1 Bathymetric Expression of the Superswell 15
    1.3.2 Midplate Swells 17
    1.3.3 Plate Boundary Features 18
    1.3.4 Off-Ridge Features 20
1.4 Conclusions 25
    Acknowledgements 26
    References 26

2 Seismicity of the Society and Austral Hotspots in the South Pacific: Seismic Detection, Monitoring and Interpretation of Underwater Volcanism 29
2.1 Introduction 29
2.2 Seismic Waves Used 30
    2.2.1 Seismic Tremors 32
    2.2.2 T Waves 34
2.3 Volcano-Seismic Activities on the Society Hotspot 35
    2.3.1 Generalities and Chronological Events 35
    2.3.2 The Over-All Seismicity of the Society Hotspot 37
    2.3.3 Seismic Detection, Magnitude and Seismic Moment 38
    2.3.4 Overview of the Swarms 39
    2.3.5 Evolution of the Swarms and Nature of the Recorded Events 45
    2.3.6 Frequency-Magnitude Relationship 52
    2.3.7 Seismic and Magmatic Activity in the Society Hotspot Volcanoes 55
2.4 Volcano-Seismic Activity of the Austral Hotspot: Macdonald Seamount 59
    2.4.1 Seismic Swarms 60
    2.4.2 Bathymetric Surveys of the Macdonald Seamount 63
2.5 Summary and Conclusions 65
    2.5.1 Society Hotspot 67
    2.5.2 Austral Hotspot 68
    2.5.3 General Conclusions 69
    Acknowledgements 70
    References 70

X Contents

3 A Global Isostatic Load Model and its Application to Determine the Lithospheric Density Structure of Hotspot Swells ... 73
3.1 Introduction ... 73
3.2 Isostasy of the Lithospheric Plate ... 74
    3.2.1 Lithostatic Load ... 74
    3.2.2 The Generalized Equation of Isostatic Load ... 76
3.3 Reference Model ... 80
    3.3.1 Compensation Depth ... 80
    3.3.2 Lithospheric Density ... 82
    3.3.3 Location of the Reference Column ... 84
3.4 Lithospheric Density Structure of Hotspot Swells ... 95
    3.4.1 Introduction ... 95
    3.4.2 French Polynesia, South Pacific Super Swell ... 96
    3.4.3 Hawaiian-Emperor Island Chain ... 102
    3.4.4 Mascarene-Réunion Hotspot Track ... 109
    3.4.5 Ascension Island ... 112
    3.4.6 The Great Meteor and Josephine Seamounts ... 116
    3.4.7 Iceland ... 120
3.5 Subsidence of Hotspot Structures ... 133
3.6 Conclusions ... 136
    Acknowledgements ... 136
    References ... 137

4 Origin of the 43 Ma Bend Along the Hawaiian-Emperor Seamount Chain . 143
4.1 Introduction ... 143
4.2 The Emperor Seamount Chain Paradox ... 145
    4.2.1 Paleomagnetic Interpretations ... 145
    4.2.2 A Simple Test ... 146
    4.2.3 The E-SMC Paradox and Solution ... 148
4.3 The Origin of the 43 Ma Bend ... 148
    4.3.1 Reasoning Towards a Preferred Model ... 148
    4.3.2 "Trench Jam" at 43 Ma Caused by the Arrival of Hawaiian Plume Head/Oceanic Plateau ... 149
    4.3.3 Evidence Versus Coincidence ... 151
4.4 Summary and Conclusion ... 152
    Acknowledgements ... 153
    References ... 153

5 South Pacific Intraplate Volcanism: Structure, Morphology and Style of Eruption ... 157
5.1 Introduction ... 157
5.2 Society Hotspot ... 158
    5.2.1 Abyssal Hill Region and Limits of Hotspot Volcanism ... 161
    5.2.2 The Sea Floor ("Bulge") Around the Hotspot Edifices ... 163
    5.2.3 The Volcanic Edifices of the Society Hotspot ... 165
5.3 Austral Hotspot ... 175
    5.3.1 The Submarine Edifices of the Austral Hotspot ... 175

Contents

XI

5.4	Pitcairn Hotspot	178
	5.4.1 Volcanic Edifices of the Pitcairn Hotspot	180
	5.4.2 The Distribution and Extent of Hotspot Volcanism	187
5.5	Hotspot Versus Non-Hotspot Volcanoes	190
	5.5.1 Sea-Floor Lineation and Seamount Distribution	191
	5.5.2 Morphological Classification of Intraplate Volcanoes	194
5.6	Style of Eruption and Formation of Hotspot Edifices	197
	5.6.1 Types of Eruption	197
	5.6.2 The Formation of a Volcanic Edifice	198
	5.6.3 Relationship Between Hotspot Volcanic Edifices	200
5.7	Summary and Conclusions	201
	Acknowledgements	203
	References	203

6	**Submarine Landslides in French Polynesia**	209
6.1	Introduction	209
6.2	Geological Setting	210
	6.2.1 Data	212
	6.2.2 Landslide Characterization	213
6.3	Landslides of the Society Islands	214
	6.3.1 Mehetia	214
	6.3.2 Moua Pihaa Seamount	214
	6.3.3 Tahiti	215
	6.3.4 Moorea	217
	6.3.5 Huahine	217
	6.3.6 Raiatea-Tahaa	217
	6.3.7 Bora Bora	219
	6.3.8 Tupai	220
6.4	Austral Island Landslides	222
	6.4.1 Macdonald	222
	6.4.2 Rapa	222
	6.4.3 Raivavae	224
	6.4.4 Tubuai	224
	6.4.5 Arago	227
	6.4.6 Rurutu	229
	6.4.7 Rimatara	231
6.5	Classification of the Society and Austral Landslides	233
	6.5.1 Geometric Characteristics	233
	6.5.2 Seismic Velocity	235
6.6	Evolution of the Mass Wasting with the Age of the Edifices	235
	6.6.1 Landslide Related to Submarine Active Volcanoes	235
	6.6.2 Landslide Related to Young Oceanic Islands (<4 Ma)	236
	6.6.3 Landslide Related to Older Oceanic Islands (>4 Ma)	236
	6.6.4 Landslide Related to Tectonic Events	236
6.7	Conclusion	236
	Acknowledgements	237
	References	237

XII   Contents

7   Mantle Plumes are NOT From Ancient Oceanic Crust 239
7.1   Introduction 239
7.2   Petrological Arguments 240
      7.2.1   Melting of Oceanic Crust Cannot Produce the
              High Magnesian Melts Parental to Many OIB Suites 240
7.3   Geochemical Arguments 240
      7.3.1   Melting of Subduction-Zone Dehydrated Residual Oceanic Crusts
              Cannot Yield the Trace Element Systematics in OIB 240
      7.3.2   OIB Sr-Nd-Hf Isotopes Record no Subduction-Zone
              Dehydration Signatures 242
7.4   Mineral Physics Arguments 246
      7.4.1   Subducted Oceanic Crusts are too Dense to Rise to the Upper Mantle 247
      7.4.2   Basaltic Melts in the Lower Mantle Conditions are Denser
              than Ambient Solid Peridotites 248
7.5   Summary 249
      Acknowledgements 250
      References 250

8   The Sources for Hotspot Volcanism in the South Pacific Ocean 253
8.1   Introduction 253
8.2   The Hotspot Chains of the South East Pacific 254
      8.2.1   Cook-Australs 256
      8.2.2   Society Islands 260
      8.2.3   Pitcairn-Gambier Chain 264
      8.2.4   Marquesas Islands 265
      8.2.5   Juan Fernandez Chain 268
      8.2.6   Foundation Seamounts 272
      8.2.7   Easter/Sala y Gomez-Nazca Chain 273
8.3   Discussion: Petrogenesis of South East Pacific Hotspots 274
      8.3.1   Location of Magma Sources: Plume, Asthenosphere or Lithosphere? 274
      8.3.2   Superswell – How Geochemically Different is It? 275
      Acknowledgements 280
      References 280

9   Plume-Ridge Interactions: New Perspectives 285
9.1   Introduction 285
9.2   Concepts 286
      9.2.1   Mantle Plumes: Deep-Rooted Hot Materials or
              Wet Shallow Mantle Melting Anomalies? 286
      9.2.2   Nature of Plume Materials 286
      9.2.3   Ocean Ridges: Ridge Suction –
              The Active Driving Force for Plume-Ridge Interactions 288
      9.2.4   Ridge Suction Increase with Increasing Spreading Rate 290
      9.2.5   The Effect of Plume-Ridge Distance 292
9.3   Examples 292
      9.3.1   "Proximal" Versus "Distal" Plume-Ridge Interactions 292
      9.3.2   Spreading Rate Directs Plume Flows 294

Contents   XIII

9.4 Summary and Conclusion 301
    Acknowledgements 304
    References 304

10 Intraplate Gabbroic Rock Debris Ejected from the Magma Chamber of the Macdonald Seamount (Austral Hotspot): Comparison with Other Provinces 309

10.1 Introduction 309
10.2 The Macdonald Seamount 312
    10.2.1 Eruptive Activity 312
    10.2.2 Morphology and Structure 313
    10.2.3 Sampling and Observations 314
    10.2.4 Volcanic Terrains 314
10.3 Petrology 315
    10.3.1 Analytical Techniques 315
    10.3.2 Rock Descriptions 320
10.4 Geochemistry 331
10.5 Discussion 336
    10.5.1 Comparison with Gabbros Recovered
           from Mid-Ocean Ridges 336
    10.5.2 Comparison with Gabbroic Ejecta
           from Other Intraplate Regions 338
    10.5.3 Origin of the Macdonald Seamount Gabbroic Clasts 341
10.6 Summary and Conclusions 343
    Acknowledgements 344
    References 344

11 The Foundation Chain: Inferring Hotspot-Plate Interaction from a Weak Seamount Trail 349

11.1 Introduction 349
11.2 Sample Preparation and Analytical Procedure 351
    11.2.1 Sample Selection and Preparation 351
    11.2.2 Dating Technique 351
    11.2.3 Irradiation and Analysis 353
    11.2.4 Data Reduction 355
11.3 Results 363
    11.3.1 Migration of Volcanism Along the Foundation Chain 363
    11.3.2 Hotspot-Spreading Center / Microplate Interaction 363
    11.3.3 Volcanic Elongated Ridges (VERs) 364
11.4 Discussion 367
    11.4.1 VERs and the Pacific-Antarctic Spreading Axis 367
    11.4.2 Foundation VERs and the Selkirk Microplate 368
    11.4.3 Pacific Plate Motion 369
    11.4.4 Implications for Plume-Hotspot Theory 370
11.5 Conclusions 371
    Acknowledgements 372
    References 372

12 Hydrothermal Iron and Manganese Crusts from the Pitcairn Hotspot Region ... 375

12.1 Introduction ... 375
12.2 Geological Setting ... 376
12.3 Sample Description ... 378
    12.3.1 Mineralogy ... 379
    12.3.2 Age Dating ... 381
    12.3.3 Biomineralization ... 382
12.4 Chemical Composition ... 385
    12.4.1 Fe Crusts ... 388
    12.4.2 Mn Crusts ... 390
    12.4.3 Rare Earth Elements (REE) ... 393
12.5 Formation of Fe and Mn Crusts ... 395
12.6 Conclusions ... 398
    Acknowledgements ... 399
    References ... 399
    Appendix ... 401

13 Methane Venting into the Water Column Above the Pitcairn and the Society-Austral Seamounts, South Pacific ... 407

13.1 Introduction ... 407
13.2 Geological Setting ... 409
13.3 Methods ... 410
13.4 Results and Discussion ... 411
    13.4.1 Water Column Characteristics and Methane Distribution ... 411
    13.4.2 Origin of Hydrothermal Methane ... 423
13.5 Conclusions ... 425
    Acknowledgements ... 426
    References ... 426

14 Petrology of Young Submarine Hotspot Lava: Composition and Classification ... 431

14.1 Introduction ... 431
14.2 Composition and Description of Oceanic Rocks ... 432
    14.2.1 Common Mineral Constituents ... 432
    14.2.2 Rock Types ... 436
14.3 Relationship Between Intraplate-Hotspot and Spreading-Ridge Magmatism ... 449
14.4 Compositional Differences Among Hotspots ... 450
    14.4.1 Relationship between Large and Small Hotspot Edifices ... 451
    14.4.2 Volcanic Stratigraphy ... 454
14.5 Summary and Conclusions ... 455
    Acknowledgements ... 457
    References ... 457

Index ... 461

Contributors

Avedik, Felix

Lesconvel
29280 Locmaria Plouzané, France

Bideau, Daniel

IFREMER
Centre Océanologique de Bretagne
Département Géosciences Marines
29280 Plouzané, France
E-mail: Daniel.Bideau@ifremer.fr

Binard, Nicolas

Christian-Rübsamen Strasse 9
35578 Wetzlar, Germany
E-mail: nicolas.binard@fresenius.de

Blanz, Thomas

Baltic Sea Research Institute Warmemünde (IOW)
Seestrasse 15
18119 Rostock, Germany
E-mail: thomas.blanz@io-warnemuende.de

Bonneville, Alain

Laboratoire de Géosciences Marines
Institut de Physique du Globe de Paris
4, Place Jussieu
75252 Paris Cedex 05, France
E-mail: bonnevil@ipgp.jussieu.fr

Botz, Reiner

Institut für Geowissenschaften, Universität Kiel
Olshausenstr. 40–60
24118 Kiel, Germany
E-mail: rb@gpi.uni-kiel.de

Caress, David

Monterey Bay Aquarium
Research Institute
Moss Landing
CA 95039, USA
E-mail: caress@mbari.org

Cheminée, Jean Louis †

Clouard, Valérie

Departamento de Geofísica
University de Chile
Blanco Encalada 2085, Casilla 777
Santiago, Chile
E-mail: valerie@dgf.uchile.cl

Devey, Colin W.

Fachbereich 5 – Geowissenschaften
Universität Bremen
Klagenfurter Starße
28359 Bremen, Germany
E-mail: cwdevey@uni-bremen.de

Fietzke, J.

Institute für Geowissenschaften
Universität Kiel
Olshausenstr. 40–60
24118 Kiel, Germany
E-mail: jf@gpi.uni-kiel.de

Garbe-Schönberg, Dieter

Institut für Geowissenschaften
Universität Kiel
Olshausenstr. 40–60
24118 Kiel, Germany
E-mail: dgs@gpi.uni-kiel.de

Haase, Karsten M.

Institut für Geowissenschaften
Universität Kiel
Olshausenstr. 40–60
24118 Kiel, Germany
E-mail: kh@gpi.uni-kiel.de

Hekinian, Roger

Keryunan
29290 Saint Renan, France
E-mail: hekinian@wanadoo.fr

Jegen, Marion D.

University of Brest, IUEM
Place Nicolas Copernic
29280 Plouzané, France
E-mail: jegen@univ-brest.fr

Jordahl, Kelsey

Monterey Bay Aquarium, Research Institute
Moss Landing
CA 95039, USA
E-mail: kels@mbari.org

Kennedy, C. B.

Department of Geology
University of Toronto
22 Russel Street
Toronto, Ontario, M5S 3B1, Canada

Klingelhöfer, Frauke

IFREMER, Centre de Brest
BP 70
29280 Plouzané, France
E-mail: frauke.klingelhoefer@ifremer.fr

Matias, Luis M.

CGUL
University of Lisbon
Campo Grande
1749-016 Lisbon, Portugal

McNutt, Marcia

Monterey Bay Aquarium, Research Institute
Moss Landing
CA 95039, USA
E-mail: mcnutt@mbari.org

Niu, Yaoling

Previously:
Department of Earth Sciences
Cardiff University
Cardiff CF10 3YE, UK
Presently:
Department of Geosciences
University of Houston
Houston, TX 77204, USA
E-mail: Yaoling.Niu@Mail.uh.edu

O'Connor, John

Department of Isotope Geochemistry
Faculty of Earth Sciences
Vrije Universiteit
De Boelelaan 1085
1081 HV Amsterdam, The Netherlands
E-mail: john.o.connor@falw.vu.nl

O'Hara, Mike J.

Department of Earth Sciences
Cardiff University
PO Box 914
Cardiff, CF10 3YE, UK

Schmidt, Mark

Institut für Geowissenschaften
Universität Kiel
Olshausenstr. 40–60
24118 Kiel, Germany

Schmitt, Manfred

Geochemische Analysen
Glückaufstr. 50
31319 Sehnde-Ilten, Germany

Scholten, Jan C.

Institut für Geowissenschaften
Universität Kiel
Olshausenstr. 40–60
24118 Kiel, Germany
E-mail: js@gpi.uni-kiel.de

Scott, Steven D.

Department of Geology, University of Toronto
22 Russel Street
Toronto, Ontario, M5S 3B1, Canada
E-mail: chair@geology.utoronto.ca

Stoffers, Peter

Institut für Geowissenschaften
Universität Kiel
Olshausenstr. 40–60
24098 Kiel, Germany
E-mail: pst@gpi.uni-kiel.de

Talandier, Jacques

Le Bourg
24240 Saussignac, France
E-mail: jtalandier@compuserve.com

Thießen, Olaf

Institut für Geowissenschaften
Universität Kiel
Olshausenstr. 40–60
24118 Kiel, Germany
E-mail: ot@gpi.uni-kiel.de

Wijbrans, J. R.

Department of Isotope Geochemistry
Vrije Universiteit
1081 HV Amsterdam, The Netherlands
E-mail: jan.wijbrans@falw.vu.nl

Introduction

Intraplate volcanism occurs in both submarine and subaerial regions on the Earth's surface. However, most of the magmatic activity responsible for the construction of the Earth's volcanic structures takes place under the sea in oceanic basins and along spreading centers. The divergent plate boundaries (Mid-Ocean Ridges) and back-arcs are relatively straight and structurally continuous features, making them easier to investigate than the more scattered and dispersed intraplate volcanic provinces within the oceanic basins.

The intraplate regions associated with hotspots provide a unique opportunity to investigate the exchange of matter between the Earth's mantle and its lithosphere. For a global understanding of the origin and evolution of the Earth's mantle, a multidisciplinary approach is often necessary. It makes sense, therefore, to combine geophysical, structural, mineralogical and geochemical studies to better understand hydrothermalism, crustal alteration, and magmatic gas and fluid discharges into seawater.

Our first awareness of hotspot activity was based on the observation that linear chains of volcanic islands and seamounts were younger than the oceanic crust on which they were built. These chains tended to be parallel while being oriented perpendicular to the magnetic anomaly pattern of the oceanic crust. Furthermore, the volcanic chains were marked by a progressive increase in the age of their volcanoes with increased distance from the nearest mid-oceanic spreading centers.

It has been about forty years since Tuzo Wilson (1963) first proposed his model linking oceanic island volcanism to deep-seated magma convection cells. Nine years later, Jason Morgan (1972) suggested that linear volcanic chains could be explained by the movement of lithospheric plates over fixed hotspots supplied by a magma source in the mantle (mantle plume).

The world's hotspots play an important role in enhancing the global magmatic budget of the planet and are also a means of indicating the variability in rigidity of lithospheric plates. The number of hotspots on the surface of the Earth is difficult to estimate, with opinions ranging from about 40 to 120 (Crough 1978; Crough and Jurdy 1980; Stefanick and Jurdy 1984) (Fig. 0.1). When looking at the satellite altimetric data of Smith and Sandwell (1997), it is evident that oceanic basins are the locations of several irregularly shaped structures – including circular edifices and ridges – all of which seem to have a volcanic origin. In addition, it appears that the majority of these structures were probably built by intraplate hotspot volcanism rather than by volcanism at spreading centers.

While marine exploration and discovery will continue to produce new information, the present volume entitled "*Oceanic Hotspots*" is intended to summarize our progress

Introduction

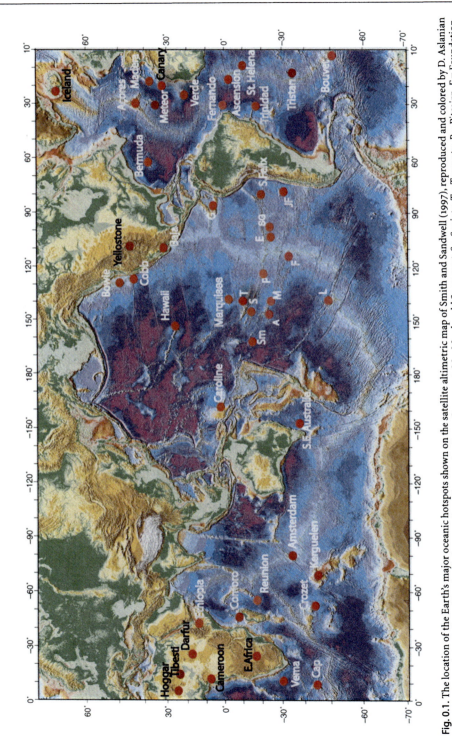

Fig. 0.1. The location of the Earth's major oceanic hotspots shown on the satellite altimetric map of Smith and Sandwell (1997), reproduced and colored by D. Aslanian (IFREMER, Brest). *L* = Inferred Louisville hotspot, *Sm* = Samoa, *A* = Arago Seamount, *M* = Macdonald Seamount, *S* = Society, *T* = Tuamotu, *P* = Pitcairn, *F* = Foundation, *E* = Easter Island, *SG* = Sala y Gómez Island, and *JF* = Juan Fernandez Island. Both spellings for the *Marquises* or *Marquesas* hotspot are found in this volume

concerning intraplate volcanic activity. With the constant growth of data acquisition, written reports and scientific publications, it has become increasingly difficult to keep up with the available information on oceanic basins and/or intraplate hotspot research. The present volume does not intend to cover all aspects of studies pertaining to the theory of hotspot volcanism. Rather, we offer a comprehensive overview and indicate on-going research concerning hotspot-generated volcanism in selected Pacific and other intraplate regions.

An international project on "intraplate volcanism" (VIP – "Volcanism Intra-Plaque") began in 1986 between French and German oceanographic institutes. This project was subsequently extended to include U.S. and Canadian colleagues, who participated in seagoing operations and shore-based studies. The project, designed to share oceanographic research, ship time and equipment, was very fruitful and led to close collaboration among the various partners. The program lasted seventeen years and resulted in more than forty publications in international journals arising from more than fifteen oceangoing expeditions using French, German, British (with side scan "GLORIA" and bathymetry coverage) and American surface ships. The French submersibles *Nautile* and *Cyana* were also used during three cruises to gather geological and geophysical information.

While the present volume is intended to be an overview, particular emphasis has been placed on data obtained between 1986 and 2000. Nonetheless, it is also our intention to broaden the scope to include other major contributions related to the Pacific hotspots. In fact, the book aims to be fairly comprehensive, since some of the authors have used a global approach when discussing such things as the gravity field implications during volcanic construction and the interaction of intraplate plumes with the divergent plates in the Atlantic and the Pacific Oceans. However, it should also be kept in mind that the Pacific Ocean is one of the world's most active volcanic regions and the location of the fastest crustal renewal during the past 170 million years.

Most available data concerning hotspots are either dispersed in the literature or unpublished for various reasons. Thus, this project has provided us with an ideal opportunity to bring together a series of fourteen comprehensive articles highlighting some major discoveries, new data and scientific thinking related to mantle plume processes and Pacific hotspot activity. Please note that throughout the volume, the terms "volcano" and "seamount" are used as synonyms and refer to geological structures formed during volcanic activity on the sea floor.

The use of bathymetry based on satellite altimetric data and research vessels' depth soundings is one of the first steps in recognizing and preparing an inventory of potential hotspot volcanism. A comprehensive approach to the study of intraplate regions by *Jordahl et al.* (Chap. 1) includes a compilation of multibeam bathymetric data collected from sixty-five oceanographic expeditions equipped with multibeam echo sounders and GPS (Global Positioning System) navigation. The maps also include all single-beam track-line data from the American "National Geophysical Data Center" (NGDC), and soundings from the French "Service Hydrographique et Océanographique de la Marine" (SHOM). These ship data were combined with multibeam data collected by other oceanographic institutions using a variety of sonar systems. In addition, predicted bathymetry from satellite altimetry was used to interpolate between the ship soundings. These maps have improved confidence in the interpretation of large-scale features such as regional depth anomalies, fracture zones, swells and chains of volcanoes, and the maps also provide new insights into the interpretation of small-scale structures. The imagery reveals that long, low, sinuous volcanic ridges are prevalent throughout the Polynesian region of the South Pacific. Their origins, which are related to regional stress in the Pacific, are also discussed.

The geophysical aspect of this volume includes detecting South Pacific earthquakes thanks to analyses based on the data collected by the Polynesian Seismic Network since the 1960s and presented in a paper by *J. Talandier* (Chap. 2). Special attention is given to the detection and interpretation of volcano-magmatic activities of the South Pacific in the regions of the Society hotspot near Tahiti and the Macdonald Seamount region in the prolongation of the Austral chain. The implication of this seismic activity in changing the morpho-structural aspect of any subsequent volcanic construction is discussed.

Thermal anomalies within the asthenosphere and the mantle plumes rising to the Earth's surface have a major impact on the lithosphere's density and consequently on its internal state of isostatic equilibrium. An "isostatic load model" discussed in the first part of the study presented by *Avedik et al.* (Chap. 3) offers a generalized description of the isostatic equilibrium in the framework of plate tectonics. Their model shows a continuous transition between the constant – and variable sub-crustal lithospheric density domain, corresponding to the Airy and Pratt compensation mechanisms in the lithospheric plate. In the second part of the study, the authors apply their "isostatic load model" to investigate the distribution of sub-crustal lithospheric density in areas of different oceanic structures associated with hotspot activity in intraplate regions and at spreading axes.

The recognition of intraplate spreading directions and the velocity of plate motion is fundamental knowledge for determining plate reconstruction and the origin of intraplate, arc and back-arc provinces. This has been possible through a detailed study of the Hawaiian-Emperor chain, which is one of the most prominent, continuous and linear volcanic structures known in the ocean. The present-day hotspot that gave rise to the Hawaiian-Emperor volcanic chain is centered underneath the island of Hawaii and its immediate vicinity and is one of the most active in the world. Its continuity is marked by a N330° orientation that changes to a more northerly direction near the middle of the chain. The origin of the 43 Myr "bend" (or direction change) along the Hawaiian-Emperor chain is discussed by *Y. Niu* (Chap. 4). His paper presents a new, alternative interpretation for explaining the origin of the 43 Ma bend that is consistent with simple physics as well as with many on-site observations by the world's oceanographic community. He suggests that the 43 Ma bend was caused by a reorientation of Pacific Plate motion at that time. This new interpretation is supported by the cessation of continental arc volcanism in the Okhotsk-Chukotka area as a result of a jamming of the trench and the ending of subduction about forty-four million years ago.

The morpho-structural study of major recent South Pacific (Society, Austral and Pitcairn hotspots) volcanic structures and their mode of construction are presented by *Binard et al.* (Chap. 5). The chapter summarizes the distribution of various types of volcanic activities such as explosive versus quiet volcanism and the sea-floor structures that have influenced the location of hotspot magmatism.

The construction of volcanic edifices is followed by their degradation during later eruptions and plate motion. This is true for submarine as well as subaerial volcanoes that constitute islands. Therefore, another aspect of a morpho-structural approach is presented by *Clouard and Bonneville* (Chap. 6) in their study of active intraplate volcanoes. These authors have shown that the edifices are subject to slope failures of fragmented material due to gravitational instabilities of the flank. Later, during their mature stage, these seamounts and volcanoes form well-developed rift zones. These structures are again modified during the subaerial volcanic stage when cataclysmic events such as debris avalanches occur at least once on most of the islands. Finally, after the cessation of volcanism, landslides produced by erosion processes are likely to take place. By determining the thickness of the debris avalanche after a landslide and combining this information with data on the seismicity of the area, *Clouard and Bonneville* have revealed that seismic velocity can be as high as 3 000 m s^{-1} during a landslide.

Magmatic processes and mantle compositions are of primary importance for understanding the evolution of mantle flow and composition, at least during the past 120–150 Ma. The nature and composition of lava erupted from a volcano varies from one hotspot to another as well as within the same hotspot. In order to decipher the origin of a magma source derived from the partial melting of a heterogeneous mantle and therefore to better understand the changes that the magma has undergone during its upwelling towards the surface, the geochemistry of a large variety of rock types has been investigated by *Niu and O'Hara* (Chap. 7). In their paper, these authors invoke the petrologic and genetic aspects of intraplate magnetism and have presented a new view on the concept of source material for the hotspot volcanoes. They show evidence that there is no genetic link between the ancient subducted oceanic crust and the source materials of OIB (Ocean-Island Basalt). The arguments presented are based on well-understood petrological, geochemical and experimental data on mineral physics.

A summary and new results based on geochemical studies concerning the nature and origin of the magma sources for the Marquises, Society, Austral, Pitcairn, Easter, Juan Fernandez and Foundation hotspots in the South Pacific are presented in a chapter by *Devey and Haase* (Chap. 8). All the hotspots they have studied show a consistent evolutionary feature, since all the volcanoes have erupted a more alkaline magma over a period of time. In most cases, except for the Marquises hotspot, the most alkaline magmas were derived from sources with the strongest time-integrated incompatible element depletion. A comparison of these South Pacific hotspots with other hotspot regions from all over the world reveals extremely high radiogenic values, perhaps indicating an exceptionally old, subduction debris-rich mantle beneath the Superswell (or "bulge") of the South Pacific oceanic basin.

The interaction between the mechanism giving rise to hotspots and the convection of the mantle related to plate motions is not well understood. In fact, we still are not sure if there is a relationship between upwelling mantle material at diverging plate boundaries and the plume-generated hotspots. *Niu and Hekinian* (Chap. 9) present, in the form of schematic models, an overview with a new perspective on the geochemical and geological consequences of plume-ridge interactions. These models aim to provide a better understanding of the expression and intensity of plume-ridge interactions. The models described include (1) what mantle plumes are; (2) the nature and composition of plume sources; (3) the actual role of ocean ridges; (4) the effect of the rate of the plate separation; and (5) the plume-ridge distance. Examples of South and North Pacific plume-ridge interaction across intraplate regions and along the Icelandic hotspot and Mid-Atlantic Ridge axis are discussed.

Up until now, how hotspot activity affects the composition of the lithosphere has merely been speculated, and has only rarely been demonstrated. With the exception of the Hawaiian xenoliths that carry information about the deep-seated composition of the lower crust and upper mantle, very little is known about the rest of the world's oceanic intraplate regions. *Bideau and Hekinian* (Chap. 10) report the occurrence of gabbroic cumulates and isotropic gabbros collected after the explosion of a crater at the summit of Macdonald Volcano located on the Austral hotspot. These accidental fragments are able to give an indication about the composition and degree of alteration of the magma reservoir.

A reliable understanding of the history and distribution of oceanic intraplate volcanism created as tectonic plates drift over hotspots is a key for understanding the processes controlling hotspot-lithosphere interaction. High-precision dating of rock samples is therefore of first-order importance when investigating the physical and geochemical development of hotspot volcanism. In an overview of ^{40}Ar/^{39}Ar dating of the Foundation Chain in the SE Pacific, *O'Connor et al.* (Chap. 11) show how a detailed understanding of the history and distribution of magmatism in a seamount chain and surrounding volcanic lineaments can distinguish long-lived plume-hotspot behavior from local lithospheric control.

It is well known that hydrothermalism is a dominant alteration process during intraplate volcanism and that hydrothermal activity is often observed before and following volcanic events. Hydrothermal fluids charged with elements leached from crustal rocks circulate and precipitate dissolved minerals within the lithosphere and on the sea floor in the upper part of intraplate seamounts where seawater penetrates and mixes with upwelling hydrothermal fluids. This type of oxygenated environment is most suitable for giving rise to Fe-, Si- and Mn-oxyhydroxide precipitates rather than sulfides. However, *Scholten et al.* (Chap. 12) have also suggested that massive flows inhibit a deep penetration of seawater so it does not become hot enough to extract metallic elements such as Cu and Zn but only Fe. They present geochemical and isotopic investigations of the Fe and Mn crusts from several Pitcairn seamounts. They found that for the Mn and Fe crusts, the ages range from 2.9 to 130 ka for the Mn crusts and from 0.29 to 3.1 ka for the Fe crusts, which indicates their very rapid formation. In addition, it is known that low-temperature iron oxyhydroxide (ferrihydrite) deposits from low-temperature hydrothermal vents are commonly accompanied by mineralized bacterial forms. Although rare in the Pitcairn region, the types and distribution of iron oxidizing bacteria give indications of the oxidized nature of the vent fluids preventing the formation of sulfides. Dating, in combination with the geochemical data, indicates that the formation of Fe and Mn crusts at the Pitcairn seamounts is not the result of a chemical fractionation of a single hydrothermal fluid. It rather seems that the Mn crusts were precipitated during past periods of relatively intense hydrothermal activity, whereas the Fe crusts were formed during more recent periods of lower activity.

The release of volatiles into seawater is also important in facilitating the development of biological communities, as well as for maintaining their concentration at the seawater-atmosphere interface. *Thießen et al.* (Chap. 13) have contributed a paper dealing with the CH_4 concentration around several intraplate volcanoes of the Society, Pitcairn and Austral hotspots. These authors show that the undersea CH_4 concentration exceeds the surface water equilibrium value of 48 nanoliters per liter of seawater by several orders of magnitude. By studying carbon-isotopic composition, the authors have been able to differentiate between a purely microbially produced CH_4 and that released during hydrothermal venting. This result has important implications for understanding the origin of some hydrothermal deposits (e.g., Fe-Mn-oxyhydroxide deposits) and the development of bacterial communities during the construction of submarine volcanoes.

To conclude the book, *Hekinian* presents a petrological approach for understanding the mineralogical and compositional variation between hotspot and East Pacific spreading-ridge volcanism. The various types of rocks are classified according to their modal and normative mineral constituents in relation to their geochemistry and lava morphology. The cyclic and sequential nature of magmatic events is identified due to *in situ* observations of the volcanic stratigraphy, which gives credence to certain hypotheses concerning the construction of undersea volcanoes.

The aim of the volume is to bring the reader up to date, and to provide an historical perspective concerning studies about intraplate volcanic activity. The authors and editors sincerely hope that these fourteen chapters will generate further interest and encourage future research on hotspots and their related phenomena in both oceanic and continental regions.

References

Crough ST (1978) Thermal origin of mid-plate, hot-spot swells. Geophys J Roy Astr S 55:451–469

Crough ST, Jurdy DM (1980) Subducted lithosphere, hotspots, and the geoid. Earth Planet Sc Lett 48:15–22

Morgan WJ (1972) Plate motion and deep mantle convection. Geolo Soc Am Mem 132:7–22

Stefanick M, Jurdy DM (1984) The distribution of hot spots. J Geophys Res 89:9919–9925

Smith WHF, Sandwell DT (1997) Global sea floor topography from satellite altimetry and ship depth soundings. Science 277:1956–1962

Wilson JT (1963) A possible origin of the Hawaiian Islands. Can J Phys 41:863–870

Chapter 1

Sea-Floor Topography and Morphology of the Superswell Region

K. Jordahl · D. Caress · M. McNutt · A. Bonneville

1.1 Introduction

The islands of French Polynesia were discovered and populated by Polynesians between 500 B.C. and A.D. 500. European exploration of the region began in the seventeenth century. The main island groups (Society, Marquesas, Tuamotu, and Austral) have long been known, but the submarine topography has only been explored in the past fifty years using conventional echo sounders, and only in the past twenty years using higher-resolution multibeam sonar systems.

Better knowledge of sea-floor bathymetry in French Polynesia has been a priority for a number of reasons, both practical and scientific. To begin with, French Polynesia encompasses the Pacific's second largest Exclusive Economic Zone (EEZ), and certainly the world's largest EEZ per capita. Improved maps of the sub-sea topography are the first step to proper inventory of economic resources, including prime fishing grounds and submarine minerals, and accurate assessment of geologic hazards, including volcanic eruptions, tsunamis, and landslides (Auzende et al. 1997; ZEPOLYF 1996a, 1996b). In addition, the nature of the sea floor within French Polynesia has been of interest for a number of scientific reasons. For example, the area is unusually shallow for its age and contains an exceptional number of islands and seamounts that do not neatly fit into hotspot theory on account of anomalous trends or age progression (McNutt 1998; McNutt and Fischer 1987; McNutt and Judge 1990). A number of prominent tectonic lineations cross the region, providing information on relative plate motions and spreading center reorganizations (Cande and Haxby 1991; Jordahl et al. 1998).

Since the early 1990s, on average there have been three expeditions each year that have mapped the sea floor of French Polynesia using multibeam sonar systems (Fig. 1.1). The availability of these new data provides an excellent opportunity to create a new, up-to-date database of depth information for the sea floor in the central Pacific that can support a number of scientific studies and other practical uses.

Fig. 1.1. Histogram showing number of multibeam expeditions per year to visit French Polynesia. Not all of these data sets were available for this analysis

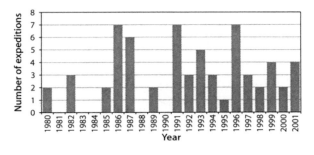

K. Jordahl · D. Caress · M. McNutt · A. Bonneville

Table 1.1. Institutions, ships, and multibeam systems

Institution	Ship	Multibeam system
France – Institut Français de Recherche pour l'Exploitation de la Mer (IFREMER)	N/O Jean Charcot	Seabeam
	N/O L'Atalante	Simrad EM12D
Scripps Institution of Oceanography (SIO)	R/V Thomas Washington	SeaBeam
	R/V Melville	SeaBeam 2000
Lamont-Doherty Earth Observatory (LDEO)	R/V Conrad	SeaBeam
	R/V Maurice Ewing	Hydrosweep DS (through 2001)
USA – National Oceanic and Atmospheric Administration (NOAA)	R/V Surveyor	SeaBeam
	R/V Discoverer	SeaBeam
Germany	F/S Sonne	SeaBeam
Japan Marine Science and Technology Center (JAMSTEC)	Kairai	SeaBeam 2112

Table 1.2. Expeditions used in the multibeam bathymetry compilation

Cruise	Institution	Year	Ship	Scientist
ETM19	IFREMER	1986	Charcot	Voisset
NIXO46	IFREMER	1986	Charcot	Voisset
NODCO1-1	IFREMER	1986	Charcot	Le Suavé
NODCO1-2	IFREMER	1986	Charcot	Voisset
NODCO2	IFREMER	1987	Charcot	Le Suavé
PAPNOUM	IFREMER	1987	Charcot	Foucher
RAPANUI2	IFREMER	1987	Charcot	Francheteau
SEAPSO5	IFREMER	1986	Charcot	Pontoise
SEARISE1	IFREMER	1980	Charcot	Francheteau
SEARISE2	IFREMER	1980	Charcot	Francheteau
TEAHITIA	IFREMER	1986	Charcot	Cheminée
FOUNDA	IFREMER	1997	L'Atalante	Maia
MANZPA	IFREMER	1999	L'Atalante	Dubois
NOUPA	IFREMER	1996	L'Atalante	Reyss
OLIPAC	IFREMER	1994	L'Atalante	Coste
PAPNOU99	IFREMER	1999	L'Atalante	Pelletier
POLYNAUT	IFREMER	1999	L'Atalante	Dubois
ZEPOLYF1	IFREMER	1996	L'Atalante	Bonneville
ZEPOLYF2	IFREMER	1999	L'Atalante	Bonneville
KR9912	JAMESTEC	1999	Kairei	–
ARIA01WT	SIO	1982	Washington	Shipley
ARIA02WT	SIO	1982	Washington	Lonsdale
CNXO01WT	SIO	1982	Washington	Robert (IFREMER)
PPTU03WT	SIO	1985	Washington	Mammerickx
CRGN01WT	SIO	1987	Washington	Winterer

Chapter 1 · Sea-Floor Topography and Morphology of the Superswell Region   11

Table 1.2. *Continued*

Cruise	Institution	Year	Ship	Scientist
CRGN02WT	SIO	1987	Washington	Natland/McNutt
CRGN03WT	SIO	1987	Washington	Cronan
RNDB16WT	SIO	1989	Washington	Guenther (transit)
TUNE01WT	SIO	1991	Washington	Tsuchiya
TUNE02WT	SIO	1991	Washington	Swift
TUNE03WT	SIO	1991	Washington	Talley
BMRG01MV	SIO	1995	Melville	Orcutt
BMRG02MV	SIO	1995	Melville	Lonsdale, Hawkins, Castillo
BMRG09MV	SIO	1995–1996	Melville	Lonsdale
WEST01MV	SIO	1993	Melville	Urabe (Geol. Surv. Japan)
WEST02MV	SIO	1993–1994	Melville	Lonsdale
WEST05MV	SIO	1994	Melville	Bryden
WEST12MV	SIO	1995	Melville	Moe
WEST13MV	SIO	1995	Melville	Coale
GLOR03MV	SIO	1992–1993	Melville	Forsyth
GLOR04MV	SIO	1993	Melville	Sandwell
GLOR05MV	SIO	1993	Melville	Sandwell
SOJN01MV	SIO	1996	Melville	*Transit*
SOJN02MV	SIO	1996	Melville	Chave
SOJN08MV	SIO	1997	Melville	*Transit*
SOJN09MV	SIO	1997	Melville	Chave
PANR05MV	SIO	1998	Melville	Hey
PANR06MV	SIO	1998	Melville	Gee
C2608	LDEO	1985	Conrad	Weissel
EW9102	LDEO	1991	Ewing	Detrick/Mutter
EW9103	LDEO	1991	Ewing	McNutt/Mutter
EW9104	LDEO	1991	Ewing	Larson
EW9106	LDEO	1991	Ewing	McNutt
EW9204	LDEO	1992	Ewing	McNutt
EW9205	LDEO	1992	Ewing	*Transit*
EW9602	LDEO	1996	Ewing	McNutt
EW9603	LDEO	1996	Ewing	*Transit*
EW0003	LDEO	2000	Ewing	*Transit*
RITS93B	NOAA	1993	Surveyor	Capt. F. J. Jones
RITS93C	NOAA	1993	Surveyor	Capt. F. J. Jones
RITS94A	NOAA	1994	Surveyor	Capt. Thomas Ruszala
DI9301	NOAA	1993	Discoverer	Capt. Robert Smart
TOGA92	NOAA	1992	Discoverer	Capt. Robert Smart
SO-47	Germany	1986	Sonne	Stoffers
SO-100	Germany	1995	Sonne	Devey
KR9912	–	–	–	–

1.2
Data Sources and Methods

We have assembled multibeam bathymetry from sixty-five oceanographic expeditions undertaken by research vessels from France, the United States, Germany, and Japan (Table 1.1). Table 1.2 lists the expeditions included in this compilation. Incorporation of these swath bathymetry data into a single map represents a significant challenge, in that they were collected by about half a dozen different sonar systems from various vendors and were archived in approximately two dozen different digital data formats. MB-System (Caress and Chayes 1996, 2003), an open source software package for the processing and display of swath mapping sonar data, is well suited for this sort of data compilation because of its modular and extensible i/o library. Most of the relevant formats were already supported, but in order to work with the French multibeam data it was necessary to create i/o modules for the raw Simrad EM12D format and the generic archival format (MBB) used by IFREMER (l'Institut Français de Recherche pour l'Exploitation de la Mer). Most of

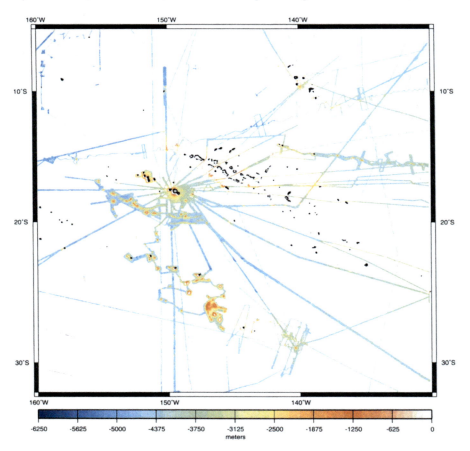

Fig. 1.2. New depth compilation for French Polynesia. Gaps between soundings have been left blank

the compiled multibeam bathymetry required little additional processing, but in a few cases interactive bathymetry and navigation editing removed significant artifacts.

Once the multibeam bathymetry database was in place, the data were used to create bathymetric grids using the MB-System. The full regional grids were produced at 1 km × 1 km resolution. In addition to the multibeam data, NGDC trackline data were included in gridding, as well as island topography from GTOPO30 and shoreline information from GSHHS (Wessel and Smith 1996). Soundings from the French Service Hydrographique et Océanographique de la Marine (SHOM) were also included, where they have been made available to the ZEPOLYF (Bonneville et al. 1995; Sichoix and Bonneville 1996) program (south of 10° S latitude).

Three separate small-scale grids were generated. The first contains data values only for grid nodes that are constrained by sonar bathymetry (Fig. 1.2) and provides an excellent representation of the true sonar coverage given the very different swath widths of the various multibeam systems. The other two grids used different strategies to fill the remaining grid nodes. Figure 1.3 was based entirely on thin-plate spline interpola-

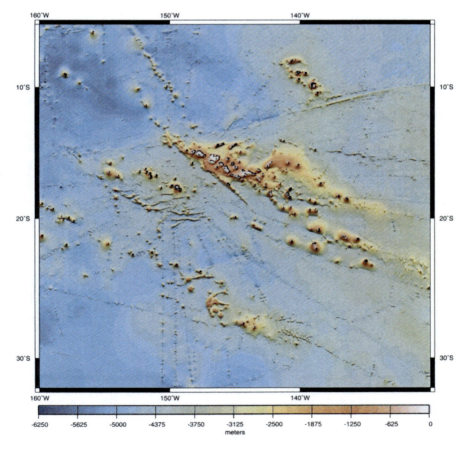

Fig. 1.3. New depth compilation for French Polynesia. Gaps between soundings were filled with spline interpolation

tion (Smith and Wessel 1990), while the final grid (Fig. 1.4) used predicted bathymetry from satellite altimetry (Smith and Sandwell 1994), incorporating the bathymetry from this database for both the long and the short wavelengths. The spline grid has the advantage of not including any data derived from gravity data, thus avoiding the problems of circular arguments in using the bathymetry grid for scientific studies that use both bathymetry and gravity as independent constraints to derive physical properties of the crust and mantle. Studies of lithospheric flexure, lateral variations in sediment thickness, sub-crustal intrusions, and subsurface loading would all be examples of studies that should not use bathymetry predicted from gravity. The spline grid is best in areas where ship track coverage is relatively dense, such as in the Society Islands. In other regions where gaps between the shiptracks can be on the order of 100 km (e.g., south of the Austral Islands), spline artifacts will degrade the utility of this grid for quantitative studies. In such areas, the predicted bathymetry grid (Fig. 1.4) provides the most accurate qualitative view of submarine features.

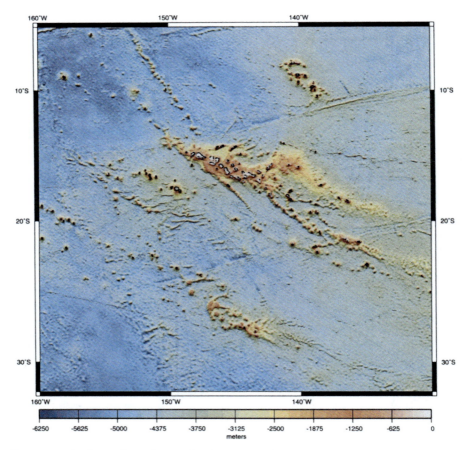

Fig. 1.4. New depth compilation for French Polynesia. Gaps between soundings were filled with depth predicted from satellite altimetry. In order to get a smooth blending between the two different data sets, their difference has been gridded (i.e., low-pass filtered) and then added to the predicted depth grid

There are a few artifacts seen in the grid that are emphasized by the illumination scheme used in Fig. 1.4. Most obvious is the pervasive "ringing" in the predicted bathymetry, where it is used to interpolate between ship soundings. This noise, with an average amplitude of ~100–200 m and spectral peak at ~50 km, is a side-lobe effect in the filtering scheme used to convert satellite altimetry to predicted bathymetry (Smith and Sandwell 1994). Other filtering schemes to suppress this noise would negatively impact the ability of the predicted bathymetry to resolve the positions and amplitudes of uncharted seamounts, and therefore this noise is tolerated.

A second type of artifact is a noticeable edge at the boundary between swaths of multibeam bathymetry and the base map. This effect is caused by local deviations between the multibeam depth and the regional average depth determined in the prediction scheme. The effect is enhanced by the differing resolutions inside versus outside the swaths.

In addition to viewing the sea-floor morphology of all of French Polynesia, this multibeam database allows us to grid the data on more local scales for a variety of purposes. For example, for small-scale morphological studies of seamounts and tectonic fabric, custom grids can be efficiently created at fine scale, because the database automatically reads only those files with soundings within the specified area of interest. The grids presented here are available via anonymous ftp from the server at ftp://ftp.mbari.org/pub/Polynesia.

1.3
Sea-floor Morphology in French Polynesia

Sea-floor depths in French Polynesia provide an excellent view of the morphology of the basaltic basement rocks on account of the low level of sedimentation, except locally near islands and shallow banks. For this reason, the bathymetry of French Polynesia has inspired a number of interpretations of sea-floor depth, ranging from regional studies of depth anomalies to fine-scale investigations of sea-floor fabric.

1.3.1
Bathymetric Expression of the Superswell

McNutt and Fischer (McNutt and Fischer 1987) first used the term "Superswell" to refer to the area in the south-central Pacific, largely coincident with French Polynesia, where the sea floor is several hundred meters to more than a kilometer shallower than typical for sea floor of the same geologic age elsewhere in the world's oceans. They noted that this area is also underlain by slow seismic velocities in the uppermost mantle (Nishimura and Forsyth 1985), has experienced an unusual volume of off-ridge volcanic activity, subsides more slowly than the global norm (Marty and Cazenave 1989), and appears to have an unusually weak lithospheric plate for its age at the time of volcanic loading based on flexure analysis (Calmant and Cazenave 1987). They argued that the South Pacific Superswell might have been unusually shallow for 100 million years or more, based on the fact that plate reconstructions place Menard's (1964) "Darwin Rise" over the same area of the mantle in the mid-Cretaceous. Menard had inferred a shallow depth for the Darwin Rise sea floor, currently located in the northwest Pacific, based on the elevation of the wavecut summits of drowned former islands (guyots) above the surrounding sea floor.

McNutt and Judge (1990) identified an additional unusual feature of the Superswell: a geoid low corresponding to the topographic high. Cazenave and Thoraval (1993, 1994) demonstrated that the seismic velocity anomalies in the upper mantle, geoid, and depth anomalies over the Superswell are all correlated at degree 6 spherical harmonic. Theoretical modeling at both labs demonstrated that the distinctive geoid/topography relationship of the Superswell could be explained by a broad dynamic upwelling of hot mantle material in a shallow low-viscosity zone beneath the region. However, the thermal anomaly in the upper mantle would not explain the unusually low values for elastic plate thickness reported for Superswell lithosphere (McNutt and Judge 1990). Therefore, any weakening of the elastic plate must be local to the island chains.

Levitt and Sandwell (1996) later questioned the existence of the South Pacific Superswell based on a modal analysis of newly collected multibeam bathymetry data from sea floor aged 15 to 35 Ma east of the French Polynesian hotspots. Their soundings were systematically deeper than the depths indicated in the older ETOPO5 gridded database (NGDC 1988) that had been used in the early studies to quantify the depth anomaly, leading them to question whether the Superswell might be an artifact of bad data. McNutt et al. (1996) re-examined the evidence for a large, regional depth anomaly spanning French Polynesia, confining the analysis to actual ship soundings. Their modal depth demon-

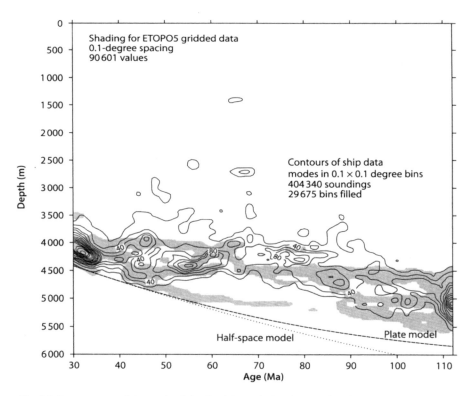

Fig. 1.5. Contour map of the mode of the depth/age relation for French Polynesia. The contour lines show the number of 0.1 × 0.1 degree latitude/longitude bins containing an original ship sounding that falls within each 100 m interval of depth and 2 Myr interval of age. Out of a total of 90 601 latitude/longitude bins, only about one third contain a ship sounding. Contour interval is 20 values beginning at 20. *Shading* represents the density of ETOPO5 values in the same bins, with contour interval 200 values starting at 100 (from McNutt et al. 1996)

Chapter 1 · Sea-Floor Topography and Morphology of the Superswell Region   17

strated remarkable agreement between the most up-to-date compilation of ship soundings and ETOPO5 in the depth/age relationship in French Polynesia, even though the number of ship soundings had increased fivefold since the release of ETOPO5 (Fig. 1.5).

The studies by Levitt and Sandwell (1996) and McNutt et al. (1996) highlight the importance of using actual depth soundings, rather than interpolated or predicted depths, in isolating depth anomalies. The new bathymetric map presented in Fig. 1.2 is based on a slightly augmented data set compared with what was available to McNutt et al. (1996) at the time, but it confirms the same general features of the Superswell. The depth anomaly grows from about 250 m on 30 Ma sea floor to approximately 1 000 m on 70–80 Ma lithosphere.

1.3.2
Midplate Swells

Broad (500–1 000 km) areas of elevated (500–2 000 m) sea floor typically surround regions of active intraplate volcanism (Crough 1978a,b, 1983). On account of their association with hotspots, these midplate swells have generally been interpreted as thermal in origin, either caused by altering the thickness/temperature in the lithosphere surrounding the hotspot (Detrick and Crough 1978; Detrick et al. 1981; McNutt and Judge 1990; McNutt 1987; Menard and McNutt 1982) or by the thermal buoyancy of a plume (Courtney and White 1986; Robinson and Parsons 1988a). One key observation in support of models that place at least some of the compensation for the swell as shallow as the mid to lower lithosphere is the ratio of the geoid anomaly to the depth anomaly, and thus accurate digital bathymetry maps have been important for constraining the origin of midplate swells.

In their broad survey of the heights and geoid-to-topography ratios of swells globally, Monnereau and Cazenave (1990) found that swell height increases and the geoid-to-topography ratio increases as the age of the lithosphere increases. Simple interpretation of the ratio in terms of a dipole compensation model would place the mass deficit supporting the swell in the mid to lower lithosphere, in apparent agreement with the lithospheric reheating model for midplate swells. However, theoretical models with realistic physical parameters had difficulty explaining the rapid rise time of midplate swells on account of the low coefficient for thermal conduction of the lithosphere. In comparison, swells surrounding hotspots in the South Pacific are generally smaller and have lower geoid-to-topography ratios than the global norm for hotspots elsewhere on lithosphere of the same age (Fig. 1.6), and therefore even shallower apparent compensation depths. Thus, the difficulty in supplying enough heat at shallow enough depth in the lithosphere to explain the low geoid-to-topography ratios is even more pronounced for the hotspots of the South Pacific.

More recent investigations have pointed out that melting models would predict a chemical component to swell relief as well as a thermal component (Jordan 1979). The extraction of a basaltic melt from more fertile mantle leaves a residuum depleted in aluminum and other elements that form the dense mineral garnet below 40–60 km depth. Therefore, the residuum is lighter than the undepleted mantle, potentially providing buoyancy to support a midplate swell depending on the viscosity and dynamic evolution of the depleted layer. For example, the depleted residuum might flow within the lower lithosphere to widths of hundreds of kilometers around the site of surface volcanism, thus matching the width of midplate swells (Phipps Morgan et al. 1995). On the other hand, it could spread more

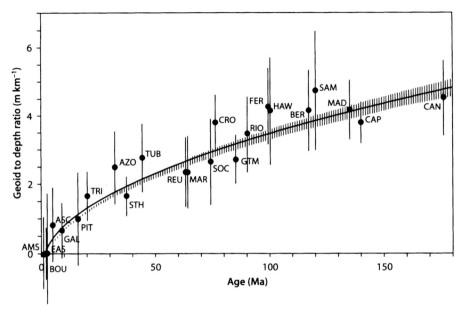

Fig. 1.6. Geoid-to-topography ratios as a function of age computed from swell height and geoid anomaly over hotspot swells. The *solid line* is the best fitting square root of age function. Note that all of the examples from the South Pacific such as Easter (EAS), Pitcairn (PIT), Marquesas (MAR), and Society (SOC) fall below the global trend, with the exception of Tubuai (TUB) (from Monnereau and Cazenave 1990)

uniformly over an even larger region, potentially halting the square-root-of-age subsidence of old oceanic lithosphere, particularly in the western Pacific. One advantage of providing at least a portion of the compensation for swell by chemical depletion is that it would have a shallow source, thus explaining the low geoid-to-topography ratios for swells without requiring substantial reheating of the lithosphere.

The observation from seismic reflection and tomographic studies that the crust immediately beneath several island chains is underlain by a presumably magmatic body lighter than the surrounding mantle offers another source for the buoyancy of swells. McNutt and Bonneville (2000) demonstrated that a seismically imaged body underplating the Marquesas Islands (Caress et al. 1995) has sufficient buoyancy to explain both the depth and the geoid anomaly over the Marquesas swell (Fig. 1.7). Bonneville and McNutt (to be published) are applying this same approach on the new bathymetric data from this paper to show that the Society and Austral swells could also have a similar origin, although the seismic tomography data for those island chains is not of sufficient quality to determine whether such underplating indeed exists. These recent investigations point out the importance of correcting for all shallower sources of anomalous crustal structure before attributing the rise of midplate swells to deeper sources.

1.3.3
Plate Boundary Features

The new bathymetric map for French Polynesia displays a number of features that record the history of plate boundaries in the South Pacific: fracture zones, abyssal hill

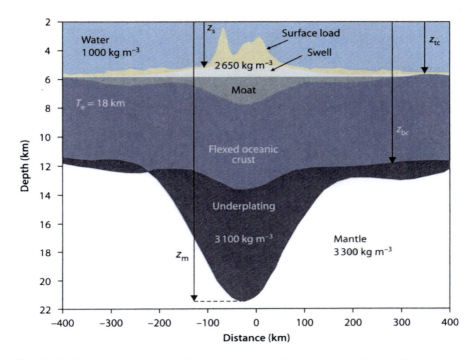

Fig. 1.7. Model for the Marquesas swell based on geophysical data from a line through the center of the chain. Only that portion of the bathymetric load above the swell is supported by elastic flexure. The swell itself is compensated by underplating. The shape of the underplating is consistent with seismic refraction data, and the depth of the moat agrees with seismic reflection data. Depths to the top of the swell, the top of the oceanic crust, the bottom of the oceanic crust, and the bottom of the underplating are indicated by z_s, z_{tc}, z_{bc}, and z_m, respectively. The small geoid-to-topography ratio results from the fact that the compensation for the swell is chemical rather than thermal and is indeed shallow (from McNutt and Bonneville 2000)

fabric, and the ancient traces of triple junctions and propagating rifts. Because some of the sea floor in this region was created during the Cretaceous superchron, such bathymetric features provide crucial evidence for past relative plate motions. Using actual bathymetry provides a substantial improvement over tectonic studies based on satellite altimetry (Okal and Cazenave 1985) on account of the higher resolution in sonar data and the fact that many of the plate boundary features are nearly isostatically compensated, and thus they have only a small signature in the altimetric geoid. For example, Kuykendal et al. (1994) used high-resolution, digitized locations of the Marquesas fracture zone from multibeam sonar to detect two subtle but important changes in Pacific-Farallon plate motion, the earlier one at 65–85 Ma (large uncertainty on account of lack of magnetic isochrons) and the more recent one between 33 and 37 Ma. The Pacific-Farallon ridge responded to a major change in the pole of rotation at about 50 Ma through the creation of a propagating rift, the trace of which has also been detected and interpreted in this data set (Cande and Haxby 1991; Jordahl et al. 1997; Jordahl et al. 1998).

Fracture zones have also been proposed as potential indicators of the strength of lithospheric plates by observing the magnitude of flexure that develops across fracture zones as the opposing plates subside at different rates depending on the age difference across the fault (Sandwell and Schubert 1982). Most studies have relied on geoid profiles recovered from satellite altimetry to determine the flexural signal from a locked fault (Sandwell 1984; Sandwell and Schubert 1982; Wessel and Haxby 1990). However, a detailed study of the bathymetry along the large-offset Marquesas Fracture Zone (Jordahl et al. 1995) demonstrated that conclusions drawn from geoid studies alone can be in error, particularly if the section of the fracture zone being used was inside the active section of the transform fault during a time when the pole of rotation was shifting. In that case, the fine-scale details concerning the formation of intra-transform relay zones and propagating rifts as the transform adjusts to a new direction of spreading are not observed in the smooth geoid data. Estimates of the fault strength based on geoid analysis incorrectly indicate a weak fault unless high-resolution bathymetry data are available to sort out the details of transform adjustments. The new map presented in Fig. 1.2 includes some high-resolution crossings of the Austral Fracture Zone that could be useful for this sort of study.

1.3.4
Off-Ridge Features

One of the most distinguishing features of the Pacific Plate in the region of Polynesia is the large volume of off-ridge volcanism in forms ranging from linear, age progressive chains of large islands and seamounts (Duncan and McDougall 1974; Duncan and McDougall 1976; Bonneville et al. 2002) to discontinuous subparallel ridges (Sandwell et al. 1995; Winterer and Sandwell 1987) to an unusually large number of isolated seamounts (Bemis and Smith 1993). The pattern of volcanism is far from random even on casual inspection. Clearly the new map presented here, by displaying what has been observed for small-scale features within the large framework, has great potential for leading to new insights on what process or processes have led to this large volume of off-ridge volcanism and whether properties associated with the lithosphere (e.g., thickness, rigidity, state of stress, pre-existing fabric, etc.) have interacted with the melt source to create the variety of volcanic expressions (ten Brink 1991; Vogt 1974).

High-resolution bathymetry reveals a variety of morphologies amongst the Polynesian volcanoes. The most common morphology is a circular or oval seamount with

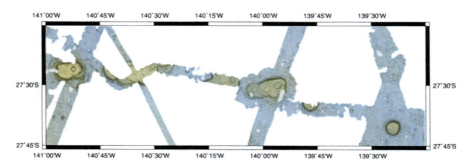

Fig. 1.8. Bathymetry of the Taukina Seamount chain in the south Austral region. The Taukina Seamounts are flat-topped, steep sided cones, 6–10 km in diameter and 1000–1500 m high. The Hydrosweep DS multibeam bathymetry is shown using a 100 m grid interval and shading by slope magnitude

Chapter 1 · Sea-Floor Topography and Morphology of the Superswell Region   21

Fig. 1.9. Bathymetry of Macdonald Seamount in the south Austral region. This volcanically active seamount rises 3 000 m above the surrounding sea floor. The SeaBeam and Hydrosweep DS multibeam bathymetry is shown using a 100 m grid interval and shading by slope magnitude

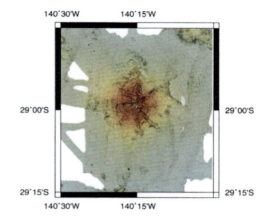

Fig. 1.10. Guyot Zep17 in the Tarava Seamounts south of the Society Islands is another example of a rift zone volcano (Clouard et al. 2003). The Simrad EM12D multibeam bathymetry is shown using a 100 m grid interval and shading by slope magnitude

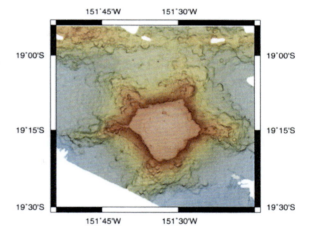

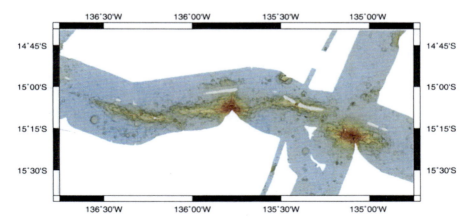

Fig. 1.11. Bathymetry of a segment of the Puka Puka Chain (Sandwell et al. 1995). These volcanic ridges are clearly constructed by the superposition of many small, flat-topped volcanic cones. The SeaBeam multibeam bathymetry is shown using a 100 m grid interval and shading by slope magnitude

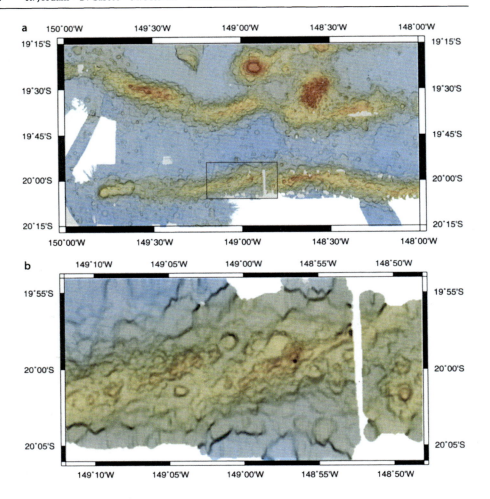

Fig. 1.12. a Bathymetry of a segment of the Va'a Tau Piti Ridges near Tahiti (Clouard et al. 2003). These volcanic ridges are clearly constructed by the superposition of many small, flat-topped volcanic cones. The Simrad EM12D, Hydrosweep DS and SeaBeam multibeam bathymetry is shown using a 100 m grid interval and shading by slope magnitude. b Detail of volcanic morphology of ridges in region shown by box in a

steep sides and flat tops, often with a prominent caldera (Scheirer et al. 1996). Most of the smaller seamounts have this form. Figure 1.8 shows examples from the Taukina Chain in the south Austral region that are 6–10 km across and 1 000–1 500 m high.

Many of the larger volcanoes develop linear zones of weakness emanating from a central peak. The tendency for magma to migrate along and erupt from these rift zones produces prominent radial ridges (Binard et al. 1991; see Sect. 5.1). Two fully mapped examples of these rift zone volcanoes are shown here: the Macdonald Seamount of the south Austral Chain in Fig. 1.9 (McNutt et al. 1997) and Guyot Zep17 in the Tarava Seamounts south of the Society Islands (Clouard et al. 2003) in Fig. 1.10.

Figures 1.2–1.4 show that en échelon volcanic ridges are also quite common in Polynesia. These ridges have been mapped in detail in the Puka Puka Chain (Fig. 1.11)

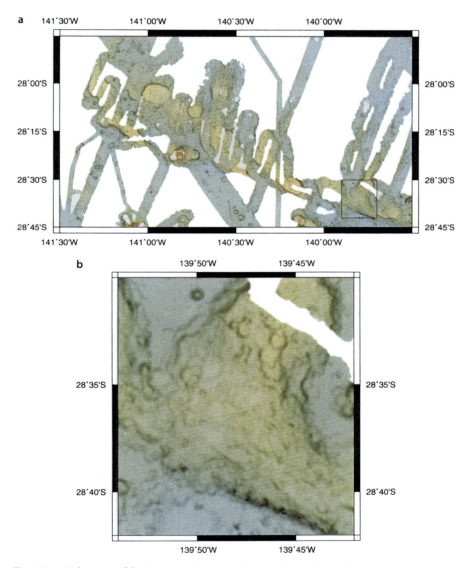

Fig. 1.13. a Bathymetry of the Ngatemato Seamount chain in the south Austral region. The Ngatemato Chain is mostly constructed of many small cones, like the Puka Puka and Va'a Tau Piti Ridges, but also includes some large, flat-topped edifices. The Hydrosweep DS, SeaBeam and Simrad EM12D multibeam bathymetry is shown using a 100 m grid interval and shading by slope magnitude; **b** detail of volcanic morphology of ridges in region shown by box in **a**

(Sandwell et al. 1995), the Austral Islands (McNutt et al. 1997) and the Va'a Tau Piti Ridges near Tahiti (Clouard et al. 2003), as seen in Fig. 1.12a,b. These ridges have generally been formed on young (<20 Ma) lithosphere and span scales from tens to hundreds of kilometers, possibly including entire island chains (e.g., the Tuamotu). High-resolution bathymetry reveals that these ridges are actually constructed of many small, flat-topped

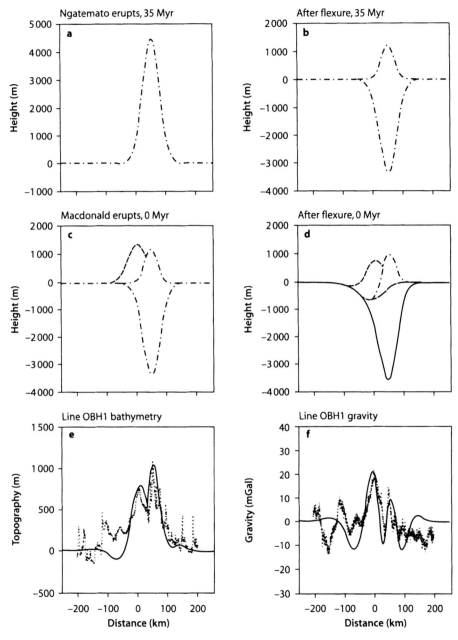

Fig. 1.14. Flexural modeling of lithospheric loading at two distinct times near Macdonald Seamount; **a** cross section showing load from older, near-ridge volcano erupting at 35 Ma; **b** net bathymetric profile resulting after load is compensated by flexure of a weak elastic plate; **c** smaller Macdonald Volcano erupts in the present; **d** net bathymetric profile after stiffer plate flexes to compensate the new, young load; **e** comparison of simple model bathymetry to multibeam data from the southern Austral Islands; **f** comparison of model gravity anomaly to the observed. From McNutt et al. (1997)

volcanic cones superimposed on one another (Fig. 1.11–1.13a and 1.13b). Apparently, these features are formed by a large number of discrete, small volcanoes forming in close proximity, presumably over a long period of time. The ridges that have been sampled and dated (Winterer and Sandwell 1997) show no evidence for a linear age progression along the feature and thus cannot be explained as hotspot features. The lack of age progression and the large-scale en échelon morphology suggests that these ridges result from lithospheric fractures that either tap ubiquitous upper mantle melt in the region or generate local decompression melting (Winterer and Sandwell 1997). Recent geochemical analyses also indicate that the Puka Puka lavas originated from the shallow upper mantle rather than from a hotspot (Janney et al. 2000). The morphological similarity of these en échelon ridge systems at both small and large scales suggests they may share a common origin.

High quality bathymetric data has also proved critical to calibrating the elastic thickness of the lithosphere (Filmer et al. 1993). As a good example, an initial estimate of the effective elastic plate thickness of Macdonald Seamount was 0 km (Calmant 1987; Calmant and Cazenave 1986), which is unexpectedly low considering that it is an active volcano resting on 43 Ma lithosphere. However, higher-resolution bathymetric surveys undertaken in the region of the Austral Islands demonstrated that Macdonald has recently erupted in the region of much older volcanoes (McNutt et al. 1997). The gravity signal used to calibrate the plate stiffness was measuring an average of the very low elastic thickness associated with the older volcanoes that erupted near the spreading center and the current, larger elastic thickness (15 km) associated with Macdonald (Fig. 1.14). While retaining the resolution of the multibeam data embedded in the context of the spline interpolation and single-beam sonar data, the new map presented in Fig. 1.3 should prove very useful for flexure studies.

Detailed analysis of multibeam bathymetry of the sea floor around Society and Austral Islands has also shown evidence of more than thirty-six submarine landslides (see Sect. 6.2.2 and 6.7). This inventory shows an evolution of the landslide type with the age of oceanic islands. Submarine active volcanoes are subject to superficial landslides of fragmental material, whereas young islands exhibit marks of mass wasting corresponding to giant lateral collapses due to debris-producing avalanches that occurred during the period of volcanic activity.

1.4
Conclusions

Bathymetric observations form the basis of nearly all geological, geophysical, and geomorphological studies of the sea floor. Unfortunately, we lack high quality bathymetry for vast areas of the sea floor, including most of French Polynesia. What data do exist are generally of uneven quality in terms of accuracy, resolution, and navigation. Furthermore, until now, there has not been one source of bathymetry data for French Polynesia that is comprehensive and in a common format. In producing the new data base and maps presented here, we have attempted to retain the information content in the original soundings, while still presenting the data at a large enough scale to be useful for regional synthesis. Our goal is to stimulate more studies of the sea floor of French Polynesia by making the highest quality data readily available.

Acknowledgements

The data synthesis efforts of Jordahl, Caress, and McNutt have been supported by the Packard Foundation. Development and support of MB-System is supported by the National Science Foundation and the Packard Foundation. Part of the database has been made available by the ZEPOLYF program funded by the French government and by the local government of French Polynesia. The authors particularly wish to acknowledge the work of all the scientists and crews involved in the expeditions that collected the data presented in this study.

References

Auzende JM, Bonneville A, Grandperrin R, Le Visage C (1997) Les programmes ZoNéCo et ZEPOLYF: Inventaire des zones économiques des Territoires Français du Pacifique, Marine Benthic Habitats. ORSTOM/IFREMER, Nouméa, pp 8

Bemis KG, Smith DK (1993) Production of small volcanoes in the Superswell region of the South Pacific. Earth Planet Sc Lett 118:251–262

Binard N, Hekinian R, Cheminée JL, Searle RC, Stoffers P (1991) Morphological and structural studies of the Society and Austral hotspot regions in the South Pacific. Tectonophysics 186(3–4):293–312

Bonneville A, Le Suavé R, Audin L, Clouard V, Dosso L, Gillot PY, Hildenbrand A, Janney P, Jordahl K, Keitapu M (2002) Arago Seamount: The Missing Hot Spot found in the Austral Islands. Geology 30:1023–1026

Calmant S (1987) The elastic thickness of the lithosphere in the Pacific Ocean. Earth Planet Sc Lett 85:277–288

Calmant S, Cazenave A (1986) The effective elastic lithosphere under the Cook Austral and Society Islands. Earth Planet Sc Lett 77:187–202

Calmant S, Cazenave A (1987) Anomalous elastic thickness of the oceanic lithosphere in the South Central Pacific. Nature 328:236–238

Cande SC, Haxby WF (1991) Eocene propagating rifts in the southwest Pacific and their conjugate features on the Nazca plate. J Geophys Res 96:19609–19622

Caress DW, Chayes DN (1996) Improved processing of Hydrosweep DS multibeam data on the R/V Maurice Ewing. Mar Geophys Res 18:631–650

Caress DW, McNutt MK, Detrick RS, Mutter JC (1995) Seismic imaging of hotspot-related crustal underplating beneath the Marquesas Islands. Nature 373:600–603

Cazenave A, Thoraval C (1993) Degree 6 upper mantle tomography and the south Pacific superswell. Earth Planet Sc Lett 57:63–74

Cazenave A, Thoraval C (1994) Mantle dynamics constrained by degree 6 surface topography, seismic tomography and geoid: Inference on the origin of the South Pacific Superswell. Earth Planet Sc Lett 122:297–219

Clouard V, Bonneville A, Gillot PY (2003) The Tarava Seamounts: A newly characterized hotspot chain on the South Pacific Superswell. Earth Planet Sc Lett 207(1–4):117–130

Courtney RC, White RS (1986) Anomalous heat flow and geoid across the Cape Verde Rise; Evidence for dynamic support from a thermal plume in the mantle. Geophys J Roy Astron Soc 87:815–867

Crough ST (1978a) Thermal origin of mid-plate hotspot swells. Geophys J Roy Astron Soc 55:451–469

Crough ST (1978b) Thermal origin of mid-plate, hot-spot swells. Eos Transactions American Geophysical Union 59(4), pp 270

Crough ST (1983) Hotspot swells. Annu Rev Earth Planet Sc Lett 11:165–193

Detrick RS, Crough ST (1978) Island subsidence, hotspots, and lithospheric thinning. J Geophys Res 83:1236–1244

Detrick RS, Von Herzen RP, Crough ST, Epp D, Fehn U (1981) Heat flow on the Hawaiian swell and lithospheric reheating. Nature 292:142–143

Duncan RA, McDougall I (1974) Migration of volcanism with time in the Marquesas Islands, French Polynesia. Earth Planet Sc Lett 21:414–420

Duncan RA, McDougall I (1976) Linear volcanism in French Polynesia. J Volcanol Geotherm Res 1: 197–227

Filmer PE, McNutt MK, Wolfe CJ (1993) Elastic thickness of the lithosphere in the Marquesas and Society Islands. J Geophys Res B 98:19565–19577

Janney PE, Macdougall JD, Natland JH, Lynch MA (2000) Geochemical evidence from the Pukapuka Volcanic ridge system for a shallow enriched mantle domain beneath the South Pacific Superswell, Earth Planet Sc Lett 181:47-60

Jordan TH (1979) Mineralogies, densities, and seismic velocities of garnet lherzolites and their geophysical implications. The Mantle sample: Inclusions in Kimberlites and other volcanics 2:1-14

Jordahl K, McNutt MK, Webb HW, Kruse SE, Kuykendall MG (1995) Why there are no earthquakes on the Marquesas Fracture Zone. J Geophys Res 100:24431-24447

Jordahl K, McNutt M, Buhl P (1997) Volcanism in the southern Austral Islands; time history, tectonic control, and large landslide events. Eos Transactions American Geophysical Union 78(46) Suppl, pp 720

Jordahl KA, McNutt M, Zorn H (1998) Pacific-Farallon relative motion 42-59 Ma determined from magnetic and tectonic data from the southern Austral Islands. Geophys Res Lett 25:2869-2872

Kuykendall MG, Kruse SE, McNutt MK (1994) The effect of changes in plate motions on the shape of the Marquesas Fracture Zone. Geophys Res Lett 21:2845-2848

Levitt DA, Sandwell DT (1996) Modal depth anomalies from multibeam data bathymetry: Is there a South Pacific Superswell? Earth Planet Sc Lett 139:1-16

Marty JC, Cazenave A (1989) Regional variations in subsidence rate of oceanic plates: A global analysis. Earth Planet Sc Lett 94:301-315

McNutt MK (1987) Temperature beneath the midplate swells: the inverse problem. In: Keating B, Fryer P, Batiza R, Boethlert G (eds) Seamounts, islands and atolls. Geophys Monogr AGU, pp 123-132

McNutt MK (1998) Superswells. Rev Geophys 36:211-244

McNutt MK, Bonneville A (2000) A shallow, chemical origin for the Marquesas swell. Geochemistry Geophysics Geosystems 1

McNutt MK, Fischer KM (1987) The south Pacific Superswell. In: Keating B, Fryer P, Batiza R, Boethlert G (eds) Seamounts, islands and atolls. Geophys Monogr AGU, pp 25-34

McNutt MK, Judge AV (1990) The superswell and mantle dynamics beneath the south Pacific. Science 248:969-975

McNutt MK, Sichoix L, Bonneville A (1996) Modals depths from shipboard bathymetry: There IS a South Pacific Superswell. Geophys Res Lett 23:3397-3400

McNutt MK, Caress DW, Reynolds J, Jordahl KA, Duncan RA (1997) Failure of plume theory to explain midplate volcanism in the Southern Austral Islands. Nature 389:479-482

Menard HW (1964) Marine geology of the Pacific. International series in the earth sciences. McGraw-Hill, New York, pp 271

Menard HW, McNutt MK (1982) Evidence and consequence of thermal rejuvenation. J Geophys Res 87:857

Monnereau M, Cazenave A (1990) Depth and geoid anomalies over oceanic hotspot swells: A global survey. J Geophys Res 95:15429-15438

N.G.D.C. (1988) National Geophysics Data Center, ETOPO5 bathymetry/topography data. Data announc. 88-MGG-02. National Oceanic and Atmospheric Administration, U.S. Department of Commerce, Boulder, CO

Nishimura CE, Forsyth DW (1985) Anomalous Love-wave phase velocitieis in the Pacific: Sequential pure-path and spherical harmonic inversion. Geophysical J Roy Astronom Soc 81:389-407

Okal EA, Cazenave J (1985) A model for the plate tectonic evolution of the east-central Pacific based on Seasat investigations. Earth Planet Sc Lett 72:99-116

Phipps MJ, Morgan WJ, Price E (1995) Hotspot melting generates both hotspot volcanism and hotspot swell? J Geophys Res 100:8045-8062

Robinson EM, Parsons B (1988a) Effect of a shallow low-viscosity zone on the formation of midplate swells. J Geophys Res 93:3144-3156

Sandwell DT (1984) Thermomechanical evolution of oceanic fracture zones. J Geophys Res 95 89: 11401-11413

Sandwell DT, Schubert G (1982) Lithospheric flexure at fracture zones. J Geophys Res 95 89:4657-4667

Sandwell DT, et al. (1995) Evidence for diffuse extension of the Pacific plate from Pukapuka Ridges and Cross-Grain gravity lineations. J Geophys Res 95 100:15087-15099

Scheirer DS, Macdonald KC, Forsyth DW, Shen Y (1996) Abundant seamounts of the Rano Rahi Seamount field near the Southern East Pacific Rise, 15 degrees S to 19 degrees S. Marine Geophys Res 18:13-52

Sichoix L, Bonneville A (1996) Prediction of bathymetry in French Polynesia constrained by shipboard data. Geophys Res Lett 23:2469-2472

Smith WHF, Sandwell DT (1994) Bathymetric prediction from dense satellite altimetry and sparse shipboard bathymetry. J Geophys Res 95 99:21803-21824

Smith WHF, Wessel P (1990) Gridding with continuous curvature splines in tension. Geophysics 55: 293-305 ten Brink US (1991) Volcano spacing and plate rigidity. Geology 19:397–400

Vogt PR (1974) Volcano spacing, fractures, and thickness of the lithosphere. Earth Planet Sc Lett 21: 235–252

Wessel P (2002) GMT Geoware CD-ROM vol. 1 version 1.2

Wessel P, Haxby WF (1990) Thermal stresses, differential subsidence, and flexure at oceanic fracture zones. J Geophys Res 95:375–391

Wessel P, Smith WHF (1996) A global self-consistent, hierarchical, high-resolution shoreline database. J Geophys Res 101:8741–8743

Winterer EL, Sandwell DT (1987) Evidence from en-echelon crossgrain ridges for tensional cracks in the Pacific plate. Nature 329:534–537

ZEPOLYF (1996a) Exploration de la zone Economique de Polynésie Française. Université du Pacifique – SMA – Ifremer – Ostom – Shom, Papeete – Tahiti

ZEPOLYF (1996b) Rapport de mission de la campagne ZEPOLYF-1. Université Française du Pacifique and IFREMER, Papeete – Tahiti

Chapter 2

Seismicity of the Society and Austral Hotspots in the South Pacific: Seismic Detection, Monitoring and Interpretation of Underwater Volcanism

J. Talandier

2.1
Introduction

Since the oceanic column provides an optical, thermal, and to a large extent, chemical shield for the remote sensing of the planet's surface, careful monitoring of seismic activity on the ocean floor remains one of the few methods of studying submarine volcanic activity. This line of research goes back to more than fifty years ago, when based on a suggestion of Ewing et al. (1946), Dietz and Sheehy (1954) obtained a detailed history of the 1952 eruption of the Myojin Volcano, south of Japan, using teleseismic T waves propagated in the oceanic column over a distance of 8 600 km to a receiving array on the California coastline.

However, this type of research is difficult due to a number of problems generally connected with the remote character of the underwater environment. Since the beginning, no permanent ocean-bottom seismographs have been deployed in oceanic basins (intraplate regions); therefore, all recording must be done using island stations. Unfortunately, islands that are equipped with instruments are few, and their seismic noise characteristics are generally poor, which results in a limited coverage and mediocre detection and location capabilities. Furthermore, the logistics and costs of seagoing expeditions prohibit the rapid deployment of portable arrays over the epicentral area of a recognized shock, a procedure routinely carried out on land. Finally, the scarcity of detailed bathymetry of ocean basins further inhibits geological interpretation.

The Polynesian Seismic Network (Réseau Sismique Polynésien, RSP) constitutes a notable exception to the generally unfavorable conditions described above for other regions of the world's oceans. The RSP is a wide-aperture seismic network centered in Tahiti and especially instrumented to provide detection characteristics comparable to those of the best continental sites. The routine magnifications for the seismic displacements are 125 000 at 1 Hz, and reach 2×10^6 at 3 Hz. This network has been previously described in a number of papers (Talandier and Kuster 1976; Okal et al. 1980; Talandier 1993). Its location on the Polynesian Island chains is favorable for detecting oceanic underwater volcanism (Fig. 2.1).

The present work includes a short review of the seismological methods used to identify volcanic sources on the ocean floor and provides examples drawn from over forty years of experience with the RSP in Polynesia. There is also a detailed look at the seismic and magmatic activity of the Mehetia/Teahitia and Macdonald Seamounts, which are respectively associated with the Society and Austral hotspots and which were directly monitored by the RSP network. The first two volcanoes are very close to the Tahiti seismic stations, whereas the Macdonald Seamount is further away; therefore, these two examples can clearly illustrate the different methods of detection and monitoring of submarine volcano-seismic activity in an oceanic environment.

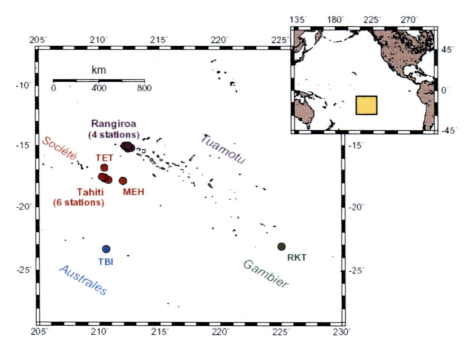

Fig. 2.1. Location map of the Polynesian Seismic Network (Réseau Sismique Polynésien = RSP) receiving stations in French Polynesia. *Inset* indicates the general location of the area investigated

2.2
Seismic Waves Used

Seismic detection of submarine volcanic activity makes use of three different kinds of seismic waves: *(i) conventional* seismic waves (mostly body waves) are individually separated in recordings as seismic phases and thus lead to the identification and location of individual seismic events (earthquakes); *(ii) seismic tremor* consists of signals of a ringing character that cannot be decomposed into individual events; and *(iii) T* waves which are propagated in the oceanic SOFAR (SOund Fixing And Ranging) channel over distances that can be teleseismic in range. It goes without saying that not all types of waves are generated by every phase of every submarine volcanic eruption; in addition, their differing propagation characteristics further affect their eventual detection by an island station. Figure 2.2 presents typical examples of each kind of seismic wave as recorded by the RSP.

▶
Fig. 2.2. a Example of small earthquakes from the 1985 swarm of the Teahitia Volcano recorded routinely by the five stations of the Tahiti subarray (distances 40-100 km). The different phases (refer to distance, *Pb* and *Sb*, or *Pn* and *Sn*) are clearly visible. This type of event can be located accurately. **b** *T* wave recorded by the same network during an explosive event, which started on December 24, 1980 on the Macdonald. Note the large amplitude at the station situated on the southern coast side of the source; **c** example of high-frequency harmonic tremor, continued noise of variable amplitude, as recorded by the same network during the 1985 Teahitia swarm; **d** example of high-frequency spasmodic tremor with repeated small earthquakes from the 1982 Teahitia swarm. Note the large amplitude at the Moorea station (*AFR*) located more 100 km from the source

Chapter 2 · Seismicity of the Society and Austral Hotspots in the South Pacific

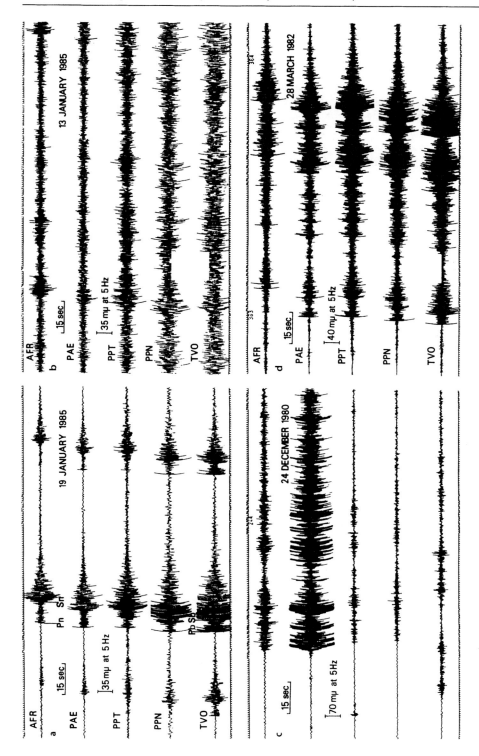

Since conventional seismic waves (mostly body waves) can be individually separated in recordings as seismic phases, we can use them to enable the identification and location of the recorded events. The strongest earthquakes of a purely volcanic origin, which are generally of low magnitude, are detected by sensitive seismic stations at regional distances of ten to hundreds of kilometers. The seismic waves that are used include direct P and S waves, which are propagated in the crust, and/or Pn and Sn waves, which are refracted at the Mohorovicic (Moho) discontinuity, depending on epicentral distance. In addition, high-frequency Rayleigh waves in the 1–2 s period range were occasionally recognized.

Earthquakes due to volcanism have usually been interpreted as related to one of two mechanisms. During the initial phases of a volcanic episode, magma is forced upwards through a series of conduits. The increased pressure in the magma is transmitted to the country rock, and the resulting stress is released through a brittle fracturing of the rock. This process obviously precedes the eruption of magma at the surface and forms the basis for the use of volcanic seismicity, which serves the purpose of forecasting eruptions of well-instrumented volcanoes (e.g., Kilauea, Mount St. Helens). These events can start as deep as 55 km (Eaton 1962; Butler 1982). After a major cycle of activity, the newly constructed volcanic edifice could become gravitationally unstable, and its mass tends to readjust itself through earthquakes. These events have occasionally been called 'tectonic' earthquakes (e.g., Klein 1982) in order to emphasize that they are not directly related to a magmatic process. Of course, they remain related to the presence and general activity of the volcano. This type of earthquake would include such phenomena as caldera collapses, and it is believed to represent isostatic compensation, which occasionally reaches very large magnitudes, such as in Hawaii.

There are two limitations for the use of conventional seismological techniques in the study of underwater volcanic seismicity. Not all episodes of a volcanic crisis give rise to individually recognizable earthquakes. In particular, the eventual eruption of lava on the ocean floor can take the form of a relatively continuous phenomenon, which may not be described properly as a succession of 'earthquakes'. In addition, the capabilities of detecting earthquakes in the largest parts of the ocean basins are still very limited, and adequate uniform coverage exists only at the level of $mb = 4.5$. Experience acquired from volcanoes such as the Kilauea, Loihi or Teahitia has shown that very few, if any, of the events at these sites have attained this threshold during a pure volcanic episode. The fast opening Mid-Oceanic Ridges, conspicuously absent from the worldwide seismic maps, are another obvious example of underwater volcanism whose seismicity totally evades detection by presently available permanent networks.

2.2.1
Seismic Tremors

Seismic tremors, related to the more or less continuous seismic agitation accompanying some stages of magma ascent toward the surface, consist of continuous resonant signals that cannot be separated into individual events. These underground tremors, not to be confused with submarine noise (T waves), which will be discussed below, are not detected at a distance of more than about 100 km. This is an upper limit for the combination of the most intense tremors and the existing capabilities of seismic sensing stations. The seismic tremors can be classified as either spasmodic tremors, with simultaneous bursts of small earthquakes, or as harmonic tremors, with a continued noise of high or low frequencies of fluctuating amplitude.

The 'high-frequency' ($\approx$7 Hz) tremor has often been described as originating in the resonance of cracks opening in the rock under magmatic pressure (Aki et al. 1977). More recently, the role of the oscillation of the fluid body of magma itself inside the cracks was recognized, and it was suggested that harmonic tremor may be dominated by Rayleigh waves of so-called 'long-period' earthquakes, excited by such an oscillation (Chouet 1985). These 'long-period' earthquakes are known to be significantly excited by the so-called 'B-type' events (McNutt 1986).

Low-frequency ($\approx$2 Hz) tremor has been reported at many subaerial sites in conjunction with the documented observation of the culminating phases of activity, involving deflation and volcanic fountaining (e.g., Einarsson and Brandsdottir 1984; McNutt and Harlow 1983). Thus, it is hypothesized that submarine eruptions were taking place at the times when low-frequency tremor was recorded from Teahitia. However, some low-frequency tremors have escaped detection, due to the relatively large distance from their sources to the recording stations.

The interpretation of tremor activity in terms of eruptive processes of a volcanic edifice is made extremely difficult by the great variability of the tremor patterns among well-instrumented subaerial volcanoes. This probably reflects the fact that the morphology of the magma plumbing and the kinematics of ascension of the magma itself have no *a priori* reason for being similar under all volcanic edifices. As an example, at Kilauea in Hawaii, Aki and Koyanagi (1981) have correlated high-frequency tremor (about 7 Hz) with deep magma progression occurring weeks to months before the eruption. They have further observed an evolution in tremor frequency with time. In the Mehetia and Teahitia edifices, high-frequency tremor is clearly absent from the earliest phase of the 1982 earthquake activity, which is interpreted as being a deeper event. This high-frequency tremor seems to have started when the seismic activity became shallower and was observed accompanying the earthquake activity observed during the 1983, 1984 and 1985 swarms, which are unrelated to the episodes of deep seismicity present in 1981 and 1982. This is, of course, in clear contrast to the case of Kilauea.

The main limitation for using tremor in order to monitor underwater volcanic activity is due to the tremors' rapid attenuation with distance. They are not observed at Kilauea at distances greater than 50 km, and the fact that we do not observe any tremor for the Mehetia Volcano at the Tahiti station may be due to its attenuation over the 120 km distance rather than to its absence at the source. In the marine environment, tremor has not been observed at distances greater than 90 km. At shorter distances, it can provide some insight into the level of magmatic activity or the eruptive process, which could be better monitored directly in the case of a subaerial volcano. Furthermore, for edifices concealed under water, the interpretation of tremor on the basis of comparison with documented cases must take into account the extreme variability of the characteristics of individual volcanoes.

2.2.2
T Waves

Hydro-acoustic waves, called *T* waves, propagated at great distances in the ocean's low-velocity SOFAR channel and recorded at seismic stations, provide another means for distant detection of volcano-seismic activity. Because of the guided nature of the *T* waves and their almost totally negligible anelastic attenuation in water, *T* waves can propagate over truly teleseismic distances, e.g., across the whole width of the Pacific Ocean, as long as no obstacle such as an island or seamount masks their path. A first category of *T* waves actually consists of submarine noise generated in the ocean at the lava/seawater interface. These waves originate directly from a degassing of lava and a possible vaporization of ambient water. As a result, they occur only for submarine eruptions at ocean depths less than 900 m. A second category of *T* waves consists of seismic waves that are converted into acoustic waves on a submarine slope near the epicenter.

Volcanic earthquakes or magmatic phenomena generate *T* waves only under favorable conditions at the source. In attempting to use *T* waves to recognize underwater volcanism, it is important to record 'magmatic' *T* waves that are not associated with earthquakes. In addition to this fundamental limitation at the source, the propagation (and hence reception) of *T* waves can be affected by a masking of the wave's path by a seamount or an island structure. A particular station or small-aperture array might be shielded from large

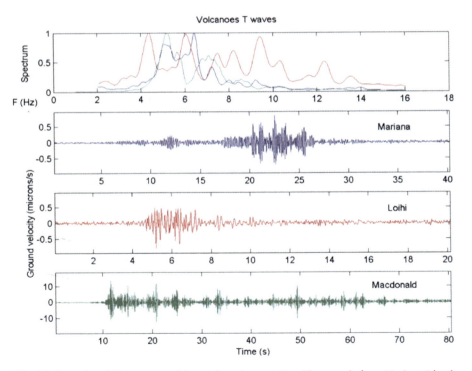

Fig. 2.3. Examples of *T* waves caused by a submarine eruption. The records from Mariana Islands, Loihi (Hawaii), and Macdonald Austral hotspots are shown. *Top*: The amplitude spectrum of the three recordings presented underneath is shown

epicentral areas. In conclusion, T waves can be most valuable for underwater volcano detection of shallow edifices located at a great distance from an array of hydrophones or island stations. This means of monitoring volcanoes is particularly important for the Pacific Ocean. Talandier and Okal (1998, 2001) have defined the identification criteria for sources of T waves and their ways and means of recording by seismic stations.

The most spectacular example of teleseismic detection of underwater volcanism by T waves is undoubtedly the recording of the eruption at the Macdonald Seamount (29° S, 140° W) on May 29, 1967 (Norris and Johnson 1969), and the subsequent discovery of the seamount by R. H. Johnson on July 20, 1969 (Johnson 1970). Since then, activity of the Macdonald has been documented by the RSP. In addition to the monitoring of Polynesian volcanoes, T waves of a volcanic origin are regularly received by the RSP from the following regions: Rumble Seamounts (north of New Zealand); Raoul Island (Kermadec); Monowaii Seamount (south of Tonga); Tori-Shima Island (south of Japan); Volcano, Bonin and Izu Island; Mariana Archipelago; Galapagos Islands; Hawaii; and Loihi. Other remarkable sources of volcanic origin detected by T waves are the monochromatic signals from the hydrothermalism of the Hollister ridge in the South Pacific Ocean (Talandier and Okal 1996). Examples of T waves recorded by the RSP for Macdonald, Mariana and Loihi Volcanoes are shown in Fig. 2.3.

2.3
Volcano-Seismic Activities on the Society Hotspot

The volcano-seismic activity of the Society hotspot presently in its early stages of evolution as well as in its next, insular form is discussed. The Society Islands are, after Hawaii, the second best developed among the Pacific Ocean island chains.

2.3.1
Generalities and Chronological Events

Multichannel bathymetry surveys performed by R/V J. CHARCOT, SONNE and L'ATALANTE indicate that the region of the Society hotspot includes about thirty submarine volcanic edifices ranging from 300 to 2 900 m high (Fig. 2.4). The island of Mehetia located at the eastern tip of the hotspot is about 3.5 km high with a maximum elevation of 435 m above the sea level (see Sect. 5.2.3.8) (Fig. 2.4). These edifices emerged from a regional swell or bulge located at a depth of 3 700–3 900 m and extending up to 100 km in diameter. This bulge is made up of low hills that are not associated with the individual volcanoes, and the bulge is covered by slightly more than 50 cm of sediment thickness. The hotspot swell rises about 300–500 m above the level of the neighboring deep ocean floor. Beyond this swell, the old oceanic crust is characterized by a sediment thickness of more than 20 m and by fault escarpments (Cheminée et al. 1989).

The volcanic complex of the Society archipelago hotspot is similar to the volcanoes of Mauna Loa, Kilauea and Loihi, which form a triangle of 50–60 km on each side, and which are contemporary evidence of the Hawaiian hotspot. However, a major difference between the Society and Hawaiian Islands is that the largest island of the Society Chain, Tahiti, is presently inactive, strongly eroded and fringed by a coral reef. On the other hand, the easternmost island in the chain, Mehetia, located about 130 km east of Tahiti, is an extremely small and steep cone, where no recent coral barrier has devel-

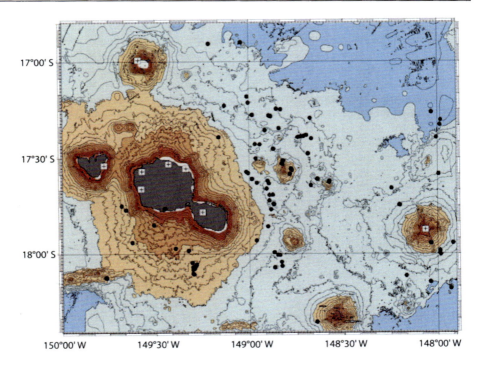

Fig. 2.4. The overall seismicity of the Society hotspot lasted for a quarter of a century, from July 1967 to March 1990. There is a sharp contrast between the relatively older edifices (Tairapu, Tahiti and Moorea), which are more or less consolidated and almost aseismic, and the more recent active volcanoes of the bulge area. However, there is a residual seismicity at the south of Tahiti and Tairapu, which suggests the presence of residual tectonic constraints in this area

oped. Thus, it is likely that the hotspot is presently in the early process of building the next island in the chain. The Society Islands and the hotspot area offer a rather unique opportunity to look at the transitional period preceding the building of a new major volcanic edifice. This episode is comparable to the early genesis of the island of Hawaii, about one million years ago. Mehetia's situation is also somewhat reminiscent of the Loihi Seamount, southeast of Hawaii.

During the last forty years, a systematic study of the seismic activity in the immediate vicinity of Tahiti was made possible thanks to the high-gain short-period stations of the Polynesian Seismic Network on Tahiti, Moorea, Rangiroa and, since 1986, on Mehetia and Tetiaroa. For the period 1962–1980, we have identified approximately thirty seismic epicenters in a 100 000 km^2 area centered about 60 km east of Tahiti (Fig. 2.4). Also, earthquakes of weak magnitude with repeated small seismic swarms, suggesting active volcanism in the vicinity of two seamounts at 180 and 2 100 m depth and named Moua Pihaa and Rocard were detected. This moderate seismicity is probably representative of the hotspot activity since 1917, the year during which numerous earthquakes were felt by the inhabitants of Tahiti, and which probably indicated one or several volcano-seismic swarms of the hotspot.

Contrasting to a quiet period, the Society hotspot area was the site of five major seismic swarms, whose intensity was by far greater than anything previously recorded in Polynesia (Talandier and Okal 1984). First, in 1981, about 4 000 earthquakes occurred on the southeastern flank of Mehetia. Then, in 1982, a swarm of more than 8 000 recorded earthquakes took place in the vicinity of the Teahitia Seamount, topping 1 600 m below sea level, located about 90 west-northwest of Mehetia. Later, in the vicinity of Teahitia, a swarm of 2 500 earthquakes was recorded in 1983. They were followed in 1984 by 8 000 earthquakes, and finally a seismic swarm of about 10 000 earthquakes was recorded in 1985. Since these five years of intense activity, this area of the Society hotspot has returned to its previous period of moderate seismicity.

2.3.2
The Over-All Seismicity of the Society Hotspot

The seismicity produced during volcanic activities is mainly due to the transfer of magma in the crust and within the volcanic edifice. Clustering swarms of numerous very small earthquakes are accompanied by volcanic tremors during some stages of the eruptive process. These are harmonic or spasmodic underground continuous noise of fluctuant intensities or in strong gusts. This noise is due to the vibrations generated by the filling and resonant oscillation of cracks under the pressure of the magma. This type of seismicity is spatially limited at the volcanic edifices where a vertical migration of the seismic events is frequently associated with horizontal migration. In general, following the effusion of lava, a diffuse seismicity, caused by a smaller number of earthquakes with a stronger magnitude that reflect crustal readjustment, will terminate the crisis.

On the other hand, an overall permanent seismicity is related to the sea floor associated with the bulge surrounding the edifices. There are individual and small numbers of earthquakes that rarely form clusters. These earthquakes are due to the isostatic underground readjustment. The new stability is intended to adjust the magmatic and thermal deep processes to the subsidence of the hotspot edifices. These activities are more pronounced in the vicinity of previously active volcanoes such as Teahitia after the crisis of 1982–1985. The overall seismicity of the Society hotspot lasted a quarter of a century (from July 1967 to March 1990) and is shown in Fig. 2.4. There is a sharp contrast between the relatively older edifices (Tairapu, Tahiti and Moorea), which are more or less consolidated and almost aseismic, and the area of the bulge with its active volcanoes. However, there is a residual seismicity at the south of Tahiti and Tairapu, which suggests greater tectonic constraints in this area.

The subsidence of the Tahitian edifices is as significant as those from Hawaii when considering their size and ages (*i.e., Maui*). In the Society hotspot, the 3 000-meter contour line tends to confirm the presence of a shearing mechanism. This contrasts with the structure of the hotspot characterizing the sea floor deeper than the 3 500 m deep region, which has more of a bulging appearance, interpreted as being a stage of magmatic upwelling (Chap. 5, Sect. 5.2.2.1). It is noticed that the seismically most prominent area is located between the Tairapu Peninsula and the region around the Teahitia Volcano, which was the site of strong seismic activity during 1981 to 1985. The transition zone between the tectonic stresses of shearing and the bulging effect of the crust should be revealed, as in the case of the Hawaiian area, by a pinching of the crust. In fact, we have no evidence to corroborate the existence of crustal pinching based on the detailed bathymetry from the area; however, it is also possible that the definition of the bathymetric map may be insufficient.

2.3.3
Seismic Detection, Magnitude and Seismic Moment

In order to detect the events of the Society hotspot area, up until 1986, the only relevant stations were the five stations on Tahiti and nearby Moorea and the four stations on the atoll of Rangiroa, 350 km to the north, and therefore too far away to detect the weaker events.

Magnitudes were estimated using the formula

$$Ml = \log A + 2.5 \log \Delta - 2.1,$$

where A is the peak-to-peak amplitude in microns at a period inferior to 1 s and Δ is the epicentral distance in kilometers. The numerical constants in this formula are designed to lock the upper end of this magnitude scale onto the teleseismic mb and are similar to those used by the Hawaii Volcano Observatory (HVO) Wood-Anderson instrument for the seismic studies of Kilauea and Mauna Loa events.

An estimate of the seismic energy involved in the individual events was obtained using *Gutemberg and Richter's* relation:

$$\log Es = 2.9 + 1.9 Ml - 0.024 Ml^2,$$

where Es is in joules. This formula was used in previous studies for the regional seismicity, where it was also found to match the relation $Es = 4.8 + 1.5 Ms$.

In addition, an estimate of the seismic moment of the earthquakes was obtained using the relation (Geller 1976)

$$\log Mo = Ml + 17.56 \quad \text{for } Ml > 4.2 \text{ (by alignment of } mb),$$

where Mo is in dynes centimeters. This formula provides a comparative basis for discussing the regional output of seismic energy and seismic moments during various seismic episodes.

The thresholds for detection of seismic activity centered at Mehetia and Teahitia is estimated at $Ml = 1.1$ and 0.8, respectively, but detection below magnitude 1.5 at Mehetia and 1.0 at Teahitia is affected by day-to-day variations in the level of background seismic noise. On particularly quiet days, events were detected with $Ml = 0.9$ at Mehetia and 0.5 at Teahitia. Although these thresholds are considered excellent in the oceanic environment, they remain higher than in the case of the densely instrumented island sites, such as at Kilauea or Mauna Loa. In particular, it is clear that smaller events, such as the type recorded at Kilauea directly on the flanks of the volcano, would totally escape detection. Any comparison between these volcanic edifices must involve either a "magnitude filtering" of the Kilauea data set or an extrapolation of the frequency-magnitude relations in Polynesia. This would suggest up to 60 000 events at Mehetia in 1981, 70 000 at Teahitia in 1982, 35 000 in 1983, 300 000 in 1984 and 60 000 in 1985 at the $Ml = 0.1$ level. These features are at least comparable to the events recorded at Mauna Loa in Hawaii during swarms previous to the eruptions after twenty-five years of quiescence.

2.3.4
Overview of the Swarms

Independent from the overall seismic crisis, and prior to the strong activity recorded in the 1980s, two small swarms were recorded in April 1969 and July 1972 in the Society hotspot region. The first swarm of 600 small earthquakes was located on Moua Pihaa and the second of 200 small earthquakes was on Rocard (Talandier and Kuster 1976). However, although this seismicity indicates magmatic activity for these two seamounts, it cannot be compared in terms of its duration, or in terms of the number of events or the energy released to those of the 1981–1985 periods (Talandier and Okal 1984). These five later crises will be analyzed in detail below.

The 1981 seismic swarm at Mehetia started abruptly on March 6, 1981, and lasted until December 1981, with some sporadic activity throughout 1982. The history of the swarm showing both the number of recorded earthquakes and the energy release, using three-day windows for the entire swarm and six-hour windows for the initial two weeks was recorded (Fig. 2.5a,b). The swarm features have several distinct episodes, and they will be more fully described in a later section. The largest event occurred on March 15, with a magnitude $Ml = 4.3$.

Fig. 2.5. Histograms of earthquake distribution of $Ml \geq 0.9$ and the square root of the seismic energy released (*top curve, blue*); **a** the total duration of the 1981 Mehetia swarm, using three-day windows, totaling 3 476 events and 8.4×10^{10} J; **b** the first eighteen days of the swarm, using six-hour windows, totaling 1 161 events and 5.3×10^{10} J. Note the pattern change of the two curves around March 15, when the number of earthquakes decreased sharply, and when the energy released reached a peak

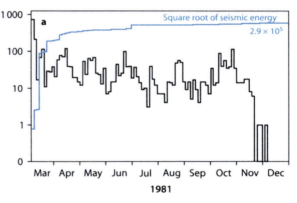

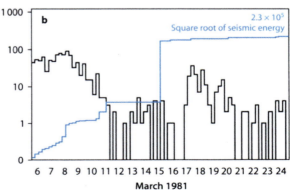

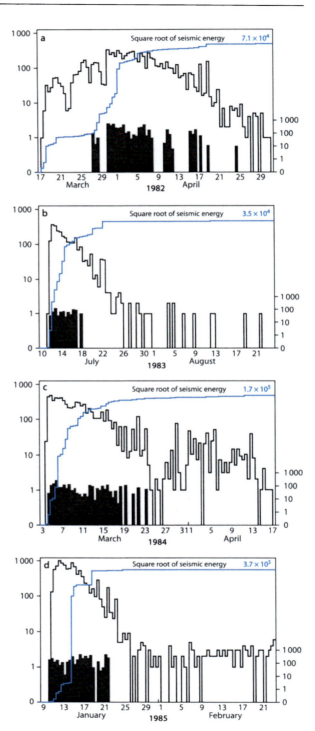

Fig. 2.6. Histograms of earthquake distribution (scale on *left*), square root of the seismic energy released (*top curve in blue*), and duration in minutes of the high-frequency tremors (scale at the *right dark field*) using twelve-hour win-dows for the first forty-six days of each Teahitia swarm; **a** swarm of 1982: 7798 earthquakes of $Ml \geq 0.9$; 5517 minutes with high-frequency tremors; **b** swarm of 1983: 2491 earthquakes; 719 minutes of tremors; **c** swarm of 1984: 7084 earthquakes; 2051 minutes of tremors; **d** swarm of 1985: 9673 earthquakes; 3990 minutes of tremors

Chapter 2 · Seismicity of the Society and Austral Hotspots in the South Pacific

The chronology of the events recorded and the energy release during the 1982 seismic swarm at Teahitia are presented (Fig. 2.6a–d). Activity started abruptly on March 16 and increased regularly until March 27, involving mostly low-magnitude earthquakes. After March 27, the earthquakes were accompanied by seismic tremors, recorded by all five stations of the Tahiti-Moorea subarray, and the tremors lasted more or less constantly until April. Then the tremor activity decreased and disappeared on April 18. Earthquake activity decreased regularly until May 19. After that date it dwindled to a number of rare, occasional events. The largest event was an $Ml = 3.4$ earthquake on April 1. Obvious differences between the Teahitia and the Mehetia swarms are the shorter duration and more homogeneous character of the Teahitia swarm compared to that of the Mehetia swarm.

The 1983 swarm of Teahitia started very abruptly on July 12 but was relatively short-lived (Fig. 2.6b). After two very active days, the intensity of the swarm decreased regularly, and it died off on July 15. The maximum magnitude reached was 2.5 on July 15. From the very start of crisis, the high-frequency tremors were present.

The 1984 and 1985 swarms also started very abruptly with numerous small earthquakes (about 1 500 to 2 000 earthquakes recorded each day) and intense high-frequency tremors (Fig. 2.6c,d). After a few days, the swarms diminished slowly over a period of about fifteen days and more rapidly for ten days, respectively. However in these two cases, as opposed to the two previous Teahitia swarms, the seismic activity continued at a variable level for several months. Practically absent from the 1985 swarm, some low-frequency tremors were sporadically observed for the swarm of 1984 (Fig. 2.6c). The largest event occurred on March 10, with a magnitude $Ml = 3.1$ for the 1984 swarm, but with an earthquake of $Ml = 3.3$ on July 2 at the end of this activity. The 1985 swarm differs from that of 1984 because it is marked by a high-energy episode with magnitudes of 4.4 and 4.0 and by several earthquakes of $Ml = 3.0$–3.5 on January 15 and one of $Ml = 4.2$ on January 19. This sequence is the most energetic of the five crisis recorded in the Mehetia-Teahitia area between 1981 and 1985.

The main differences among the developments of Teahitia's four swarms are shown in Fig. 2.6 and consist of several points:

i. In 1982 there were sequences of fourteen days without tremors prior to the main swarm with intense high-frequency tremors. Then the three following swarms started abruptly in 1983–1985 with the presence of these high-frequency tremors;
ii. The high-frequency tremors were more intense for the 1982 and 1985 swarms;
iii. Significantly low-frequency tremors were only observed for the swarms of 1982 and 1984, but this recorded result may be due to the fact that the events had different locations, involving a more or less favorable path around the magmatic structure;
iv. The last two swarms of 1984 and 1985 were followed by a large amount of seismic activity over a period of several months;
v. The swarm of 1985, the most energetic of the five Mehetia-Teahitia swarms, shows one sequence with high energy lasting five days after the start of the crisis.

The main characteristics of these five intense swarms of the Society hotspot are presented in Table 2.1.

Table 2.1. Main characteristics of the Mehetia-Teahitia swarms of 1981–1985

Location	Year	Duration (d)[a]	Number of earthquakes		Maximum Ml	High frequency tremors (min)[b]	b-value[c]	Deep sequence	Tectonic episode
			Total	Ml 4					
Mehetia	1981	280	3536	1	4.3	–	1.14 (1.42)	Yes	Yes
Teahitia	1982	80	8047	0	3.4	5517 (29)	1.40	Yes	No
Teahitia	1983 July	15	2487	0	2.8	727 (7)	1.49	No	No
Teahitia	1983 Dec.	7	328	0	2.6	–	–	No	No
Teahitia	1984	45 (307)	7795	0	3.7	2051 (20)	(1.17+)	No	No
Teahitia	1985	16 (80)	9670	3	4.4	3990 (12)	1.05 (1.33)	No	Yes

[a] Number shows duration of main period of activity. The number in parentheses indicates the total duration of the swarm.
[b] First number is cumulative duration of high-frequency tremors in minutes. The numbers in parentheses indicate the number of days with high-frequency tremors.
[c] Number is average b-value for whole swarm (including 'tectonic' episode if present). The number in parentheses are b-value for early phase of swarm if tectonic episode is absent. (+) Poorly defined b-value (see Sect. 3.6).

2.3.4.1
Location Techniques

Location techniques are based on Klein's (1978) HYPOINVERSE program, using an adapted version. Four different crustal models, each involving three layers over a halfspace, are used for (1) oceanic crust in the Tahiti-Mehetia (Society hotspot) area, (2) the northwestern Tuamotu Plateau, (3) the Tahiti and Mehetia Volcanic edifices, and (4) the edifice of the Rangiroa Atoll. These models were obtained from seismic refraction experiments (Talandier and Okal 1987). Accordingly, the first to arrive from the Mehetia swarm are Pn waves at all stations. From the Teahitia area they are Pn or Pg waves, depending on the epicentral distances where Pg is used for the crustal phase of basaltic composition. For impulsive signals, the errors in reading of less than ±0.025 s are negligible when compared to other sources of uncertainty such as station anomalies and the accuracy of the crustal models used.

The accurate determination of hypocenters in the vicinity of Mehetia suffers both from the relatively large distance to the closest station (about 120 km) and from the repartition of all stations in two subarrays concentrated around the azimuths N10°E and N275°W from the epicenter. In particular, hypocentral depths could not be constrained by travel times alone. The only available depth constraint came from the portable station operating for two days on the island of Mehetia in late March 1981. This station recorded only one event of low magnitude, which unfortunately went undetected by the stations of the permanent network. The $S-P$ interval for this record suggests a depth of 13 km, which according to the crust model, could be representative of the Mohorovicic discontinuity (Moho). We have chosen to use this figure as a starting value for all Mehetia relocations. Significantly, HYPOINVERSE then failed to adjust the focal depths and has opted to keep their values constrained. Similarly, in the case of Teahitia, we used the same starting depth of 13 km; therefore, the station repartition was insufficient for further constraints. This situation contrasts to the case of Loihi, where the numerous stations of the HVO network, by providing homogeneous azimuthal coverage over 90° and at distances as close as 35 km, have made it possible to resolve focal depths to a precision of about 5 km.

One hundred forty events were detected from the Mehetia swarm, with clear arrivals from both Tahiti and Rangiroa, to epicenter relocation using the HYPOINVERSE routines. Since reading errors are negligible, station residuals consist of a station correction, resulting from a local deviation of the crustal thickness under the stations to that used in the model and that takes into consideration the possibility of a path effect. For clustered epicenters recorded at a greater distance compared to the size of the cluster, these parameters will not vary significantly for individual events and can be modeled as a single station correction. These corrections were obtained by averaging residuals from the initial locations of the events and then using these averages in relocating the 140 events. The final locations obtained for the 140 events were computed using between six and nine Pn arrivals as well as the Sn arrival on the horizontal short-periods at the central station of PPT (on Tahiti), since the other stations were not equipped with horizontal instruments. This technique of using two successive locations, where the second employed the station corrections obtained by the first, has served to minimize the errors:

- First, standard residuals, already low for the initial locations (average value 0.068 s) are significantly improved by the relocations (average 0.042), confirming that the most random parameters (reading errors) are of a negligible nature;
- Second, large semi-axes of the horizontal ellipses are again significantly reduced by the relocations going from an average value of 8.6 km to an average of 5.4 km. Despite the general orientation of these axes along a bisector of the vectors pointing to the Tahiti and Rangiroa subarrays, this last value of 5.4 km is definitely smaller than the horizontal extent of the epicentral area. This indicates that the source area of the swarm is truly elongated in the NNW-SSE direction;
- Finally, despite a similar reduction in their absolute value, the vertical semi-axes remain about twice as large as the horizontal ones. They are much less meaningful, since the program did not readjust the depths, and thus the relocations of the earthquakes lack depth resolution.

The influence of the use of S times at the PPT station on the relocation of the earthquakes can be discussed as follows: The simultaneous use of P and S at PPT is equivalent to fixing the total distance traveled by the waves to PPT. For hypocenters located at/or below the Moho, this will result in constraining the epicentral distance for this station because of the very low inclination from the horizontal for any mantle wave. On the other hand, for sources whose true location is above the Moho, the observed S-P depends on both epicentral distance and depth. Since we have constrained the depth at the Moho, the inclusion of S wave data improves the relocation of earthquakes below the Moho, while it degrades the epicentral relocations of events whose true depth is above it; however, it will still improve the relative location of events whose true depths are comparable. Therefore, if two clusters are apparent, it is probable that the apparent distance between the clusters primarily reflects differences in focal depths. A decrease in true focal depth would tend to increase the apparent distance from Tahiti. The two clusters correspond to temporally separated events, with the northwestern ones occurring during the first weeks of the swarm, while the southeastern cluster became active later.

It is possible to conclude that the relative locations of the epicenters of Mehetia events at comparable depths are probably accurate to better than ± 3 km. On the other hand, the relative position of the whole ensemble of epicenters is somewhat less well constrained, probably no better than ± 6 km.

A similar procedure was used for the four Teahitia swarms of 1982–1985. A second set of station corrections was selected independently, as was done for Mehetia, since these corrections are also affected by the seismic wave paths to the station, which is shorter for Teahitia. As was done for Mehetia, the initial locations were used to infer station corrections, which were then used to relocate the 713 events (164 in 1982, 146 in 1983, 256 in 1984 and 147 in 1985). These signals with clear first arrivals could be picked at all five stations of the Tahiti subarray and by at least one station (usually PMO) of the Rangiroa subarray. The improvement provided by the relocations for events is clear. However, the epicentral distances are closer to the stations than in the case of Mehetia and consequently the first arrivals are Pn or Pg waves. Therefore, the values of the residuals and the precision in terms of the relative relocations of epicenters are better, on the order of ± 2 km (assuming common depths) for the same swarm. Uncertainty about the position of the simultaneous swarms may be greater, but their distribution in relationship to the area of Teahitia and to other smaller edifices, which are active volcanoes, tends to prove that the absolute location is accurate.

For Teahitia, the solutions obtained exclusively from P wave data are accurate. However, the use of S wave arrivals could give an indication of the event's depth according to the principle exposed previously. If the depth is equivalent to or greater than the Moho, there will be no significant modification of the epicenter. However, the location of the epicenter will change if the focal depth is clearly higher than the 13 km depth of the Moho. A somewhat more quantitative measure of an event's depth could come from the occasional observation of surface waves following the *coda* of the *Sn* wave trend on vertical short-period seismograms. We interpret those as being 1–2 s Rayleigh waves; the order of magnitude calculations for a point-source double-couple in the layered medium as defined by the seismic refraction experiments indicates that their excitation could become negligible at a depth below the sea floor comparable to one-half of their wave-length, or in the present case, 5 km. Thus, earthquakes for which a crustal depth is suggested from S times and which do not feature Rayleigh waves are probably confined to the lower crust. Those earthquakes that do show the surface waves are probably within a few km of the top of the edifice.

2.3.4.2
Correlation with Bathymetry

The swarm of 1981 at Mehetia is located on the southeast-elongated volcanic apron at a distance of 5 to 8 km from the summit of the island, between 1 400 to 1 700 m depth (Fig. 2.7a). It is possible that the eruptions took place on a small structure that was presumed to be a crater (17°57' S, 148°04' W) and which had been previously identified by several surveys (French Navy patrol-ship LA PAIMPOLAISE; R.V. MELVILLE, Craig 1983). Although it is not visible on the map of Fig. 2.7b on the close-up of the south flank of Mehetia, this crater might be as small in size as the one that covers the summit of the island (Fig. 2.7c).

In the Teahitia area, the swarms of 1982 and 1983 are closer to the main volcanic edifice, but those of 1984 and 1985 extend over an area further to the north (Fig. 2.8). The 1984 swarm largely extends to the southeast on a small volcanic edifice at 2 900 m depth, and the 1985 swarm is distributed on two small edifices at about 3 100 m depth with the highest energy episode being concentrated on the northern edifice. About 1 400 km² of the Teahitia area, including the small edifices to the north, are affected by this extensive volcano-seismic activity.

2.3.5
Evolution of the Swarms and Nature of the Recorded Events

Seismic activity developing in the form of swarms, rather than of classic fore-shock, main shock, and after-shock patterns, was observed in areas of active volcanism or extensional tectonism. Swarms directly associated with major eruptions are most often short-lived (lasting from one week to a few months); the seismic activity, which preceded the eruption of Kilauea and Mauna Loa in Hawaii, would be a typical ex-

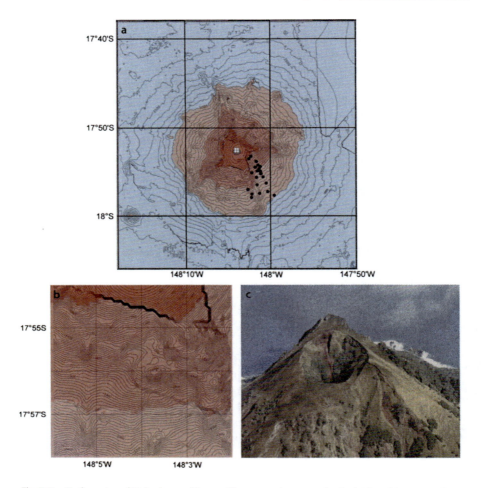

Fig. 2.7. a Bathymetry of Mehetia area (Bonneville, personal communication). The *white square* shows the seismic station on the rim of the crater at the summit of the small island about 400 m above sea level. The 1981 swarm is concentrated on the southeastern apron of the volcanic edifice. b A close-up of Mehetia's southern flank is shown. It is possible that the eruptions arose through a presumed small crater at 17°57' S, 148°04' W identified during several surface ship surveys as well as by submersible. Because of the insufficient resolution, the crater is not visible on this map. c An aerial photograph of Mehetia Island shows a small crater of about 200 m in diameter, covering the summit of the island. The presumed crater identified by several surveys on the southeastern submarine flank of the island could be of similar dimensions

ample of a volcanic swarm. On the other hand, swarms not known to be associated with eruptive volcanism tend to have a longer duration, with only a few exceptions.

The chronological development of the five crises of 1981–1985 on Mehetia and Teahitia is summarized below (Table 2.2). As a way of illustrating the type of recordings of the volcano-seismic activity that occurred in the Society hotspot area, the first swarms of 1981 on Mehetia and of 1982 on Teahitia will be described in some detail. Those, which have followed from 1983–1985, will be more succinctly described.

Chapter 2 · **Seismicity of the Society and Austral Hotspots in the South Pacific**   47

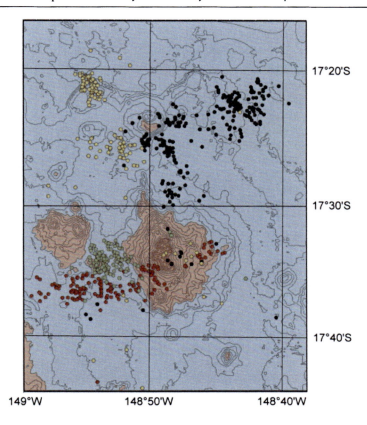

Fig. 2.8. Bathymetry of the Teahitia area using SeaBeam surveys by the R/V J. CHARCOT and R/V SONNE. The epicenters of the main earthquakes of four swarms of this volcano are shown. *Red*: 1982; *Green*: 1983; *Black*: 1984; *Yellow*: 1985. Note that the active zone extends forward about 1 400 km^2 beyond the main edifice of Teahitia

Table 2.2. Summary of the development of Mehetia and Teahitia crises

Episode	Magmatic activity			Tectonic activity
	Deep	Intermediate	Superficial[a]	
Mehetia 1981	March 6–9	March 10–25	March 26 to May 30	June 1 to Dec.
Teahitia 1982	March 17–22	March 23–27	March 28 to April 20	None
Teahitia 1983	None	None	July 11–25	None
Teahitia 1984	None	None	March 4–23	March 24 to Dec.
Teahitia 1985 (1)	None	None	Jan. 10–14	–
Teahitia 1985 (2)			Jan. 15–23	Jan. 24 to Dec.[b]

[a] Presumed for Mehetia and observed for Teahitia intense high-frequency tremors are associated with earthquakes during all this episode.
[b] This tectonic activity post crisis concerns the entire area of Teahitia.

2.3.5.1
Mehetia 1981 Swarm

A systematic analysis characterizing the evolution of events of the Mehetia swarm (location, frequency of occurrence, magnitude and spectral content) reveals four different periods of activity (Table 2.2, Fig. 2.5a,b).

- *Episode 1:* The swarm starts abruptly, with a large number of earthquakes (average of more than 240 events per day) of relatively low magnitude ($Ml \leq 3.3$). This activity is concentrated about 6 km southeast of the island. The seismograms show a high repeatability, with P and S waves featuring high frequencies and simple wave shapes (Fig. 2.9a–9d). As discussed below, these events are interpreted as having their epicenters at least as deep as the Moho. Therefore, the epicentral locations are probably accurate.
- *Episode 2:* The number of events decreases considerably (only twenty-seven per day on the average), but the activity remains substantial, with the two largest events in the swarm occurring on March 15 ($Ml = 4.3$) and March 25 ($Ml = 4.0$), respectively. The March 15 event is not truly representative of this seismic episode, being both shallower and with lower frequency. Apart from these events, the seismograms remain homogenous in their characteristics, while the epicenters move towards a crater presumed to exist in the area.
- *Episode 3:* The activity continues to decrease (on the average only thirteen events per day) and is characterized by higher-magnitude earthquakes. The signature of the seismograms becomes significantly different: their spectrum evolves toward lower frequencies, and the duration of Pn and Sn increases, leading to occasional ringing and in certain cases to the occurrence of two distinct arrivals of Pn separated by about 1.3 s (Fig. 2.9c). The regularity of this situation precludes the development of several sources but rather suggests a multipathing phenomenon. The multipathing of Pn was indeed observed in Rangiroa for seismic refraction arrivals originating in the Mehetia area (Talandier and Okal 1987). Finally, high-frequency surface waves develop after the Sn on the seismogram. All these characteristics suggest that foci are becoming shallower and are now located in the crust, which would mean that the epicenter locations might be less accurate. S waves are generally of smaller amplitude than P. Epicenters remain clustered until April 17 and then become dispersed. The cluster is clearly located southeast of the crater; and this migration from the summit to flank is directly comparable to the pattern observed on Loihi by Klein (1982).
- *Episode 4:* The level of seismicity decreases with some renewed activity in November and December (on the average eleven events per day). Characteristics of the seismograms vary widely, as do the epicenters, falling in any of the previous three zones as well as outside. The spectral content of the signals is generally of lower frequency, with occasional signals duplicating those of the preceding periods; in such occurrences, the epicenters also coincide.

In addition, the temporary station operating on Mehetia at the end of March recorded repeated puffs of seismic noise, whose characteristics could be compared to the high-frequency tremors such as those recorded from Teahitia during the crisis of 1982–1985; however, these puffs were unable to be recorded by the permanent stations

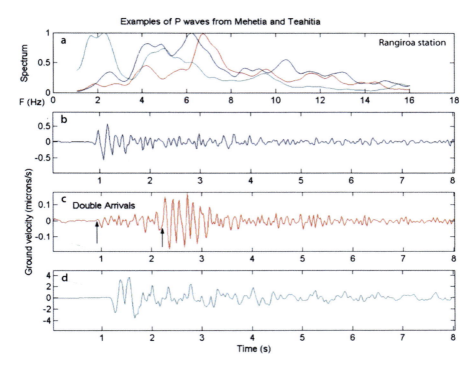

Fig. 2.9. Comparison of *Pn* waves for three events recorded at station PMO (Rangiroa); **a** amplitude spectrum of each recording presented underneath; **b** early stage of the swarm, presumed deep (Mehetia, March 9, 1981: *Ml* 2.6). Note the simple waveform and short duration of signal; **c** swarm presumed to be shallower (Teahitia, July 15, 1983: *Ml* 2.2). Note the second arrival, separated by 1.3 s; **d** *Ml* 4.3 event of March 15, 1981 at Mehetia. At this time, the low-frequency components of *Pn* and the ringing *P* waves lasted for several seconds with a characteristic frequency of 2 Hz. Not shown here: strong surface wave following *Sn* waves. This event is interpreted as being very shallow, approximately 1–2 km below sea level

on Tahiti. Nevertheless, our experience of the Teahitia swarm as well as data reported for Kilauea indicates that volcanic tremors rarely propagate over distances greater than 100 km. Thus it is probable, as suggested by the data from the temporary station, that seismic tremors accompanied the Mehetia swarm at least during its third phase. In particular, earthquakes similar to those accompanied by tremors on Teahitia were observed on Mehetia after March 25.

2.3.5.2
Teahitia 1982 Swarm

Figure 2.6 shows a chronological evolution of the swarm, both in terms of numbers of events detected, and of seismic energy released. The most intriguing pattern concerning the Teahitia swarm is the apparent migration of the seismicity (Fig. 2.8). During a period of about five weeks, the relocated epicenters moved regularly from east to west along the southern flank of the Teahitia Seamount. However, the relocations used in this figure did not adjust hypocentral depth, and the apparent epicentral migration may merely be a reflection of the change in true depth of the hypocenters (Fig. 2.8).

Thus, the pattern observed could be a true epicentral migration to the west, an ascent of the hypocenters toward the surface, or a combination of both. However, starting on approximately March 29, a complex pattern appears in the Pn seismograms recorded at the PMO station (Rangiroa), featuring two successive arrivals, separated by 1.3 s (Fig. 2.9a–c). Just as in the case of Mehetia, this suggests that the source is becoming shallower. It is worth noting that this date coincides with the development of strong tremor. Whether or not a vertical component of migration is present, the order of magnitude on the rate of seismic migration remains 1 km d^{-1}. This is comparable to the progression of seismicity during slow intrusions in moderately active rift zones, such as Loihi or the southwest rift of Kilauea. On more active systems, such as the eastern rift at Kilauea, the progression would be somewhat faster (Klein 1982).

The 1982 Teahitia swarm has several differences from that of Mehetia. The whole swarm is much shorter-lived, with the activity becoming practically negligible after only six weeks. The swarm is much more homogeneous in character, with the number of earthquakes growing steadily for about two weeks, and then decaying regularly. Finally, and most importantly, the two curves shown in Fig. 2.6 are similar to each other, indicating that there are no drastic changes in the magnitude of the earthquake distribution, which contrasts with episode 2 at Mehetia.

In addition, a large amount of seismic tremors were recorded from Teahitia during the period March 27 to April 17. Basically, two types of these tremors exist: (1) high-frequency ones, with seismic energy peaked in the 7 Hz range, of a rather spasmodic nature, following sequences of small but sharp earthquakes, and (2) low-frequency tremors, peaked at 2–3 Hz (Figs. 2.10a,b and 2.11).

For this seismic crisis, the largest event recorded at Teahitia is the $Ml = 3.4$ earthquake of April 1, 1982. A few inhabitants in and outside the peninsula felt this particular earthquake as well as reporting three more in early April. This relatively low level of maximum seismicity is in sharp contrast with the case of the 1981 Mehetia and of the 1971–1972 Loihi swarm, during which events as large as $Ml = 4.3$ were recorded. T waves were recorded only from the largest Teahitia events and could not be used to identify otherwise unsuspected activity.

2.3.5.3
Teahitia 1983 Swarm

As compared to the 1982 crisis, the 1983 Teahitia swarm had very short-lived seismic activity (only thirteen days), lower magnitudes (maximum $Ml = 2.4$), and the seismic signature of its events was quite homogeneous. In particular, the pattern of double Pn arrivals at the Rangiroa stations was present for all events of the 1983 swarm (Fig. 2.9c). It was interpreted that the whole sequence took place at very shallow depths, probably within a few kilometers of the sea floor. The activity started abruptly with about 500 events per day during the first days and decreased gradually until July 26 (Fig. 2.6a–d). The seismic activity was concentrated on the western flank of the seamount (Fig. 2.8) with high-frequency and a few low-frequency tremors accompanying the earthquakes during the whole seismic episode.

Finally, a mini-swarm of about 300 earthquakes, preceeded by an $Ml = 2.6$ event, took place between December 18 and 21, 1983.

Chapter 2 · Seismicity of the Society and Austral Hotspots in the South Pacific   51

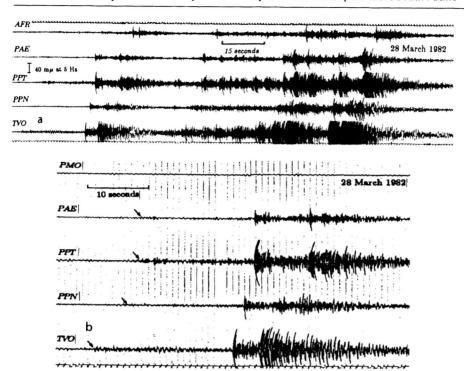

Fig. 2.10. a Typical examples of repeated small earthquakes intermixed with high-frequency tremor, as recorded in Tahiti during the Teahitia 1982–1985 swarms. *Tick marks* indicate the seconds; **b** detail of the beginning of the earthquake sequence shown in *Part a*. Note that the phenomenon starts as an apparent increase in the amplitude and frequency of the ambient noise (*arrows*), followed by more substantial events. The seismic stations are indicated in the records, AFR = (Afareaitu, Moorea), PAE = (Paea, Tahiti), PPT = (Papeete), PPN = (Papenoo, Tahiti) and TVO = (Taravao)

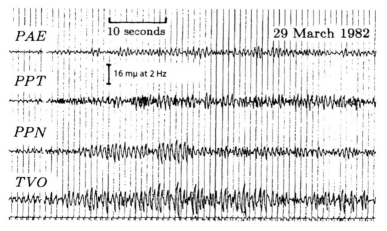

Fig. 2.11. Typical example of low-frequency tremor from the Teahitia 1982–1985 swarms. Note that the decay of the amplitude of low-frequency tremors with station distance is stronger, suggesting that they have originated at shallow depths. The abbreviations on the records are the same as in Fig. 2.10

2.3.5.4
Teahitia 1984 Swarm

As was true for the previous event in 1983, the 1984 crisis started abruptly with a large number of earthquakes and simultaneous high-frequency tremors. This high seismic activity was maintained for twenty days and was accompanied by several sequences of low-frequency tremors. The cumulative seismic energy release was greater than for the 1983 swarm with several earthquakes of Ml 3.0 to 3.6. Other differences include the fact that the seismic events were not concentrated but had migrated over an area of 20 km in a southwest-northeast direction (Fig. 2.8). In addition, although the seismic activity was decreasing, it still lasted five months, up to August. The pattern of double Pn arrivals at the Rangiroa stations suggests a very shallow depth.

2.3.5.5
Teahitia 1985 Swarm

As was observed for the 1983 and 1984 swarms, the seismic crisis of 1985 started suddenly with numerous earthquakes (about 2 000 per day) and intense high-frequency tremors. The seismic activity was characterized by two distinct sequences: (1) a first swarm from January 10 to 14 located in an area southeast of a small seamount, and (2) another very energetic swarm beginning January 15, which was concentrated on this seamount (Fig. 2.8). There were thirty earthquakes of $Ml \geq 3.0$, including three of $Ml = 4.0$ to 4.4, all of which were felt by the inhabitants of the Tairapu Peninsula, Tahiti, and for the last, on Moorea. This sequence is the most energetic of the five seismic crises of Mehetia and Teahitia between 1981–1985. As was observed during the second period of the 1982 swarm and during those of 1983 and 1984, the characteristic of the seismograms suggests a shallow-depth hypocenter.

2.3.6
Frequency-Magnitude Relationship

For each seismic crisis, a number of frequency-magnitude (b value) investigations were carried out, using all the recorded earthquakes for Mehetia (during March 6 to October 28, 1981) and Teahitia (during March 18 to April 28, 1982; July 12–26, 1983; March 4 to September 1, 1984 and January 10 to September 23, 1985). This technique models the number (N) of earthquakes with their magnitude (M) in a relationship using the formula:

$$\log N = a - bM$$

Values of b that are significantly larger than the worldwide average ($b = 0.9$) are used for rocks that are undergoing thermal weakening or excessive fracturing. In particular, documented volcanic seismicity has been associated with b values varying from 1.4 to more than 3 (McNutt 1983). We have reported a b value of 1.0 outside the swarms, and this value ranged from 1.5 to 3.2 during the smaller swarms of 1969 and 1972 at Moua Pihaa. In the case of Mehetia in 1981, using a magnitude window of 0.2 units, the results indicate an average b value of 1.14 ±0.04 for the whole sequence. However, if the data set is restricted to the first two days of the swarm, the b value increases to

1.42 ±0.15. This change of b value illustrates the dramatic increase in the number of larger earthquakes occurring after March 8 (Fig. 2.12a). In the case of the 1982 swarm at Teahitia, an investigation for the entire period between March 17 to April 30 yields a well constrained b value of 1.41 ±0.03 (Fig. 2.12c). An attempt to use shorter sampling periods failed to unveil significant variations of this coefficient with time. The figure $b = 1.41$ is in excellent agreement with the value found for the first two days of the 1981 Mehetia swarm. A remarkably similar figure of $b = 1.49$ ±0.13 was obtained for the 1983 swarm (Fig. 2.12d). For all of the swarm of 1984, the figure of $b = 1.17$ ±0.07 was determined; however, it was also noticed that the magnitude of repartition was irregular and when limited to two partial sequences, large b values were observed (Fig. 2.12e). This effect is probably due to the larger area affected that resulted in a heterogeneous distribution of the earthquake magnitude determination. Finally for the entire swarm of 1985, a b value of 1.05 ±0.02 is obtained (Fig. 2.12f). However, this crisis was clearly separated into two sequences: the first low energy swarm of January 10–14 with a b value of 1.33 ±0.06, and the second high energy sequence which started January 15 with numerous earthquakes of Ml 3.5 to 4.4, for which the b value is only 0.94 ±0.02 (Fig. 2.12g,h).

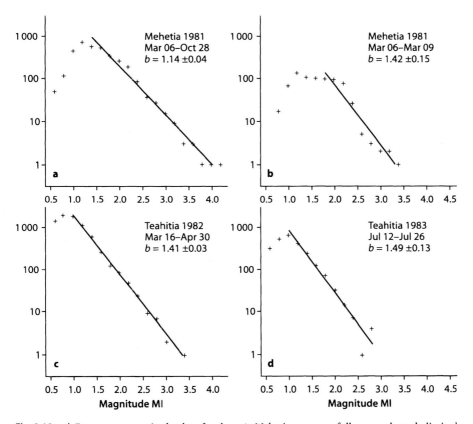

Fig. 2.12a–d. Frequency-magnitude plots for the 1981 Mehetia swarm; **a** full swarm; **b** study limited to the first 3 days of the swarm. Windows of 0.2 units of magnitude used in all cases. Frequency-magnitude plots for the Teahitia full swarms; **c** 1982; **d** 1983

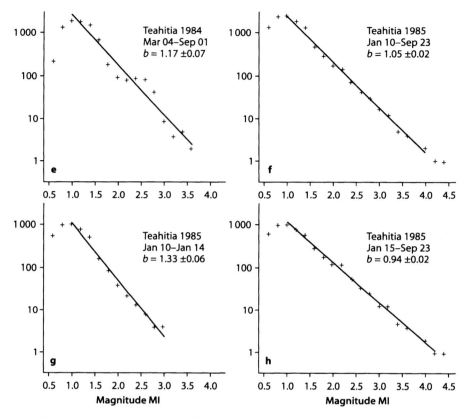

Fig. 2.12e-f. Frequency-magnitude plots for the Teahitia full swarms; e 1984; f 1985. Frequency-magnitude plots for the two sequences of the 1985 Teahitia swarm. Note the *b* value for the first one with weak energy (g) and second (h) with numerous earthquakes of relatively high magnitude

The range of *b* values obtained during the Teahitia swarms of 1982, 1983, 1984, and during the first phases of activity in 1985 at Teahitia and 1981 at Mehetia are comparable to those obtained in the Hawaiian hotspot (Klein 1982; McNutt 1983). In particular, the *b* values reached during the five Polynesian swarms are significantly higher than those found at sites of recurrent tectonic intraplate seismicity such as Regions A and C in the south-central Pacific (Okal et al. 1980).

While high *b* values have generally been recognized as indicative of volcanic seismicity, Okada et al. (1981) have cautioned against the use of *b* values for short-term sampling and recording in the study of the evolution of volcanic swarms. For example, their data at Mount Usu show the late development of large earthquakes and the disappearance of smaller ones, which lead to a negative *b value*. Thus, it may not be warranted to attribute a sudden decrease in *b* values to a variation in the physical properties of the rocks involved. However, a general trend towards fewer but larger earthquakes is clearly present at Mehetia after March 9, which is similar in character to the phenomena reported by Klein (1982) at Loihi and Kilauea, and which he interpreted as an evolution of the activity of rift zones compared to the flanks of the edifices.

In conclusion, the volcanic nature of the seismicity observed during the 1981 to 1985 swarms in the Society hotspot (Tahiti-Mehetia area) is determined on the basis of several factors: (1) the short-lived swarm-like character of the activity, (2) the large b values, (3) the identification of hydrothermal activity on Teahitia, (4) the evolution of the swarms reminiscent of those at Hawaii and Loihi, and (5) the abundant numbers of intense high-frequency tremors recorded from Teahitia that are also suspected to have occurred on Mehetia.

2.3.7
Seismic and Magmatic Activity in the Society Hotspot Volcanoes

Since the Teahitia and Mehetia sites are located under the ocean and are therefore inaccessible, we must infer the magmatic processes that may have accompanied the swarms, using only the variation in the characteristics of the seismicity. Unfortunately, two problems hamper our potential insight into these processes. First, our lack of hypocentral depth resolution obscures one of the crucial parameters in the evolution of the swarms. Secondly, the characteristics and evolution of volcanic swarms are known to vary substantially from one volcano to another. We will nevertheless attempt to draw a parallel between our observations and other examples of volcanic seismicity.

One of the characteristics of volcanic seismicity most crucial to placing the seismic swarm in the context of magmatic processes, namely hypocentral depth, could not be constrained by our relocations. The only travel time data that can be used to gain an estimate of depth are S wave arrival times. Nevertheless, their use tends to improve relocations for earthquakes truly located at or below the Moho, while it degrades the veracity of our interpretations when the source is within the crust. On this basis, the third period in the 1981 activity of Mehetia (March 26 to May 30) and the large March 15 event are interpreted as being shallower than the earlier events. Similarly, the first and second periods of earthquake activity on Teahitia (March 17–27, 1982, until the appearance of high-frequency tremors) are located below or on the Moho. These observations differ from the Teahitia swarms of 1983–1985, which also started suddenly with numerous earthquakes but where high-frequency tremors were registered simultaneously.

For the two first crises in Mehetia and Teahitia, the evolution of the waveform of P waves confirms the vertical migration of the seismicity. The Pn waves of two events of Mehetia and Teahitia, which were recorded at the PMO station on Rangiroa Atoll at a distance of 320 km north of the two volcanoes, are shown in Fig. 2.9. In the first case, the form of the wave is extremely sharp and simple, while in the second case, double arrivals separated by 1.3 s and having a much slower decaying amplitude are observed. The former type of earthquake was seen only in the early stages of the 1981 Mehetia and 1982 Teahitia swarms. This type was absent from the 1983–1985 activity at Teahitia. The first type is interpreted as having originated below the Moho discontinuity, and the second type within the crust. Thus, the sharper, deeper events would correspond to the initial opening of the magmatic conduits at depth, while the more complex, shallower ones could represent magmatic transfer within the crust and the volcanic edifice. On the other hand, the development of substantial surface waves within the 1 Hz range is controlled by the structural layering of the volcanic edifice, which would constrain the focal depth to about $\lambda/4$ or approximately 1–2 km below sea level. This means that the entire series of the 1983–1985 swarms at Teahitia are certainly no more than a few kilometers deep.

Volcanic tremors, consisting of more or less continuous seismic agitation accompanying some phases of volcanic activity, have been reported and studied extensively. The tremor could be spasmodic, featuring a repeated number of individual, identifiable seismic events, or it could be harmonic, in which case the frequency content of the signal is predominantly monochromatic (Figs. 2.2 and 2.10). Deep seismic tremors of a spasmodic nature originating about 50 km below Mauna Loa were identified by Eaton and Murata (1962) as representing the filling of magma conduits, and this took place three months before the 1959 eruption. Harmonic tremors located at 30–35 km depth were found by Aki and Koyanagi (1981) to be caused by magma oscillation in longitudinal cracks. These harmonic tremors represent a continuous, ongoing aspect of the volcano's activity, unrelated to any given eruption. These authors also noticed an evolution of the dominant frequency of the tremor with time (from about 7 to 3 Hz) during an episode of tremor (typically a few hours), which they interpreted as cracks joining with each other, thus increasing the characteristic length of the oscillator. The sequence of repeated small earthquakes followed by tremors observed in 1982–1985 at Teahitia are believed to represent a succession of small crack openings, followed by movement and oscillation of the magma in the cracks (Chouet 1981) (Fig. 2.10a). The initial waveforms, which started as an increase in the amplitude and frequency of background noise, are reminiscent of the harmonic tremors observed at Kilauea by Aki and Koyanagi (1981). The amplitude of the ground motion (about 0.01 µm) is also comparable. On the other hand, the pattern of a decrease in the wave frequency during a tremor episode is absent from the Teahitia tremors.

The total duration of the episodes of high-frequency tremor following small earthquakes as recorded for the single month period of March 17 to April 17, 1982, is 5 500 minutes. If these high-frequency tremors are interpreted as representing crack openings and magma transport according to the theory of Aki and Koyanagi's (1981), they represent a "reduced displacement" of 130 m^2, which is only four times less than the cumulative value over eighteen years of the Hawaiian Volcanoes. By adding 720, 2 050, and 4 000 minutes of high-frequency tremors for the 1983, 1984 and 1985 swarms respectively, the cumulative reduced displacements of four years of activity in Teahitia is on the same order as that observed during eighteen years at Hawaii.

However, a comparison of volcano-seismic activity for the two areas is difficult, because the characteristics of the swarms are different and the tremors are not directly comparable. In particular, the Kilauea tremors were not related to a given eruption. Also, Aki and Koyanagi (1981) and Chouet (1981) have argued that the quantification of seismic tremors has seriously underestimated the amount of lava ejected at Mauna Loa and Kilauea. These authors suggest that part of the transport of the magma through the lithosphere escapes seismic detection. Since the Hawaiian volcanoes are the only ones for which this kind of quantification has been performed, it is not clear that exactly identical situations could exist at other locations. In particular, and as argued below, the volcanic system in the Tahiti-Mehetia area is probably in a much earlier stage of its development than is Kilauea, where the plumbing is well established and the shield-building stage is steady. In this respect, the opening of cracks under Teahitia could involve a higher density of resistive barriers (Aki 1979), leading to the puffs of earthquakes that started the high-frequency tremor sequences.

While high-frequency tremors have also been reported simultaneously with ejection during eruption at Kilauea, the fountaining phases of volcanic activity are usually accompanied by intense lower-frequency harmonic tremors, located only a few kilometers deep and which peak at 2–3 Hz (Eaton and Murata 1962). At the same time, seismicity and spasmodic tremor activity were reported as strongly decreasing. This suggests that the high-frequency tremors are related to magma upwelling through the volcanic edifice, while the low-frequency tremors have accompanied the venting and fountaining processes. Fountaining probably also acts like the valve of a pressure-cooker, allowing the sudden release of pressure, at least from the shallowest parts of the plumbing system, and leading to the opening of the cracks generating the high-frequency tremors. Puffs of activity similar to low-frequency tremors were also occasionally observed during each Teahitia swarm, although their origin could not be positively localized.

The strong decay of the amplitude of low-frequency tremors in relationship to the station's distance (Fig. 2.11) suggests that they originated at shallow depths. They may be associated with eruptive processes, although a direct comparison with the case of Kilauea suggests their disappearing into the background noise beyond a distance of 90 km from the recording station. The 1982–1985 crises lasted for 53, 18, 374 and 58 minutes of low-frequency tremors, and this total duration of 503 minutes is just a small fraction of the 12 300 minutes recorded for the high-frequency tremors. This difference is probably due to the effect of propagation.

2.3.7.1
Volcano-Seismic Scenarios

On the basis of the above discussion concerning volcano-seismicity, the following paragraphs present scenarios for interpreting the Teahitia and Mehetia crises.

Scenario for Mehetia, 1981

The swarm started on March 6 with a series of small earthquakes, whose epicenter was located deep under the island, presumably below the Moho. By March 10, fewer but larger earthquakes took place, and the epicenters moved closer to the underwater crater (Fig. 2.7b). In addition, a unique large shallow earthquake occurred on March 15, which may have been due to a tectonic stress release under the pressure of the magmatic intrusion. After March 25, the poor fit of S wave travel times and the presence of surface waves and ringing Pn waves all suggest that the seismic activity had moved to shallower depths, certainly above the Moho. This ascent of the seismic activity was probably associated with the upward progression of the magma. Low-frequency events suggest that tremor was taking place at the end of March and submarine eruptions may have occurred. These eruptions would not have generated T waves because of the impossibility of magma degassing at 1 700 m depth. At least part of the later seismicity may have been due to the release of the intraplate tectonic stress accumulated in the plate, as indicated by the quiescence of the area prior to the swarm. These events are spatially more spread out, which would correspond to "tectonic" classes of seismicity, and their presence could lower the average b value of the whole swarm.

Scenario for Teahitia, 1982

The swarm began abruptly on March 16, 1982, with numerous small and simple events, possibly as deep as the Moho. These might have included a fracturing of the country rock under the increased magma pressure. Also, it is likely that the east-west migration of the seismic activity included a component of decreasing depths, as suggested by the later development of ringing phases. By March 28, a pattern started to form numerous cracks within the seamount, followed by magma filling, which gave rise to high-frequency tremor. An eruption took place, occasionally accompanied by low-frequency tremor. The pressure release temporarily shut off the process of fracturing, as noticed by the disappearance of the high-frequency tremors. Similarly to the previous Mehetia crisis, no T waves were generated, because the submarine eruption occurred at depths that were too deep for magma degassing. During this phase, three major earthquakes were felt on Tahiti. On April 17, after a three-week duration, the tremors stopped, marking the end of the eruptive process, and since previous seismic activity had released the intraplate tectonic stress, the seismic swarm died off quickly during the month of May.

Scenario for Teahitia, 1983

The volcano awoke again with a series of small, shallow earthquakes concentrated on its western flank. This activity may have taken the form of a lateral intrusion from the 1982 plumbing, and thus the swarm was able to move directly into the final seismic stage of shallow events accompanied by high-frequency tremor. This pattern lasted only two weeks, but eruptions that were probably more intense than in 1982 continued afterwards. In December 1983, a submarine survey found evidence of ongoing magmatic and hydrothermal activity. An interpretation concerning the small burst of seismic activity in December is difficult, but having the same location as that of the July crisis, it may correspond to a short reactivation of magmatic activity of this volcano.

Scenario for Teahitia, 1984

Located to the north of the Teahitia Volcano, the 1984 crisis presented several distinct episodes that seem to correspond to a seismic event migrating for about twenty km from the southwest to the northeast. Similarly to the previous crisis of 1983 and to those which will follow in 1985, this 1984 swarm started suddenly with a large number of earthquakes and intense high-frequency tremors accompanied here by a few sequences of low-frequency tremors. This activity was only superficial and decreased until it disappeared on March 23. However, there was also some residual seismic activity that slowly decreased until the end of the year. This was similar to the superficial sequence of the 1982 events and seemed to be an indication of an eruptive process about twenty days long, which was most likely followed by several tectonic readjustments.

Scenario for Teahitia, 1985

The 1985 crisis, like the two previous ones, started with a large number of small earthquakes accompanied by intense high-frequency tremors; however, this final

Chapter 2 · Seismicity of the Society and Austral Hotspots in the South Pacific 59

crisis is clearly composed of two distinct episodes: (1) During the first, from January 10–14, the swarm was located to the north of Teahitia between two small volcanic edifices, and the epicenters were dispersed. (2) The second episode, from January 15 to about January 23 at the end of the swarm, was notable for its high-energy earthquakes having a large magnitude of $Ml \geq 3.5$ up to and including four earthquakes of a $Ml \geq 4.0$. These earthquakes were concentrated on the small volcanic edifice to the north, at a distance of about 10 km from the first episode (Fig. 2.8). The presence of high-frequency tremors confirmed the eruptive process, but the cause of the high-energy seismic activity corresponding to the most important crisis of this five-year period (1981–1985) is still not clear. Nevertheless, the location of this activity was close to the northern limit of the hotspot bulge and most likely corresponded to a more rigid geological structure, which could explain this high level for the release of constraints.

2.4
Volcano-Seismic Activity of the Austral Hotspot: Macdonald Seamount

Contrary to the Society hotspot, the Macdonald Seamount is extremely far away from the seismic stations; therefore, it is a good example for testing the detection and monitoring of intraplate volcanic activity by means of T waves. The Macdonald Seamount (28.99° S, 140.26° W) was discovered in May 1967, following a strong seismic swarm detected by the hydrophones of the Hawaiian Institute of Geophysics Network (Norris and Johnson 1969). Macdonald is generally considered to be the active expression of a hotspot having generated the Cook and South Alignment of the Austral Islands.

Macdonald has been frequently explored and mapped, and its eruptions were even directly observed on two occasions. The report of the E.V. HENRY on July 1983 noted a discoloration spot located approximately 2 km east (leeward) from the summit, oriented NNW-SSE and about 700 m long. This was noticed seventy-five days after the May 1983 seismic swarm. Further observations were made on October 11, 1987 by the N/V MELVILLE, when bubbles, steam and ash ejecta occurred. The explosion of large bubbles with a rising gas column and the formation of large green stains were observed on the ocean's surface (Talandier et al. 1988). In January 1989, the N/O LE SUROIT and the submersible *Cyana* recorded the second set of in situ observations. While *Cyana* was diving near the summit (200 m depth), the divers, looking through a porthole, observed magmatic gas release associated with the eruption of pyroclastic debris. On the surface, glowing red gases, probably caused by burning H_2 and colored by high-temperature oxidation of FeO to Fe_2O_3 included in volcanic ash, were also frequently observed. Hydrogen sulfide vapors occurred, especially during episodes of intense turbulence in the water including bursts of steam and gas discharge. Also, a green discoloration of the surface water appeared, spreading over an area at least one nautical mile in diameter (Cheminée et al. 1989). The hydro-acoustic waves (T waves) detected during this eruption only gave a weak signal. This is an indication of the size of an eruption that must be taking place during a strong crisis, as exemplified by the sequence shown in Fig. 2.13. This will be further discussed in the next subchapter (Sect. 2.4.1).

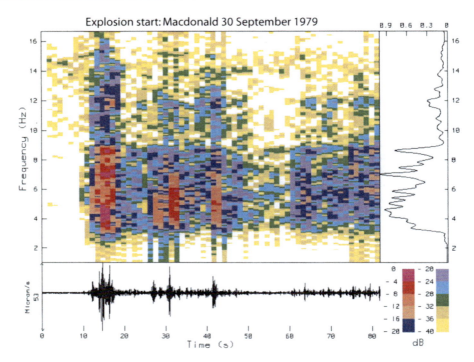

Fig. 2.13. *T* waves were recorded at the Tuamotu seismic station located at 800 km from the source. The beginning of explosive event during the Macdonald swarm in 1979 was recorded. The figure is composed of three frames: The *bottom diagram* shows an eighty-second time series of the ground velocity. Amplitude is marked at the *left* in μ s⁻¹ peak to peak. The frame on the *right-hand side* is a plot of the amplitude spectrum of the ground velocity record. The *main color frame* is a spectrogram representation of the distribution of spectral amplitude in the record, as a function of time and frequency. The color coding (or gray shading) is logarithmic, with the key (in dB relative to the most energetic pixel) shown at bottom right side of diagram. *White pixels* correspond to spectral amplitudes below −40 dB. Note the very high intensity (53 μ s⁻¹) of the impulsive start of the signal

2.4.1
Seismic Swarms

From 1977 to 1988, a total of twenty-nine seismic swarms on Macdonald were identified through detection of *T* waves by the RSP stations (Talandier and Okal 1982, 1984). This network used routinely high magnification (2×10^6 at 3 Hz) at the seismic receiving stations, which allowed a regular, systematic monitoring of the Macdonald area by the Polynesian stations, including Tubuai (1 110 km), Rikitea, (Gambiers 838 km), Afareaitu (Moorea 1 599 km), and Vaihoa (Rangiroa 1 704 km). The location of the most explosive events has a precision of ≈5 km, leaving no doubt as to the origin of the observed seismic swarms. On the other hand, no earthquakes at Macdonald have ever been detected using conventional seismic waves, because the level of detection in this remote area is about Ml = 3.5, and our experience studying the Society hotspot shows that the major seismic swarms emanating from volcanic areas could take place below this level. Also, seismic tremors could not be detected either, since they are only rarely known to be propagated over distances greater than 100 km (see Sect. 2.2.1).

Examples of T waves recorded during the start of Macdonald's explosive activity have shown that their principal characteristic is the long duration of the wave-train, which can be as long as an hour. Figures 2.2, 2.3 and 2.13 show the hydro-acoustic signal recorded by the seismic stations of the Polynesian network for the beginning of the Macdonald explosive crises of December 24, 1980, February 15, 1981, and September 30, 1979. They are interpreted as being generated by the acoustic pulse resulting from a degassing and a boiling of seawater on the ocean floor during the extrusive phase of submarine eruptions. This mechanism is made possible by the shallow character of the source. In a deeper environment (e.g., Mehetia and Teahitia in the immediate vicinity of Tahiti), the higher ambient pressure will prevent the existence of a gas phase, and T waves will therefore be absent (see Sect. 2.2). Frequently, after a period of inactivity, the impulsive arrivals correspond to explosive events accompanying the start of the main phase of some (but not all) of the swarms.

From its discovery in 1967 until December 1977, no activity was detected on the Macdonald. Since then, twenty-four swarms have taken place. The detection capabilities of the Polynesian network have not improved since 1967; thus, the quiescence of the volcano from 1967 to 1977 is real, and similarly, since 1989, no activity has been detected on the Macdonald by this network. The characteristics of the 1977–1983 swarms are summarized in Table 2.3. The 1977 swarm, which started suddenly with explosive sequences characterized by a high level of strongly modulated noise amplitude, was associated with numerous and short-lived explosive phases. The shortest events occurred in 1979, except for the event of 1967, which started with very high intensity explosive phases followed by regularly decreasing noise intensity over several hours. With a sustained noise level marked by the absence of explosive phases, the two swarms of December 1980 and February 1981 are comparable. The November 1980 swarm is reminiscent of that of December 1977, with very numerous explosive phases having a short duration, and a noise level that is strongly modulated in amplitude. Although it is without an explosive character, the crisis of December 1980 is marked by a stronger amplitude fluctuation associated with puffs of noise having a short duration.

Table 2.3. Summary of volcano-seismic 1977–1983 swarms on Macdonald Seamount

Date	Origin time	Duration	Characteristics
29 May 1967	03:22	4 hours	Weak, Macdonald discovered
11 Dec 1977	02:30	5 days	Strong, explosive sequences
30 Sep 1979	12:46	5 hours	Intermediate, short-lived; intense explosions
12 Feb 1980	23:30	12 hours	Intermediate, sustained noise; few explosions
10 Nov 1980	11:09	21 hours	Strong, many explosive sequences
24 Dec 1980	16:10	13 hours	Intermediate, explosions
15 Feb 1981	16:11	13 hours	Intermediate, few explosions
01 Mar 1982	22:07	11 hours	Intermediate, few explosions
05 Jun 1982	04:18	43 hours	Strong, explosive sequences
14 Mar 1983	17:37	9 days	Long and strong, large number of explosions
17 May 1983	05:15	4.5 days	Intermediate, few explosions
27 Oct 1983	14:12	15 hours	Weak swarm, but strong explosive sequences
24 Dec 1983	21:10	9.5 days	Very strong, longest swarm; no explosions

The following swarms were generally more intense. The March 1982 swarm started with a progressive increase in the noise level, followed by puffs of stronger amplitude and a few explosive sequences. The June 1982 sequence started with sudden, explosive events followed by nineteen impulsive sequences, lasting up to eighty seconds with sustained noise of variable intensity during the whole swarm period. The March 1983 swarm started slowly through a series of small explosions with a large number of explosive sequences (about 320) lasting up to 100 seconds, and accompanied by strong puffs of high amplitude noise, some lasting up to an hour. The May 1983 swarm began more slowly with an increase in the noise level, and showed very few explosive sequences; it consisted mostly of noise, whose amplitude is strongly modulated, and which eventually disappears through slow decay. The October 1983 swarm is a short-lived series of twenty-two explosive sequences lasting up to ten minutes, accompanied only by low-amplitude noise. The December 1983 swarm is marked by a long and intense sequence characterized by the absence of explosive events, and the presence of sporadic noise occurring in puffs, which decreases towards the end of the swarm until it dies out. A quasi-continuous noise level was recorded on Christmas Day. After 2.5 years of an apparently quiescent period, the Macdonald activity started violently on May 20, 1986. With explosive starts, the three swarms of May and August 1986, and June 1987 are of relatively short duration. Then, a stronger activity period of nearly two years, during which a total of 200 days of continuous noise separated by only a few quiet sequences, took place.

Figure 2.14 summarizes the history of activity detected on the Macdonald Seamount since 1977. It is clear that the level of activity of the Macdonald Seamount picked up sharply in 1987 and 1988. Nevertheless, the total duration of the swarm activity reported

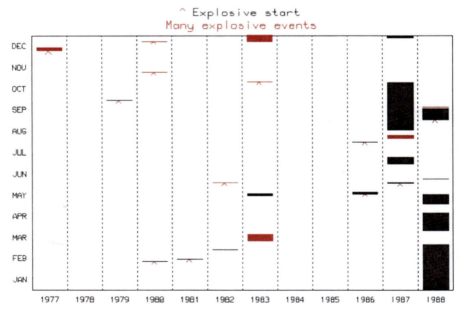

Fig. 2.14. History of the seismic swarm duration at the Macdonald Seamount. The beginning of the explosive events and several small explosions (in *red*) are indicated. The period from 1967 to 1977 was totally quiet

here (about 250 days over eleven years) remains low when compared to other Hawaiian-type volcanoes (e.g., Kilauea, Loihi; Klein 1982) or when compared to Teahitia in the Society hotspot area. It should also be emphasized, however, that if the extrusion took place on the southern flank of the seamount, it would probably go undetected by our stations to the north. Moreover, the activity concentrated inside the crater might inhibit the development of oceanwide T waves. In addition, activity at a weak level following or preceding swarms could also have gone undetected.

The evolution observed in the last few months of 1988 has gone from swarms due to explosive sequences to that of the latest swarm of only sustained noise without explosions, and this is reminiscent of the pattern observed at other volcanic sites. This is the result of a general weakening of the seismic signature after the final fountaining phases of lava extrusion has taken place. An example of this would be the development of a seismic swarm with high and low frequency tremors on the Teahitia Seamount. The intensity and characteristics of the activity during 1987 and 1988 on the Macdonald suggest that an intense phase of volcanic extrusion was then taking place. In conclusion, the seismic detection of hydro-acoustic waves from the Macdonald Seamount has revealed an intense activity between 1979 and 1988, and especially in 1987 and 1988.

2.4.2
Bathymetric Surveys of the Macdonald Seamount

The bathymetric surveys of the Macdonald Seamount comprise at least six expeditions by R. H. Johnson on R.V. HAVAIKI, ARGO, KAWAMEE in 1969-1975, four visits by French ships: the Navy-patrol boat LA PAIMPOLAISE on June 7, 1981, the R.V. MARARA on January 21, 1982, the Navy escort-ship ENSEIGNE DE VAISSEAU (E.V.) HENRY on July 30, 1983, the R.V. MARARA in April 1986, and three cruises by the F.S. SONNE in 1987 (Stoffers et al. 1989) and the N.O. LE SUROIT in 1989 (Cheminée et al. 1991) (Fig. 2.15). Johnson's (1970) surveys identified a shallow pinnacle topping at 49 m below the surface, which was explored by scuba divers during the 1975 expedition.

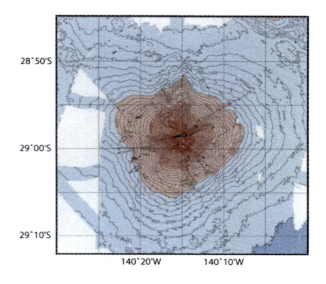

Fig. 2.15. Bathymetry of Macdonald Seamount with 100 m contour lines (courtesy of A. Bonneville)

64 J. Talandier

In 1981, LA PAIMPOLAISE reported a 27 m sounding; however, this ship has a rather poor navigation system. In 1982, the satellite-navigated R.V. MARARA mapped the summit of the seamount as a plateau extending approximately 100 × 150 m, at depths ranging from 34 to 50 m below the surface. A pinnacle at 29 m depth with an elliptical shape of about 30 × 50 m was recognized on the NW side of the plateau (28°59'5" S, 140°15'10" W) (Fig. 2.16a). The horizontal precision of the site location was estimated at 200 m. The general bathymetry of the area surveyed by R.V. MARARA is in good agreement with that obtained by the R.V. KAWAMEE (1973 and 1975). Nevertheless, a

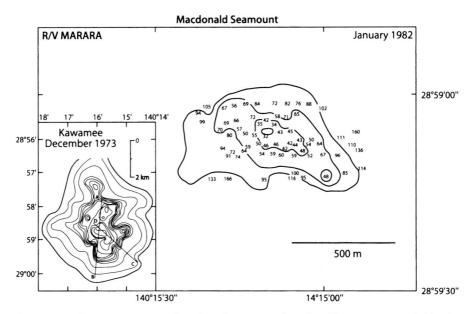

Fig. 2.16. a Bathymetric survey conducted on the summit of Macdonald Seamount recorded by the R.V. MARARA (January 1982). The coordinates were obtained by the ship's satellite navigation system. The *inset at the lower left* is a reproduction of KAWAMEE's 1973 survey (Johnson 1980)

Fig. 2.16. b photograph of a spatter cone located on the summit of the Macdonald taken by scuba divers from the E.V. HENRY in July 1983, at a depth of 40 m

comparison of the two surveys shows fundamental differences. For example, the precise location of the submittal plateau has moved approximately 500 m to ESE. This figure is approximately twice as precise as the figure presented by the KAWAMEE's navigation system but may not be significant. More importantly, the R.V. MARARA data shows the submittal plateau to be extending over a distance of about 150 m for an average depth of 40 m (with many soundings being shallower than 49 m), and this information was not shown in the KAWAMEE's survey. The pinnacle depth reported by Johnson may correspond to the 48 m depth and is located on the SE end of the plateau (Fig. 2.16a). However, it is unlikely that Johnson's repeated surveys would have consistently overestimated the depth of the submittal plateau by a factor of 2. It is also unlikely that he would have missed an existing pinnacle at 29 m depth, given the density of tracks in the immediate vicinity of the summit (Johnson 1980). In addition, the echogram profiles taken across the structure (Johnson 1980) are incompatible with MARARA's chart. We are thus led to propose that the pinnacle at 29 m depth did not exist at the time of Johnson's last survey in 1975, and similarly, that the submittal plateau rose significantly between 1975 and 1982.

In 1983, E.V. HENRY confirmed the presence of the pinnacle, within a 27 m sounding depth. A team of scuba divers explored the central part of the summit plateau at an average depth of 40 m. They identified a fissure with fresh walls and spatter cones made up of scoria-like lava on either side of this rift, about 3 m in diameter and 6 m high (Fig. 2.16a,b). These are not covered with algae, and their summits do not exhibit a cavity. It is probable that they were created by lava ejected from the nearby fissure. No such formations were described by the scuba divers on the 1975 expedition. The absence of glass from lava dredged on the pinnacle and from the summit plateau and a dredging of coral fossils have suggested that both structures were formed in the Pleistocene period, above sea level, and were later sunk by a combination of erosion and eustatic sea-level rise. This would not be possible for the fresh spatter cones observed in 1983.

In 1986, divers from the R.V. MARARA found a spatter cone several meters high made up of a friable, scoria-like, vesicular rock in the area of the 27-meter-deep pinnacle previously observed (Figs. 2.16a,b and 2.17a,b). The submarine photographs taken by divers showed numerous recent structures with hydrothermal deposits (Fig. 16b) and spatter cones in the process of edification (Reymond 1986) (Fig. 2.17a). The spatter cone observed in 1986 was 42 meters deep (Figs. 2.16a and 2.17a), and its vertical vent has emerged on top of a conical pile of blocks and debris that covers an area of more than 10 meters in diameter. The blocks and debris at the base seem to have originally been part of the cone's peak, so the cone has apparently slumped. However, the volume of debris does not seem to coincide with the tens of meters missing from the top (Reymond 1986). It is obvious that there has been a rapid evolution of this structure, and we are left with questions as to the original height of this cone and its underlying plateau (Figs. 2.16b and 2.17a).

2.5
Summary and Conclusions

The volcano-seismic activity of the Society hotspot area has been monitored more closely than that of the Austral hotspot. This is due to the proximity of the Society hotspot to the Tahiti and the Rangiroa stations of the Polynesian Seismic Network

Fig. 2.17. a Sketched cartoon of the volcanic and hydrothermal morphology observed by the IRD scuba divers at the summit of Macdonald in 1986; **b** detail of the highest spatter cone was hand drawn (Reymond 1986); **c** bottom photograph of a collapsed tubular flow associated with ocher Fe-Si-Mn hydrothermal precipitates

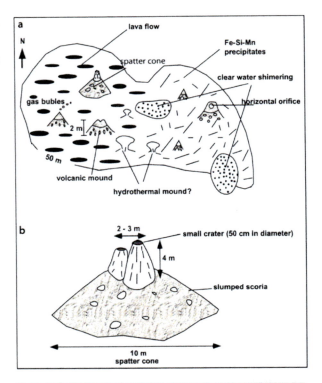

(RSP). Nevertheless, and in spite of the large distance from the receiving stations, T waves still allow an efficient and effective, although less complete, monitoring of Austral hotspot. However, and unlike the Mehetia and Teahitia Volcanoes for which the different volcano-seismic crises can be analyzed in detail, only the last stage of the eruptive process is recorded for the Macdonald Seamount. Unfortunately, the ascent of magma and its transit in that particular volcanic edifice has, up to now, completely escaped observation.

2.5.1
Society Hotspot

The seismic swarms at Mehetia and Teahitia are representative of magmatic phenomena, which have culminated for each crisis with volcanic submarine eruptions. This volcanic activity and the much weaker swarms at Moua Pihaa and Rocard Seamounts in 1969 and 1972 prove that the Society hotspot is alive and active to the east of Tahiti Island.

These seismic crises clearly indicate eruptive processes. The ascent of magma comes from deep regions, where the viscous nature of the medium prevents a strong stress release, or in other words, where no earthquakes are possible. Within the context of the Society hotspot, the first two crises in 1981 at Mehetia and in 1982 at Teahitia took place after a long quiet period. Since they were very close to one another in time and space, they are therefore likely to have a common origin. We can summarize the common characteristics of these two crises as follows: (1) The initial seismic swarm is composed of a large number of small, deep earthquakes which are spatially concentrated, and this seismicity then becomes more widespread as it moves upwards towards the surface; (2) the migration of seismic activity is nearly vertical; (3) the seismic crises occur suddenly, and they are relatively short lived and isolated without any advance seismic manifestations. An ascent of pockets of magma that have separated themselves from the viscous mantle and become more diffuse towards shallower depths could explain this process of seismic migration and its temporal and spatial distribution. Finally, the fact that the first two crises occurred within a one-year interval (at Mehetia in March, 1981 and Teahitia in March, 1982) suggests that the magmatic origin is much deeper than the horizontal distance of 90 km that separates these two volcanoes.

The recent activity on Mehetia and Teahitia is interpreted as being an episode in the ongoing process of building the next major volcanic edifice along the Society Island chain. Despite the fact that only Mehetia has succeeded in rising above sea level, three other sites (Moua Pihaa, Rocard, and especially Teahitia) are active, and extend over an area of about 4 800 km^2 (Fig. 2.4). The spatial distribution of volcanism in the Windward group of the Society Islands is arranged along two lines that are roughly parallel to the absolute motion of the Pacific Plate. (1) The northern line passes through Mehetia, Rocard, Teahitia, and Tetiaroa. Also, the small atoll of Tetiaroa located 50 km north of Tahiti may represent an edifice constructed during the previous stage of activity of the Society hotspot, and it has later been moved by lithospheric sinking under the load of the neighboring and larger Tahiti Volcano. (2) The second line, about 60 km to the south, comprises Moua Pihaa, the Tairapu Peninsula (Tahiti-Iti), Tahiti, and Moorea. This situation is similar to that of the presently active Hawaiian volcanoes, where two volcanic lines 30 km apart were reported. These lines extend from Haleakala to Kilauea and from Kahoolawe to Loihi, as reported by Jackson et al. (1972). Although an explanation has yet to be found for this fascinating pattern, it could be a common property of Hawaiian type hotspot constructional island chains.

In the absence of systematic archives, it is extremely difficult to compile a long-term history of the seismicity of the Society hotspot, and it is even more difficult to estimate the possible recurrence rate of its volcanic activity. There have been several accounts of earthquakes felt by the people residing on the island of Tahiti, as well as in the Polynesian legends, which mention large fires on the island of Mehetia. However, an isolated account of experiencing an earthquake could also be due to a distant high magnitude earthquake such as the one that occurred in Tonga and was felt in Tahiti through its *T* waves (Talandier and Okal 1979). On the other hand, Lespinasse (1919) provided a report of an earthquake swarm that was felt on Tahiti more than 80 years ago:

> Numerous earthquakes of variable intensity were felt starting November 21st [1918], and up to the end of the year. Some days, seismic tremors were felt every hour.

Such a swarm seems to have similar characteristics to the 1985 Teahitia activity, where numerous earthquakes of magnitude $Ml \geq 3.5$ were felt. However, only a few earthquakes were felt during the 1981–1984 crises. This suggests that the Tahiti-Mehetia (Society hotspot) area is experiencing a large-scale period of volcanic activity. A sixty-three-year long period of quiescence, as suggested by this report and by the lifetime of the seismic network, falls within the broad range of observed recurrence of eruptions on the moderately active Hawaiian volcanoes.

The volcano-seismic activity of the Society hotspot would have gone undetected if seismic instrumentation on the nearby islands did not exist. None of the numerous events reported here were able to be detected at teleseismic distances. In the absence of systematic monitoring of lower magnitudes, the events that were felt and reported might have been mistaken for isolated earthquakes of tectonic origin. This raises the question of the real level of underwater volcanic activity in remote ocean basins. However, in situations where the active seamount has grown to shallow depths (e.g., Macdonald Seamount) and when the seamount has penetrated the SOFAR channel, adequate detection is possible through *T* waves even at large distances.

However, if the seamount is small or if eruptions have occurred at large depths, degassing or water vaporization will be absent and no acoustic wave will penetrate the SOFAR channel. Thus, volcanic episodes can only be detected if the seismicity is above the worldwide detection level of $mb = 4.5$. Fast-spreading ridges such as the East Pacific Rise are an example of extremely active underwater volcanism; nevertheless, the events occurring there are not routinely detected. Hence, it is likely that our knowledge about the distribution of active volcanoes on the ocean floor is still quite limited, which leads us to the conclusion that intraplate volcanism is probably more prominent than thought.

The above discussion suggests that unsuspected numbers of active volcanoes on the floor of the world's oceans must exist. For instance, it is unknown if the Marquises hotspot archipelago is currently active. This is a distinct possibility, but its distance from any seismic station (about 800 km away from the Rangiroa Atoll stations) is too great to record even small earthquakes. Furthermore, if the submarine eruptions are too deep, as was the case for Mehetia in 1981 and Teahitia in 1982–1985, they will be undetectable because no *T* waves will be generated.

2.5.2
Austral Hotspot

A combination of seismic detection and geological exploration of the Macdonald Seamount has provided the following information: *(i)* there has been intense volcano-seismic activity from 1977–1988, picking up significantly during 1987 and 1988; *(ii)* the highly probable swelling and upwelling of the submittal plateau as well as the presence of a central pinnacle now reaching to the ocean's surface was observed; *(iii)* recent fissuring and spatter cones, probably due to violent ejections, have been identified;

Chapter 2 · Seismicity of the Society and Austral Hotspots in the South Pacific   69

(iv) the occurrence of hydrothermal activity during the quiescence of the volcano (Figs. 2.17a and 2.17b) as well as the absence of explosive phenomena at the start of the most recent swarms was noted. Both events suggest the permanence of superficial magmatic activity; *(v)* the possible formation of a future dome or crater at the location of the central pinnacle is likely.

At this stage, it is possible to speculate about the eventual emergence of the Macdonald Seamount in the near future. The evolution of volcano-seismic activity, as evidenced by the various surveys made at seven-year intervals, suggests that the seamount's emergence could occur relatively rapidly. However, if a small edifice such as the spatter cones (Figs. 2.16b and 2.17a) should emerge, it is likely that such a fragile construction would be rapidly swept away by the oceanic swell. The durable formation of an island would involve a large, massive and continuous underwater magmatic upwelling and eruption in a relatively short period of time. This is always possible, but the absence of a seismic crisis from 1967 to 1977 and then again from 1989 to 2002 indicates that at present, Macdonald's magmatic activity is rather sporadic. Finally, the dredging of coral fossils at depths of 151 and 283 m (R.V. MELVILLE, Helios expedition, Dredge #2, H. Craig 1987, personal communication) suggests that the edifice has previously emerged during the last glacial period. This, also, suggests that the growth of Macdonald's main edifice is relatively slow.

2.5.3
General Conclusions

The detailed study of the volcano-seismic activity of the Society and Austral hotspots during the last forty years shows the large diversity of manifestations associated with, resulting in, or caused by magmatic intrusions. Extending our study to include what is observed at Hawaii, Pitcairn and on Réunion Island, we have obtained an overview that confirms the specificity of the external manifestations for each of these hotspots. On the basis of approximately fifty years of observations, we have attempted to summarize (Table 2.4) the main features of the development of activity for these hotspots. We

Table 2.4. Generalities on the development and seismic activity of intra oceanic hotspots

	Magmatic activity	Origin seismicity[a]			
		Magmatic	M	Tectonic	M
Hawaii[b]	Quasi continuous	Considerable	≤6.0	Considerable	≤6.5
Society[c]	Sporadic by strong and short crisis	Important	≤4.0	Moderate	≤4.4
Pitcairn[d]	Sporadic by strong and short crisis	Important	≤4.6	Moderate	≤4.7
Austral[e]	Sporadic by crisis of mean duration	Weak	<3.5	Weak	<3.5
Réunion	Quasi continuous	Weak	<2.5	Weak	<3.5

[a] The distinction between 'magmatic' and 'tectonic' origin is sometimes ambiguous in particular in Hawaii. M is the magnitude scale mb or local Ml generally recalled on mb.
[b] Essentially for Kilauea and Loihi from 1973 to 1990: 278 earthquakes of $M \geq 3.0$; $24 \geq 5.0$ and $3 > 6.0$.
[c] For Mehetia and Teahitia region from 1973 to 1990: 100 earthquakes of $M \geq 3.0$ and $6 \geq 4.0$.
[d] The 2001/2002 crisis is considered.
[e] The limit of magnitude is based on the detection threshold of Macdonald earthquakes by the RSP. It is probably that magnitude of 3.5 is not attained.

can distinguish between the so-called 'magmatic' seismicity that is the direct result of magmatic transfers from that of the so-called 'tectonic' seismicity associated with isostatic readjustments. Tremors, which must be distinguished from earthquakes, are also systematically and directly associated with the eruptive process; however, due to a lack of sufficient documentation, the tremors are not taken into account on this table (Table 2.4).

To conclude, although the volcano-seismic activity of the Society hotspot has been documented in far more detail than for other regions of the world's oceans, our present observations are still limited to a relatively short contemporary time period. It is obvious that we are unable to predict the behavior of hotspots over a more extensive time period.

Acknowledgements

I am extremely grateful to Emile Okal who has extensively participated in the different studies that were the basis of this article. Many thanks to Roger Hekinian for his advice and assistance with the manuscript and especially to Ginny Hekinian who corrected this manuscript with a great deal of patience. I also wish to thank Alain Bonneville, who kindly drew and compiled the maps, as well as Dominique Reymond and Olivier Hyvernaud for their research in the archives of the "Laboratoire de Géophysique" in Tahiti.

References

Aki K (1979) Characterization of barriers on an earthquake fault. J Gephys Res 84:6140–6148
Aki K, Koyanagi RY (1981) Deep volcanic tremor and magma ascent mechanism under Kilauea, Hawaii, Characterization of barriers on an earthquake fault. J Gephys Res 86:7095–7109
Aki K, Fehler M, Das S (1977) Source mechanism of volcanic tremor: Fluid-driven crack models and their application to the 1963 Kilauea eruption. J Volcanol Geotherm Res 2:259–287
Butler R (1982) The 1973 Hawaii earthquake: A double earthquake beneath the volcano Mauna Kea. Geophys J Roy Astr Soc 69:173–186
Cheminée JL, Hekinian R, Talandier J, Albarede F, Devey CW, Francheteau J, Lancelot Y (1989) Geology of an active hot spot: Teahitia-Mehetia region in the South Central Pacific. Marine Geophys Res 11:27–50
Cheminée JL, Stoffers P, McMurtry G, Richnow H, Puteanus D, Sedwick P (1991) Gas-rich submarine exhalations during the 1989 eruption of Macdonald Seamount. Earth Planet Sci Let 107:318–327
Chouet B (1981) Ground motion in the near field of a fluid-driven crack and its interpretation in the sudy of shallow volcanic tremor. J Gephys Res 86:5985–6016
Chouet B (1985) Excitation of a buried pipe: A seismic source model for volcanic tremor. J Gephys Res 90:10237–10247
Dietz RS, Sheehy MJ (1954) Transpacific detection of Myojin volcanic explosions by underwater sound. Geol Soc Amer Bull 65:941–956
Eaton JP, Murata KJ (1962) How volcanoes grow. Science 132:925–938
Einarsson P, Brandsdottir B (1984) Seismic acitivity preceding and during the 1983 volcanic eruption in Grimsvotn, Iceland. Jokull 34:13–23
Ewing M, Woollard GP, Vine AC, Worzel JL (1946) Recent results in submarine geophysics. Geol Soc Amer Bull 57:909–934
Jackson ED, Silver EA, Dalrymple GB (1972) Hawaiian-Emperor Chain and its relation to Cenozoic circum-Pacific tectonics. Geol Soc Am Bull 83:601–617
Johnson RH (1970) Active volcanism submarine in the Austral Islands. Science 167:977–979
Johnson RH (1980) Seamounts in the Austral Islands region. National Geographic Society Research Reports 12:389–405
Klein FW (1978) Hypocenter location program HYPOINVERSE. U.S. Geol Surv Open File Rep 78: 694–735

Chapter 2 · Seismicity of the Society and Austral Hotspots in the South Pacific

Klein FW (1982) Earthquakes at Loihi submarine volcano and the Hawaiian hot spot. J Gephy Res 87: 7719–7726

McNutt SR (1983) A review of volcano seismicity. EOS Trans Am Geophys Union 64:265

McNutt SR (1986) Observations and analysis of B-type earthquakes, explosions and volcanic tremor at Pavlov Volcano. Bull Seismol Soc Amer 76:153–175

McNutt SR, Harlow DH (1983) Seismicity at Fuego, Pacaya, Izalco and San Cristobal Volcanoes. Central America Bull Volcanol 46:283–297

Norris A, Johnson RH (1969) Submarine volcanic eruptions recently located in the Pacific by SOFAR hydrophones. J Geophys Res 74:650–664

Okada H, Watanabe H, Yamashita H, Yokoyama I (1981) Seismological significance of the 1977–1978 eruptions and the magma intrusion process of Usu Volcano, Hokkaido. J Volcanol Geotherm Res 9:311–334

Okal EA, Talandier J, Sverdrup KA, Jordan TH (1980) Seismicity and tectonic stress in the southcentral Pacific. J Geophys Res 85:6479–6495

Reymond D (1986) Mission de reconnaissance du Macdonald. Rapport CEA/LDG/PAC

Stoffers P, Botz R, Cheminee JL, Devey CW, Froger V, Glasby G, Hartmann M, Hekinian R, Kogler F, Laschek D, Larque P, Michaelis W, Muhe R, Putanus D, Richnow HH (1989) Geology of Macdonald Seamount: Recent submarine eruption in the South Pacific. Mar Geophys Res 11:101–112

Talandier J (1989) Detection, monitoring and interpretation of submarine volcanic activity. EOS Trans Am Geophys Union Transactions American Geophysical Union 70:560–568

Talandier J (1993) French Polynesia Tsunami Warning Center (CPPT). Natural Hazards 7:237–256

Talandier J, Kuster GT (1976) Seismicity and submarine volcanic activity in French Polynesia. J Geophys Res 81:936–948

Talandier J, Okal EA (1979) Human perception of T waves: the June 22, 1977 Tonga earthquake felt on Tahiti. Bull Seismol Soc Amer 69: 1475–1486

Talandier J, Okal EA (1982) Crises sismiques au volcan Macdonald (Océan Pacifique Sud). C R Acad Sci Paris Sér II 295: 195–200

Talandier J, Okal EA (1984a) The volcanoseismic swarms of 1981–1983 in the Tahiti-Mehetia area, French Polynesia. J Geophys Res 89:11216–11234

Talandier J, Okal EA (1984b) New surveys of Macdonald Seamount, Southcentral Pacific, following volcanoseismic activity, 1977–1983. Geophys Res Lett 11:813–816

Talandier J, Okal EA (1987a) Seismic detection of underwater volcanism: the example of French Polynesia. Pure Appl Geophys 125:919–950

Talandier J, Okal EA (1987b) Crustal structure in the Tuamotu and Society Islands, French Polynesia. Geophys J Roy Astr Soc 88:499–528

Talandier J, Okal EA (1996) Monochromatic T waves from underwater volcanoes in the Pacific Ocean: Ringing witnesses to geyser processes? Bull Seismol Soc Amer 86:1529–1544

Talandier J, Okal EA (1998) On the mechanism of conversion of seismic waves to and from T waves in the vicinity of island shores. Bull Seismol Soc Amer 88:621–632

Talandier J, Okal EA (2001) Identification criteria for sources of T waves recorded in French Polynesia. Pure Appl Geophys 158:567–603

Talandier J, Okal EA, Craig H (1988) Seismic and in situ observations of Macdonald Seamount eruption. 11 October 1987. EOS Trans Am Geophys Union Trans Amer Geophys Un 69:258–259

Chapter 3

A Global Isostatic Load Model and its Application to Determine the Lithospheric Density Structure of Hotspot Swells

F. Avedik · F. Klingelhöfer · M. D. Jegen · L. M. Matias

3.1
Introduction

The concept of "continental drift" advocated by A. Wegener from about 1912 until his death in 1930 was based on geological observations and (for those times) modern principles of isostasy. His idea about the "wandering continents" was complemented and strengthened later on by important notions, among which were the rejuvenation of the oceanic lithosphere and its absorption in the inner Earth after subduction that later evolved as the theory of "plate tectonics" and gained general acceptance in the 1970s. Plate tectonics introduced the radically new notion in geodynamics of the large horizontal motion of about 100 km thick lithospheric plates that were gliding on their substratum, the asthenosphere. The dynamics of the plates, thought to be driven by convection currents in the asthenosphere, determines the relief of the Earth's crust.

The particularly well-observed, large-scale vertical motions of the crust in the oceanic basins shed a new light on the thermal control of the density changes in the lithosphere. T. Crough, dealing with the topography of the ocean floor, wrote in 1983, "Density is apparently so sensitive to temperature that an average change of 600 °C in a 100 km thick lithosphere can change the surface elevation by 3 km, thus causing the observed sea-floor subsidence from the ridge crest to the old oceanic basins." Moreover, numerous hotspot-generated regional elevations of the crust and active volcanoes are frequent in the oceanic area but also present on land, and these provide further testimony to the importance of the thermal control of the density in the lithosphere.

In view of the multiplicity of thermo-mechanical stresses that the lithospheric plates are subject to, it is of primary interest to examine what the scheme of their internal equilibrium is, particularly when responding to these stresses.

Considering the geodynamics of the continental and oceanic lithosphere, it appears plausible to assume that the isostatic compensation scheme is not uniform. The recognition of the fact that the Earth's crust rigidity also contributes to compensate for smaller elements of the topography adds to the complexity in elucidating the mechanisms of isostasy.

The isostatic equilibrium of large-scale topographic features of the oceanic lithosphere and the search to relate and link the different compensation mechanisms acting in the continental and oceanic part of the lithospheric plate have led us to propose a generalized notion of isostasy and an isostatic load model, which combines the Airy and Pratt compensation schemes in the frame of plate tectonics.

The application of this model allows us in turn to investigate and to describe how the density distribution in the lithospheric plate is associated with the major tectonic features that control the regional geology at land and in the oceanic areas.

The following discussion is divided into two parts. In the first part (see Sects. 3.2 and 3.3), we shall present our model and discuss the choice of model parameters. In the second part (see Sect. 3.4), applying this model, we determine and comment on the density distribution in the lithospheric plate associated with tectonic features of the ocean floor, with a special emphasis on structures generated by hotspot activity.

3.2
Isostasy of the Lithospheric Plate

3.2.1
Lithostatic Load

The pressure exerted by a column of rock and/or water at any depth H is the lithostatic pressure and expressed by an n-layered model is

$$P = g \sum_{i=1}^{n} \rho_i h_i \tag{3.1}$$

where ρ_i and h_i are the ith layer density and thickness.

The densities can be conveniently determined from empirical seismic velocity and density relationships and from high-pressure/high-temperature laboratory experiments such as, for example, the work of Ludwig, Nafe and Drake (1970) and Birch (1961). In this paper the polynomial approximation used is

$$\rho = -0.6997 + 2.2302\alpha - 0.598\alpha^2 + 0.07036\alpha^3 - 0.0028311\alpha^4$$

where α is the P-wave velocity (in km s^{-1}) derived from refraction and wide angle reflection experiments and the resulting density is in g cm^{-3}.

As the gravity acceleration "g" can be considered constant down to 200 km depth with an error less than 1%, the lithostatic load (L) in the lithosphere may be defined as the product of layer density and thickness only.

To avoid additional conversion factors, we will express density and thickness in g cm^{-3} and km, which gives the lithostatic load in 100 kg cm^{-2} units. Multiplying the lithostatic load by 9.8 m s^{-2}, one obtains the lithostatic pressure in MPa.

Let us now look briefly at the differences of lithostatic load and gravity potential field, which are both governed by the density of the subsurface.

For a 1D model, the gravity effect and the lithostatic load are identical except for a constant. The gravity effect of a layered model

$$\Delta g = 2\pi\gamma \sum_{i=1}^{n} \rho_i h_i$$

is thus proportional to the sum of the loads.

The sensitivity for both the 1D gravity and load model to the density of a layer are proportional to the thickness of the layer and independent of its depth.

For a two-dimensional or three-dimensional model, the response of the two methods is different. For the isostatic load model, we assume that the load is determined by the underlying vertical density variation only. Thus, the horizontal changes in density can be represented by a series of one-dimensional models. This assumption is reasonable, since the Earth is plastic and horizontal forces are small when compared to vertical forces. The isostatic balance means that rock columns at different locations must exert the same pressure at a given depth.

For the gravity methods, the response depends on the two-dimensional or three-dimensional density variation, i.e., it depends on the density distribution underneath and around the station. A multidimensional density anomaly will cause a response that depends not only on the density contrast, but also on the distance and depth from the observation point.

It is not only of historical interest to remember the origin of the theory of isostasy and the concepts of crustal mass compensation. Beginning in 1850, gravity measurements were made on the Himalayan mountains over a large range of elevations. The corresponding calculated Bouguer anomalies were largely negative over high elevations. Moreover, their magnitude appeared to be a systematic function of the terrain elevation. These results corroborated the findings of earlier plumb line deflection measurements in the Andes and Himalayas, which showed that the observed deflections were smaller than the calculations, because they assumed a uniform density for the earth.

Later, both series of observations led to the suggestion that the Earth's crust must be less dense under the mountains and by analogy under the continents than in oceanic areas where the surface elevation is low. This systematic variation of the Earth's crustal density related to elevation was termed "isostasy".

These discoveries led Airy and Pratt to publish their rivaling theories of isostasy in 1855.

Airy postulated that the Earth's rigid crust of constant average density floats on a denser fluid substratum of constant density. In his concept the surface elevations, the mountains, are supported by the buoyancy of the thickened lower crust forming crustal "roots" (Fig. 3.1).

Considering the hydrostatic type equilibrium of the lithostatic load, Airy's isostatic compensation is expressed by

$$\rho_c h = R(\rho_L - \rho_c) \tag{3.2}$$

where h is the surface elevation, R the thickness of the compensating "root", and ρ_c and ρ_L are the densities of the crust and the substratum, respectively. In oceanic areas where the topography is covered with water, the equilibrium is expressed by

$$h(\rho_c - \rho_w) = R(\rho_L - \rho_c) \tag{3.3}$$

Fig. 3.1. Principles of isostasy after Airy and Pratt

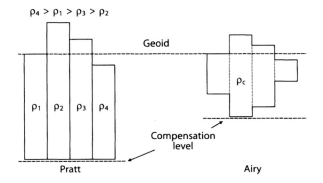

Thus, Airy's type of isostatic compensation concept suggests that the compensation depth is variable and is considered generally to correspond to the base of the deepest "root" in the area of interest.

In Pratt's vision of isostasy, the lighter crust also floats on a denser substratum, but contrary to Airy, the lower boundary of the crust is horizontal and at a uniform depth, the "depth of compensation". To achieve equilibrium of the lithostatic load, Pratt postulates a lateral variation of density in such a way that under crustal elevation, the material between the depth of compensation and the surface is assumed to be lighter than under areas of topographic depression (Fig. 3.1). The Pratt type of isostatic compensation is then expressed by

$$\rho(H_0 + h) = \rho_0 H_0 \qquad (3.4)$$

where ρ is the variable density, H_0 the compensation depth, h is the surface elevation and ρ_0 is the reference density. To define the most probable depth of compensation, the isostatic gravity anomalies were calculated for stations with a wide range of elevation using different compensation depths. The smallest calculated isostatic anomalies were considered to indicate the accurate compensation depth, which was found to be approximately 100 km.

For the past 150 years, the Airy and Pratt hypotheses have offered principles that can explain gravity observations, and based on gravity results alone, one could hardly discriminate between the pertinence of one or the other theory. However, as time passed, doubts emerged about the general validity of the individual concepts. Airy's hypothesis, for example, appears to be more realistic in explaining the geological structures observed in the case of collisional tectonics, while Pratt's concept better explains the higher density of the oceanic substratum deduced from the higher velocity of earthquake waves. On the other hand, Pratt's "about 100 km thick crust" was not supported by the emerging results of seismic experiments.

Because of today's vision of geodynamics brought up by plate tectonics and the knowledge we have gained about the thermal structure in oceanic basins as well on land, the established opinion nowadays is that neither of these hypotheses alone fully explains the observations.

Let us close this section with Alfred Wegener's statement, written in 1915: "... die richtige Deutung dürfte in einer Verbindung beider Vorstellungen zu finden sein ..." (... the correct explanation would probably be a combination of both concepts ...).

3.2.2
The Generalized Equation of Isostatic Load

Since Airy and Pratt formulated their hypotheses concerning the hydrostatic equilibrium of the crust approximately 150 years ago, the concept of plate tectonics and the importance of the thermal structures of the lithosphere related to these dynamics have considerably changed our perception of geodynamics.

These discoveries raise the question about the isostatic equilibrium in a lithospheric plate and the transition between the different mechanisms of isostasy.

To illustrate our approach, let us consider a mature cold lithospheric plate with a thickness H situated at sea level. The crustal part of the plate with the thickness h_c has a block with an elevation (h) compensated by a "root" R according to the Airy compensation model. We assume that the base of the lithospheric plate is a compensation level. The terminology used in the following discussion is illustrated in Fig. 3.2.

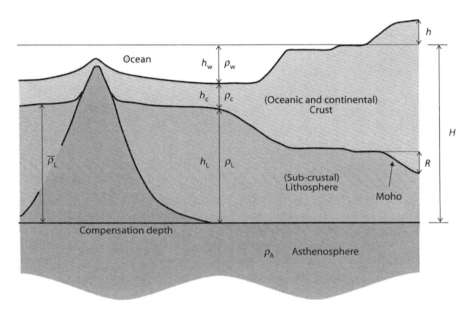

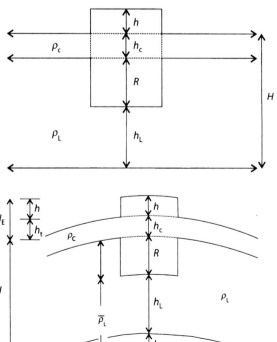

Fig. 3.2a. Schematic diagram of a lithospheric plate

Fig. 3.2b. Explanation of the symbols and terms used in this study

The Airy compensation requires that

$$h = \left(\frac{\rho_L}{\rho_c} - 1 \right) R \tag{3.5}$$

and

$$R = \left(\frac{\rho_c}{\rho_L - \rho_c} \right) h \tag{3.6}$$

where h and R are the surface elevation and the compensating "root"; ρ_c and ρ_L are the crust's and cold sub-crustal lithosphere's density.

Then,

$$H\rho_0 = (h + h_c + R)\rho_c + h_L\rho_L \tag{3.7}$$

expresses the Airy type of compensation of the crust embedded in the lithospheric plate. When the plate passes over a thermally anomalous part of the asthenosphere such as a hotspot, the temperature within the sub-crustal lithosphere increases (reheating) and/or part of the lithosphere is replaced by the upwelling asthenosphere (lithospheric thinning, Crough 1978). Both processes lead to a decrease of the average density of the sub-crustal lithosphere from ρ_L to $\overline{\rho_L}$. We assume that the doming of the crust passively follows the sub-crustal lithospheric thermal expansion h_t and that isostatic load balance is maintained. Thus, the thermal expansion of the sub-crustal lithosphere must be associated with an elevation h_t of the surface.

Such a compensation scheme is described by Pratt's equation. Applying this relation to the lithosphere gives

$$\overline{\rho_L} = (h_L + h_t)\rho_L h_L \tag{3.8}$$

and

$$h_t = h_L \left(\frac{\rho_L - \overline{\rho_L}}{\rho_L} \right) \tag{3.9}$$

where $\overline{\rho_L}$ is the average sub-crustal lithospheric density and h_t the thermally induced surface elevation. They represent the variable parameters, while ρ_L and h_L represent the lithospheric density and thickness and are the reference parameters in Pratt's equation.

The total thickness of the plate including the tectonic and thermal surface elevations is

$$H + (h + h_t) = H + H_E \tag{3.10}$$

and

$$H + H_E = (h + h_c + R) + (h_L + h_t) \tag{3.11}$$

Including the total plate thickness $H + H_E$ in Pratt's equation, we have

$$(H + H_E)\rho_0 = (h + h_c + R)\rho_c + (h_L + h_t)\overline{\rho}_L \tag{3.12}$$

thus, the generalized equation of the isostatic load is

$$H_c\rho_c + H_L\overline{\rho}_L = H_0\rho_0 = \text{const.} \tag{3.13}$$

where
- $H_c = h + h_c + R$ is the total crustal thickness (including the water layer when applicable);
- $H_L = h_L + h_t$ is the total sub-crustal lithospheric thickness;
- ρ_c is the crustal density;
- $\overline{\rho}_L$ is the variable average sub-crustal lithospheric density;
- ρ_0 is the constant reference density; and
- H_0 is the uniform depth of compensation level relative to sea level.

This equation of isostatic load expresses the combination of the Airy and Pratt compensation schemes in the frame of plate tectonics and offers a general description of the isostatic compensation mechanisms in the lithospheric plate. These mechanisms operate in such a way that the sum of the crustal and lithospheric load remains constant at a uniform compensation depth. The important role the crustal load plays in the evolution of the average sub-crustal lithospheric density is implicitly expressed by the isostatic load equation.

Since the general acceptance of plate tectonics in the 1970s, the thermal state of the lithosphere is viewed as a first order parameter. Since the density of the sub-crustal lithosphere is directly related to its temperature, the determination of the density distribution in the lithosphere is obviously an important objective when dealing with geodynamics. Fortunately, the average density of the sub-crustal lithosphere from the isostatic load equation is readily obtained, once the seismic velocity structure (thus the density) and thickness of the crust is determined. From Eq. 3.13, the average sub-crustal lithospheric density is

$$\overline{\rho}_L = \left(\frac{H_0\,\rho_0 - H_c\,\rho_c}{H_L} \right) \tag{3.14}$$

In order that Eqs. 3.13 and 3.14 can be applied to the determination of isostatic balance and sub-crustal lithospheric densities, it is necessary to define

1. The depth of the uniform compensation level H_0;
2. The standard or reference sub-crustal lithospheric density ρ_0;

which are the parameters necessary to determine the constant reference lithostatic load $H_0\rho_0$ at the compensation level.

In the equation of isostatic load equilibrium that integrates both Airy's and Pratt's models of isostasy, we have separated the crustal lithosphere from the sub-crustal lithosphere essentially because of the active role the former plays in the evolution of the average lithospheric density. As most of the discussion in the following sections is devoted to the evaluation and interpretation of this sub-crustal lithospheric density, we propose omitting the word "sub-crustal" from the classification from here on.

3.3
Reference Model

3.3.1
Compensation Depth

The fluid Earth model serves as a reference to describe the Earth's gravity potential field. It can be considered with a good approximation that the Earth's interior of increasing density must have equipotential levels, along which the density and pressure are constant. According to the plate tectonics concept, the Earth's outer shell is composed of lithospheric plates lying on and moving on the asthenosphere. The lithosphere/asthenosphere transition is therefore considered to be the first level having approximately constant physical characteristics. Therefore, it is plausible to assume

1. That the local and regional density- and lithostatic pressure variations observed in the Earth's crust are compensated for in the lithospheric plate itself; and
2. That the lithostatic pressure is constant along the lithosphere/asthenosphere transition level.

On the other hand, the effect of large-scale processes such as, for example, the subduction of a lithospheric plate and the "recycling" of the slabs in the asthenosphere will involve their compensation in the deeper levels of the Earth.

The implication that the lithosphere/asthenosphere boundary is taken as a compensation level raises the question of the nature and depth of this transition.

In dealing with this question, we are reminded by Vogt (1974) that since "there are several possible definitions of lithospheric plate, plate thickness is not necessarily identical in those definitions and this fact should be noted when relevant empirical and theoretical data are compared." There is a great variety of physical parameters that are used to define lithospheric plate thickness. This is also true for plate models, which range from rheological, thermal to seismological models. Nevertheless, there is a marked concordance between different approaches identifying a zone in the 100 km depth range, which may be considered as being the lithosphere asthenosphere transition.

It is of historical interest that it was Wegener himself (1915) who first attempted to address the question of the plate thickness of the moving continental blocks ("Schollen"). His approach (assuming isostatic equilibrium of the continental and oceanic blocks) resulted in an estimate of a thickness of 91 km. He also noted that this value is only approximate because of the very loosely constrained density parameters at that time.

Walcott (1970) calculated a thickness of about 110 km for the normal continental lithosphere and 74 km or more for the oceanic lithosphere using its flexural rigidity.

The advent of exploration seismics and their rapid development in the second half of the 20th century initiated the determination of crustal structures in almost all areas of the world, including the oceanic regions. However, tentative exploration of deeper parts of the lithospheric plate by long-range explosion seismology remained very uncommon because of the considerable technical and financial challenges.

Earthquake body wave analyses frequently show a marked decrease of shear wave velocities in the 100 to 150 km depth range, which is interpreted as the lithosphere-asthenosphere transition.

In modern plate tectonic models, the key feature is the thermal state of the lithosphere. Increasing or decreasing temperatures govern density changes in the lithosphere and result in regional elevation or subsidence. The systematic subsidence of the ocean floor and the decrease of heat flow with age are the primary constraints of models dealing with the thermal evolution of the oceanic lithosphere.

The oceanic lithosphere is created at the Mid-Ocean Ridge, then it cools and subsides as it moves away from the ridge crest. During this process, the lithospheric thickness increases and asymptotically approaches the thickness of a cold mature lithosphere.

One of the first thermal models (Turcott and Oxburgh 1967) considers the newly formed lithosphere as a cooling halfspace. The Parson and Sclater model (1970) views the lithosphere as a cooling plate with a lower isothermal boundary at about 1 350 °C. This model suggests a plate thickness of about 125 km for the cold plate. Both models predict an increasing sea-floor depth (h_w) with age (t) according to

$$h_w = h_0 + a t^{1/2} \tag{3.15}$$

where $h_0 = 2.5$ km to 2.6 km, $a = \sim 0.35$ km Ma$^{-1/2}$ and t is expressed in Ma. The cooling halfspace as well as the plate model fail to match the observed abrupt flattening of the sea-floor depth for ages older than about 70 Ma (Stein and Stein 1992; Smith and Sandwell 1997). It is to be noted, however, that in some oceanic basins this "flattening" of the sea-floor depth with age has not been observed.

For ages greater than about 70 Ma, the sea-floor depth (sediment load corrected) levels off to a depth approximately 5.3 to 5.4 km and stays nearly constant to about 120 Ma when it deepens again (Fig. 3.3). The constant depth of the sea floor and the absence of significant heat flow variations in lithosphere older than about 70 Ma suggest that the lithosphere is at thermal equilibrium in these oceanic basins. Using improved heat flow and depth data, Stein and Stein (1992) found that a somewhat reduced plate thickness of 95 km and a slightly higher isothermal base of 1 450 °C best satisfy the observations.

Fig. 3.3. Relationship between sea-floor depth and age after Smith and Sandwell (1997) and Stein and Stein (1992) (*dashed line*)

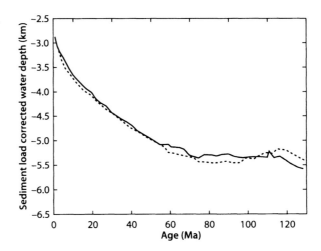

As density is one of the primary parameters of our isostatic load model, it is of interest to compare the above depth estimates for the base of the lithospheric plates with the "depth of compensation" established from gravity observations. Based on these, the most probable compensation depth is obtained when the calculated average isostatic anomalies are minimized. In this way, a depth range of 96 to 122 km was obtained from stations situated over a wide range of elevations in continental areas. A depth of 113.7 km was used by the U.S. Geological Survey as its "standard" for the compensation depth.

The good agreement between the depth ranges suggested by the thermal and gravity models for the base of the lithospheric plate and the compensation depth respectively, leads us to consider the same depth range for the definition of the compensation depth for our isostatic load model. However, the effect of deepening the compensation depth from 95 to 125 km results in a reduction of about 14% of the range of the lithospheric density variations derived from Eq. 3.14. Therefore, in order to keep the maximum span of possible density variations, we have chosen 95 km, the shallowest compensation depth that is compatible with our isostatic load model.

3.3.2
Lithospheric Density

In order to determine the reference lithospheric density for our model, theoretical values may be used as a first approach. Since the Earth is considered a self compressed body with increasing pressure and temperature towards its center and is thought to be composed of essentially homogeneous layers, the increase of the inner Earth's density with depth may be expressed as a function of its radius. Taking also the bulk modulus into consideration, the variation of density with the radius can be further refined by seismic velocities. This nevertheless simplified approach shows that the Earth's density in thermally stable areas is globally constant down to a depth of approximately 400 km.

Using a large and global data set from continental and oceanic areas, H. Hotta (1970) investigated the relationship between the depth of crust/mantle interface (Mohorovicic interface or "Moho") and lithostatic load at this interface. Layer thicknesses were obtained from seismic refraction measurements and layer densities from the widely used seismic velocity versus density relationships cited earlier. Hotta showed that there is a linear relationship between Moho depth and lithostatic load in vast areas of the Earth's surface (Fig. 3.4). However, as Hotta noted, this relationship deviates from linear when the Moho depth is shallower than about 12 km.

Hotta distinguished between data from geodynamically active and inactive or stable areas. The latter are characterized by low-level natural seismicity and by isostatic equilibrium based on gravity studies. Using data from these stable, mainly continental areas, Hotta established the following empirical relationship between lithostatic load and "Moho" depth:

$$L = A + \rho_L h_M = -15.38 + 3.31 h_M \tag{3.16}$$

where L is the lithostatic load in 100 kg cm^{-2} units, ρ_L is the lithospheric density in g cm^{-3} and h_M is the depth to the Moho in kilometers. According to Hotta's definition, Eq. 3.16 represents the "isostatic pressure-depth relationship" of the Moho. From this relationship, Hotta concluded that: *(a)* the lithospheric density in geodynamically stable areas is constant from the Moho down to at least the deepest observations; *(b)* its constant average value is 3.31 g cm^{-3}; (c) that the isostatic compensation of the crustal structures in stable areas must be the same type as described by Airy in his theory.

Adding to Hotta's data some more recent seismic soundings and restricting the original data set to Moho depths greater than 15 km (Fig. 3.5), the parameters in Eq. 3.16 change slightly to ρ_L = 3.332 g cm^{-3} and A = -16.00 kg cm^{-2}, which are adopted for our reference model.

It is noteworthy that in 2000, Darbyshire et al. used a regional data set from Iceland, and in a similar fashion determined a mean mantle density of 3.18 g cm^{-3} for this active region. The high magmatic activity within the Icelandic hotspot may explain the low value of the upper mantle density derived in this area.

Recq (1983), while investigating the isostatic anomalies in the Indian Ocean, observed the increase of sub-crustal (Moho) seismic velocities with increasing distance from the East Indian Ocean spreading center. The lithospheric densities that he determined from these velocities also showed a distance-dependent increase expressed as ρ_L = 4.58 - $L^{1/2}$, where L is longitude. Recq attributed this increase in density to the cooling of the lithosphere moving away from the ridge and proposed a density of

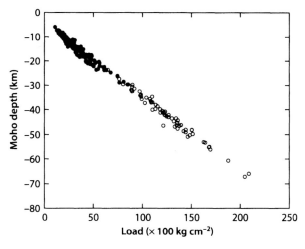

Fig. 3.4. Relationship between Moho depth and lithostatic load (including also the water layer where applicable) in continental (*white dots*) and oceanic (*black dots*) areas (Hotta 1970)

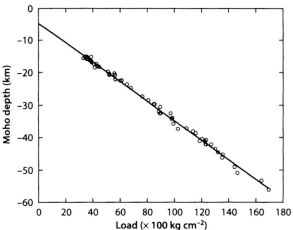

Fig. 3.5. Relationship between Moho depth and crustal load (including the water layer where applicable) for tectonically stable areas (Hotta 1970). A least square linear fit to the data is shown

3.3.3
Location of the Reference Column

Extending Hotta's conclusions, we suggest that in oceanic areas the observed age independence of sea-floor depth of 5.4 km (sediment load corrected) for ages between about 70 Ma and 120 Ma may well signify that these oceanic basins reached thermal conditions corresponding to an average constant density lithosphere which characterizes the geodynamically stable areas discussed previously.

To test the validity of this assumption, we calculate the lithostatic load L at the Moho level as a function of age by assuming average oceanic crustal parameters with crustal thickness $h_c = 7.1$ km, crustal density $\rho_c = 2.83$ g cm^{-3} (White et al. 1992), and the observed water depth variation h_w with age (see Fig. 3.3):

$$L = h_w\rho_w + h_c\rho_c \tag{3.17}$$

where ρ_w is the seawater density.

Using Hotta's "isostatic pressure-depth" relationship (Eq. 3.16) and the calculated lithostatic load L above, a water depth h_{wp} can be predicted:

$$h_{wp} = \frac{L - 16.0}{\rho_L} - h_c \tag{3.18}$$

If our assumption is correct, then the observed and the predicted water depth must agree for ages of 70 Ma to 120 Ma.

The results of our calculations are presented in Fig. 3.6 and show a large divergence of the observed and predicted water depth for young ages, their convergence when the ages increase, and finally, their excellent agreement for ages older than approximately 75 Ma. This agreement suggests that the lithosphere is thermally stable under oceanic basins where the specific depth is 5.4 km.

Because the oceanic crust has a globally constant thickness and density, the divergence of the observed and predicted sea-floor depths must result from a lateral change of lithospheric density. Therefore, the observed sea-floor elevation at ridge crests and on other thermal structures such as hotspot swells is considered to be induced by increased temperature and corresponding decreased densities in the lithosphere.

The importance of the 5.4 km areas of deep oceanic basins (sediment load corrected) is that they represent a smoothly continuous transition zone between the two domains of dissimilar isostasy: the first characterized by the dynamic of a dominantly laterally variable lithospheric density, the second by an essentially constant density lithosphere. The merit of the generalized equation of isostatic load (Eq. 3.14) is that it expresses this continuity and interlacing of the two isostatic compensation mechanisms acting in the lithospheric plate. Thus, the singular characteristics of these oceanic basins at 5.4 km leads us to locate the reference column of our isostatic load model here.

In summary, our reference model has the parameters shown in Table 3.1 (see also Figs. 3.2 and 3.7).

Fig. 3.6. *1:* Observed sea-floor depth versus age after Smith and Sandwell (1997) (*solid line*) and Stein and Stein (1992) (*dashed line*); *2:* predicted sea-floor depth from the crustal load versus Moho depth relationship (Hotta 1970) based on the observations above. See discussion in the text

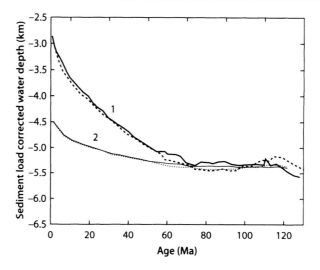

Table 3.1. Parameters of the reference model

Symbol	Parameter	Calculation	Value	Unit
h_w	Water depth		5.4	km
ρ_w	Seawater density		1.03	g cm^{-3}
h_c	Thickness of oceanic crust		7.1	km
ρ_c	Crustal density		2.83	g cm^{-3}
h_M	Moho depth	$h_M = h_w + h_c$	12.5	km
ρ_L	Lithospheric density		3.332	g cm^{-3}
Z_R	Compensation depth, base of plate		95	km
h_L	Thickness of lithosphere	$h_L = Z_R - h_M$	82.5	km
L_R	Lithostatic reference load	$L_R = L_w + L_c + L_L$	300.54 × 10^2	kg cm^{-2}
P_R	Lithostatic reference pressure	$P_R = gL_R$	2 948 × 10^6 29.48	Pa kbar

After substitution of the numerical reference load L_R in Eq. 3.14, we can apply this equation to determine the density distribution in the lithosphere.

The average lithospheric density is

$$\bar{\rho}_L = \frac{L_R - (\rho_c h_c + \rho_w h_w)}{h_L} = \frac{300.54 - (\rho_c h_c + \rho_w h_w)}{h_L} = \frac{L_L}{h_L} \quad (3.19)$$

($\rho_w h_w$ if applicable), when the load is in 100 kg cm^{-2} units and the other units used are in km and g cm^{-3}.

Now that our isostatic load model is complete, we shall compare the results obtained with it to geological and geophysical evidence and show that the isostatic load equation is valid for the different types of compensation as illustrated in the following sections. In the subsequent discussion, our attention will be focused on oceanic areas only.

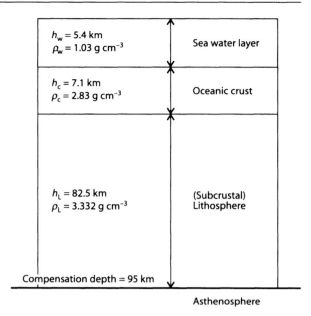

Fig. 3.7. Isostatic load model: reference column. See discussion in the text

3.3.3.1
The Constant Lithosphere Density and Variable Crustal Thickness Domain

Observations on passive continental margins, typically close to or in isostatic equilibrium, show that for given lithospheric, crustal and seawater densities (ρ_L, ρ_c and ρ_w), there is a linear relationship between changes in crustal thickness h_c and in water depth h_w (for example: Charvis et al. 1995). This is expressed and adapted to our model by the following equation:

$$\Delta h_c \frac{\rho_L - \rho_c}{\rho_L - \rho_w} = -\Delta h_w \qquad (3.20)$$

where Δh_c and Δh_w represent crustal thickness and water depth differences with respect to $h_c = 7.1$ km and $h_w = 5.4$ km, the reference values of the isostatic load model.

During the transition from our reference oceanic basin depth to the continental shelf at sea level, the water depth decreases by 5.4 km. According to Eq. 3.20 and assuming that the oceanic and continental crustal densities are closely the same, the predicted increase in crustal thickness will be 24.8 km, which gives (for the average continental crust at sea level) a total thickness of 24.8 + 7.1 = 31.9 km for an average density of 2.83 g cm^{-3}. Thus, for example in the continental shelf region, a three layered crust composed of 5.4 km, 7.1 km and 19.4 km thick layers having 5.7 km s^{-1}, 6.6 km s^{-1} and 6.9 km s^{-1} seismic velocities would satisfy this model. In the same way, a two layered crust, with a seismic velocity of about 6.2 km s^{-1} for the upper 12.5 km and 6.7–6.8 km s^{-1} for the lower layer down to 31.9 km depth could also be representative of a continental crust at sea level. These crustal models not only agree with the observations but also emphasize the validity of our reference model in both the marine and continental environments.

When using published seismic data from various sources, the densities calculated for the continental crust increase with increasing crustal thickness, confirming earlier observations (see, for example, Woolard 1959). An average of 2.82 g cm^{-3} was obtained for a crust about 30 km thick, often taken as a reference in gravity studies for crust "at zero elevation" (sea level). This density value is very similar to the 2.83 g cm^{-3} average obtained for the oceanic crust. For crustal thickness $h_c \sim 40$ km, the average density value increases slightly to 2.86 g cm^{-3}.

Using experimental data from tectonically stable areas in the isostatic load equation (Eq. 3.13), the values obtained for lithospheric densities are grouped around a value of 3.33 g cm^{-3} despite the changes in water depth and crustal thickness, illustrated in Fig. 3.8. With the previously derived sea level crustal thickness, the calculated average density of the entire lithospheric plate is close to 3.16 g cm^{-3}.

Figure 3.9 is a good illustration of the continuity of the constant lithospheric density domain on passive margins despite the great geological change that takes place between the oceanic and continental lithosphere.

We shall briefly outline here the procedure used to calculate the average lithospheric densities, using crustal velocity models in order to illustrate the proportions of cold and hot lithosphere in the plate. First, the layer densities were derived from the layer velocities, using the polynomial relation discussed above (Sect. 3.2.1). The lithostatic load, due to the water layer and the crust, were calculated along each km of the model using the multi-layer equation (Eq. 3.1). The average lithospheric density was then obtained using Eq. 3.19. When the average lithospheric density found is lower than the reference model value $\overline{\rho}_L = 3.332$ g cm^{-3}, we assume that the sub-crustal lithosphere is divided into a cold layer with a reference density (ρ_L) and a hot layer with a density of ρ_{hot}. Then the approximate depth of the boundary between the hot and the cold lithosphere is calculated by $H_{hot-cold} = 95 - (h_L - h_t)$ in km, where $h_t = (L_L - \rho_{hot}h_L) / (\overline{\rho}_L - \rho_{hot})$ and $L_L = \overline{\rho}_L h_L$, which represents the lithostatic load of the lithosphere, as previously obtained. When this boundary is shallower than 60 km, the ρ_{hot} density value will vary according to the average lithospheric density obtained. When the boundary is deeper than 60 km, we have arbitrarily assigned a constant 3.25 g cm^{-3} density to the hot lithosphere.

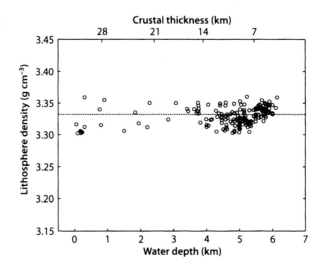

Fig. 3.8. Relationship between water depth, crustal thickness and lithospheric density for tectonically stable areas derived from the isostatic load model. Predicted lithospheric density (*dashed line*) and what is calculated from observations (*circles*)

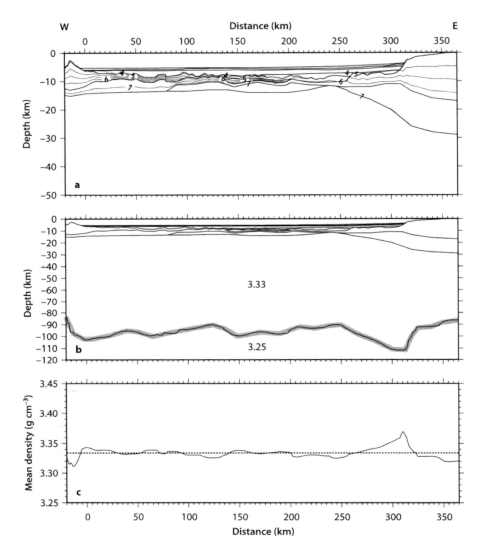

Fig. 3.9. a Crustal structure of the ocean continent transition on the Atlantic coast of the Iberian Craton at 40° N (Dean et al. 2000); **b** approximate depth of the boundary between hot and cold lithosphere; **c** mean lithospheric density

The term "hot lithosphere" used in this study could also designate the "upwelling asthenosphere", whenever required by the geodynamic context.

Throughout this work, the figures will follow the same order:

a   crustal seismic velocity structure; derived from the isostatic load model;
b   approximate depth of the hot/cold lithosphere boundary;
c   average lithospheric density (Eq. 3.19).

3.3.3.2
Variable Lithospheric Density – Constant Crustal Thickness Domain

The creation of the Earth's crust and lithosphere takes place mainly in the oceanic areas at spreading centers or ridges, which represent the longest active volcanic chain on the Earth's surface. Also, the large number of isolated volcanoes as well as the numerous hotspot chains suggest that the oceanic areas belong to the Earth's thermally most dynamic regions. Excluding the volcanic edifices themselves, considerable sea-floor elevations or swells characterize these active areas. As the oceanic crust exhibits globally nearly constant thickness, thermally induced density changes in the lithosphere must characterize the dominant isostatic compensation scheme. Figure 3.10 shows the evolution of the lithospheric density as a function of water depth and age as calculated according to Eqs. 3.17 and 3.19 using average oceanic crustal parameters, $h_c = 7.1$ km and $\rho_c = 2.83$ g cm^{-3}. Approximately 2.5 km water depth is typical over the crests of fast spreading ridges such as for example the East Pacific Rise. As the upwelling asthenosphere here rises close to the sea floor, a density $\overline{\rho}_L$ of 3.25 g cm^{-3} represents its average density. As the crustal thickness at the ridge crest is thinner than the standard crust, the density value calculated at 3.23 to 3.24 g cm^{-3} appears to be a realistic value for the upwelling asthenosphere under steady-state spreading centers. Seismic velocities in the 7.6–7.8 km s^{-1} range are usually found at shallow depths under the ridge crest and correspond to the density range predicted by the isostatic load model.

When the sea-floor elevation reaches sea level, as do some volcanic structures associated with thermal asthenosphere anomalies such as ridge centered hotspots, the predicted average lithosphere/asthenosphere density is 3.19 g cm^{-3}. From this density value combined with the values for oceanic crustal parameters, the average density of the entire active oceanic plate can be calculated. The average value is 3.16 g cm^{-3}, which is the same as the one previously obtained for the continental part of the lithospheric plate at sea level.

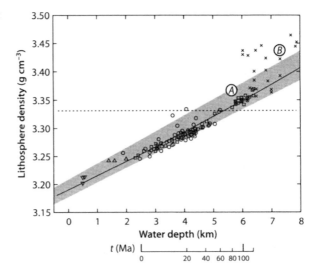

Fig. 3.10. Average lithospheric density versus water depth. Predicted value using a crustal thickness (h_c) of 7.1 km and a crustal density (ρ_c) of 2.83 g cm^{-3} (*solid line*). *Circles, squares, triangles* and *crosses*: experimental data from Hotta (1970), White et al. (1992), Navin et al. (1998) and Weir et al. (2001), respectively. *Shaded area* represents crustal thickness from 4 to 10 km. Time-scale based on sea-floor depth versus age relationship of Parson and Sclater (1970). "A" and "B": see discussion in the text

Using experimental data obtained for the oceanic crust in the 4 to 10 km thickness range in the Atlantic, Pacific and Indian Oceans, the calculated lithospheric densities derived from the isostatic load model are plotted as a function of the sea-floor depth in Fig. 3.10. Again, experimental and theoretical values are in good agreement, confirming the pertinence of the "isostatic load model".

However, there is a group of experimental data (labeled "A" in Fig. 3.10), which, although well aligned with the general trend representing the typical oceanic crust, nevertheless exceeds the limits of the cold lithospheric density (3.33 g cm^{-3}) and water depth (5.4 km) for oceanic basins older than 75 Ma for which the "flattening" of the water depth versus age relationship (Fig. 3.3) is explained by their proximity to hotspot or mantle plumes in these areas in the past. This "A" data material probably comes from areas that are sufficiently distant from the hotspot tracks, such as the Argentine Basin, analyzed by Hohertz et al. (1998), where the flattening of the age vs. depth relationship is not observed. The conclusion of their investigation is that the bathymetry in the Argentine Basin may be explained by the half space cooling model and possibly some additional dynamic effects such as induced flow in the asthenosphere.

A second group of experimental data is distributed in our model over a wide range of lithospheric densities, water depth and crustal thickness (labeled "B" in Fig. 3.10). All these data are for regions of subduction where isostasy, the basic criterion of our isostatic load model, breaks down. To illustrate such a scenario, an example (Fig. 3.11a) that shows the subduction of the Juan de Fuca Plate beneath North America (Gerdom et al. 2000) is used. At the origin of the profile, to the west, the plate age is about 5 to 6 Ma. It is remarkable that while isostatic conditions are maintained, the isostatic load model realistically describes the lithosphere's density and its gradual thickening as it cools when aging. From about 100 km eastwards from the origin of the profile, the process of the mechanical depression of the oceanic crust leads to an increasing discrepancy between Moho depth (when considered incorrectly as the base of the subducted oceanic crust for the whole crustal complex) and the lithostatic load exerted by the overburden. The response of the "isostatic load model" to this setting is an increase of lithospheric densities in order to keep the load on the compensation level constant. Figure 3.11b,c illustrates this process.

Other examples of the evolution of lithospheric densities for different oceanic areas are shown in Figs. 3.12 and 3.13: across the Mid-Atlantic Ridge at 46° N and along the Mid-Atlantic Ridge axis from the Reykjanes Ridge to the Kane Fracture Zone.

It is tempting to use the average lithospheric density values that our generalized isostatic load model predicts, coupled with the temperature of the lithosphere obtained from other independent estimates, in order to assess the off-axis temperature of the lithosphere. The density difference our model predicts between the cold lithosphere under oceanic basins and the ascending asthenosphere at steady-state spreading centers is $\Delta\rho = 0.09$ g cm^{-3}. The associated temperature difference is expressed by $\Delta T = \Delta\rho / (\overline{\rho_L}\alpha)$ and gives a value of about 700 °C difference between the cold and hot lithosphere (where the volume coefficient of thermal expansion $\alpha = 3.8 \times 10^{-5}$ °C). Thus, the average lithosphere temperature estimated for the cold oceanic reference basins is about 600 to 700 °C, depending on the temperature attributed to the upwelling asthenosphere at the spreading center (1 300 to 1 400 °C).

Also it appears from the lithospheric density difference compared to our reference lithosphere (3.332 g cm^{-3}) that the average lithosphere temperature in the area of the "deep oceanic basins" previously discussed (where the water depth is >5.4 km) is about 140 °C lower.

3.3.3.3
Variable Lithospheric Density and Variable Crustal Thickness Domain

Mid plate swells and their associated volcanic areas are the surface manifestations of ascending mantle plumes and hotspots. They are also a typical example of a domain where isostasy is governed by the combined effects of lithospheric density variation (including the role played by the lithosphere's rigidity) and crustal thickness accre-

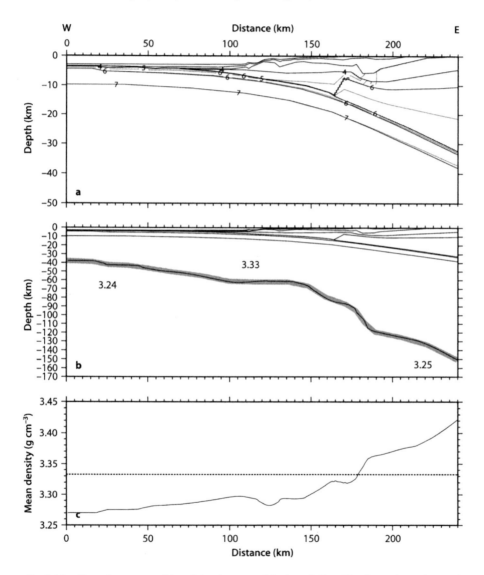

Fig. 3.11. a Crustal structure of the subduction zone of the Juan de Fuca Plate beneath North America (Gerdom et al. 2000); **b** approximate depth of the boundary between hot and cold lithosphere; **c** mean lithospheric density

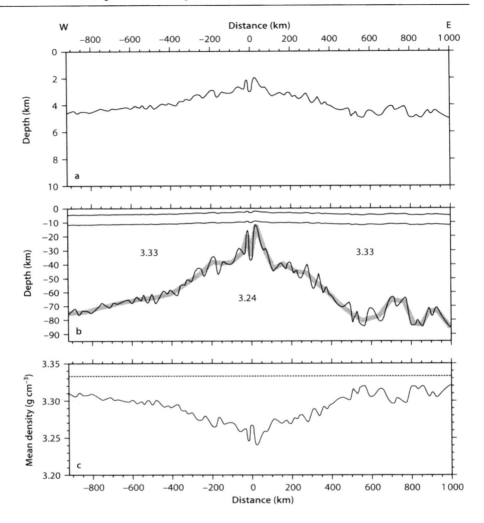

Fig. 3.12. a Bathymetry profile across the Mid-Atlantic Ridge at 46° N (Keen and Tramontini 1970); b approximate depth of the boundary between hot and cold lithosphere; c mean lithospheric density, calculated using a constant crustal thickness of 7.1 km and density of 2.83 g cm^{-3}

tion. Schematically, these hotspot swells and their volcanic structure in fact represent the elevation of the sea floor due to several factors:

- A thermally induced vertical doming (swell) of the crust, which leads to an average depth of 4.25 km (Crough 1978), corresponding to an approximate thermal age of 25 Ma, or alternatively, the sea-floor depth is proportional to the square root of crustal age (Menard and McNutt 1982);
- In addition, a further elevation of the sea floor caused by crustal accretion. The accretion proceeds "from the top" by the accumulation of volcanic material on the existing crust. This additional volcanic load may cause a flexing of the crust. The

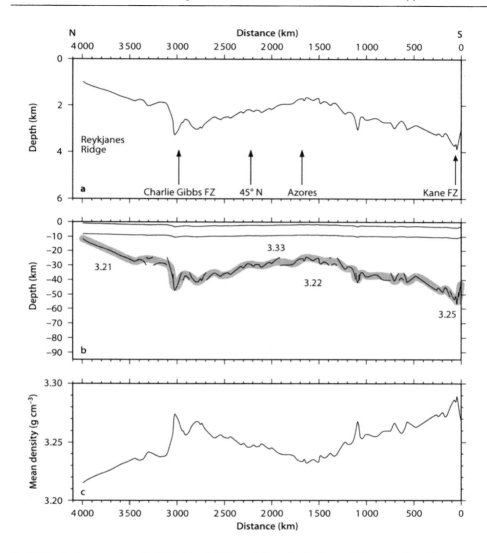

Fig. 3.13. a Bathymetry along the Mid-Atlantic Ridge axis from the Reykjanes Ridge to the Kane Fracture Zone from Mello (1999); **b** approximate depth of the boundary between hot and cold lithosphere; **c** mean lithospheric density, calculated using a constant crustal thickness of 7.1 km and density of 2.83 g cm^{-3}

total crustal thickness is further amplified by ascending asthenospheric material underplating the crust from below in such a way that a crustal accretion of more than 20 km may be attained.

Because of the their particularity, the isostatic compensation of the hotspot-generated structures should have a variable lithospheric density component (also including the lithospheric rigidity) and a variable crustal thickness component depending on the amount of the crustal accretion.

Figure 3.14 shows the schematic evolution of typical hotspot structures in the framework of the "isostatic load model": the "hotspot track".

As a lithospheric plate of age "t" moves towards the hotspot, the increasing thermal flux and the upwelling asthenosphere reheats and rejuvenates the lithosphere, resulting in a decrease of lithospheric densities from $\bar{\rho}_1$ to $\bar{\rho}_2$ (path labeled "A" in Fig. 3.14a). Due to the thermal expansion of the lithosphere, the oceanic crust progressively rises and the "swell" forms for which we have derived a characteristic average of 3.28 to 3.30 g cm^{-3} lithospheric density from the "isostatic load model". Since no crustal accretion occurs ($h_{c1} = h_{c2}$), once the heat input ceases, the subsidence of the thermally rejuvenated lithosphere follows the depth versus age relationship of the original lithosphere on a parallel path, labeled "C" in Fig. 3.14a.

In a further stage of this hotspot-induced process, the melt produced by the progressive decompression of the ascending asthenosphere while the upper lithosphere is also reheated is discharged on the sea floor. Parallel to the beginning of the extrusive volcanic activity and the crustal accretion through the extrusives, an internal lower crustal accretion is thought to begin through underplating (labeled "A" in Fig. 3.14b). Thus the evolution of the "hotspot track" in our isostatic load model follows an intermediate path between the constant- and variable lithospheric density domain corresponding to a state of equilibrium between the crustal growth (crustal load) and the

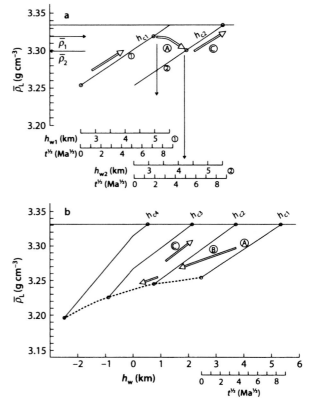

Fig. 3.14. Hotspot structures: mean lithospheric density ($\bar{\rho}_L$) versus water depth (elevation) (h_w) and time t. **a** Hotspot swells: scheme of thermal rejuvenation and subsidence of the oceanic lithosphere (*arrows*). Crustal thickness remains unchanged ($h_{c1} = h_{c2}$). See discussion in the text. **b** Hotspot tracks (*arrows*): crustal accretion and subsidence of hotspot generated structures. Subsidence represents a time span of 75 Ma. The crustal structures subside until the density of the cooling lithosphere increases to 3.33 g cm^{-3}. Association of hotspots and spreading centers (*dots*). Crustal thickness: $h_{c1} = 7$ km, $h_{c2} = 14$ km, $h_{c3} = 21$ km, $h_{c4} = 28$ km. Negative values of water depth represent elevation above sea level. See discussion in the text

lithospheric density and rigidity (labeled "B" in Fig. 3.14b). In this scheme, both the underplated material and the lower crustal "bulge" formed when the crust flexes due to the additional load of extrusives represent the accretion of the lower crust. This crustal "root" is expected to play an important role in the compensation of the whole crustal structure.

The accretionary phase of structures generated by an association of spreading center and hotspot would follow the path illustrated by the dotted line in our scheme (Fig. 3.14b). When the thermal input of the hotspot ceases, the lithosphere will cool and subside again at a rate depending on its effective thermal age (Crough 1978 and Menard and McNutt 1978). In the isostatic load model, the subsidence phase of these structures is labeled "C" in Fig. 3.14b. The subsidence will cease when the average lithospheric density reaches 3.332 g cm^{-3}, the value corresponding to the normal, cold lithosphere.

However, other investigators (Ito and Clift 1998) suggest that in addition to the cooling and thermal subsidence, the uplift caused by the prolonged late stage magmatism and subsequent lower crustal accretion (underplating) also plays an active role in the subsidence of hotspot-generated structures.

In the following sections, we shall discuss in more detail these different phases of the "hotspot track" and the lithospheric density distribution associated with them.

3.4
Lithospheric Density Structure of Hotspot Swells

3.4.1
Introduction

Parallel to the first-order large-scale variation of the sea-floor depth such as mid-ocean ridge crests and oceanic basins, for example, intermediate scale, anomalous elevations of the sea floor on the regional scale were referred to as sea-floor swells, which are generally associated with active or recent volcanism.

Since the middle of the 20th century, the hypothesis for explaining the origin of the swells evolved rapidly. Betz and Hess (1942) proposed an elongated pile of extrusive volcanic rocks for the Hawaiian swell. Dietz and Menard (1953) considered the role of convection cells in the mantle to interpret the swell's characteristics. Wilson (1963) and Morgan (1971, 1972) introduced the notion of spatially stationary thermal anomalies resulting in ascending mantle plumes. Accordingly, swells and the associated volcanic activity are generated when lithospheric plates pass over such thermal anomalies. The concept of lithospheric reheating, rejuvenation thinning and its subsequent subsidence represented a further evolution of the hypothesis about the thermal control of swells (Crough 1975, 1978). Consequently, thermally induced density changes in the lithosphere appear as first order parameters to explain the hotspot-generated swells' morphology and their dynamics.

In the following sections, we shall apply our "isostatic load model" to investigate the density distribution in the lithosphere-asthenosphere system associated with some active hotspot swells as well as the ancient structures that were witnesses of past activity.

3.4.2
French Polynesia, South Pacific Superswell

The Pacific, at about 5 to 30° southern latitudes, is considered to be one of the thermally most active areas of the world's oceans. In addition to the South East Pacific Rise (SEPR), which accounts for one of the most vigorous spreading centers of the Earth, the regional lithospheric density map (Fig. 3.15) shows two prominent low-density zones, oriented approximately in the direction of the actual movement of the Pacific Plate, from the rise westwards:

a   The South Pacific Superswell (1) a triangle shaped zone of elevated sea floor with respect to its age which prolongs the higher western flank of the SEPR (2) towards the Marquesas Fracture zone and islands (3);
b   Numerous volcanic island chains neighboring the Tuamotu plateau (4) in the south are thought to be the superficial expression of the Polynesian hotspots (Society islands – 5, Austral islands – 6, Gambier island – 7).

The northern zone extends the axial low-density area in the direction of the Marquesas Islands, and the southern one evolves towards the Pitcairn-Gambier and Society Island chains. It is interesting to note that the eastward prolongation of the Tuamotu plateau also corresponds to an eastward retreat of the lithosphere of low density.

3.4.2.1
The South Pacific Superswell

Several features are observed along a track running median to the Marquesas and Austral fracture zones, starting east of the SEPR on oceanic crust less than 10 Ma old (Figs. 3.15 and 3.16). The profile clearly shows the topographic asymmetry of the eastern and western ridge flanks. This anomaly cannot merely be explained by the coupling of the western flank to the faster-moving Pacific Plate and to the westward migrating ridge axis. Investigations now suggest an asthenospheric flow from the hotspots to the ridge axis to explain the observations (Toomey et al. 2003).

In the area of the Polynesian hotspot swell with crustal ages going from about 35 Ma to more than 80 Ma, the sea floor remains at an average depth of 4.5 km, which is also the approximate basal depth of the volcanic edifices on the swell. In order to calculate the average lithospheric densities along the profile, a typical crustal thickness for the Pacific Ocean of 6.0 km was used, and only the upper crustal structure was taken into account for the islands (see Fig. 3.16b). The calculated lithospheric densities (Fig. 3.16c) increase from the usual ridge crest values of about 3.25 to 3.30 g cm^{-3} at 30 Ma and remain approximately constant in the region of the Polynesian hotspot swell until passing the Society Island chain. In the area of the Austral Islands the lithospheric density reaches about 3.31–3.32 g cm^{-3}. West-Southwest of the Austral Island chain, both the density and the water depth reach values predicted by the isostatic load model for the cold lithosphere. In order to illustrate the depth reached by the hot lithosphere or the upwelling asthenosphere (Fig. 3.16b), we have assigned a value of 3.25 g cm^{-3} for its effective density. The cold lithosphere is represented by a density of 3.33 g cm^{-3}. It

Chapter 3 · **A Global Isostatic Load Model and its Application**   97

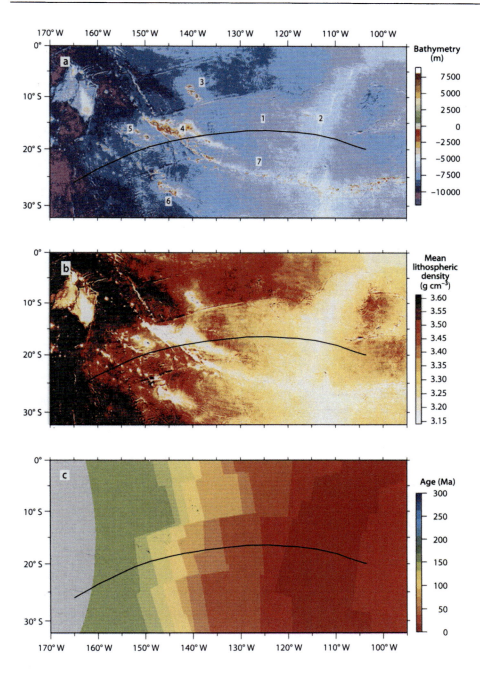

Fig. 3.15. a Bathymetry from Sandwell and Smith (2000); *1:* Superswell; *2:* South East Pacific Rise; *3:* Marquesas Fracture Zone and Islands; *4:* Tuamotu Plateau; *5:* Society Islands; *6:* Austral Islands; *7:* Gambier Islands; **b** mean lithospheric density calculated with a uniform crustal thickness of 6 km; **c** crustal age from Mueller et al. (1997)

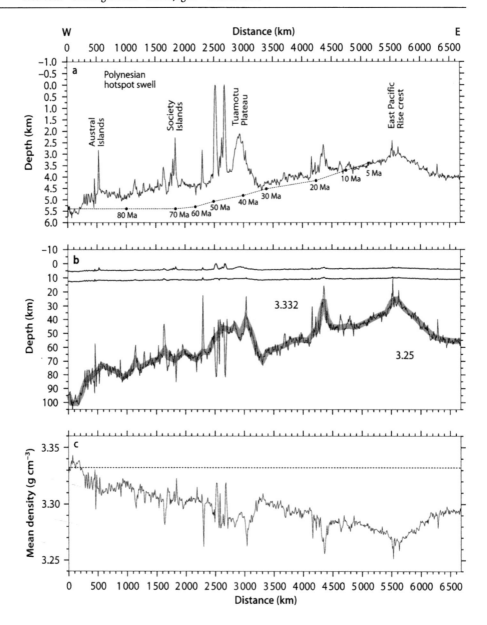

Fig. 3.16. a Bathymetry along the profile indicated in Fig. 3.15 and corresponding sea-floor depth calculated from depth versus age relationship (*dotted line*); **b** approximate depth of the boundary between hot and cold lithosphere; **c** mean lithospheric density, calculated using a constant crustal thickness of 6 km and density of 2.83 g cm^{-3}

is noteworthy that when using these density parameters, the transition between hot and cold lithosphere occurs at about 60 to 70 km depth in the region of the Polynesian hotspot swell.

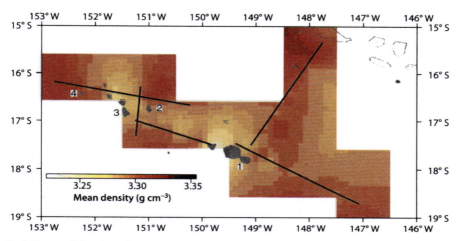

Fig. 3.17. Mean lithospheric density in the region of French Polynesia. Wide-angle seismic profiles (*solid lines*) acquired during the Midplate II cruise (Grevemeyer et al. 2001; Patriat et al. 2002). *1:* Tahiti; *2:* Huahine; *3:* Raiatea; *4:* Maupiti

3.4.2.2
Society Island Chain

One of the best-studied Polynesian hotspot tracks is the Society Island chain, situated on 70–80 Ma lithosphere and oriented about 300° N, which is the present motion of the Pacific Plate. The age of the island chain spans 4.3 Ma. The youngest island associated with the present hotspot is Mehetia, southeast of Tahiti, and Maupiti, to the northwest, is the oldest. Danobeitia et al. (1995), Grevemeyer et al. (2001) and Patriat et al. (2002) established the crustal structure of the island chain along several seismic wide-angle reflection and refraction profiles (Fig. 3.17). These data were used to calculate the regional lithospheric density map of the Society Island chain, which clearly shows (as far as the resolution of the seismic investigation of the crustal structure allows) that the individual islands are all associated with low-density lithospheric roots. It is noteworthy that the low-density lithospheric root shown by the regional map between Tahiti and the western island group does not appear to be associated with any notable extrusive island-building volcanic activity. Thus, this low-density lithosphere root may be the remnant of an asthenospheric upwelling of limited thermal intensity. Figure 3.18 shows the presence of the low-density lithosphere under Tahiti in detail, and consequently the shallow depth of the hot lithosphere, source of the ongoing hotspot-related volcanic activity east of Tahiti.

3.4.2.3
Tuamotu Plateau

The Tuamotu plateau rests on an oceanic lithosphere aged about 35 to 65 Ma. Samples from the northwestern extremity of the plateau show radiometric ages in the 50 Ma range, suggesting an ancient origin for this volcanic edifice. Unfortunately, the crustal base of the plateau has not yet been investigated in detail, a circumstance that prevents us from determining the lithospheric density distribution beneath this structure.

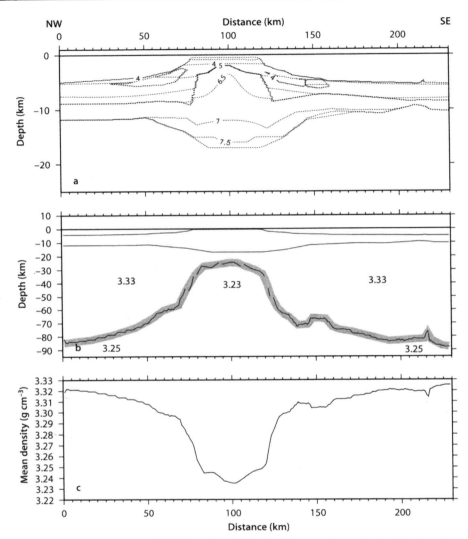

Fig. 3.18. a Crustal structure of Tahiti from Danobeitia (1995); b approximate depth of the boundary between hot and cold lithosphere; c mean lithospheric density

3.4.2.4
Marquesas Island Chain

The age of the oceanic crust in the area of the Marquesas Island chain is about 50 Ma. Despite the location of this chain on the northern side of the Marquesas fracture zone, the regional map of lithospheric densities suggests its connection to the South Pacific Superswell (Fig. 3.15) and to the region of Polynesian hotspots. The 4.5 km average seafloor depth surrounding the islands is slightly deeper than is typical for a hotspot swell, and the lithospheric density is in the 3.31 g cm^{-3} range, somewhat higher than is characteristic for hotspot swells in active areas (Fig. 3.19). The depth of the hot/cold litho-

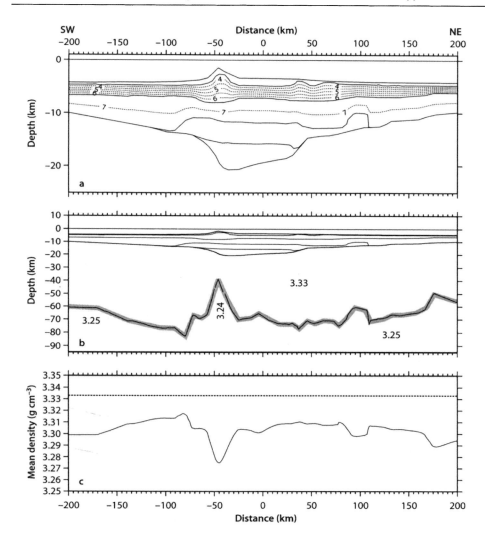

Fig. 3.19. a Crustal structure of the Marquesas from Caress et al. (1995); **b** approximate depth of the boundary between hot and cold lithosphere; **c** mean lithospheric density

sphere boundary is also rather deep (on average ≥70 km) when compared to its depth in the area of the Society Islands. The crustal structure across the island chain shows a large accretion of the lower crust (Caress et al. 1995), which appears to characterize the structure of the island chain as a whole, rather than individual volcanic edifices only. It is noteworthy that the ratio of the higher structures above the surrounding sea-floor level (H_e) and the corresponding lower crustal "bulge" or root (H_R) below the average Moho-depth of the area is approximately 0.3. Considering the usual range of lithosphere to crust density ratios, this particular type of H_e/H_R ratio characterizes, according to Eq. 3.3, crustal structures in an Airy type of isostatic equilibrium, which corresponds to the constant lithosphere domain in our isostatic load model.

The rather high lithospheric density environment coupled with the modest development of the low-density lithosphere beneath the Marquesas Islands do not favor the vision of a dynamic hotspot with associated volcanic activity in this area. The regional lithospheric density map shows, however, some channeling of low-density lithosphere across the Marquesas fracture zone. It is interesting to note in this respect the arguments of Gutscher et al. (1999) for an ancient origin of the Marquesas Plateau overprinted by recent, rather small-scale volcanism.

3.4.3
Hawaiian-Emperor Island Chain

The Hawaiian-Emperor Island chain is clearly representative of an intraplate hotspot-generated structure. This hotspot trace, more than about 5 500 km long, has a volcano-magmatic history spanning as much as 75 Ma. Like some other hotspot tracks in the Pacific, it has a characteristic bend to the north at about 43 Ma of its life, signifying a marked change in the direction of the motion of the Pacific Plate with respect to a fixed hotspot reference frame (Fig. 3.20).

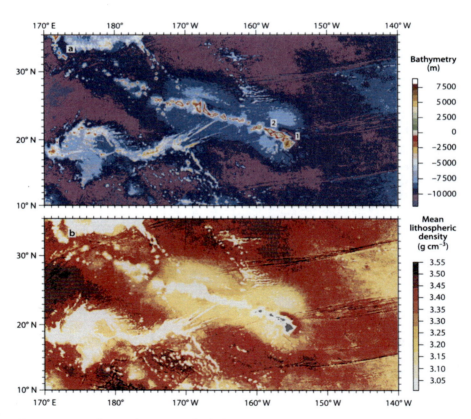

Fig. 3.20. Regional map of Hawaiian Island chain; **a** bathymetry from (Sandwell and Smith 2000), *1:* Hawaii, *2:* Oahu; **b** mean lithospheric density derived from the isostatic load model, calculated with constant crustal thickness of 6.0 km

The younger structures of the seamount chain are built on top of a sea-floor swell, about 1200 km wide, which rises at Hawaii, the youngest element at the southeastern extremity of the hotspot trace, approximately 1.2 km above the depth of the surrounding abyssal plains. The age of the oceanic lithosphere there is about 80 Ma. Due to the large amount of geological and geophysical data accumulated from numerous studies conducted in the region, the Hawaiian island chain is particularly well suited to document the dynamics associated with hotspot structures.

Passing over the ascending hot mantle plume, the old oceanic lithosphere is reheated and thermally rejuvenated. The new, shallower sea floor generated by reheating relates to the "effective thermal age" (t_{eff} in Ma) of the lithosphere by (McNutt 1984):

$$t_{eff} = ((h_w - h_{w0}) / 0.35)^2 \qquad (3.21)$$

where h_w is the actual sea-floor depth in km and h_{w0} = 2.5 to 2.6 km, the reference depth of the well-known age/depth relationship of Parsons and Sclater (1970).

To illustrate the process of thermal rejuvenation of the lithosphere and the subsequent subsidence of the swell in the framework of the isostatic load model, the bathymetric data collected on the Hawaiian swell (Fig. 3.21a) (von Herzen 1989) was used. Figure 3.21b shows the subsidence of the Pacific oceanic crust (h_c = 6 km) from 0 to 70 Ma, accord-

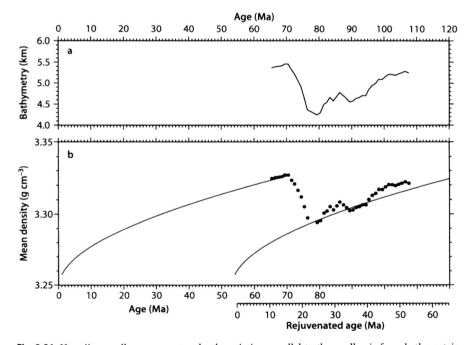

Fig. 3.21. Hawaiian swell; **a** mean water depth variation parallel to the swell axis from bathymetric profiles at distances from the island axis of about 150 km and 300 km to the south and 150 km, 300 km and 450 km to the north (von Herzen et al. 1989); **b** mean lithospheric density-age relationship (*solid lines*) calculated from the isostatic load model assuming a uniform plate velocity of 9 cm yr^{-1} using bathymetry above (*dots*). Passing over the hotspot area leads to a thermal rejuvenation of the lithosphere to an age of about 25 Ma. The subsequent subsidence follows the depth-age relationship (Parson and Sclater 1970) on a curve parallel to the original one

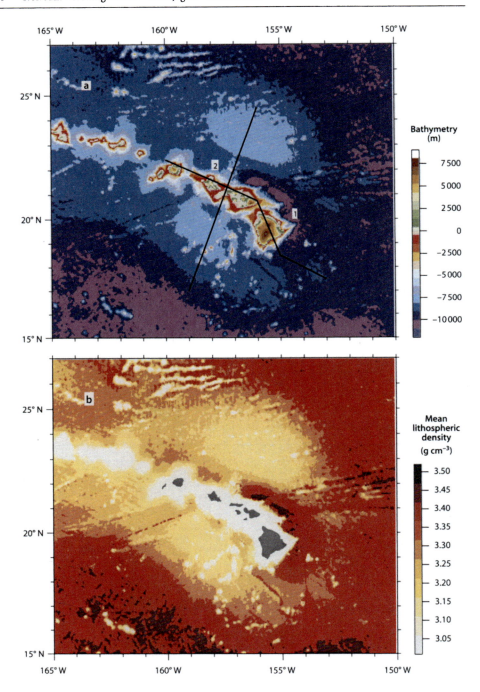

Fig. 3.22. Regional map of Hawaii; **a** bathymetry from Sandwell and Smith (2000); location of seismic profiles (*solid lines*); *1:* Hawaii, *2:* Oahu; **b** mean lithospheric density calculated with a uniform crustal thickness of 6.0 km

ing to the age versus sea-floor depth relationship (also called the time "$t^{1/2}$ law"), and the increase of the lithospheric density from 3.25 to 3.33 g cm^{-3}. The thermal rejuvenation of the lithosphere passing over the hotspot, from about 72 Ma to the effective thermal age of approximately 25 Ma (h_w = 4.3 km) takes place in about 7 Ma. The lithospheric density decreases to 3.29 g cm^{-3}, in close agreement with the value predicted by the isostatic load model for hotspot swells. The subsequent subsidence phase of the swell conforms to its increasing thermal age. At Midway, approximately 2 600 km away (and about 29 Ma later, assuming a Pacific Plate velocity of 9 cm yr^{-1}), the litho-

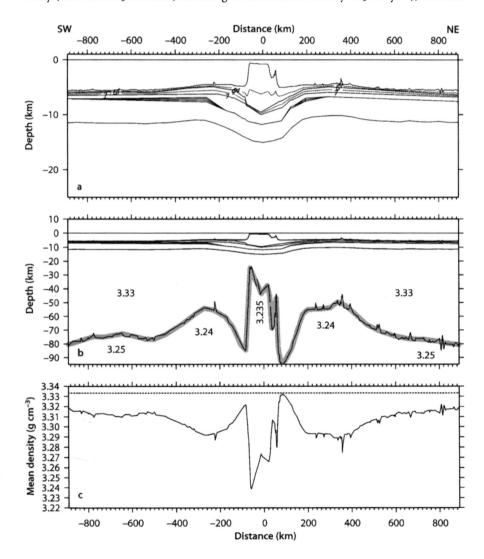

Fig. 3.23. a Bathymetric profile (Watts 1976) and crustal structure (Lindwall 1988) over the Hawaiian hotspot swell; see location in Fig. 3.22; **b** approximate depth of the boundary between hot and cold lithosphere; **c** mean lithospheric density. See discussion in text

spheric density increased again to approximately 3.32 g cm^{-3}, and the swell subsided to 5.2 km, where the lithosphere's thermal age is about 55 Ma.

Seismic investigations were carried out along and across the Hawaiian swell and Island chain (crossing the seamount chain in proximity to Oahu Island (3 Ma)) to explore the crustal structure (Fig. 3.22). The crustal structures and the density distribution in the associated lithosphere are illustrated by Figs. 3.23-3.26. The lithospheric densities were calculated for the across axis profile by using typical Pacific crustal parameters (h_c = 6.0 km, ρ_c = 2.83 g cm^{-3}) and the Hawaiian crustal structure from

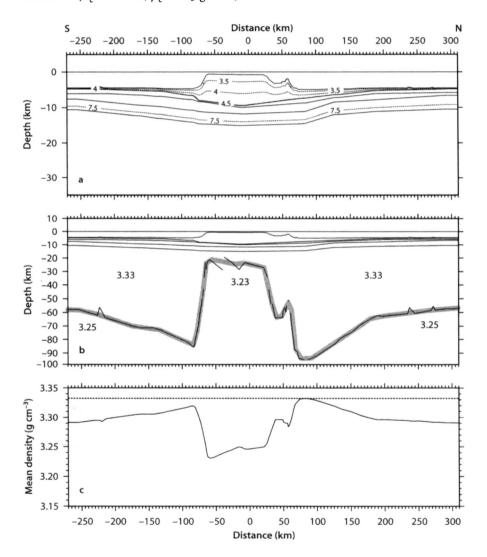

Fig. 3.24. a Crustal structure of Hawaii from Lindwall et al. (1988), derived from the isostatic load model; b approximate depth of the boundary between hot and cold lithosphere; c mean lithospheric density

Lindwall (1988) (Figs. 3.23, 3.24) and Watts (1985) (Fig. 3.25). The profile (Fig. 3.23 shows high lithospheric densities for the abyssal plains at the NE and SW extremities of the profile, about 800 km away from the hotspot track axis. Parallel to the gradual thermally induced elevation of the sea floor, the lithospheric densities decrease to about 3.29 g cm^{-3}, where the sea-floor depth is approximately 4.3 km. Both values, as we have already seen, are typical for hotspot swells.

The flexing of the oceanic crust due to the load of the volcanic edifices starts symmetrically at a distance about 200 km away from the structural axis of the edifices and

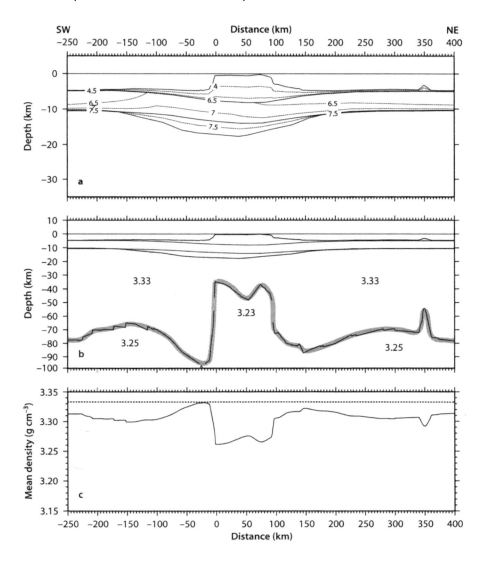

Fig. 3.25. a Crustal structure of Hawaii from Watts et al. (1985); see location in Fig. 3.22; derived from the isostatic load model; **b** approximate depth of the boundary between hot and cold lithosphere; **c** mean lithospheric density

causes a depth increase of its original surface of almost 5 km. The moats, formed essentially above the flanks of the depressed crust, are filled with mainly unconsolidated, light volcanoclastics and sediments. Due to the flexural rigidity of the lithosphere, the depth of the lower crustal boundary (Moho) at the flanks corresponds essentially to the central load exerted by the volcanic edifice and not to the lithostatic load above them. Due to the disproportionate depth of the Moho with respect to the lithostatic load exerted by the unconsolidated moat infill, the lithospheric density calculated from

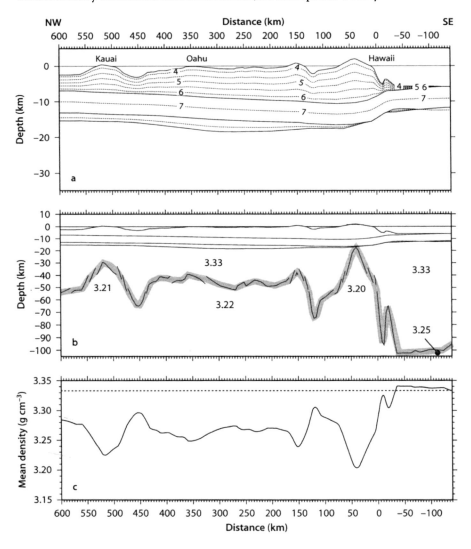

Fig. 3.26. a Crustal structure of the Hawaiian island chain from 0 to approximately 5 Ma age from Watts et al. (1985); see location in Fig. 3.22; **b** approximate depth of the boundary between hot and cold lithosphere; **c** mean lithospheric density

Chapter 3 · A Global Isostatic Load Model and its Application   109

our isostatic load model will tend to compensate by increasing the lithospheric densities in order to insure a constant lithostatic load at the compensation depth. Therefore, the sudden increase of densities on both sides of the prominent low-density root under the volcanic edifice is a response of our model to the crustal geometry and may also be considered as a measure of the lithosphere's rigidity. Moreover, the mean lithospheric density profile has a shape that is similar to the observed gravity signal.

The pronounced low lithospheric densities under the axis of the seamount chain (close to Oahu) of about 3.25 g cm^{-3} are very similar to those observed under highly active oceanic spreading centers and seem to indicate the young thermal age of this volcanic structure (Figs. 3.23–3.25). The essential difference in the crustal models shown is the importance of the lower crustal accretion from underplating. A crustal- and lithospheric density profile along the axis of the young part (0 to about 5 Ma) of the Hawaiian island chain shows (Fig. 3.26) that after the sudden decrease of lithospheric densities to about 3.21 g cm^{-3} at the Hawaiian end of the structure, corresponding to zero age and the beginning of the superficial volcanic activity, the lithospheric densities remain low, for an average 3.27 g cm^{-3}. The boundary of the hot/cold lithosphere is also shallow along the structure corresponding approximately to the 5 Ma interval considered here. The average shallow depth of the low-density lithosphere suggests that even after the extrusive volcanism ceases (after the equilibrium between the ascending magma column and surrounding lithostatic load has been reached), the process of underplating may go on for several million years.

Moreover, the rising of the hot/cold lithosphere boundary under the islands of Maui-Molokai and Kauai suggests that a similar phenomenon must have taken place under the island of Oahu too, but it is not depicted as such by the crustal structure (Watts et al. 1985) (Fig. 3.26a). Lithospheric densities decreasing to approximately 3.23 to 3.24 g cm^{-3} are associated with the periodically rising hot/cold lithosphere boundary. This observation may express the periodic recrudescence of hotspot dynamics along the hotspot track. Assuming a plate velocity of 9 cm yr^{-1} and based on the 5 Ma period considered here, the periodicity of the renewed hotspot activity appears to be 2.0 Ma on average.

A comparison of the approximate 3.21 g cm^{-3} lithospheric density value obtained for the volcanic chain's zero age at Hawaii to the average 3.24 g cm^{-3} calculated for active, steady-state fast spreading centers suggests that the potential temperature of the asthenosphere at the Hawaiian hotspot should be about 200 to 250 °C higher.

3.4.4
Mascarene-Réunion Hotspot Track

The Réunion volcanic island is located on the southwestern extremity of the aseismic Mascarene ridge, which is thought to represent the approximate 65 Ma history of activity of the Réunion-Mascarene hotspot. This history attributes an age span of 65 to 45 Ma to the northern Mascarene Ridge, 7 to 8 Ma to Mauritius Island and 0 to 2 Ma to Réunion Island (Bonneville et al. 1997; Schlich 1982) (Fig. 3.27). The Réunion-Mascarene hotspot is now located beneath an oceanic lithosphere that is approximately 68 Ma old. The hotspot swell characterizing the environment of Réunion Island reaches more than 1 km above the depth of the surrounding oceanic basins.

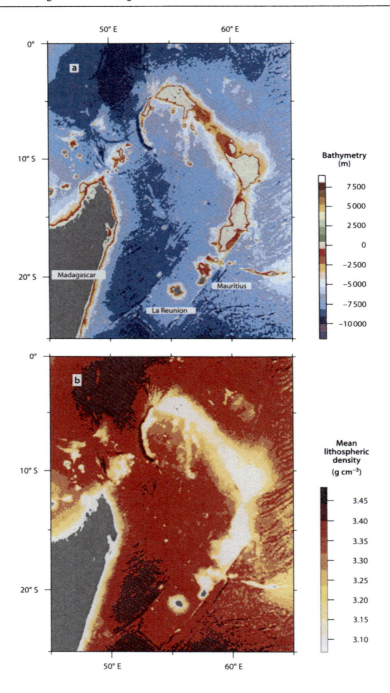

Fig. 3.27. Regional map of the Mascarene Ridge area; a bathymetry from Sandwell and Smith (2000); b mean lithospheric density calculated with a uniform crustal thickness of 7.1 km

The ridge bordered by the Mahahoro Fracture zone to the NW and the Mauritius Fracture zone to the SE separates two major basins of the Indian Ocean (Fig. 3.27). The Mascarene Basin, adjacent to Madagascar Island, was created between magnetic anomalies 34 to 27 (83 to 60 Ma) (Schlich 1982). The Madagascar Basin was formed by the Central Indian Ridge from 66 to 45 Ma (Sclater et al. 1981; Schlich 1982). More recent interpretations suggest the existence of an extinct spreading center south of Réunion Island (Schlich et al. 1990).

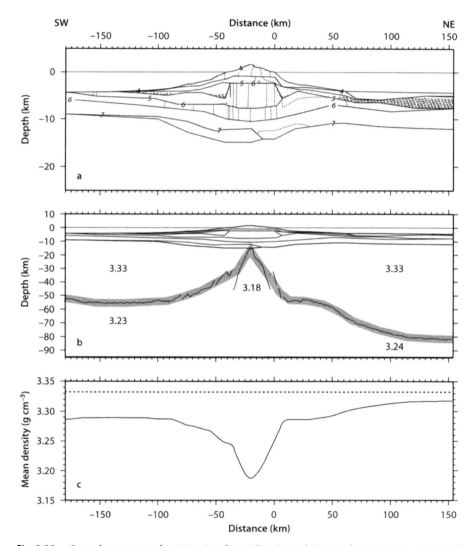

Fig. 3.28. a Crustal structure of La Réunion from Charvis et al. (1999); **b** approximate depth of the boundary between hot and cold lithosphere; **c** mean lithospheric density. See discussion in the text

Gallart et al. (1999) and Charvis et al. (1999) have investigated the crustal structure of Réunion Island's volcanic edifice and of the surrounding oceanic crust (Fig. 3.28a). Summarizing the essence of their findings,

a   The typical oceanic crust under and around the island has a rather constant thickness of about 5 to 6 km;
b   No volcanic load induced flexing of the crust under the edifice was observed;
c   An underplated body has been detected beneath the active volcanic island, which does not seem to be elongated in the SSW to NNE direction of the hotspot track.

The lithospheric densities determined using the published crustal structures led to a typical value about 3.29 g cm^{-3} for the lithospheric density beneath the sea-floor swell south of the volcanic edifice (Fig. 3.28c). Under Réunion Island, the lithospheric densities are as low as 3.19 g cm^{-3}. However, this very low value relates to the peak elevation of the volcanic structure, which is approximately 1.6 km above sea level. Distributing the island's load evenly over its surface, a scale more compatible with the isostatic compensation, the elevation of the island above sea level decreases to 0.9 km and the lithospheric density increases to approximately 3.21 g cm^{-3}. This value is comparable to the lithospheric density determined for Hawaii, suggesting similar hotspot dynamics at both sites. It is important to note that the lithospheric density increases notably towards the NE in the direction of the axis of the hotspot track. Furthermore, the boundary of the hot lithosphere deepens from about 55 km SW of Réunion to 80 km at the NE. These observations tend to show the gradual relative movement of the hotspot plume center to the SW, which is consecutive to the northeasterly plate drift. Therefore, the higher lithospheric densities as well as the pronounced deepening of the hot lithosphere's boundary to the NE of Réunion indicate the presence of colder lithosphere here and consequently, a low-level activity or even a hiatus in hotspot dynamics for a period.

3.4.5
Ascension Island

Ascension Island is a 4 km high volcanic edifice with a basal diameter of 60 km, located on a 7 Ma old oceanic crust in the equatorial Atlantic (8° S), 90 km west of the Mid-Atlantic Ridge between the Ascension Fracture Zone (FZ) and the Bode Verde FZ (Figs. 3.29 and 3.30).

From the anomalously shallow bathymetry of the nearby ridge segment as well as trace element signatures and isotopic ratios of basalt between MORB and enriched basalt from hotspot islands of the nearby ridge axis, it has been concluded that Ascension Island might be of plume/hotspot origin (Brozena 1986; Hanan et al. 1986). Schilling et al. (1985) suggest that the plume location lies close to Circe Seamount, with asthenospheric flow channeling towards the ridge axis or alternatively resulting from the melting of small pockets of heterogeneous mantle that have risen into the melting region at or close to the ridge axis (Minshull et al. 1998).

Radiometric dating of surface lava flows gave ages between 0.6 and 1.5 Ma, and the presence of fresh lava flows suggests that the volcano has probably been active in the last few hundred years. Three-dimensional gravity modeling (Minshull and Brozena 1997) predicts a significant lower elastic thickness (3 ±1 km) of the lithosphere under-

Chapter 3 · **A Global Isostatic Load Model and its Application** 113

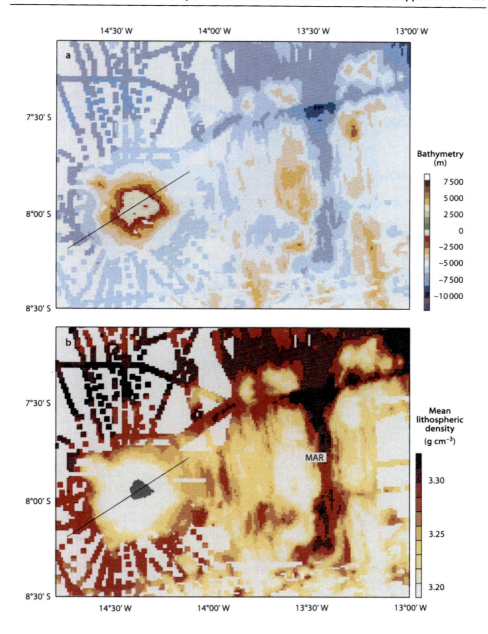

Fig. 3.29. Regional map of Ascension Island; **a** from different bathymetric shipboard measurements; **b** mean lithospheric density calculated with a uniform crustal thickness of 7.1 km

neath the island than the 12 km predicted from thermal cooling-lithospheric thickness relations. Due to the absence of a flexed two-or-three layer boundary, the absence of a thick layer of underplate, the results of the previous gravity modeling, and the existence of a seamount (proto-Ascension) close to the Mid-Atlantic Ridge axis,

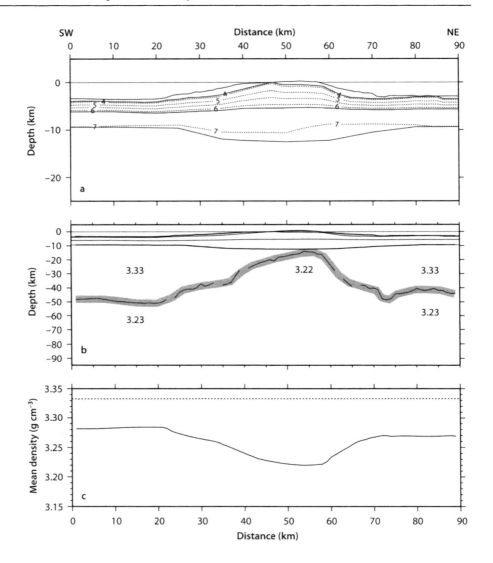

Fig. 3.30. a Crustal structure of Ascension Island from Klingelhöfer et al. (2001); see location in Fig. 3.29; **b** approximate depth of the boundary between hot and cold lithosphere; **c** mean lithospheric density

Klingelhöfer et al. (2001) conclude that part of the island might in fact be older than previously thought and may have been formed in a Mid-Oceanic Ridge environment.

Figure 3.30 shows the crustal model (Klingelhöfer et al. 2001) and the resulting mean lithospheric densities predicted from the isostatic load model. The mean lithospheric densities on the eastern side of the island are slightly lower than those on the western side away from the ridge, which gives some support to the hypothesis of asthenospheric flow from the ridge towards the island as the cause of recent volcanism. The lowest values are found underneath the eastern part of the island close to the most recent volcanism.

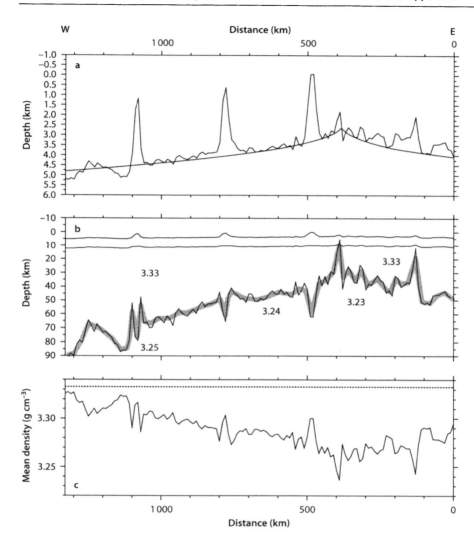

Fig. 3.31. a Bathymetric profile over the Mid-Atlantic Ridge (MAR), Ascension Island and two seamounts mentioned in Klingelhöfer et al. (2001); b approximate depth of the boundary between hot and cold lithosphere; c mean lithospheric density calculated with a uniform crustal thickness of 7.1 km

Generally, the mean lithospheric density is low, due to the young age of the island and the vicinity of the Mid-Atlantic Ridge (MAR) axis. A comparison between the predicted sea-floor depth (Fig. 3.31) and bathymetric data (Brozena 1985) shows that the western side of the MAR follows the theoretical age vs. relationship of Parsons and Sclater, while the eastern side is elevated and thus is evidence of anomalous low density lithosphere.

No lithospheric rejuvenation can be derived from the lithospheric densities determined, a fact which could be interpreted as evidence that part of the island was built directly on the ridge.

3.4.6
The Great Meteor and Josephine Seamounts

Not all seamount chains necessarily have a "hotspot" origin, in the sense that the genesis of these characteristic groups of volcanic edifices is described today. Clusters of seamounts or rises such as the Atlantis-Plato-Cruiser-Great Meteor group or the Madeira-Tore Rise in the eastern North Atlantic represent a particular product of an oceanic lithospheric/asthenospheric system, despite the fact that thermal anomalies of the asthenosphere contributed to their origin.

3.4.6.1
The Great Meteor Seamount

The Great Meteor Seamount, the southern end-member of the seamount group, is located at about 30° N and 8° E, on an oceanic lithosphere approximately 80 to 84 Ma (Fig. 3.32). The seamount, discovered in 1938 by the R/V Meteor, is a large guyot, culminating at a water depth of about 275 m. Two basalt samples dredged from the seamount had radiometric ages of about 11 Ma and 16 Ma (Wendt et al. 1976). The crest of the seamount is thought to have been created under subaerial to shallow water conditions. Foraminiferical limestone samples from the summit suggest ages older than 7 Ma (Pratt 1963). Verhoef and Collette (1987) estimated a Late Oligocene/Early Miocene age (about 22 Ma) for the bulk of the Great Meteor complex.

The seismic refraction exploration of the seamount was conducted onboard the R/V Meteor II in 1990. The results studied here for our analysis were published by Weigel and Grevemeyer (1999) (Fig. 3.33). It is important to note that the seismic investigations did not reveal the flexing of the original oceanic crust, one of the features of the structural analysis of Watts et al. (1975), based essentially on gravity and bathymetric data. Instead of a flexed crust, Weigel and Grevemeyers' study identifies the intrusive core of the seamount and a lower crustal accretion of about 6 km thickness, composed of material with high seismic velocities representing densities in the 3.0 g cm^{-3} range. It is remarkable that in both cases, the calculated gravity signal from the models satisfies the observed gravity field. Let us note that the characteristic moat and its low-density infill, which are generally associated with the presence of flexed crust, are not observed here. The present overall crustal thickness of the structure is approximately 17 km.

The lithospheric densities determined by applying the "isostatic load model" (Fig. 3.33) are an average 3.32 g cm^{-3} surrounding the volcanic edifice, beneath an undisturbed oceanic crust about 7 km thick and a sediment blanket averaging approximately 0.5 km. The water depth, (sediment load corrected) is 4.9 km. Both the shallower sea floor, which is about 0.5 km, and the lithospheric density values, which were lower than were expected for an oceanic crust that is approximately 80 to 84 Ma suggest that the magmatic process that built the seamount must have been associated with a significant thermal rejuvenation of the surrounding lithosphere. The present thermal age of this lithosphere is around 45 Ma.

The average lithospheric density found beneath the volcanic complex is approximately 3.24 g cm^{-3}, in the range of lithospheric densities associated with steady-state spreading centers. This low-density value as well as the shallow depth of the hot/cold lithospheric boundary may explain the youth of the radiometric ages of basalt dredged

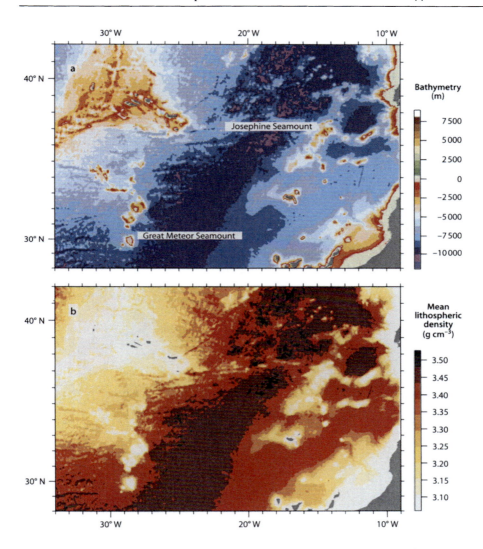

Fig. 3.32. Regional map of the Great Meteor and Josephine Seamount area; **a** bathymetry from Sandwell and Smith (2000); **b** mean lithospheric density calculated with a uniform crustal thickness of 7.1 km

from the seamount, witnesses of a rather recent and only gradually decreasing magmatic dynamic that built the bulk of the volcanic edifice (Verhoef and Collette 1987).

3.4.6.2
The Josephine Seamount

The Josephine Seamount occupies the northern tip of the Madeira-Tore rise and lies across the Azores-Gibraltar fracture zone, at about 38° N and 14° W (Fig. 3.32). The age of the oceanic lithosphere in the area is estimated to be about 120 Ma, whereas the

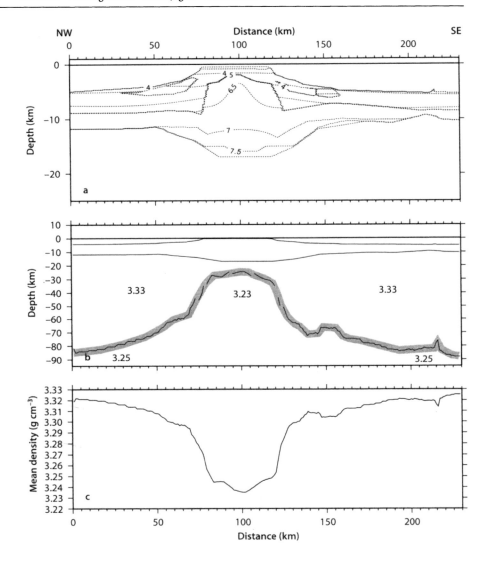

Fig. 3.33. a Crustal structure of the Great Meteor Seamount from Weigel and Grevemeyer (1999); **b** approximate depth of the boundary between hot and cold lithosphere; **c** mean lithospheric density

emplacement of the Madeira-Tore rise must have occurred on young lithosphere at, or adjacent to, the Mid-Atlantic Ridge (Tucholke and Ludwig 1982). The extrusive structure of the Josephine Seamount rises about 3.4 km above the surrounding, approximately 5.2 km deep ocean floor, where the oceanic crust is about 6.3 km thick and shows the typical crustal seismic velocity and density structure (Fig. 3.34). The seismic image does not reveal any flexing of the crust under the seamount. Beneath its extrusive structure, the lower crustal accretion reaches a thickness of about 5.8 km.

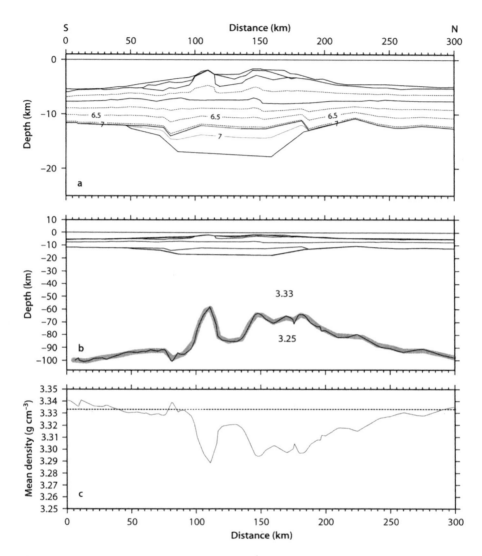

Fig. 3.34. a Crustal structure of the Josephine Seamount from Peirce and Barton (1991); b approximate depth of the boundary between hot and cold lithosphere; c mean lithospheric density

The lithospheric densities calculated for the area around the seamount show high densities characterizing an old and cold lithosphere. Under the seamount, only a slight decrease of the lithospheric density to about 3.30 g cm^{-3} on average is observed, which suggests that the Josephine Seamount, after its emplacement 120 Ma ago, was probably reactivated in its early history and it has been cooling ever since. Alternatively, the position of the seamount on the Azores-Gibraltar Fracture Zone may be responsible for this slight renewal of thermal activity in the lithosphere.

3.4.7
Iceland

The continued interaction between the Mid-Atlantic Ridge spreading center and the Icelandic mantle plume began in early tertiary times following the opening of the North Atlantic, and it still plays a major role in the geodynamics of the region. A substantially higher potential temperature of the asthenosphere causes a large melt production. The thick crust resulting from the increased melt production is present not only in Iceland, but also along the Greenland-Iceland and Faeroe-Iceland ridge, thought to represent the hotspot traces on the American and Eurasian lithospheric plates, respectively. Moreover, bathymetric anomalies generated by the hotspot swell are observed more than 1000 km away from Iceland (Fig. 3.35).

The subaerial locations of the Mid-Atlantic Ridge in Iceland are (Fig. 3.38) in the south, the Western Volcanic Zone, which is a continuation of the Reykjanes Ridge spreading center and further southeast, the Eastern Volcanic Zone, which appears to take over the spreading and transfer it to the Northern Volcanic Zone. The latter links the subaerial Icelandic rift system to the active underwater Kolbeinsey Ridge spreading center. The center of the hotspot plume is currently thought to be located between these two volcanic zones, towards the southeastern shore of Iceland in the area of the Vatnajökull highlands, where the topography has an of average height of 1 km and peaks at about 2 km. The surface rocks on Iceland span ages from 0 to 16 Ma.

The nature of the Icelandic crust is subject to debate and has already generated a considerable amount of literature. One of the two current hypotheses proposes a thin, 10 to 15 km thick crust and a mantle with an anomalous low seismic velocity, indicating possible partial melting at the base of the crust (Gebrande et al. 1980). The other model suggests a crustal thickness of 20 to 40 km under Iceland (Darbyshire et al. 2000). Moreover, the results of recent seismic investigations indicate that the crustal thickness and the elevation decrease with distance from the hotspot center (Darbyshire et al. 2000).

Recently, the results of several surveys contributed to more detailed knowledge about the crustal structure of Iceland and the adjacent areas. To comment on the lithospheric density distribution in the Icelandic region, the crustal models determined by the following surveys will be used:

a "Reykjanes Axial Melt Experiment: Structural Synthesis from Electromagnetic and Seismics (RAMESSES)" (Navin et al. 1998), aiming at the investigation of the crustal accretionary processes on the Reykjanes Ridge at 58°45' N;

b "Reykjanes Ridge Iceland Seismic Project (RRISP)" (Gebrande et al. 1980; Goldflam et al. 1980): an exploration of the crustal structure along an approximate NE-SW profile across Iceland and the southeastern flank of the Reykjanes Ridge;

c "Reykjanes-Iceland Seismic Experiment (RISE)" (Weir et al. 2001), linking crustal structures of the Reykjanes Ridge and Iceland.

d "ICEMELT" survey (Darbyshire et al. 1998), probing the crust and upper mantle beneath central Iceland;

e "Faeroe-Iceland Ridge Experiment (FIRE)" (Staples et al. 1997 and Smallwood et al. 1999), which explores the crustal structure of northeastern Iceland in order to link Faeroe Island to a present-day spreading center.

Figure 3.38 gives a schematic view of the location of the studies.

Chapter 3 · **A Global Isostatic Load Model and its Application** 121

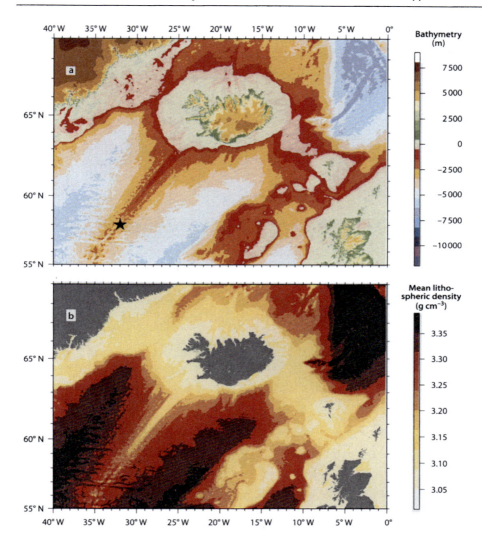

Fig. 3.35. Regional map of Iceland; **a** bathymetry from Sandwell and Smith (2000); **b** mean lithospheric density calculated with a uniform crustal thickness of 7.1 km. *Star* indicates the location of the RAMESSES seismic and electromagnetic experiment

3.4.7.1
"RAMESSES" Study

A seismic and electromagnetic survey has been performed on the Reykjanes Ridge spreading center (half-spreading rate 10 mm yr^{-1}) at about 59° N, approximately 900 km to the south of the Iceland hotspot plume (Fig. 3.35). The results show an anomalously shallow average of 1.8 km water depth of the ridge axis here, clearly the result of the additional thermal influence of the hotspot. However, no notable accretion of the crustal thickness is observed, and the thickness of 7.5 km found here is similar to that of aver-

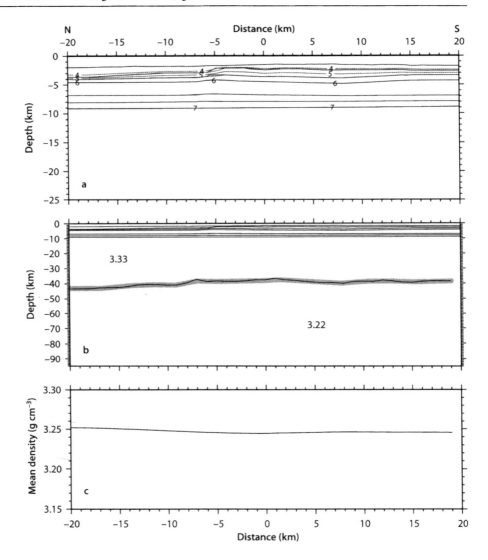

Fig. 3.36. a Crustal structure of along the Reykjanes Ridge at 57°45' N from Navin et al. (1998); see location of experiment in Fig. 3.35; b approximate depth of the boundary between hot and cold lithosphere; c mean lithospheric density

age oceanic crust. The crust-lithosphere boundary is characterized by a 7.8 km s^{-1} velocity, a rather high value in this context, and a narrow well-defined transition zone (Fig. 3.36). The lithospheric densities calculated from the foregoing crustal structure show an approximately uniform value of 3.24 to 3.25 g cm^{-3} along the length of the 60 km on-axis profile and a hot/cold lithosphere boundary about 40 km deep.

It is interesting to note that a cross-axis profile (Fig. 3.37) shows average lower lithospheric densities on the southeastern flank of the Reykjanes Ridge than the along-axis profile, possibly associated with the off-axis extensional tectonics indicated by

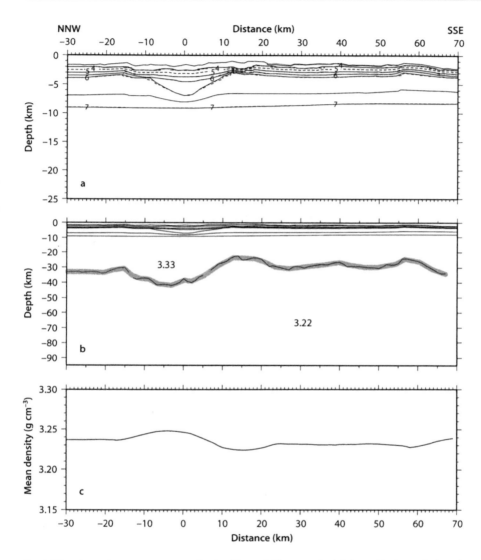

Fig. 3.37. a Crustal structure across the Reykjanes Ridge at 57°45' N from Navin et al. (1998); see location of experiment in Fig. 3.35; **b** approximate depth of the boundary between hot and cold lithosphere; **c** mean lithospheric density

Navin et al. (1998). The rise of the hot lithosphere boundary of about 15 km from the ridge axis towards the southeast associated with low lithospheric densities having an average 3.23 to 3.24 g cm^{-3} also suggest a renewal of magmatic activity that is linked here to the extensional tectonics mentioned previously. Despite the presence of an intra-crustal magma chamber interpreted on the basis of seismic and electromagnetic parameters, the higher lithospheric densities and a deeper lithospheric-asthenospheric boundary under the ridge axis suggest that the investigated segment of the Reykjanes Ridge does not appear to be in an active magmatic phase at the present time.

3.4.7.2
"RRISP" Profile

The combined land-sea seismic survey was carried out in 1977 along a profile that is approximately 800 km, in order to investigate the deep structure beneath Iceland and its transition towards the eastern flank of the Reykjanes Ridge to the south (Fig. 3.38). The marine part of the survey (from about 61°30' to 63°30' N) investigated the oceanic lithosphere along magnetic anomaly 5 (10 Ma) (Goldflam et al. 1980). The study revealed the presence of an oceanic crust that is approximately 9 to 10 km thick and a well-defined Moho discontinuity with a 7.7 km s^{-1} seismic velocity. The anomalously high velocity zone (8.3 to 8.6 km s^{-1}) found in the depth range from about 20 to 50 km may be explained by anisotropy (Goldflam et al. 1980). The lithospheric density calculated along the profile averages at 3.24 g cm^{-3} and is similar to those found at about 59° N (see "RAMESSES" study previously discussed).

According to Gebrande et al. (1980) (Figs. 3.38 and 3.39), a sharp transition in the velocity structure occurs crossing the shoreline to the north. The velocity structure of the mainland of Iceland is characterized by a two-layered crust of variable thickness (10 to 15 km), showing similar seismic characteristics to the one found beneath Reykjanes Ridge. However, the major differences in the sub-crustal structure are the low velocities (7.0 to 7.6 km s^{-1}) reaching down to more than 50 km depth. They are interpreted to be caused by a diapiric upwelling of the asthenosphere in a state of partial melt. Furthermore, considering RRISP profile II and earlier investigations (Bath 1960), the low-velocity lithosphere/asthenosphere appears to extend beneath all Iceland.

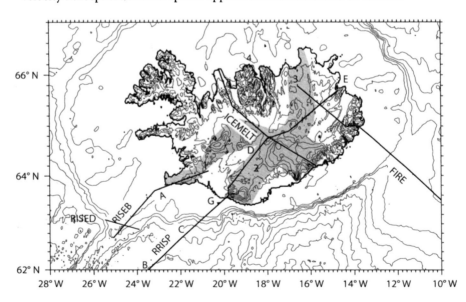

Fig. 3.38. Altimetry and bathymetric chart of Iceland and surrounding ocean (from Sandwell and Smith 2000) showing the location of seismic lines and shot-points (*B, C, G, D, E*) of the RRISP 77 project (from Gebrande 1980). The neo-volcanic zone is outlined in *gray*; *1:* western volcanic zone, *2:* eastern volcanic zone, *3:* northern volcanic zone, *4:* Kolbeinsey Ridge spreading center

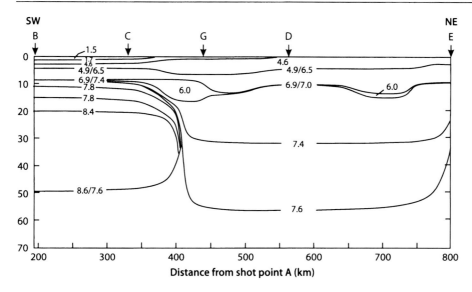

Fig. 3.39. Final velocity model of RRISP profile 1 (from Gebrande 1980); see location in Fig. 3.38

The results of longitudinal (P) and shear (S) wave parameter analyses suggest about 13% melt for the depth range of 20 to 50 km in the area. This would lead to a mean density change of about 0.07 g cm^{-3} for the lithosphere between the Reykjanes Ridge flank at 62° N (crustal age ~10 Ma, lithospheric density about 3.24 g cm^{-3}, see RRISP marine profile, Figs. 3.38 and 3.39) and the active hotspot plume area beneath Iceland (Gebrande et al. 1980). Since velocities in the 7.0 to 7.4 km s^{-1} range are often interpreted as belonging to the lower crust, it is difficult to draw lower crustal boundaries in active magmatic provinces. In the case of Iceland, this boundary may vary between 10 and 45–50 km. Its exact depth is a question of how this boundary is defined.

The 10 to 15 km thick crust suggested by the RRISP model represents, according to today's terminology, the "thin hot crust" hypothesis. Using this crustal structure, the lithospheric density calculated from the isostatic load equation results in an average value of 3.22 g cm^{-3} for the active hotspot plume beneath Iceland. In order to simulate the currently accepted "thick cold crust" solution, we extended the crustal thickness to an average of 30 km using appropriate seismic velocities. The resulting lithospheric density then comes to a value of about 3.26 g cm^{-3}.

3.4.7.3
RISE: Reykjanes-Iceland Seismic Experiment

The study is a combined land-sea survey along the axis of Reykjanes ridge and its extension on the mainland (see profile B and land profile A, Fig. 3.38). A short profile (D) crosses the Reykjanes ridge spreading center at the southern end of the sea profile B. The approximately 150 km long sea profile (Fig. 3.40) shows a typical oceanic crustal structure. However, the crustal thickness is approximately 10.5 km in the south and thickens to 14 km

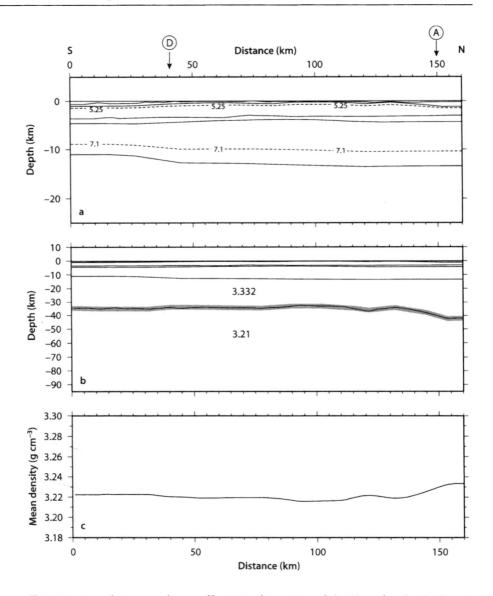

Fig. 3.40. a Crustal structure along profile RISE B from Weir et al. (2001); see location in Fig. 3.38; **b** approximate depth of the boundary between hot and cold lithosphere; **c** mean lithospheric density

beneath the Icelandic shore to the north and thus is about 2.5 to 5 km thicker than the crust found at 59° N by the RAMESSES survey. Using the published crustal structure, the lithospheric density derived from profiles B, A and D (Figs. 3.38 and 3.40–3.42) averages approximately 3.22 g cm^{-3} and is therefore lower than the 3.24 to 3.25 g cm^{-3} densities obtained for the RAMESSES study located further south.

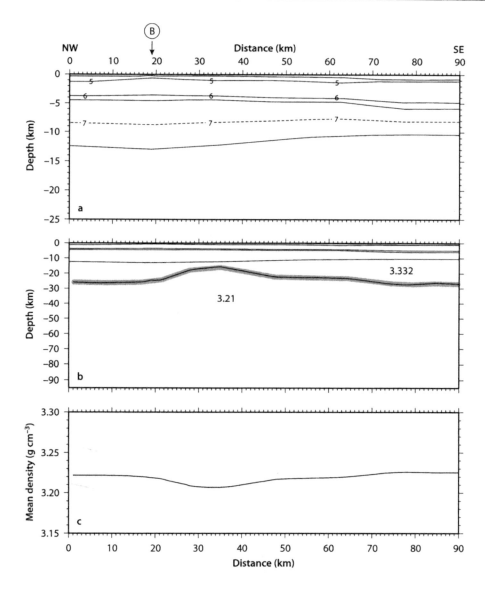

Fig. 3.41. a Crustal structure along profile RISE D from Weir et al. (2001); see location in Fig. 3.38; b approximate depth of the boundary between hot and cold lithosphere; c mean lithospheric density

This lithospheric density difference between these two sites represents an approximate 150 °C temperature increase in the lithosphere/asthenosphere's temperature between the Reykjanes ridge axis at 59° N and adjacent Iceland. The land profile "A", oblique to the off-shore profile "B", follows the shift of the spreading center on land. The crustal thickness increases towards the northeastern end of the profile to about 20 km (Fig. 3.42).

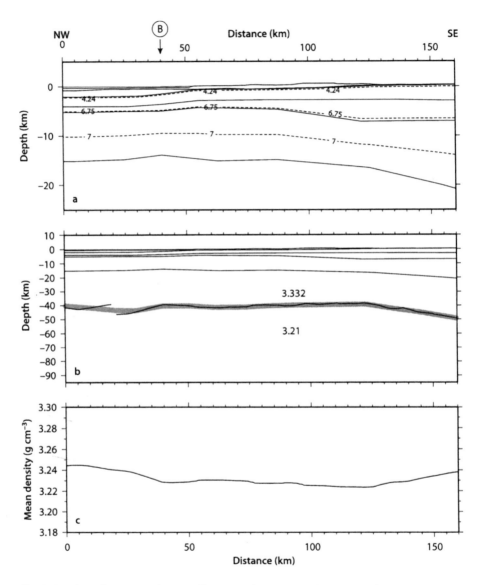

Fig. 3.42: a Crustal structure along profile RISE A from Weir et al. (2001); see location in Fig. 3.38; **b** approximate depth of the boundary between hot and cold lithosphere; **c** mean lithospheric density

3.4.7.4
"ICEMELT" Survey

Another seismic refraction survey was set up to investigate the crustal structure in central Iceland. The profile (Fig. 3.38) starts on the northwestern coast where the terrain is about 1 to 8 Ma, then crosses the Mid-Atlantic spreading center and the Vatnajökull area

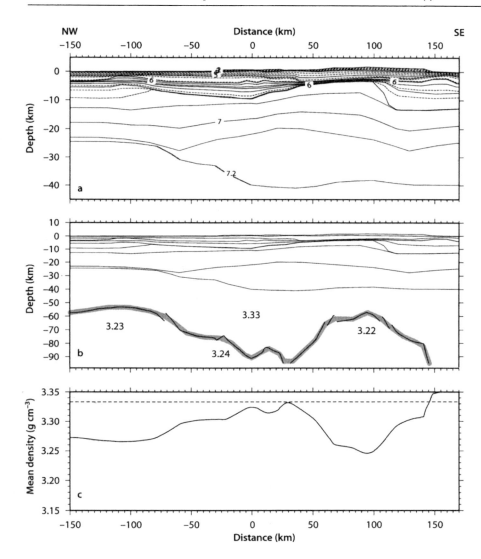

Fig. 3.43. a Crustal structure of central Iceland from the ICEMELT study (Darbyshire et al. 1998); see location in Fig. 3.38; **b** approximate depth of the boundary between hot and cold lithosphere; **c** mean lithospheric density

with mountains that are 1.7 to 2 km high and thought to be above the plume center. The profile ends at the southeastern shore of Iceland on terrain approximately 3 to 8 Ma.

The lower boundary of the upper crust (Fig. 3.43) is defined by the 6.9 km s^{-1} isovelocity, and is approximately 13 to 15 km thick at the northwestern and southeastern end of the profile and decreases gradually to about 9 km below the highlands. The 7.2 km s^{-1} isovelocity in the crustal model is thought to represent the base of the lower crust. The total crustal thickness increases from about 25 to 40 km in the area of the plume center and remains at this level to the southeastern end of the profile. The mean

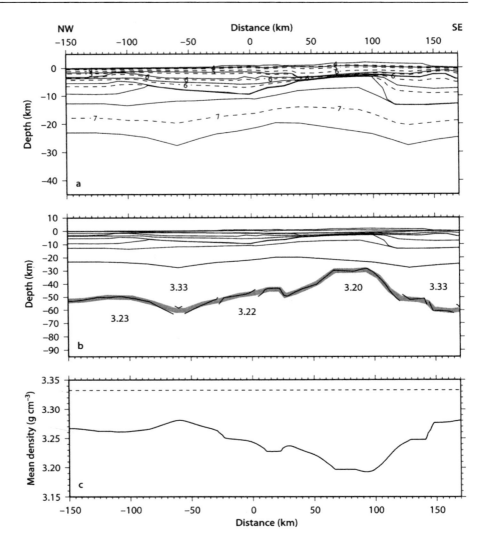

Fig. 3.44. a Crustal structure of central Iceland from the ICEMELT study; crust/lithosphere boundary defined by the 7.1 km s^{-1} isovelocity; modified after Darbyshire et al. (1998); see location in Fig. 3.38; **b** approximate depth of the boundary between hot and cold lithosphere; **c** mean lithospheric density

lithospheric densities show values of 3.27 g cm^{-3} in the northwest region of the profile, roughly corresponding to the Icelandic average obtained from the RRISP study assuming a 30 km thick crust. The lithospheric density then increases to surprisingly high values, which encompass a rather narrow low-density zone in the area of the presumed plume center. The obtained results are difficult to reconcile with the observations, since they show a high-density lithosphere underneath the active subaerial Mid-Atlantic Ridge (at the center of the profile) close to the hotspot plume center. A more plausible lithospheric density model is obtained if the 7.1 km s^{-1} isovelocity is designated as the lower crustal base (Fig. 3.44). This reduces the crustal thickness to more uniform, smaller values between 20 and 25 km. The lithospheric density at the plume center decreases to a value around 3.20 to 3.22 g cm^{-3} and is similar to the lithospheric density obtained beneath the active part of Hawaii for example. However, with the modified model of thinner crust (Fig. 3.44), a gradually decreasing density lithosphere (towards the southeast along the profile) replaces the former high densities associated with the subaerial Mid-Atlantic spreading center to the northwest of the Icelandic plume center.

In light of this modified crustal geometry, the lithospheric density for terrain of similar crustal age at the northwestern and southeastern extremities of the ICEMELT profile are now also similar and in the 3.27 to 3.28 g cm^{-3} range.

3.4.7.5
FIRE: Faeroe-Iceland Ridge Experiment

The combined land-sea seismic project was carried out to explore the crustal structure from the Mid-Atlantic spreading center in Iceland along the Faeroe Ridge (thought to represent the Icelandic hotspot track) to the Faeroe Island block (Figs. 3.35 and 3.38). The geological nature of the Faeroe Island block is controversial and considered alternately as having an oceanic origin or as being a fragment of the Eurasian continent. The FIRE survey was preceded by the Faeroe-Iceland Ridge Project – FIR (Bott et al. 1971) and the North Atlantic Seismic Project – NASP (Zverev et al. 1976; Bott and Gunnarsson 1980).

The Faeroe-Iceland Ridge was created at the Mid-Atlantic spreading center from 54 Ma to about 26 Ma (Bott 1985), when a westward jump of the ridge to the present-day Western Fjords area occurred (Hardarson and Fitton 1993). Spreading from this center generated a thick volcanic crust and lasted to about 7 Ma before the present, when successive eastward jumps transferred the spreading to the Northern volcanic zone.

At the northwestern end of the profile in Iceland, a pronounced crustal thinning from about 30 km to 20 km is interpreted across the Mid-Atlantic spreading center (Northern Volcanic Zone – NVZ) at 0 km on the distance scale of Fig. 3.45. Along the Iceland-Faeroe ridge, the crustal thickness is 25 to 35 km and increases to more than 40 km beneath the Faeroe block. Here, the overall crustal density is close to 3.0 g cm^{-3}. The upper crustal thickness remains rather constant along the whole profile, its base (7 km s^{-1} isovelocity) averaging about 15 km in depth. The lithospheric density value is 3.23 to 3.24 g cm^{-3} for the Mid-Atlantic spreading center, similar to the one obtained for the modified "ICEMELT" profile (Fig. 3.44). Beneath the lower crustal "bulge", which extends southeast from the Mid-Atlantic spreading center to the Icelandic shore, the lithospheric density increases to about 3.30 g cm^{-3} on average. This portion of about 35 km thick crust is considered a remnant of the Western Fjord spreading center discussed previously. From the Icelandic shore to the southeast along the Faeroe-Iceland ridge, the lithosphere's thickness and density values increase gradually from 3.24 to 3.30 g cm^{-3} under the Faeroe block, reflecting the cooling and aging of the lithosphere.

Summarizing the results obtained, it can be seen that low lithospheric densities (lower than the 3.24 g cm^{-3} on average associated with steady-state spreading centers) appear to extend far beyond the limits attributed to the present-day hotspot plume center. Derived

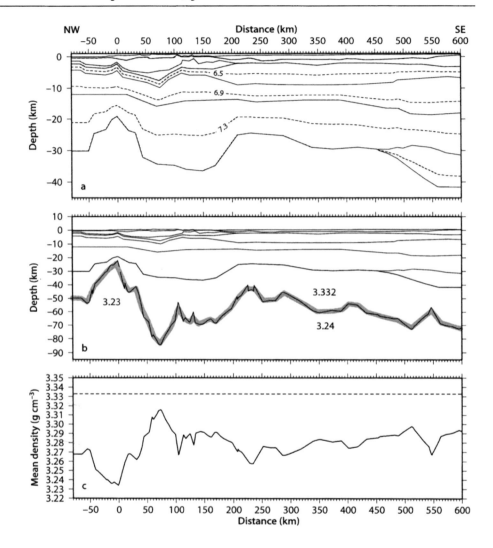

Fig. 3.45. a Crustal structure along profile FIRE from Smallwood et al. (1999); see location in Fig. 3.38; b approximate depth of the boundary between hot and cold lithosphere; c mean lithospheric density

from the density differences, the potential temperature of the asthenosphere appears to be approximately 100 to 150 °C higher than normal, not only beneath the Mid-Atlantic spreading axis, but also under its flanks. Moreover, the temperature difference calculated for the plume center can be as high as 200 to 250 °C. Therefore, from the distribution of the low lithospheric densities in the Icelandic region, one can conclude that a considerably larger volume of the lithosphere/asthenosphere system than merely a hotspot plume contributes to the creation of the thick Icelandic crust. On the other hand, the temperatures derived from the lithospheric densities suggest that this crustal thickness is in the 15 to 25 km range, on the average.

3.5
Subsidence of Hotspot Structures

Following the reheating and thermal rejuvenation of the lithosphere, the hotspot-generated swells will subside again when the thermal input ceases at a rate depending on the lithosphere's effective thermal age (Crough 1987; Menard and McNutt 1978). We discussed and illustrated this process in the framework of the isostatic load model (Sect. 3.3.3.3 and Fig. 3.14) and showed the Hawaiian swell as an example (see Sect. 3.4.3 and Fig. 3.21).

Detrick et al. (1977) have analyzed the subsidence of "aseismic ridges". These volcanic structures, such as the Rio Grande Rise, the Walvis Ridge or the Mascarene Plateau, have been formed in shallow water to subaerial environments. It was found that these aseismic ridges' subsidence rate is comparable to that of the oceanic lithosphere and can therefore be described by a depth-vs.-age relationship ("time $t^{1/2}$ law"). The isostatic load model predicts a crustal thickness in the 15–20 km range for these structures and the associated lithospheric densities should be approximately 3.23 g cm^{-3} during their genesis and youth (Fig. 3.14b). Their subsidence corresponds to the "time $t^{1/2}$ law" and ceases after approximately 75 Ma, when the cooling lithosphere's density reaches 3.33 g cm^{-3}. The predicted water depth should then be in the 2.5 to 3 km range, and this agrees well with the observations.

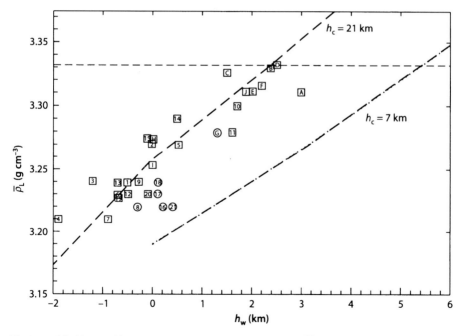

Fig. 3.46. Subsidence of hotspot structures: lithospheric density ($\bar{\rho}_L$) versus water depth (h_w). For comparison, subsidence of oceanic crust ($h_c = 7$ km) and of a crustal structure ($h_c = 21$ km); crustal density: 2.83 g cm^{-3} (*dotted lines*). *Squares:* crustal thickness more than 15 km; *circles:* crustal thickness from 10 to 15 km. *1–6:* Hawaii; *7:* La Réunion; *8:* Ascension; *9:* Great Meteor; *10:* Josephine; *11:* Marquesas; *12:* Tahiti; *13–21:* Iceland; *A:* Agulhas Plateau; *B–D:* Mozambic Ridge; *E–G:* Madagascar Ridge; *H–I:* Kerguelen Island; *J:* Crozet Rise

As a snapshot of their evolution at a given moment, Fig. 3.46 represents the position of hotspot structures discussed in the study as well as others from the Indian Ocean as a function of the water depth (or elevation) and lithospheric density. It is remarkable that the average trend follows the subsidence curve of an oceanic-type crust that is approximately 18 to 20 km thick until it reaches the density value of the cold lithosphere (3.33 g cm^{-3}) at a water depth average of 2.5 km. It is noteworthy that the trend observed here also compares well with the predictions of the isostatic load model (Fig. 3.14b).

Considering the same structures as above, the observed H_e/H_R ratios can now be examined as a function of the lithospheric density, where H_e represents the elevation of the extrusive structure above the surrounding topography and H_R the lower crustal "root" below the average depth of the unaltered crust-lithosphere boundary, the "Moho". It appears from Fig. 3.47 that while high H_e/H_R ratios are associated with low lithospheric density (hot lithosphere), decreasing H_e/H_R ratios are quite the reverse, as they are linked to the gradual increase of lithospheric densities. The lowest H_e/H_R ratios, on an average of 0.2–0.25, also represent structures with well-developed crustal "roots" embedded in a high density, already matured and cold lithosphere. The possible and also plausible explanation for the decreasing H_e/H_R ratios of these volcanic structures while the lithospheric density increases might be a lower crustal growth from flexing ("bulging") of the crust and/or the progressive "freezing" to the crust or underplating of the melt accumulated near the base of the crust, which accompanies the cooling of

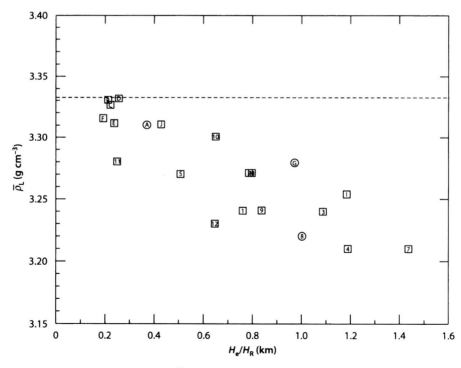

Fig. 3.47. Mean lithospheric density ($\bar{\rho}_L$) for different ratios of crustal elevation to lower crustal root, H_e/H_R. Symbols as in Fig. 3.46

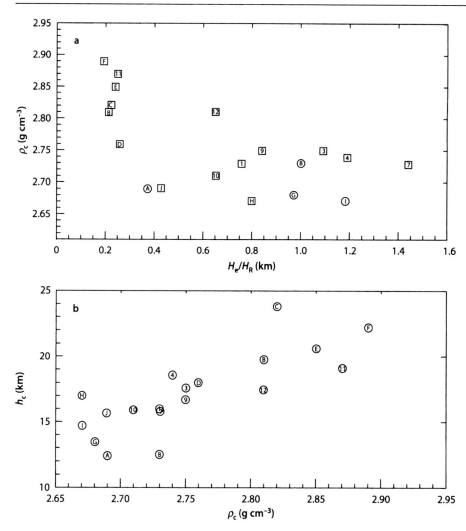

Fig. 3.48. a Crustal density (ρ_c) versus the ratio of crustal elevation to lower crustal root, H_e/H_R; **b** crustal thickness h_c versus crustal density ρ_c. Symbols as in Fig. 3.46

the lithosphere. The increasing density of the crust while the H_e/H_R ratio decreases and the crustal thickness increases may well document this process (Fig. 3.48a,b).

Both the subsidence of the volcanic structures observed and the trend of decreasing H_e/H_R ratios while the lithospheric density values approach the constant lithospheric density domain ($\rho_L = 3.33$ g cm^{-3}) suggest that these structures gradually tend towards an Airy type of isostatic equilibrium, described by the $H_e/H_R = (\rho_L - \rho_c)/(\rho_c - \rho_w)$ relation. This gradual transition between the two isostatic compensation schemes, the variable- and constant lithospheric density domains (Pratt- and Airy type of compensation), characterizes the evolution of the isostatic equilibrium of the hotspot generated structures.

3.6
Conclusions

1. The isostatic load equation and model discussed in the first part of the study represents a combination of the Airy and Pratt compensation schemes in the frame of plate tectonics, and it offers a generalized description of isostatic compensation mechanisms in the lithospheric plate. The transition between these mechanisms is continuous, and they operate in such a way that the sum of the crustal load (also including the water layer when applicable) and the lithostatic load of the sub-crustal lithosphere (termed "lithosphere" in this study) remains constant at a uniform compensation depth. This depth is set at 95 km and coincides approximately with the base of lithospheric plate suggested by the thermal plate model. It is also compatible with the reference depth in Pratt's isostatic compensation model.

   It is noteworthy that the essential physical parameters of the isostatic load model are based on observations and experimental results. The role that the crustal load plays in the evolution of the average lithospheric density is implicitly expressed by the isostatic load equation.

2. Our investigations show that the lithosphere's density is the same in tectonically stable areas on land and in the regions of the oceanic basins that are approximately 70 to 120 Ma, where the sea floor remains constant at a 5.4 km depth (sediment load corrected). This density here is 3.332 g cm^{-3} and is used as a reference value for our isostatic load model. In some deeper oceanic basins where the age independence of the sea-floor depth is not observed, the lithospheric density can reach higher values.

3. The average lithospheric densities derived for the steady-state spreading ridges are in the 3.23–3.25 g cm^{-3} range and most likely indicate the average density of the upwelling asthenosphere.

4. Our study shows that an average 3.28–3.30 g cm^{-3} lithospheric density characterizes areas of intraplate hotspot swells.

5. Low average lithospheric densities in the 3.20 to 3.22 g cm^{-3} range are associated with high-intensity hotspot dynamics, such as observed in Hawaii, Iceland or on Réunion Island. Here, the average lithospheric density difference compared to steady-state spreading centers indicates a potential temperature of the asthenosphere that is approximately 200 to 250 °C higher.

6. The crustal thickness of the hotspot-generated structures appears to average 18 to 20 km. The depth versus age ("time $t^{1/2}$") law describes their subsidence. This subsidence ceases when the cooling lithospheric density reaches 3.33 g cm^{-3}. The corresponding water depth is then in the 2.5 to 3 km range.

7. A gradual transition between the two isostatic compensation schemes, the variable and constant lithospheric density domain (Pratt and Airy type of compensation), characterizes the evolution of the isostatic equilibrium of hotspot-generated structures.

Acknowledgements

The authors are grateful to H. D. Needham, A. Hirn and J. Francheteau for their critical and helpful comments on the original manuscript. We would like to thank I. Grevemeyer and Ph. Charvis for providing their crustal velocity data used in our analysis.

We extend our thanks to V. Hekinian and C. Avedik for their help in editing this work. Warmest thanks also to R. Avedik for accepting that the 'home atmosphere' be temporarily replaced by 'hot lithosphere'.

References

Angenheister G, Gebrande H, Miller H (1979) First results from Reykjanes Ridge Iceland Seismic Project 1977. Nature 279:56–60

Angenheister G, et al. (1980) Reykjanes Ridge Iceland Seismic Experiment (RRISP 77). J Geophys 47:228–238

Asada T, Shimamura H (1976) Observation of earthquakes and explosions at the bottom of the Western Pacific: Structure of oceanic lithosphere revealed by Longshot Experiment. In: The geophsysics of the Pacific Ocean basin and its margins. Geophysical Monograph 19:135–154 American Geophys Union, Washington D.C.

Avedik F, Howard D (1975) Preliminary results of a seismic refraction study in the Meriadzek-Trevelyan area, Bay of Biskay, In: Montadert L, Roberts DG (eds) Initial report of Deep Sea Drilling Project, Vol. XLVIII. Washington D.C., 48:1015–1023

Bath M (1960) Crustal structure of Iceland. J Geophys Res 65:1793–1807

Beblo M, Björnsson A (1978) Magnetotelluric investigation of the lower crust and upper mantle beneath Iceland. J Geophys 45:1–16

Binard N, Hekinian R, Cheminée JL, Searle RC, Stoffers P (1991) Morphological and structural studies of the Society and Austral hotspot regions in the South Pacific. Tectonophysics 186:293–312

Birch F (1961) The velocity of compressional waves in rocks at pressures to 10 kbar. J. Geophys Res 71:3459–3556

Bjarnason IT, Menke W, Flovenz OG, Caress D (1993) Tomographic image of the Mid-Atlantic plate boundary in southwestern Iceland. J Geophys Res 98:6607–6622

Bodine JH, Steckler MS, Watts AB (1981) Observations of flexure and the rheology of the oceanic lithosphere. J Geophys Res 86(B5):3695–3707

Bott MHP (1985) Plate tectonic evaluation of Icelandic transverse ridge and adjacent regions. J Geophys Res 90:9953–9960

Bott MHP, Gunnarsson K (1980) Crustal structure of the Iceland-Faeroe Ridge. J Geophys 47:221–227

Bourdon E, Hémond C (2001) Looking for the 'missing end-member' in South Atlantic Ocean mantle around Ascension Island. Miner Petrology 71:127–138

Bram K (1980) New heat flow observations on the Reykjanes Ridge. J Geophsy 47:86–90

Brandsdottir B, Menke W, Einarsson P, White RS, Staples RK (1997) Faroe-Iceland Ridge Experiment 2. Crustal structure of the Krafla central volcano. J Geophys Res 102:7867–7886

Bruck R, Carbotte SM, Mutter C (1997) Controls on extension of mid-atlantic ridges. Geology 25: 935–938

Calmant S, Cazenave A (1986) The effective elastic lithosphere under the Cook-Austral and Society Islands. Earth Planet Sci Lett 77:187–202

Calmant S, Cazenave A (1987) Anomalous elastic thickness of the oceanic lithosphere in the south-central Pacific. Nature 328:236–238

Calmant S, Francheteau J, Cazenave A (1990) Elastic layer thickening with age of the oceanic lithosphere: A tool for prediction of the age of volcanoes or oceanic crust. Geophys J Int 100:59–67

Cannat M, et al. (1999) Mid-Atlantic Ridge-Azores hotspot interaction: Along axis migration of a hotspot derived event of enhanced magmatism 10 to 4 Ma ago. Earth Plan Sci Lett 173:257–269

Caress DW, McNutt MK, Detrick RS, Mutter JC (1995) Seismic imaging of hotspot related crustal underplating beneath the Marquesas islands. Nature 373:600–603

Carlson RL, Raskin GS (1984) Density of ocean crust. Nature 311:555–558

Carlson RL, Herrick CN (1990) Densities and porosities in the oceanic crust and their variations with depth and age. J Geophys Res 95:9153–9170

Case JE, Ryland SL, Simkin T, Howard KA (1974) Gravitational evidence for a low density mass beneath the Galapagos Island. Nature 181:1040–1043

Charvis P, Recq M, Operto S, Brefort D (1995) Deep structure of the Northern Kerguelen Plateau and hotspot related activity. Geoph J Int 122:899–924

Charvis P, Laesanpura A, Gallart J, Hirn A, Lepin JC, de Voogt B, Minshull TA, Hello Y, Pontoise B (1999) Spatial distribution of hotspot material added to the lithosphere under La Réunion, from wide-angle data. J Geophys Res 104:2875–2893

Cheminée JL, Hekinian R, Talandier J, Albarede F, Devey CW, Francheteau J, Lancelot Y (1989) Geology of an active hotspot: Teahitia-Mehetia region in the south central Pacific. Mar Geophys Res 11:27–50

Chocran JR (1979) An analysis of isostasy in the World's oceans. 2. Mid-Ocean Ridge crests. J Geophys Res 84 B9:4713-4729

Christensen N (1974) Compressional wave velocities in possible mantle rocks to pressures of 30 kbar. J Geophys Res 79:407-412

Clouard V, Bonneville A, Barsczus HG (2000) Size and depth of ancient magma reservoirs under atolls and islands of French Polynesia using gravity data. J Geophys Res 105:8173-8191

Collette BJ, Slootweg AP, Verhoef J, Roest WR (1984) Geophysical investigation of the floor of the Atlantic Ocean between 10° and 38° N (Kroonvlag-project). Proc Ned Akad Wet 87(B):1-76

Crough ST (1975) Thermal model of the oceanic lithosphere. Nature 256:388-390

Crough ST (1978) Thermal origin of mid-plate hotspot swells. Geophys J R Astron Soc 55:451-469

Crough ST (1983) Hotspot swells. Annu Rev Earth Planet Sci 11:165-193

Darbyshire FA, Bjarnason IT, White RS, Flovenz OG (1998) Crustal structure above the Iceland mantle plume imaged by the ICEMELT refraction profile. Geophys J Int 135:1131-1149

Darbyshire FA, White RS, Priestly KF (2000) Structure of the crust and uppermost mantle of Iceland from a continental seismic and gravity study. Earth Planet Sci Lett 181:408-428

Detrick RS, Crough ST (1978) Island subsidence, hotspots, and lithospheric thinning. J Geophys Res 83:1236-1244

Detrick RS, Sclater JG, Thiede J (1977) The subsidence of aseismic ridges. Earth Planet Sci Lett 34:185-196

Doin MP, Fleitout L (2000) Flattening of the oceanic topography and geoid: Thermal versus dynamic origin. Geophys J Int 143:582-594

Escartin J, Cannat M, Pouliquen G, Rabain A, Lin J (2001) Crustal thickness of V-shaped ridges south of the Azores: Interaction of the Mid-Atlantic Ridge (36°-39° N) and the Azores hotspot. J Geophys Res 106(B10):21719-21735

Feighner MA, Richards MA (1994) Lithospheric structure and compensation mechanisms of the Galapagos Archipelago. J Geophys Res 99:6711-6729

Flovenz OG, Gunnarsson K (1991) Seismic crustal structure in Iceland and surrounding area. Tectonophysics 189:1-17

Foucher JP, Le Pichon X, Sibuet JC (1982) The ocean continent transition in the uniform lithospheric stretching model: Role of partial melting in the mantle. Phil Trans R Soc London A305:27-43

Gallart J, Driad L, Chauris P, Sapin M, Hirn A, Diaz J, de Voight B, Sachpazi M (1999) Perturbation to the lithosphere along the hotspot track of La Réunion from an offshore-onshore seismic transect. J Geophys Res 104:2895-2908

Gebrande H, Miller H, Einarsson P (1980) Seismic structure of Icealnd along the RRISP profile 1. J Geophys 47:239-249

Gerdom M, Trehu AM, Flueh ER, Klaeschen D (2000) The continental margin off - Oregon from seismic investigations. Tectonophysics 329:79-97

Goldflam P, Weigel W, Loncarevic BD (1980) Seismic structure along RRISP - Profile I on the southeast flank of the Reykjanes Ridge. J Geophys 47:250-260

Goodwillie AM, Watts AB (1993) An altimetric and bathymetric study of elastic thickness in the central Pacific Ocean. Earth Planet Sci Lett 118:311-326

Goslin J, Beuzart P, Francheteau J, Le Pichon X (1972) Thickening of the oceanic layer in the Pacific ocean. Mar Geophys Res 1:418-427

Gradstein FM, Agterberg FP, Ogg JG, Hardenbol J, van Veen P, Thierry J, Huang Z (1994) A Mesozoic timescale. J Geophys Res 99:24051-24074

Grevemeyer I, Flueh ER (2000) Crustal underplating and its implications for subsidence and state of isostasy along the Ninetyeast Ridge hotspot trail. Geophys J Int 142:643-649

Grevemeyer I, Weigel W, Whitmarsh RB, Avedik F, Deghani AG (1995) The Aegir Rift: Crustal structure of an extinct spreading axis. Mar Geophys Res 19:1-23

Grevemeyer I, Flueh ER, Reichert C, Bialas J, Klaeschen D, Kopp C (2001) Crustal architecture and deep structure of the Ninetyeast Ridge hotspot trail from active-source ocean bottom seismology. Geophys J Int 144:414-431

Grevemeyer I, Weigel W, Schüssler S, Avedik F (2001) Crustal and upper mantle seismic structure and lithospheric flexure along the Society Island hotspot chain. Geophys J Int 147:123-140

Gudmundsson O, Brandsdottir B, Jacobsdottir S, Stefansson R (1994) The crustal magma chamber of the Katla Volcano in south Iceland revealed by two-dimensional seismic undershooting. Geophys J Int 119:227-296

Gutscher MA, Olivet JL, Aslanian D, Eissen JP, Maury R (1999) The "lost Inca Plateau": Cause of flat subduction beneath Peru? Earth Plan Sci Lett 171:335-341

Hayes DE (1988) Age-depth relationship and depth anomalies in the Southeast Indian Ocean and Atlantic Ocean. J Geophys Res 93:2937-2954

Hekinian R, Bideau D, Stoffers P, Cheminée JL, Mühe R, Puteanus D, Binard N (1991) Submarine intraplate volcanism in the south Pacific: geological setting and petrology of the Society and Austral region. J Geophys Res 96:2109–2138

Hirn A (1988) Features of the crust mantle structure of Himalayas-Tibet: A comparison with seismic traverses of Alpine, Pyrenean and Variscan orogenic belts. Phil Trans R Soc London A 326:17–32

Hirn A, et al. (1984) Crustal structure and variability of the Himalayan border of Tibet. Nature 307:23–25

Hohertz WL, Carlson RL (1998) An independent test of thermal subsidence and asthenosphere flow beneath the Argentine Basin. Earth Plan Sci Lett 161:73–83

Hotta H (1970) Stability of the crust mantle structures and tectonics of Island arc and trench systems. J Phys Earth 18(1):79–113

Hughes DS, Maurette C (1956) Variation of elastic wave velocities in granites with pressure and temperature. Geophysics 11(2):277–284

Hughes DS, Maurette C (1957) Variation of elastic wave velocities in basic igneous rocks with pressure and temperature. Geophysics 12:23–31

Ito G, Clift PD (1998) Subsidence and growth of Pacific Cretaceous Plateaus. Earth Plan Sci Lett 161: 85–100

Ito G, McNutt M, Gibson RL (1995) Crustal structure of the Tuamotu Plateau, 15° S, implications for its origin. J Geophys Res 100:8097–8114

Keen C, Tramontini C (1970) A seismic refraction survey on the Mid-Atlantic Ridge. Geoph J R Astr Soc 20:473–491

Kern H, Popp T, Gorbatsevitch F, Zharikov A, Lobanov KV, Smirnov YP (2001) Pressure and temperature dependence of V_p and V_s in rocks from the superdeep well and from surface analogues at Kola and the nature of velocity anisotropy. Tectonophysics 338:113–134

Klingelhöfer F, Minshull TA, Blackmann DK, Harben P and Childers V (2001) Crustal structure of Ascension Island from wide angle seismic data: Implication for the formation of near-ridge volcanic islands. Earth Plan Sci Lett 190:41–56

LADLE Study Group (1983) A lithospheric seismic refraction profile in the western North Atlantic ocean. Geophys J R Astr Soc 75:23–69

Leeds AR (1975) Lithosphere thickness in the western Pacific. Phys Earth Plan Interiors 11:61–64

Le Pichon X (1969) Models and structure of the oceanic crust. Tectonophysics 7:385–401

Lindwall DA (1988) A two-dimensional seismic investigation of crustal structure under the Hawaiian Island near Oahu and Kauai. J Geophys Res 93:12107–12122

Louden KE (1980) The crustal and lithospheric thickness of the Philippine Sea as compared to the Pacific. Earth Plan Sci Lett 50:275–288

Ludwig WJ, Nafe JE, Drake CL (1970) Seismic refraction. In: Maxwell AE (ed) The sea, 4, Part 1: New concepts of sea floor evolution. John Wiley & Sons, Inc., New York, pp 53–84

Marechal JC (1981) Uplift by thermal expansion of the lithosphere. Geoph J R Astr Soc 66:535–552

McNutt MK (1984) Lithospheric flexure and thermal anomalies. J Geophys Res 89:11180–11194

McNutt MK (1988) Superswells. Rev Geophys 36:211–244

McNutt MK, Fischer KM (1987) The South Pacific superswell. In: Keating BH, Fryer P, Batiza R, Boehlert GW (eds) Seamounts, islands, and atolls. Geophysical Monograph 43, American Geophys Union, Washington D.C., pp 25–34

McNutt MK, Judge AV (1990) The Superswell and mantle dynamics beneath the South Pacific. Science 248:969–975

McNutt MK, Menard HW (1978) Lithospheric flexure and uplifted Atolls. J Geophys Res 83:1206–1212

Mello SLM, Cann JR (1999) Anomalous mantle at 45° N Mid-Atlantic Ridge. J Geophys Res 104(B12):29335–29349

Menke WB, Brandsdottir B, Einarsson P, Bjarnason IT (1996) Reinterpretation of the RRISP-77 Iceland shear wave profiles. Geophys J Int 126:166–172

Menke W, West M, Brandsdottir B, Sparks D (1998) Compressional and shear velocity structure of the lithopshere in northern Iceland. Bull Seism Soc Am 88:1561–1571

Minshull TA, Brozena JM (1997) Gravity anomalies and flexure of the lithosphere at Ascension Island. Geophys J Int 31:347–360

Montagner JP (1986) First results on the three-dimensional structure of the Indian Ocean inferred form long period surface waves. Geophys Res Lett 13:315–318

Morgan WJ (1971) Convection plumes in the lower mantle. Nature 230:42–43

Morgan WJ (1971) Plate motion and deep mantle convection. Geol Soc Am Memoir 132:7–22

Morgan WJ (1983) Hotspot tracks and the early rifting of the Atlantic. Tectonophysics 94:123–139

Nafe DA, Drake CL (1963) Physical properties of marine sediments. In: Hill MN (ed) The sea. Wiley Interscience, New York, 794–815

Nagumo S, Ouchi T, Kasahara J, Koresawa S (1986) P-wave velocity in the lower lithosphere in the Western North West Pacific basin observed by an ocean bottom seismometer long range array. Bull of the Earthquake Res Inst University of Tokyo 61:403–414

Nagumo S, Ouchi T, Kasahara J, Koresawa S (1987) P-wave velocity structure of the lithosphere-asthenosphere beneath the western Northwest Pacific basin determined by an ocean seismometer array observation. Bull of the Earthquake Res Inst University of Tokyo 62:15–22

Navin DA, Peirce C, Sinha MC (1998) The RAMESSES experiment-II. Evidence for accumulated melt beneath a slow spreading ridge from wide-angle refraction and multi channel reflection seismic profiles. Geoph J Int 35:746–772

Nolasco R, Tarits P, Filloux JH, Chave AD (1998) Magnetotelluric imaging of the Society Island hotspot. J Geophys Res 103(B12):30287–30309

Operto S, Charvis P (1996) Deep structure of the southern Kerguelen Plateau (southern Indian Ocean) from ocean bottom seismometer wide-angle data. J Geophys Res 101:25077–25103

Oxburgh ER, Parmentier EM (1977) Compositional and density stratification in oceanic lithosphere – Causes and consequences. J Geol Soc London 133:343–355

Parsons B, McKenzie D (1978) Mantle convection and thermal structure of the plates. J Geophys Res 83:4485–4496

Parsons B, Sclater JG (1977) An analysis of the variation of ocean floor bathymetry and heat flow with age. J Geophys Res 82:803–827

Pautot G (1975) Analyse structurale de l'archipel des Tuamotu: Origine volcano-tectonique, paper presented at 3éme Colloque des Science de la Terre, Univ. Montpellier, Montpellier, France

Peirce C, Barton JP (1991) Crustal structure of the Madeira-Tore Rise, eastern North Atlantic – Results of a DOBS wide angle and normal incidence seismic experiment in the Josephine Seamount region. Geophys J Int 106:357–378

Phipps Morgan J, Smith WHF (1992) Flattening of the sea floor depth-age curve as a response to asthenospheric flow. Nature 359:524–527

Phipps Morgan J, Morgan WJ, Price E (1995) Hotspot melting generates both hotspot volcanism and a hotspot swell. J Geophys Res 100:8045–8062

Pollack HN (1980) On the use of the volumetric thermal expansion coefficient in models of ocean floor topography. Tectonophysics 64:45–47

Pudjom-Djomani YH, O'Reilly SY, Griffin WL, Morgan P (2001) The density structure of subcontinental lithosphere through time. Earth Plan Sci Lett 184:605–621

Purdy GM (1983) The seismic structure of 140 myr old crust in the western Atlantic ocean. Geophys J R Astron Soc 72:115–137

Putirka K (1999) Melting depth and mantle heterogeneity beneath Hawaii and the East Pacific Rise: Constrains from Na/Ti and rare earth element ratios. J Geophys Res 104(B2):2817–2829

Recq J, Goslin J (1981) Etude de l'equilibre isostatique dans le sud-ouest de l'ocean Indien a l'aide des resultats de refraction sismique. Marine Geology 41:M1–M10

Recq M (1983) Anomalies isostatiques sous le basin de Crozet et la dorsale est-indienne. Bull Soc Geol de France XXV.6:963–972

Recq M, Charvis P (1986) A seismic refraction survey in the Kerguelen Isles, southern Indian Ocean. Geophy J R Astr Soc 84:529–559

Renkin ML, Sclater JG (1988) Depth and age in the North Pacific. J Geophys Res 93(B4):2919–2935

Sandwell DT (1982) Thermal isostasy: Response of a moving lithosphere to a distributed heat source. J Geophys Res 87:1001–10014

Sandwell DT, MacKenzie KR (1989) Geoid height versus topography for oceanic plateaus and swells. J Geophys Res 94:7403–7418

Sapin M, Hirn A (1997) Seismic structure and evidence for eclogitization during the Himalayan convergence. Tectonophysics 273:1–16

Schlich R (1982) The Indian Ocean: Aseismic ridges, spreading centres and oceanic basins. In: Nairns AEM, Stehli FG (eds) The oceans basins and margins, vol 6: The Indian Ocean. Plenum, New York, pp 51–147

Schlich R, Dyment J, Munschy M (1990) Structure and age of the Mascarene and Madagascar basins. Paper presented at Colloque International Volcanisme intraplaque: Le point chaud de la Réunion. Inst. de Phys. du Globe de Paris

Schubert G, Sandwell D (1989) Crustal volumes of the continents and of oceanic and continental submarine plateaus. Earth Planet Sci Lett 92:234–246

Shimamura H, Asada T (1976) Apparent velocity measurements on an oceanic lithosphere. Phys Earth Plan Interior 13:15–22

Shimamura H, Asada T, Suyehiro K, Yamada T, Inatani H (1983) Longshot experiments to study velocity anisotropy in the oceanic lithosphere of the Northwest Pacific. Phys of the Earth and Plan Interiors 31:348–362

Sibuet JC, Veyrath-Peinet B (1980) Gravimetric model of the Equatorial fracture zone. J Geophys Res 85(B2):943–954

Sibuet JC, Le Pichon X, Goslin J (1974) Thickness of the lithosphere deduced from gravity edge effects across the Mendocino fault. Nature 252:676–679

Sinha MC, Constable SC, Peirce C, White A, Heinson G, MacGregor LM, Navin DA (1998) Magmatic processes at slow spreading ridges: Implications of the RAMESSES experiment of 57°45' N on the Mid-Atlantic Ridge. Geoph J Int 135:731–745

Sleep NH (1975) Formation of oceanic crust: Some thermal constrains. J Geophys Res 80(29):4037–4042

Sleep NH (1990) Hotspots and mantle plumes: Some phenomenology. J Geophys Res 95:6715–6736

Sleep N (1994) Lithosphere thinning by mid-plate plumes and thermal history of hot plume material ponded at sublithospheric depths. J Geophys Res 99:9327–9343

Smallwood JR, White RS, Minshull TA (1995) Sea-floor spreading in the presence of the Iceland plume: The structure of the Reykjanes Ridge at 61°40' N. J Geol Soc 152:1023–1029

Smallwood JR, Staples RK, Richardson KR, White RS, FIRE Working Group (1999) Crust generated above the Iceland mantle plume: from continental rift to oceanic spreading centre. J Geophys Res 104:22885–22902

Smith WHF, Sandwell DT (1997) Global seafloor topography from satellite altimetry and ship sounding. Science 277:1956–1962

Staples RK, White RS, Brandsdottir B, Menke W, Maguire PKH, McBride JH (1997) Faroe-Iceland Ridge Experiment 1. Crustal structure of northeastern Iceland. J Geophys Res 102:7849–7866

Stein CA and Stein S (1992) A model for the global variation in oceanic depth and heat flow with lithospheric age. Nature 359:123–128

Su WJ, Woodward RL, Dziewonski AM (1992) Deep origin of mid-ocean ridge seismic velocity anomalies. Nature 360:149–152

Toomey DR, Wilcock WSD, Couder JA, Forsyth DW, Blundy JD, Parmentier EM, Hammond WC (2003) Asymmetric mantle dynamics in the MELT region of the East Pacific Rise. J Geophys Res 108(B4): NIL-1-NIL16

Tucholke BE, Houtz RE, Ludwig WJ (1982) Sediment thickness and depth to basement in western North Atlantic basin. AAPG Bull 66:1384–1395

Verhoef I, Collette BJ (1987) Lithospheric thinning under the Atlantis-Meteor Seamount complex (North Atlantic) In: Keating BH, Fryer P, Batiza R, Boehlert GW (eds) Seamounts, islands, and atolls. Geophys Monograph 43, American Geophysical Union, Washington D.C., pp 391–404

Vogt PR (1974) Volcano height and plate thickness. Earth Plan Sci Lett 23:337–348

Von Herzen RP, Cordery MJ, Detrick RS, Fang C (1989) Heat flow and the thermal origin of the hotspot swells: The Hawaiian swell revisited. J Geophys Res 94:13783–13799

Walcott RI (1970) Flexure of the lithosphere at Hawaii. Tectonophysics 9:435–446

Watts AB (1976) Gravity and bathymetry in the central Pacific Ocean. J Geophys Res 81:1533–1553

Watts AB (1979) An analysis of isostasy in the world's oceans, 1. Hawaiian-Emperor seamount chain. J Geophys Res 83:5985–6004

Watts AB, Cochran JR, Selzer G (1975) Gravity anomalies and flexure of the lithosphere: a three-dimensional study of the Great Meteor Seamount, northeast Atlantic. J Geophys Res 80:1391–1398

Watts AB, Bodine JH, Ribe NM (1980) Observations of flexure and the geological evolution of the Pacific Ocean basin. Nature 283:532–537

Watts AB, tenBrink US, Buhl P, Brocher TM (1985) A multichannel seismic study of the lithospheric flexure across the Hawaiian-Emperor seamount chain. Nature 315:105–111

Wegener A (1915) Die Entstehung der Kontinente und Ozeane. Sammlung Vieweg: Tagesfragen aus den Gebieten der Naturwissenschaften und Technik, Braunschweig

Weigel W, Grevemeyer I (1999) The Great Meteor Seamount: Seismic structure of a submerged intraplate volcano. In: Charvis P, Danobeitia JJ (eds) Hotspot and oceanic crust interaction. J Geodyn 28:27–40

Weir NRW, White RS, Brandsdottir B, Einarsson P, Shimamura H, Shiobara H, RISE fieldwork team (2001) Crustal structure of the northern Reykjanes Ridge and Reykjanes Peninsula, southwest Iceland. J Geophys Res 106(B4):6347–6368

Wendt I, Kreuzer H, Müller D, von Rad U, Raschka H (1976) K-Ar age of basalts from the Great Meteor and Josephine Seamounts (eastern North Atlantic). Deep Sea Res 23:849–862

White RS (1993) Melt production rates in mantle plumes. Phil Trans Roy Soc London Ser A 342:137–153

White RS, McKenzie DP, O'Nions RK (1992) Oceanic crustal thickness from seismic measurements and rare earth element inversions. J Geophys Res 97:19683–19715

White RS, Brown JW, Smallwood JR (1995) The temperature of the Iceland plume and origin of outward propagating V-shaped ridges. J Geol Soc London 152:1039–1045

Wolfe CJ, McNutt MK, Detrick RS (1994) The Marquesas archipelagic apron: Seismic stratigraphy and implications for volcano growth, mass wasting, and crustal underplating. J Geophys Res 99:13591–13608

Woolard GP (1959) Crustal structure from gravity and seismic measurements. J Geophys Res 64(B10):1521–1544

Worzel JL, Shubert GL (1982) Gravity interpretation from standard oceanic and continental crustal sections. Geol Soc Am Special Paper 62:87–100

Yoshii T (1975) Regionality of group velocities of Rayleigh-waves in the Pacific and thickening of the plate. Earth Plan Sci Lett 25:305–312

Zhao D (2001) Seismic structure and origin of hotspots and mantle plumes. Earth Plan Sci Lett 192: 251–265

Zverev SM, Kosminskaya IP, Krasilstchikova GA, Mikhota GG (1976) The crustal structure of Iceland and the Iceland Färoe-Shetland region. Soc Sci Isl 5:73–93

Chapter 4

Origin of the 43 Ma Bend Along the Hawaiian-Emperor Seamount Chain: Problem and Solution

Y. Niu

4.1
Introduction

The Hawaiian-Emperor Seamount chain (H-E SMC) on the Pacific Plate (Figs. 4.1 and 4.2a,b) is the best-defined hotspot track on the Earth. If hotspots are surface manifestations of deep, fixed sources of mantle plumes (Morgan 1971, 1981), then the along-track volcanic age progression away from Hawaii (e.g., Clague and Dalrymple 1989) must record the direction, absolute velocity, and possible changes of the Pacific Plate motion. This would suggest that the prominent ~43 Ma Bend along the H-E SMC reflects a sudden change in Pacific Plate motion direction by ~60°. However, the actual cause of the 43 Ma Bend is unknown. A leading hypothesis is that the collision between India and Eurasia some ~45 Ma ago might have triggered the sudden reorientation of the Pacific Plate motion from northward to northwestward, hence the 43 Ma Bend (Dalrymple and Clauge 1976; Patriat and Achache 1984). This collision, however, is shown to have had no effect on the Pacific Plate motion (Lithgow-Bertelloni and Richards 1998). The lack of apparent mechanism for such a sudden change in Pacific Plate motion direction led to the speculation (Norton 1995) that the ~43 Ma Bend may have resulted from a southward drift of the Hawaiian hotspot prior to ~43 Ma. Indeed, recent paleomagnetic studies (Tarduno and Gee 1995; Tarduno and Cottrel 1997; Christensen 1998; Sager 2002), plate reconstructions (Acton and Gordon 1994; Norton 1995, 2000; DiVenere and Kent 1999; Raymond et al. 2000), mantle flow models (Steinberger and O'Connell 2000), and statistical analysis of plate motions using seamount geochronology (Koppers et al. 2001) all indicate that hotspots are not fixed, but they move individually or in groups at speeds up to 60 mm yr^{-1}. Specifically, using paleomagnetic data-derived paleolatitudes for the Suiko (~64.7 Ma) and Detroit (81.2 Ma) Seamounts along the Emperor Seamount chain (E-SMC), Tarduno and Cottrel (1997) suggest that the Hawaiian hotspot had drifted southward at a speed of ~30–50 mm yr^{-1} from 81 to 43 Ma. Sager (2002) suggests that this southward drift must have been even more rapid. In fact, ODP Leg 197 (July 1 to August 27, 2001) was devoted to verifying the paleomagnetic interpretations (Tarduno et al. 2001).

If the Hawaiian hotspot had indeed drifted in speed as rapidly as the lithospheric plate motion, then we have lost the best hotspot reference with which to reconstruct plate tectonic history in the Pacific and elsewhere. This would be a revolution, and we would be left with no alternative but to reconsider the tectonic history of the Pacific Plate in particular and global tectonics in general during the period from ~81 to 43 Ma. Caution is necessary!

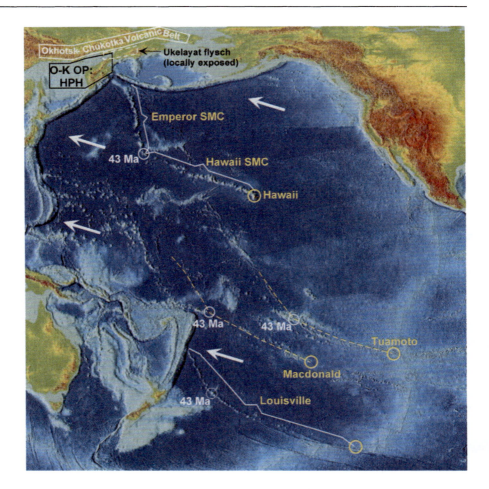

Fig. 4.1. Portion of the world's topographic map (from http://www.ngdc.noaa.gov/mgg/image/) showing Hawaiian hotspot, Hawaiian-Emperor Seamount chain, and the 43 Ma bend. Highlighted are also several other present-day hotspots (Louisville, Macdonald, Tuamotu) and the 43 Ma bends along these respective hotspot tracks on the Pacific Plate (e.g., Norton 2000). "*O-K OP: HPH*" outlines the suggested Okhotsk-Kamchatka oceanic plateau: the Hawaiian mantle plume head material. Note that the Emperor Seamount chain is perpendicular to the Cretaceous-Tertiary Okhotsk-Chukotka active continental margin, which the Pacific Plate moved towards and subducted beneath prior to 43 Ma. Note also the area where the Kamchatka forearc Ukelayat flysch sandstones are exposed and studied by Garver et al. (2000)

In this chapter, I accept that the Hawaiian hotspot is not as fixed as was previously thought, but emphatically point out that interpretations based on paleomagnetic data will have difficulties in reconciling important observations. I then present an alternative interpretation for the origin of the 43 Ma Bend, which is consistent with simple physics as well as many observations. It is my intention that this contribution will offer a stimulus to the community for a better understanding of fundamental tectonic problems, such as this one, which we should understand, but which we still have not yet mastered.

Fig. 4.2. a A simplified map of Hawaiian-Emperor Seamount chain in the Pacific after Clauge and Dalrymple (1989). K-Ar age dates of seamount lavas are also from these authors except for Detroit Seamount for which an Ar-Ar age date is from Keller et al. (1995). **b** Assuming the Hawaiian hotspot is fixed, the age data indicate a significant spreading rate change from ~5.69 cm yr^{-1} during the Emperor Seamount volcanism (~81 to 43 Ma) to 8.76 cm yr^{-1} of the present-day Hawaiian Seamount volcanism since 43 Ma

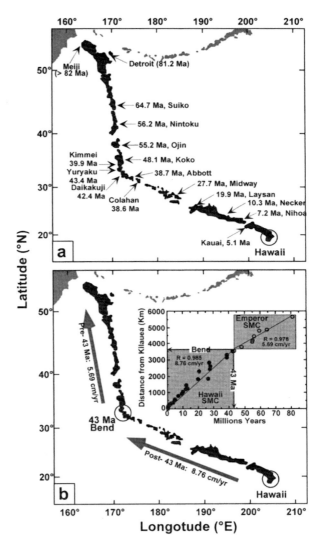

4.2
The Emperor Seamount Chain Paradox

4.2.1
Paleomagnetic Interpretations

If the Hawaiian hotspot has remained fixed, then ancient lavas erupted on the Emperor Seamounts should record paleolatitudes similar to the present-day latitude of Hawaii at ~19.5° N. This does not seem to be the case. Lavas from the Suiko Seamount (64.7 Ma) give a paleolatitude of 27.5° (±2°) (Kono 1980), and lavas from the Detroit

Seamount (81.2 Ma) give a paleolatitude of 36.2° (+6.9°/−7.2°) (Tarduno and Cottrell 1997). Very recently, using more samples from different drill holes of the Detroit Seamount, Sager (2002) obtained an even higher paleolatitude of 42.8° (+13.2°/−7.6°). If the effect of true polar wander (TPW) can indeed be ruled out (Tarduno and Cottrell 1997), then these high paleolatitudes of seamounts during eruption can only be explained by the southward drift of the Hawaiian hotspot. Using the paleolatitudes derived from paleomagnetic data of the Detroit Seamount lavas, the Hawaiian hotspot would have drifted southward during the 38 Myr period (from 81 to 43 Ma) at a speed of ~50 mm yr^{-1} (between 30 to 70 mm yr^{-1}, Tarduno and Cottrell 1997) or at a speed of ~68 mm yr^{-1} (between 46 to 107 mm yr^{-1}, Sager 2002). Such drift is remarkable, as the speeds are on the order of quite fast plate motion.

4.2.2
A Simple Test

We can carry out some simple exercises to see (1) whether such high speed drift of the Hawaiian hotspot is possible, and (2) whether the 43 Ma Bend is due to the cessation of the hotspot drift without changing the direction of the Pacific Plate motion. Figure 4.3 shows the results of the exercises. From the subtle and more Hawaii-like "trend" between the Meiji and Detroit Seamounts at the northern end of the Emperor chain (Figs. 4.1 and 4.2a,b), we can assume that the Hawaiian hotspot started the drift from the time of Detroit Seamount volcanism at ~81 Ma. We can also assume, for this testing purpose, that the Pacific Plate had spread both in direction and speed during the 38 Myr period (from 81 to 43 Ma) the same as it does today. Figure 4.3b gives several scenarios. If the Hawaiian hotspot had not drifted, the present-day E-SMC would simply be a western extension of the Hawaiian Seamount chain (H-SMC) as labeled A. If the Hawaiian hotspot had indeed drifted southward at speeds of 30 mm yr^{-1} and 60 mm yr^{-1}, respectively, then the present-day E-SMC would be in the positions B and C, respectively. Obviously, none of the three scenarios can be practically correct, as they do not match the present-day location of the E-SMC. In order to reproduce the E-SMC, we must consider several variables: (1) possible variations in both direction and velocity of the Pacific Plate motion, and (2) possible changes in the speed of southward drift of the Hawaiian hotspot. Although paleomagnetic data cannot resolve longitudinal changes, as the E-SMC is essentially north-south (~350° NNW; see Fig. 4.3a), we thus know that the Pacific Plate must have had limited motion to the west (~293° NWW; Fig. 4.3a) prior to 43 Ma. The meaningful variables to be considered are thus limited to (1) southward drift of the Hawaiian hotspot, (2) northward motion of the Pacific Plate, and/or (3) their combination.

These considerations lead to an essentially unique solution that best fits the E-SMC (Fig. 4.3c). The velocity component of the Pacific Plate motion in the 293° NWW direction is limited to 10 mm yr^{-1}. The combination of southward Hawaiian hotspot drift and the northward velocity component of the Pacific Plate motion gives a net effect of 50 mm yr^{-1} along the E-SMC. This best-fit result is in fact very close to the lava age progression rate of ~57 mm yr^{-1} along the E-SMC (Fig. 4.2b). If neither the Hawaiian hotspot nor the Pacific Plate were stationary during the formation of E-SMC, then both should have moved at a speed less than ~57 mm yr^{-1} (say, ~60 mm yr^{-1}). In other words, if the Hawaiian hotspot had stayed fixed, the Pacific Plate must have moved north-

Fig. 4.3. a Plot of the Hawaiian-Emperor Seamount chain in a Cartesian coordinate system in kilometers (recalculated in terms of "great circle distance" from latitude and longitude values) with the 43 Ma Bend chosen as the origin. **b** Several hypothetical scenarios (*A*, *B* and *C*) illustrating where and in what direction the Emperor Seamount chain would be located today, each symbol point corresponding to a seamount of the Emperor Seamount chain. I assume for this testing purpose that the Pacific Plate had spread in both direction and speed during the ~81–43 Ma period the same as it does today. Scenario A: With Hawaiian hotspot fixed, the Emperor Seamount chain would simply be an extension of the Hawaiian Seamount chain. Scenarios B and C: Hawaiian hotspot had drifted south-ward at speeds of 30 mm yr^{-1} and 60 mm yr^{-1}, respectively. This exercise indicates that one cannot reproduce the Emperor Seamount chain with Hawaiian hotspot drift only without changing the speed, direction or both of the Pacific Plate; **c** Keeping Pacific Plate motion direction as it is today, but varying speeds of both Pacific Plate motion and Hawaiian hotspot drift due south, the Emperor Seamount chain can be reproduced, essentially and uniquely, in both the orientation and length (*shaded squares*). The Pacific Plate moves to the NWW (292.5°) at a speed of only 10 mm yr^{-1}, while the Hawaiian hotspot drifts southward at a speed of 50 mm yr^{-1}. The latter is essentially the same as the 57.1 mm yr^{-1} northward motion motion of the Pacific Plate prior to 43 Ma, assuming the fixed Hawaiian hotspot (Fig. 4.2). This suggests that if the interpretations based on paleomagnetic data were correct, the Pacific Plate would have been essentially stationary during the ~38 Myr period of Emperor Seamount chain formation

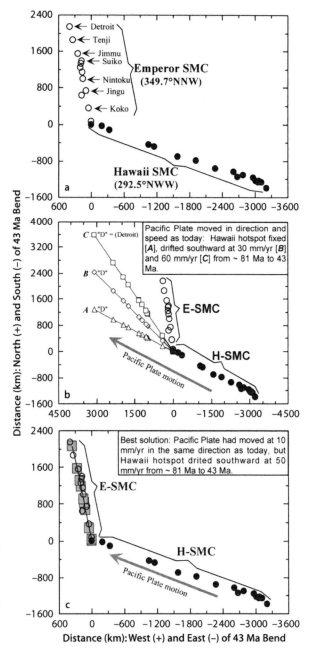

ward at a speed of ~60 mm yr^{-1}. Alternatively, if the Hawaii hotspot had indeed drifted continuously southward at a speed of ~60 mm yr^{-1}, the Pacific Plate would have to have been nearly stationary for 38 Myr (from 81 to 43 Ma).

4.2.3
The E-SMC Paradox and Solution

If we accepted the paleomagnetic interpretations that the Hawaiian hotspot had drifted continuously at a speed of 50 mm yr^{-1} (Tarduno and Cottrell 1997) or 70 mm yr^{-1} (Sager 2002) during E-SMC formation, we would also have to accept that the Pacific Plate had been essentially stationary (only 10 mm yr^{-1} drift to ~293° NWW; Fig. 4.3c) during this 38 Myr period. This reasoning is simply wrong, because it is inconsistent with observations. The well-established magnetic anomalies on the Pacific Plate at its present-day latitude of 20–35° N shows that from 81 Ma (Chron 34) to 43 Ma (Chron 18), the Pacific Plate spread at a speed of ~54 to 63 mm yr^{-1} in a relative motion direction due west (Atwater 1989). Furthermore, during this same period, the Pacific Plate continued to subduct to the NW and NNW beneath northeast Asia (Zonenshain et al. 1990; Norton 2000). This situation is indeed a paradox! To resolve the paradox means that we must make a choice between observations and interpretations. I choose to accept the observations based on well-established magnetic anomalies that the Pacific Plate was not stationary, but spread at ~54–63 mm yr^{-1} in the relative plate motion direction due west. But as "absolute" plate motion to the west (e.g., 293° NWW) is limited to ~10 mm yr^{-1} (Fig. 4.3c), the Pacific Plate would have had to move in an "absolute" direction due north. Considering that the Hawaiian hotspot may indeed have drifted southward and regarding the ~57 mm yr^{-1} vector (Fig. 4.2b) along the E-SMC, it is difficult to know how much contribution came from a northward Pacific Plate motion and how much from the southward Hawaiian hotspot drift. However, if we consider continuous Pacific Plate production and consumption (Atwater 1989; Norton 2000) regardless of the net growth in size during that 38 Myr period, we must accept that the Pacific Plate motion due north is significant, and must be greater than the Hawaiian hotspot drift to the south. I do not doubt the motion of the Hawaiian hotspot, but request careful re-evaluation of the paleolatitudes of E-SMC derived from lava paleomagnetic data. The errors associated with the current paleolatitude estimates are too large (>25%) to allow a quantitative evaluation of the actual speed of the Hawaiian hotspot drift. Furthermore, precise evaluation of the effects of both TPW and apparent polar wander of the Pacific in the Cretaceous is required.

4.3
The Origin of the 43 Ma Bend

4.3.1
Reasoning Towards a Preferred Model

Having argued above that the northward motion of the Pacific Plate must be significant during E-SMC formation, I consider that the reorientation of the Pacific Plate motion at 43 Ma best explains the origin of the 43 Ma Bend. This is supported by the expression of other hotspots and their tracks on the Pacific Plate (Fig. 4.1) (Morgan 1971, 1981; Norton 1995, 2000). In particular, the Louisville Seamount chain in the southern Pacific (Hawkins et al. 1987; Lonsdale 1988) behaves essentially the same as the H-E SMC (Harada and Hamano 2000; Norton 2000; Raymond et al. 2000; Koppers et al. 2001) and has fairly well defined along-chain age progression with a recognized 43 Ma bend (Lonsdale 1988;

Watt et al. 1988). If the Pacific Plate had not changed its direction of motion at 43 Ma, these observations would require a simultaneous multi-hotspot source swing in the Pacific deep mantle. Given the size of these mantle hotspots/plumes, their probable derivation from the core-mantle boundary (e.g., Richards and Griffiths 1988; Zhao 2001) and the very high viscosity in the deep mantle (Richards and Griffiths 1988; Davies and Richards 1992), it is indeed difficult to imagine that all these hotspot sources moved at the same time, in the same direction and by the same amount. It is, however, physically straightforward to envision that the Pacific Plate, a single unit, may have reoriented itself.

The question is, what might have caused the reorientation of the Pacific Plate motion at that period? To answer this question, we need to understand what controls the direction and speed of a moving oceanic plate. This is the same question as what causes the plate motion. The answer is straightforward – the forces that act on plate boundaries, such as subducting slab pull, ridge push etc. (Forsyth and Uyeda 1975; Gordon et al. 1978). As slab pull is the dominant driving force for an oceanic plate (Forsyth and Uyeda 1975; Davies and Richards 1992) and accounts for >90% of all the possible forces (Lithgow-Bertelloni and Richards 1998), our task is to identify a paleo-trench/subduction zone to the north towards which the Pacific Plate had moved prior to 43 Ma and to understand why the Pacific Plate failed to continue due north, but suddenly moved towards a northwest direction at 43 Ma.

4.3.2
"Trench Jam" at 43 Ma Caused by the Arrival of Hawaiian Plume Head/Oceanic Plateau

A topographically prominent feature in the far-east northeast Asia is the Late Cretaceous-Early Tertiary Okhotsk-Chukotka Andean-type active continental margin with a well-developed volcanic arc (Fig. 4.1) and subduction zone dipping gently (20°) northwest (Zonenshain et al. 1990). It was inferred, based on this shallow dipping, that the trench must have been ~500 km from the present edge of the volcanic belt within the Sea of Okhotsk (Zonenshain et al. 1990). This is the only subduction system known to exist in the region prior to the present Kamchatka-Aleutian subduction system. It is important to note that the elongation of the Okhotsk-Chukotka continental margin is essentially perpendicular to the orientation of the E-SMC. We can infer that the Pacific Plate may have moved towards and subducted beneath this continental arc prior to 43 Ma. Also note that if the subduction zone's dipping is gentle (20°), the slab-pull force, the rate of subduction and Pacific Plate motion may not be very fast, although the actual rate cannot be constrained.

The reorientation of the Pacific Plate at ~43 Ma requires that the northward motion and subduction stopped at this time. Physically, the cessation of a plate subduction can only be caused by "trench jam", e.g., the arrival of buoyant and unsubductable terranes (Ben-Avraham et al. 1981). Figure 4.4a–c illustrates the concept and physical likeliness of trench jam, and also gives reasons why the "trench jam" can lead to the reorientation of the Pacific Plate motion. Mantle plumes are probably derived from a deep thermal boundary layer, which is most likely at the core-mantle boundary as predicted theoretically (e.g., Griffiths and Campbell 1990; Davies and Richards 1992) and detected seismically (e.g., Zhao 2001). Mantle plumes ascend because of thermal buoyancy and because of the growth of the more buoyant plume heads (Whitehead and Luther 1975;

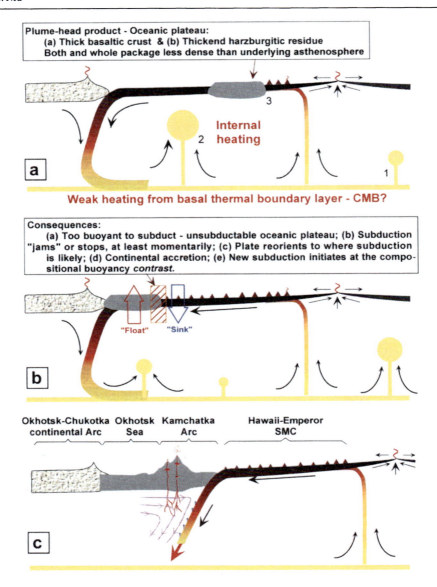

Fig. 4.4. Cartoon illustrating the consequences when a buoyant oceanic plateau (mantle plume head) collides with a subduction zone. **a** The initiation, thickening and subduction of an oceanic lithosphere. Initiation and rise of a mantle plume from a basal thermal boundary layer (*1*), development of plume head (*2*), and formation of oceanic plateau by decompression melting of plume head. **b** The plateau moves with the plate leaving a hotspot track on the younger sea floor. This plateau, when reaching the trench, has important consequences as indicated. If the trench jam leads to the cessation of the subduction, the subducting plate will reorient its motion to where subduction is likely. A large compositional buoyancy contrast at the plateau edge becomes the focus of the stress within the plate in favor of the initiation of new subduction zones (Niu et al. 2001, 2003). **c** Initiation and subduction of the dense oceanic lithosphere soon leads to dehydration-induced mantle wedge melting for arc magmatism. Note that *C* is meant to illustrate the concept, which is simplified and exaggerated to schematically describe the present-day H-E SMC, Kamchatka arc, Okhotsk Sea and abandoned Andean-type Okhotsk-Chukotka continental arc (see text for details)

Richards et al. 1989; Griffiths and Campbell 1990). When the plume head reaches a shallow level, it melts by decompression and produces thick basaltic crust and thickened, highly depleted residues (Campbell and Griffiths 1990; Hill et al. 1992; Herzberg and O'Hara 1998; Herzberg 1999; Niu et al. 2001, 2003), which have low Fe/Mg ratios and low aluminum, preventing the formation of dense garnet minerals (Niu 1997). As a result, the whole assemblage is less dense than the underlying asthenosphere (Niu and Batiza 1991), thus forming the buoyant oceanic plateau (Burke et al. 1978; Ben-Avraham et al. 1981; Abbott et al. 1997; Niu et al. 2003). This plateau moves with the plate leaving a hotspot track on the younger sea floor (Fig. 4.4a). When this buoyant plateau reaches a subduction zone (Fig. 4.4b), (1) it is too buoyant to subduct and thus will become part of a newly accreted continent (e.g., Ben-Avraham et al. 1981; Abbott et al. 1997; Albarède 1998; Herzberg 1999; Niu et al. 2003); (2) subduction stops (or in other words, the "trench jams") at least momentarily; and (3) the plate reorients its motion to where subduction is more likely (Niu et al. 2001, 2003).

In the context of this model, I hypothesize that the arrival of a buoyant Hawaiian plume head/oceanic plateau at ~43 Ma caused the trench jam, which stopped the subduction of the Pacific Plate beneath the Okhotsk-Chukotka active continental margin, and consequently led the Pacific Plate to reorient its motion in the direction where subduction was more likely, i.e., the present-day western Pacific, where subduction zones had formed some ~7 Myr earlier (Moberly 1972; Taylor 1993). This readily explains the 43 Ma Bend along the H-E SMC, and also explains why this 43 Ma Bend is less sharp along hotspot tracks further to the south such as the Tuamotu, Macdonald, and Louisville Seamount chains on the Pacific Plate (Fig. 4.1). Note that since subducting slab pull is the major drive force for plate motion, a trench jam can easily lead to the reorientation of the plate motion as is the case of the Pacific Plate during late Neogene as a result of the collision between the Ontong Java Plateau and the northern Australian Plate (Wessel and Kroenke 2000). The question is, where is the unsubductable Hawaiian plume head/oceanic plateau? Niu et al. (2001, 2003) hypothesized that the "exotic Okhotia terrane" (Zonenshain et al. 1987), now beneath the northern Okhotsk Sea, and at least part of the Kamchatka arc lithosphere are the best candidates for the Hawaiian plume head (Figs. 4.1 and 4.4c). The Okhotia terrane has been speculated to be a continental fragment (Zonenshain et al. 1987), but its long march carried by the Pacific/Kula Oceanic Plates away from any continents since some ~130 Ma ago makes it more like a buoyant oceanic plateau rather than a continental fragment. In fact, recent geological and geophysical work by Bogdanov and Dobretsov (2002) confirms the "volcanic oceanic plateau" nature of the Okhotsk Sea terrane.

4.3.3
Evidence Versus Coincidence

The above interpretation is strongly supported by recent work of Garver et al. (2000). These authors established that ~44 Ma is the youngest fission track grain age (FTGA) of primary igneous zircons in the far-east Kamchatka forearc Ukelayat flysch sandstones derived from the Okhotsk-Chukotka continental arc (Garver et al. 2000) (Figs. 4.1 and 4.5a). This minimum age strongly suggests the termination of the Chukotka continental arc volcanism – the provenance of the FTGA zircons. This termination of the arc volcanism as a result of subduction cessation is likely to be the consequence of the col-

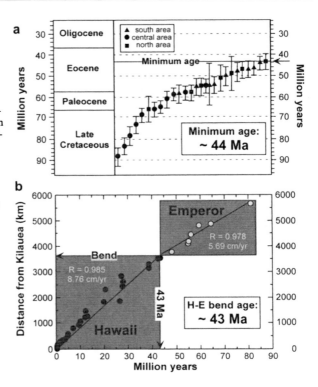

Fig. 4.5. a Simplified from Garver et al. (2000) to show the fission track grain age (FTGA) of primary igneous zircons in the Cretaceous-Tertiary Ukelayat flysch. The provenance of the zircons is the Andean-type Okhotsk-Chukotka continental magmatic arcs. Note that the minimum zircon age of ~44 Ma is essentially the same as the age of the ~43 Ma Bend along the Hawaiian-Emperor Seamount chain; b as the *inset* in Fig. 4.2b, used to compare with the minimum FTGA age in Fig. 4.5a

lision of the Hawaiian mantle plume head. I believe that the collision at ~44 Ma is the actual cause of, not coincidental with, the sudden reorientation of the Pacific Plate at ~43 Ma, marked by the bend along the H-ESC (Fig. 4.5b). Assuming that the oldest Meiji Seamount along the H-E SMC is ~82 Ma, the Hawaiian mantle plume would be ~125 Ma old, which is an interesting age as it is not significantly different from the ~122 Ma of the first phase of Ontong Java Plateau volcanism (Mahoney et al. 1993).

4.4
Summary and Conclusion

While we have known for a long time that hotspots are not really fixed (e.g., Molnar and Atwater 1973), the advertisement "Fixed Hotspots Gone with the Wind" (Christensen 1998) has stirred up our way of thinking in an unusual way – all that we thought we knew about plate motions must be wrong or at least in huge error! In particular, the suggestion based on interpretations of paleomagnetic data that the E-SMC resulted from southward drift of the Hawaiian hotspot (Tarduno and Cottrell 1997; Sager 2002) provides us with an even more severe challenge. The suggested speed of Hawaiian hotspot drift is huge, 50–70 mm yr^{-1}, which is on the order of fast plate motion velocity. If this were indeed correct, this would represent a revolution. Caution is thus necessary before we accept this suggestion. Because the total apparent velocity along the E-SMC is ~57 mm yr^{-1} (Figs. 4.2b and 4.5a,b), if the Hawaiian hotspot drifted southward at this speed, the Pacific Plate would have had to be stationary for 38 million years from 81 to 43 Ma. The latter cannot be correct, because well-established magnetic anomalies on the Pacific Plate indicate its relative motion due west at a speed of ~54 to 63 mm yr^{-1} during that period (Chron 34, ~81 Ma to Chron 18, ~43 Ma). Because significant "absolute" motion to the west is unlikely (Fig. 4.3c), mass conservation requires that the Pacific Plate moved to the north at a significant speed, probably much more so than the southward drift of the Hawaiian hotspot. This analysis requires a reconsideration of the paleo-magnetic data interpretation about the speed of hotspot drift. I suggest that the 43 Ma Bend along the E-SMC is caused by the reorientation of the Pacific Plate motion at that time. The arrival of a buoyant Hawaiian mantle plume head/oceanic plateau to the Andean-type Okhotsk-Chukotka active continental margin led to a "trench jam", and subsequent cessation of the Pacific Plate motion to the north. The Pacific Plate reoriented its motion to the northwest, where the present-day western subduction zones had formed some ~7 Myr earlier. This interpretation is supported by the fact that the Okhotsk-Chukotka continental arc volcanism stopped, as a result of trench jam and subduction cessation, at ~44 Ma, the same time as the age of the 43 Ma Bend. I further suggest that the "exotic Okhotia terrane" beneath the northern Okhotsk Sea and perhaps a significant portion of the Kamchatka arc lithosphere are the best candidates for the buoyant Hawaiian mantle plume head – an unsubductable oceanic plateau. The geology in the broad Kamchatka region is complex. Drilling at ideal sites of the Kamchatka peninsula or into the northern Okhotsk Sea lithosphere is required to verify this hypothesis.

Acknowledgements

The author thanks Cardiff University for support and acknowledges the support of UK NERC for a Senior Research Fellowship. Discussions with Rodey Batiza, Ian Campbell, Jack Casey, Jon Davidson, Phillip England, Roger Hekinian, Kaj Hoernle, Jim Natland, John O'Connor, Mike O'Hara, Julian Pearce, Marcel Regelous, Peter Stoffers, David Waters, Tony Watts, Tim Worthington, and many others at various stages were also very helpful. Y. Niu also acknowledges the support from University of Houston for completing the research.

References

Abbott DH, Drury R, Mooney WD (1997) Continents as lithological icebergs: The importance of buoy-ant lithospheric roots. Earth Planet Sci Lett 149:15–27
Acton GD, Gordon RG (1994) Paleomagnetic tests of Pacific plate reconstructions and implications for motion between hotspots. Science 263:1246–1254
Albarède F (1998) The growth of continental crust. Tectonophys 296:1–14
Atwater T (1989) Plate tectonic history of the northeast Pacific and North America. In: Winterer EL, Hussong DM, Decker RW (eds) The geology of North America – The eastern Pacific and Hawaii. Geol Soc Amer vol N:21–72
Ben-Avraham Z, Nur A, Jones D, Cox A (1981) Continental accretion: From oceanic plateaus to allochthonous terranes. Science 213:47–54
Bogdanov NA, Dobretsov NL (2002) The Okhotsk volcanic oceanic plateau. Geologiya I Geofizika 42: 1011–114
Burke K, Fox PJ, Sengör MC (1978) Buoyant ocean floor and the origin of the Caribbean. J Geophys Res 83:3949–3954
Campbell IH, Griffiths RW (1990) Implications of mantle plume structure for the evolution of flood basalts. Earth Planet Sci Lett 99:79–93
Christensen U (1998) Fixed hotspots gone with wind. Nature 391:739–740

Clague DA, Dalrymple GB (1989) Tectonic, geochronology and origin of the Hawaii-Emperor Chain. In: Winterer EL, Hussong DM, Decker RW (eds) The geology of North America – The eastern Pacific and Hawaii. Geol Soc Amer vol N:188–217

Dalrymple GB, Clague DA (1976) Age of the Hawaiian-Emperor bend. Earth Planet Sci Lett 31:313–329

Davies GF, Richards MA (1992) Mantle convection. J Geol 100:151–206

DiVenere V, Kent DV (1999) Are the Pacific and Indo-Atlantic hotspots fixed? Testing the plate circuit through Antarctica. Earth Planet Sci Lett 170:105–117

Forsyth DW, Uyeda S (1975) On the relative importance of the driving forces of plate motion. Geophys. J R Astr Soc 43:163–200

Garver JI, Solovier AV, Bullen ME, Brandon MT (2000) Towards a more complete record of magmatism and exhumation in continental arcs, using detrital fission-track thermochrometry. Phys Chem Earth A25:565–570

Gordon RG, Cox A, Harter CE (1978) Absolute motion of an individual plate estimated from its ridge and trench boundaries. Nature 274:752–755

Griffiths RW, Campbell IH (1990) Stirring and structure in mantle starting plumes. Earth Planet Sci Lett 99:66–78

Harada Y, Hamano Y (2000) Recent progress on the plate motion relative to hotspots Geophys Monogr 121:327–338

Hawkins JW, Lonsdale PF, Batiza R (1987) Petrologic evolution of the Louisville Seamount Chain. Geophys Monogr 43:235–254

Herzberg C (1999) Phase equilibrium constraints on the formation of cratonic mantle Geochem Soc Spec Publ 6:241–258

Herzberg C, O'Hara MJ (1998) Phase equilibrium constraints on the origin of basalts, picrites, and komatiites. Earth Sci Rev 44:39–79

Hill RI, Campbell IH, Davies GF, Griffiths RW (1992) Mantle plumes and continental tectonics. Science 256:186–193

Keller RA, Duncan RA, Fisk MR (1995) Geochemistry and $^{40}Ar/^{39}Ar$ geochronology of basalts from ODP Leg 145. ODP Sci Results 145:333–344

Kono M (1980) Paleomagnetism of DSDP Leg 55 basalts and implications for the tectonics of the Pacific plate. Init Rep Deep Sea Drilling Project 55:737–752

Koppers AAP, Phipps Morgan J, Morgan JW, Staudigel H (2001) Testing the fixed hotspot hypothesis using $^{40}Ar/^{39}Ar$ age progressions along seamount trails. Earth Planet Sci Lett 185:237–252

Lithgow-Bertelloni C, Richards MA (1998) The dynamics of Cenozoic and Mesozoic plate motions. Rev Geophys 36:27–78

Lonsdale P (1988) Geography and history of the Louisville hotspot chain in the Southern Pacific. J Geophys Res 93:3078–3104

Mahoney JJ, Storey M, Duncan RA, Spencer KJ, Pringle M (1993) Geochemistry and age of the Ontong Java Plateau. Geophys Monogr 77:233–262

Moberly R (1972) Origin of lithosphere behind island arcs with reference to the western Pacific. Geol Soc Amer Mem 132:35–55

Molnar P, Atwater T (1973) Relative motion of hotspots in the mantle. Nature 246:288–291

Morgan WJ (1971) Convection plumes in the lower mantle. Nature 230:42–43

Morgan JW (1981) Hotspot tracks and opening of the Atlantic and Indian Oceans. In: Emiliani C (ed) The sea. Wiley New York, vol 7, pp 443–487

Niu Y (1997) Mantle melting and melt extraction processes beneath ocean ridges: Evidence from abyssal peridotites. J Petrol 38:1047–1074

Niu Y, Batiza R (1991) In-situ densities of silicate melts and minerals as a function of temperature, pressure, and composition. J Geol 99:767–775

Niu Y, O'Hara MJ, Pearce JA (2001) Initiation of subduction zones: A consequence of lateral compositional buoyancy contrast within the lithosphere. Eos Trans AGU 82:(47) Fall Meet Suppl F10

Niu Y, O'Hara MJ, Pearce JA (2003) Initiation of subduction zones as a consequence of lateral compositional buoyancy contrast within the lithosphere: A petrologic perspective. J Petrol 44:851–866

Norton IO (1995) Plate motions in the Pacific: The 43 Ma non event. Tectonics 14:1080–1094

Norton IO (2000) Global hotspot reference frame and plate motion. Geophys Monogr 121:339–358

Patriat P, Achache J (1984) India-Eurasia collision chronology has implications for crustal shortening and driving mechanism of plate. Nature 311:615–621

Raymond CA, Stock JM, Cande SC (2000) Fast paleogene motion of the Pacific hotspots from revised global plate circuit constraints. Geophys Monogr 121:359–376

Richards MA, Griffiths RW (1988) Deflection of plumes by mantle shear flow: Experimental results and a simple theory. Geophys J Int 94:367–376

Richards MA, Duncan RA, Courtillot VE (1989) Flood basalts and hotspot tracks: Plume heads and tails. Science 246:103–107

Sager WW (2002) Basalt core paleomagnetic data from Ocean Drilling Program Site 883 on Detroit Seamount, northern Emperor Seamount chain, and implications for the paleolatitude of the Hawaiian hotspot. Earth Planet Sci Lett 199:347–358

Steinberger B, O'Connell RJ (2000) Effects of mantle flow on hotspot motion. Geophys Monogr 121: 377–398

Tarduno JA, Cottrel RD (1997) Paleomagnetic evidence for motion of Hawaiian hotspot during formation of the Emperor Seamounts. Earth Planet Sci lett 153:171–180

Tarduno JA, Gee J (1995) Large-scale motion between Pacific and Atlantic hotspots. Nature 378: 477–480

Tarduno JA, Duncan RA, Cottrell RD, Scholl DW, ODP Leg 197 Shipboard Scientific Party (2001) Motion of Hawaiian hotspot during formation of the Emperor Seamounts: Initial results of ODP Leg 197. Eos Trans AGU 82:(47), Fall Meet Suppl F1116

Taylor B (1993) Island arcs, deep sea trenches, and back-arc basins. Oceanus 35:17–25

Watt AB, Weissel JK, Duncan RA, Larson RL (1988) Origin of the Louisville Ridge and its relationship to the Eltanin Fracture zone system. J Geophys Res 93:3051–3077

Wessel P, Kroenke LW (2000) Ontong Java Plateau and late Neogene changes in Pacific plate motion. J Geophys Res 105:28255–28277

Whitehead JA, Luther Jr PS (1975) Dynamics of laboratory diapir and plume models. J Geophys Res 80:705–717

Zhao D (2001) Seismic structure and origin of hotspots and mantle plumes. Earth Planet Sci Lett 192:251–265

Zonenshain LP, Kononov MV, Savostin LA (1987) Pacific and Kula/Eurasia relative motions during the last 130 Ma and their bearing on orogenesis in northeast Asia. Geodynamic Ser 18:29–48

Zonenshain LP, Kuzmin MI, Natapov LM (1990) Foldbelts of the Northeast USSR, Taimyr and the Arctic. Geodynamic Ser 21:121–146

Chapter 5

South Pacific Intraplate Volcanism: Structure, Morphology and Style of Eruption

N. Binard · R. Hekinian · P. Stoffers · J. L. Cheminée †

5.1
Introduction

Intraplate volcanic activity in the Pacific (Fig. 5.1) has been observed and/or seismically recorded in the following areas: (1) Samoa Islands with the Rockne Seamount (Johnson 1984); (2) the Society archipelago with the Teahitia Seamount (Cheminée et al. 1989; see Sect. 2.3.5) and seismically active Mehetia Island (Talandier and Okal 1987; see Sect. 2.3.5.1); (3) the Austral archipelago with Macdonald Seamount (Johnson 1970; see Sect. 2.4.1); (4) the Pitcairn hotspot with the Bounty Volcano (Stoffers et al. 1990); and (5) Hawaii. Numerous geological and geophysical investigations have been carried out in the Pacific, including the Line Islands (Schlanger et al. 1984), Marquesas Islands (McNutt 1989; Duncan and McDougall 1974), Austral Society Islands (Duncan and McDougall 1976) and the Louisville chain (Lonsdale 1988; Watts et al. 1988). Nevertheless, except for Hawaii (Macdonald and Abbott 1970; Fornari et al. 1979, 1980, 1988; Lonsdale 1989), little is known about the morphology and structure of submarine intraplate volcanoes in the Pacific.

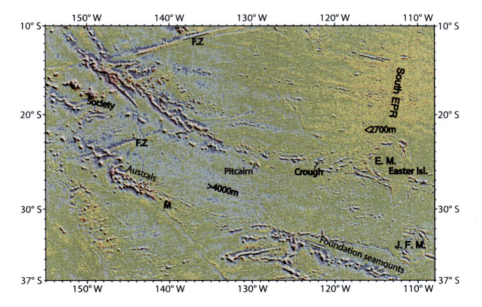

Fig. 5.1. General altimetry map after Smith and Sandwell (1997) of the South Pacific Ocean shows the main volcanic alignments, the location of hotspots and the traces of the major fracture zones (F.Z.); M = Macdonald Seamount, J.F. = Juan Fernandez Microplate, E.M. = Easter Microplate

Are hotspot activities taking place along preferential directions that represent tectonic stresses developed along the weaknesses of the oceanic lithosphere, and/or could they have been formed during the reactivation of ancient structures (faults, fissures, fracture zones)? Or, instead, are hotspots the result of lithospheric thinning due to the tectonics of spreading or a combination of all of these phenomena? Of the several hypotheses for the origin of hotspots presented by Crough (1978), the most attractive are the reheating of the lithospheric plate by an ascending mantle plume (Detrick and Crough 1978) or the underplating of the lithosphere by the less dense residue of the hotspot melting (Morgan 1972; Phipps Morgan et al. 1995).

The present study is focused on the Society, the Austral and the Pitcairn hotspots as well as on the two volcanic islands (Mehitia and Pitcairn) associated with them. In order to evaluate the submarine activity on the various morpho-structural provinces observed in these hotspot regions, bathymetric data, bottom imagery and field observations have been used. The field observations were carried out by scientists aboard the submersibles *Cyana* and *Nautile*; the bottom imagery was done using a deep-towed television camera system OFOS (Ocean Floor Observation System), side-looking sonar (GLORIA) and sea-floor back-scatter imagery data collected by the F.S. SONNE, N.O. L'ATALANTE, N.O. J. CHARCOT, and RRS CHARLES DARWIN (1988). The multichannel bathymetric coverage includes data collected by SeaBeam (N.O. J. CHARCOT 1986), Hydrosweep (F.S. SONNE 1987), and Simrad EM12D systems (N.O. L'ATALANTE 1999). The seamounts' shape, degree of flatness and lava compositions are considered in order to differentiate between hotspot and spreading ridge magmatism and to explain the formation of submarine hotspot volcanoes. Also, the study of two islands, Mehetia and Pitcairn, will document the first geological steps of seamount formation in a subaerial environment over a period of less than 1 Ma.

5.2
Society Hotspot

The Society Island chain and hotspot region covers a total area of about 65 000 km^2. The volcanic edifices increase in age from the active Mehetia Volcano (148°04' W, 17°52' S), located at the border of the hotspot surrounded by the abyssal hill region, to the Maupiti Volcano (152°15' W, 16°27' S), located at 440 km to the northwest, and dated as 4.34 Ma old (Duncan and McDougall 1976) (Fig. 5.2a). The area covered by the Society hotspot itself is only about 15 000 km^2. The contemporaneous activity of several edifices such as Teahitia and Mehetia, which are 90 km apart from each other, suggests that their eruptions might have occurred during the last 900 000 years (Fig. 5.2a) assuming that the Pacific Plate moves at a speed of 11 cm yr^{-1} over a fixed hotspot

▶

Fig. 5.2. a General location map of the Society archipelago showing the ten major islands. The hotspot region with the major submarine edifices is shown (*CY* = Cyana, *MP* = Moua Pihaa, *RO* = Rocard, *TH* = Teahitia, *TU* = Turoi). The K-Ar ages of the volcanic islands (Duncan et al. 1976) are plotted against their distances from Mehetia. The Pacific Plate velocity is inferred to be about 11 cm yr^{-1}. **b** Compiled (Bonneville, personal communication) bathymetry from several cruises of the R.V. J. CHARCOT (1986), F.S. SONNE (1987, 1989), N.O. L'ATALANTE (1999), showing the locations of the large and small seamounts forming the Society hotspot. The two boxes (#1 and #2) indicate the tracks of the GLORIA side-scan sonar (Fig. 5.3a–d). The track of a seismic profile (*A–B*) is also indicated (after Binard et al. 1991) (see Fig. 5.4)

Chapter 5 · South Pacific Intraplate Volcanism: Structure, Morphology and Style of Eruption 159

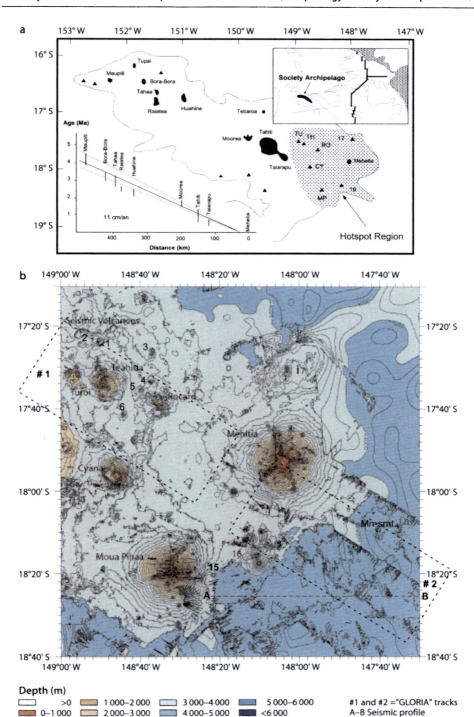

Table 5.1. Major morphological features of the volcanic edifices from the Society and Austral hotspot regions

	Latitude	Longitude	Minimum depth (m)	Basement depth of edifices (m)	Height[a] (m)	Basal surface (km^2)	Basal diameter (km)	Total volume (km^3)	Summit plateau Surface (km^2)	Summit plateau Diameter (km)	Flatness[b]	Calculated edifice slope[c] (°)
Flatness <0.25												
Volcano 4	17°33'S	148°37'W	3450	2950	500	7	3.0	0.9	0.25	0.6	0.20	22.6
Volcano 5	17°36'S	148°40'W	3450	3050	400	11	3.7	1.5	0.38	0.7	0.19	14.9
Volcano 6	17°41'S	148°44'W	3350	2900	450	8	3.2	1.1	0.25	0.6	0.19	19.0
Turoi	17°31'S	148°57'W	2226	3150	924	60	8.7	16	2	1.6	0.18	14.6
Teahitia	17°34'S	148°49'W	1456	3300	1844	225	16.9	104	3.2	2.0	0.12	13.9
Rocard	17°39'S	148°35'W	2300	3500	1200	120	12.4	33	1.5	1.4	0.11	12.3
Cyana	17°56'S	148°45'W	1435	3250	1815	210	16.4	102	3.3	2.1	0.13	14.2
Moua-Pihaa	18°20'S	148°32'W	143	3600	3457	1100	34.7	760	5.1	2.6	0.07	11.2
Mehetia Island	17°53'S	148°04'W	–435[d]	3600	4035	1050	36.6	985	–	–	–	–
Volcano 16	18°17'S	148°10'W	2750	3800	1050	80	10.1	23	1.4	1.3	0.13	13.5
Macdonald	28°59'S	140°15'W	40	3750	3710	1600	45.1	820	4.5	2.4	0.05	9.9
Rà	28°46'S	141°05'W	1040	3650	2610	260	18.2	210	2.1	1.6	0.09	17.5
Flatness >0.25												
Seismic 1	17°24'S	148°50'W	2890	3500	610	37	6.9	8.2	11	3.7	0.54	21.1
Seismic 2	17°22'S	148°55'W	3100	3500	400	24	5.5	5.3	5	2.5	0.46	15.0
Rocald	17°40'S	148°30'W	3100	3500	400	7.5	3.1	1.9	1.4	1.3	0.43	24.4
Mn-Seamount	18°24'S	147°16'W	2850	4400	1550	210	16.4	116	37.5	6.9	0.42	18.1
Macdocald	29°17'S	140°15'W	3150	4000	850	62	8.9	22	10	3.6	0.40	17.7

[a] Minimum depth minus basement depth of edifices.
[b] Summit plateau diameter/basal diameter.
[c] Calculated average slope of the edifice (arctg (2 × height / basal diameter – summit plateau diameter)).
[d] Height above sea level.

(Herron 1972; Duncan and McDougall 1976; Minster and Jordan 1978). Also, the presence of hydrothermal activity observed on the Moua Pihaa Seamount extends the influence of a magmatic upwelling zone to a presumed width of at least 70 km. The hotspot in *sensu stricto* consists of several provinces, which include different sized edifices (small: <500 m high, intermediate: 500–2000 m high and large: >3000 m high) built on top of a sea floor that is shallower (3750–4200 m depth) than the surrounding abyssal hill region (>4200 m depth) (Fig. 5.2a,b).

A compilation of the multichannel bathymetric data (Bonneville, personal communication; see Sect. 1.3.1) shows the existence of at least eight major volcanic edifices located between the Taiarapu peninsula (Tahiti island) and the island of Mehetia (Cheminée et al. 1989; Stoffers et al. 1989; Hekinian et al. 1991) (Table 5.1, Fig. 5.2b). Also using previous SeaBeam data, Cheminée et al. (1989) and Hekinian et al. (1991) have shown that hotspot volcanoes are surrounded by topographic lows (3600–3950 m depth) bordered by deeper (>4000 m depth) abyssal hill regions formed on an ancient oceanic crust. From the monitoring of the seismicity in the area, it was inferred that some of these volcanoes are still active (Talandier and Okal 1984; Okal et al. 1980). Petrological work on samples collected from these volcanoes has revealed the presence of two major magma types: (1) a low-K tholeiitic basalt associated with ancient edifices, and (2) a more alkali-rich suite related to the more recent volcanic events of the Society hotspot (Hekinian et al. 1991). Other geochemical studies (Devey et al. 1990, 2003; Hémond et al. 1994) have also shown the heterogeneous nature of the magma sources and their relationship to the alkali-basalt suites.

5.2.1
Abyssal Hill Region and Limits of Hotspot Volcanism

Detailed bathymetric studies and the GLORIA sonar survey, which were conducted across the Society hotspot and its immediate surroundings, gave an overview of the major structural and morphological domains (Fig. 5.3a and 5.3d). The GLORIA swath extends from the Taiarapu peninsula (Tahiti) (149°12' W, 17°21' S) across the Society hotspot up to the abyssal hill region near 147°30' W, 18°43' S (Searle et al. 1995) and shows the variation in the fabric of the ocean floor (Fig. 5.3a–d). The abyssal hill region comprises an area deeper than 4200 m, with a general slope less than 0.1°, and is characterized by several structural lineations (oriented N170°) representing faulted terrain (whose orientations correspond to that of the ancient Farallon ridge axis). The change in slope corresponds to that observed in most *swells* found in association with volcanic ridges (Crough 1978). There are a few eruptive lines (fault, fissures) suggested by the alignments of small volcanic cones (100–600 m high) associated with an elongated abyssal hill. Some broad, flat low-reflectivity areas are found towards the west near the 4200 m contour line and are bordered by N170° trending structures.

The seismic reflection profile taken across the abyssal hill and hotspot regions shows a succession of breaks in the sub-bottom reflectors, which correlate with the linear structures observed on the GLORIA sonographs and the bathymetric maps (Binard et al. 1992a) (Fig. 5.4). These reflectors correspond to pelagic sediments on horst and graben structures. Note that the ocean-floor depth in this area is practically uniform (slope < 0.1°). In contrast, the western end of the same seismic profile shows a flat and uniform reflector characterizing the hotspot region (Fig. 5.4). The slope increases gen-

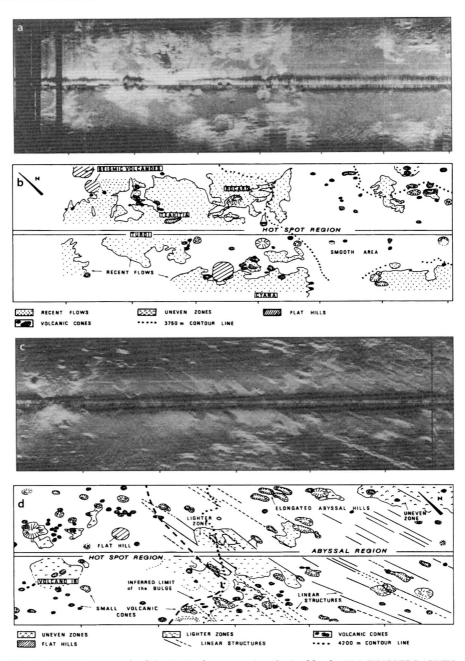

Fig. 5.3. GLORIA sonograph of the Society hotspot region obtained by the RRS CHARLES DARWIN (courtesy of R. Searle 1988); **a,b** sonograph and morpho-structural interpretations covering the area of the hotspot at less than 3750 m depth; **c,d** show the GLORIA data of the abyssal hill (>4000 m depth) region and part of Society hotspot including the transition zone at 3750–4000 m depth. The surface covered is 40 km wide, and the schematic interpretation of the sonograph is at the same scale. The lightest color in the sonograph indicates high surface reflectivity (recent flow)

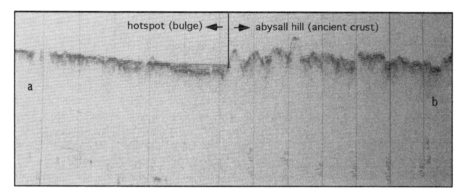

Fig. 5.4. Seismic reflection (water gun system 1.3 liter, SODERA – Sociétée pour le Développement de la Recherche Appliquée, courtesy of F. Avedik 1989) profile across the Society hotspot and its surrounding abyssal hill region. Two different morphological features, a smooth and a rougher sea floor express the transition zone between the hotspot and the abyssal hill region

tly westward and may be composed of a dense material (lava flow and/or hyaloclastites) masking the underlying structures which are overlain by a thin sediment blanket. Also, seismic profiles near 147°40' W and 18°30' S show a sediment pond, more than 10 m thick, near the hotspot boundary.

The limit of the hotspot is diffused, and is inferred as being where the crustal linear fabric disappears. In the vicinity of the boundary between the abyssal hills (faulted terrain) and the hotspot, the sea floor becomes smoother and structural lineations due to faulting have disappeared, while small hills have emerged (Fig. 5.3c,d). No fresh lava flows are distinguishable in this transition zone. The limit of the hotspot as seen on the GLORIA sonograph is the same as that which was inferred from SeaBeam data and roughly corresponds to the 4 200 m isobath. In the hotspot itself, there is a morphological contrast between the major volcanoes (Cyana, Rocard, Teahitia, Turoi) and the smoother deeper area (>3 750 m), referred to as the "bulge" (Cheminée et al. 1989; Hekinian et al. 1991). The fresh lava flows (lighter and more reflective areas, Fig. 5.3a,b) observed from a submersible during the Teahitia II cruise cover a large area above 3 750 m depth, which also includes the base and flanks of the major seamounts (Moua Pihaa and Mehetia) (Cheminée et al. 1989) (Fig. 5.2b). The rough topography is due to recent flows produced by a multitude of small vents constructed during lateral eruptions. These flows may often form ridges or rift zones with variable orientations.

5.2.2
The Sea Floor ("Bulge") Around the Hotspot Edifices

The "bulge", previously thought to be an "archipelago apron" (Menard 1956), includes the topographic lows comprising the 3 750–4 200 m contour lines. Also, it includes small conical volcanic edifices less than 500 m high and a 400–500 m shallower regional elevation than the 4 200 m isobath marking the approximate boundary between the abyssal hill area and the hotspot region (Figs. 5.2b and 5.3c,d). The Society hotspot "bulge" has a slope of more than 0.5°. It is morphologically well developed in the hotspot area where the most prominent edifices such as Teahitia, Moua Pihaa and

Mehetia are built. Its geometry is roughly elongated along a N110° direction and is oriented along the strike of the Society Island chain, especially between the 3750 m and the 4200 m contour lines. The origin and nature of the "bulge" is still somewhat uncertain. If the bulge is a sedimentary feature (turbidites and volcanic shards), the observed boundaries with the ancient abyssal hill region should be marked by a smooth topography (3750–4200 m depth) (Figs. 5.3a–d and 5.4). On the other hand, if the difference is due to thermally induced stress, we should be able to observe relict abyssal hill faulting. At less than 3750 m depth, the contour lines become more curvilinear and are centered around the Taiarapu Peninsula of Tahiti Island, which represents an ancient shield volcano that extends to 3200 m depth. The eastern limit of the bulge is marked by the bases of Moua Pihaa, Mehetia, and by volcanoes number 16 and 17 (Figs. 5.2b and 5.4). There are no parasitic cones beyond these major volcanoes, and the smooth topography of the floor is probably the result of sedimentary and/or volcanoclastic fills. The fact that the GLORIA sonographs do not show any back-light scattering as observed on active volcanic edifices, suggests that no recent flows have occurred on the sea floor forming the bulge (Fig. 5.3a). Volcanic glass interlayered by thin sediment cover was recovered from piston cores taken on the topographic lows (3750–3800 m depth) in the Society hotspot (Hekinian et al. 1991). The topographic elevations characterizing the various provinces are listed below.

- *At 4200–3750 m depth,* a large area of the sea floor surrounding the major volcanic edifices consists of gentle rolling hills and small volcanic cones (50–300 m in height), which are likely to represent short-termed events during the hotspot magmatism. The sea floor around the larger edifices has an average slope of <0.5° on several profiles (Binard et al. 1991). The presence of a few small, elongated volcanic cones (about 500 m high), which are oriented parallel to the abyssal hills, suggests the influence of the ancient crustal fabric at this depth level throughout the "bulge". This is also indicated by the north-south flow directions near 3750 m depth at the foot of Rocard, which may be influenced by underlying abyssal hill structure.

- *At 3750–3200 m depth,* the sea floor shows a sharp increase in average slope (up to 1.35°) together with an increase in the number of volcanic edifices more than 500 m high. All the small and intermediate size edifices within this area might not be directly related to hotspot activity but could represent ancient volcanoes associated with the former abyssal hill region, which may have subsequently been reactivated. Indeed, N-MORBs similar in composition to those found in the surrounding abyssal hill region were found at 3500 m depth on the top of the *Seismic Volcanoes,* at 148°50'W and 17°25'S (Hekinian et al. 1991) (Figs. 5.2b and 5.3a). A NE-SW profile, which starts near the border of the bulge in the abyssal hill region around the 4200 m depth contour line and terminates on the Taiarapu peninsula, shows that there are continuously increasing slopes.

- *At less than 3200 m depth,* the topography corresponds to the base of the Taiarapu Peninsula, which has a slope of 10°, similar to that of a Hawaiian shield volcano. The slope of the edifice southeast of Tahiti is probably formed by a succession of talus pile (slumped rock) debris (see Sect. 6.3.3) and lava step flow fronts formed during the various constructional stages of volcano building.

The difference in slope between 0.5 and 1.35° also corresponds to the more circular shape of the bathymetric contours and is due to a sudden increase in small volcanic

Chapter 5 · South Pacific Intraplate Volcanism: Structure, Morphology and Style of Eruption

cones having diverse structural orientations (Binard et al. 1991). The part of the bulge located between the depths of 4 200 m and 3 200 m is the area where the most recent constructional features prevail, and some of them represent major volcanic centers whose distribution and setting are controlled by the sea-floor fabric.

5.2.3
The Volcanic Edifices of the Society Hotspot

The Society hotspot comprises at least eight major volcanoes varying in volume between about 16 and 1 000 km^3 (Table 5.1, Fig. 5.2b). They are found in an area equivalent to an equilateral triangle with sides about 120 km long. The largest structure is Mehetia, which is 4 000 m high and presently 435 m above sea level. A common characteristic among the six largest edifices studied is the presence of lateral spurs, which form prominent bathymetric features. Such spurs, also called *rift zones* by Vogt and Smoot (1984), are commonly found on large submarine intraplate edifices formed on the Pacific crust (Fornari et al. 1988; Smoot 1982).

5.2.3.1
Turoi

Turoi Seamount has an intermediate size (about 1 500 m high with a basal diameter of about 9 km) and is located west of the Teahitia Seamount. It consists of three volcanic cones with summits culminating at 2 230 m depth (Table 5.1, Fig. 5.6a). The morphology of this seamount consists of at least three eruptive centers coalescing together and forming edifices whose sizes (600–900 m high) are larger than that of the abyssal hill volcanoes (500 m). This morphology, with its lack of well-defined rift zones and/or parasite cones, suggests that the Turoi Volcano is at an early stage of growth. Despite this, we observe that the directions of the main structural alignments due to flank eruptions are identical to those of the more mature edifices. A dive profile was made along the flank of the northern edifice crossing two eruptive cones (Fig. 5.6a, dive TH03). The northern peak is covered by fragmented pumice flow coated by indurate hydrothermal crusts (>10 cm thick) and has an elliptic collapsed crater about 10 m deep and 50 m in diameter (Fig. 5.8h). The southern peak consists of a trachytic spur made up of lightly colored massive lava. This morphology is comparable to that of a subaerial volcanic neck representing an ancient magmatic conduit. On the western flank of the eastern cone at 2 600 m depth, fresh bulbous and tubular pillows dusted by pelagic sediment occur associated with a fissure (gjà), showing that the Turoi Seamount was recently active even if the main part of the edifice is ancient. Low-K bearing dolerite (MORB), and silica-enriched volcanics (trachyte, trachy-andesite) were recovered (Binard 1991).

5.2.3.2
Teahitia

Teahitia Seamount is a large (>2 000 m high and a basal diameter of about 20 km) edifice culminating at 1 400 m depth with three summit cones (Table 5.1, Fig. 5.5). The eruptions have also taken place along lateral rift zones comparable to other submarine volcanoes

(Smoot 1982; Vogt and Smoot 1984). Along these rift zones there are a succession of small conical vents <200 m in height. Small volcanic mounds and hornitos overlaying bulbous pillows and tubular flows are observed on the flank of the volcano near 2600 m depth (Fig. 5.8b and 5.8d). Other lava includes sheet flows (draped and flat flows), giant tubes and pillows, generally with a smoother surface than typical EPR spreading ridge lavas (Ballard and Moore 1977; Hekinian et al. 1985; Choukroune et al. 1984) (Fig. 5.8c and 5.8e). They abound on the flank's eruptive sites (2600-3000 m depth) and are often associated with fissures near the base of the edifice (2900 m depth) (Fig. 5.8a). The rift zones show the following characteristics as observed from dives (Binard et al. 1991) (Fig. 5.6b):

1. bulbous and tubular pillows forming steep slopes >50-60°;
2. brecciated flow fronts giving rise to vertical scarps, up to 50 m in height;
3. poorly-sorted talus material at the foot of these scarps with a slope <30-40°.

The summits of lateral vents (30-50 m in diameter) are flattened and consist of large pillows, giant lava tubes, and haystacks <3-5 m in height. Flat flows occur near the east-

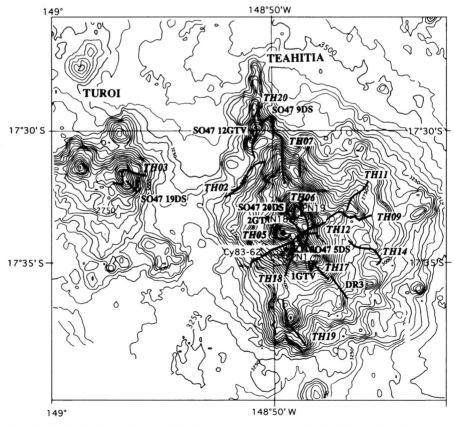

Fig. 5.5. Detailed bathymetric map of the Society hotspot showing the Teahitia and Turoi Seamounts where most surface ship (F.S. SONNE = SO...) and submersible operations (*Cyana* and N.O. SUROIT = TH) were conducted

Chapter 5 · South Pacific Intraplate Volcanism: Structure, Morphology and Style of Eruption   167

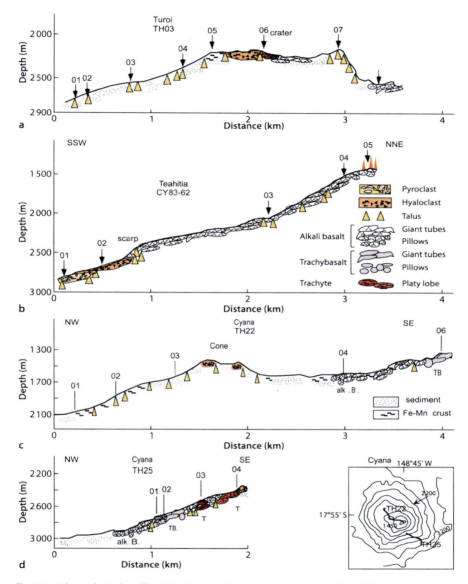

Fig. 5.6. a The geological profile (TH03) (no vertical exaggeration) across Turoi Volcano shows abundant talus coated with manganese crust. The summit of the volcano is made up of an old pumiceous flow and fresh pillows (2 200 m depth) on its the western flank. The edge of a crater consists of layered flows with interbedding of pumiceous material and blocky/tabular flows (Fig. 5.8h). The interbedded layering suggests rhythmic explosive/effusive sequences. **b** Profile along the southern flank of Teahitia (Cy83-62) at 2 800–1 450 m depth shows pyroclast-hyaloclastite association, giant lava flows and active hydrothermal chimneys (2–5 m high) near the top of the edifice (Fig. 5.8f). The hydrothermal field estimated at about 1 km in diameter is made up essentially of Fe-oxihydroxide. **c,d** Cyana Seamount (TH22 and TH25 profiles) is covered by talus, abundant pelagic sediments coated with a thin manganese crust and hyaloclastites (Fig. 5.8g). Large lava tunnels were observed near the summit. *Alk. B.* = alkali basalt, *T* = trachyte, *TB* = trachybasalt (also see Sect. 14.2.2)

ern base of the edifice (about 2 800 m depth). The summit of the edifice (1 400 m depth) consists of an extensive Fe-Si-Mn oxyhydroxide hydrothermal fields with active and colorless low-temperature (<30 °C) venting (Fig. 5.8f). These deposits are found forming tall and small (<20 cm) high hydrothermal cones discharging low-temperature fluids charged with Fe, Si and Mn and very low amounts of other transitional metals (Hoffert et al. 1987; Puteanus et al. 1991; Michard et al. 1989). The Teahitia Volcano consists of a variety of volcanics, including alkali basalt (most abundant), trachybasalt, trachy-andesite, picritic basalt, basanite and ankaramite (Hekinian et al. 1991).

5.2.3.3
Moua Pihaa

Moua Pihaa is an example of a late stage volcanism (>3 500 m high and a basal diameter of about 34 km) in the Society hotspot region (Table 5.1, Fig. 5.2b). The volcanic activity is concentrated in the summit crater zone, where several vents occur. The starfish shape is well expressed by rift zones up to 20 km in length. At greater depths (<1 500 m), interbedded intrusive rocks (more abundant at <800 m depth), pillows, and volcanic ejecta occur on a summit plateau (300–200 m depth), having a gentle topography where circular craters (<15–20 m in depth) are located. In the summit area, the rift zones are radial and the seamount presents a starfish morphology (Vogt and Smoot 1984). The slopes with a smooth topography between the rift zones are covered by unsorted talus from landslides and sediment such as reported from Hawaii (Fornari et al. 1988; Moore et al. 1989). It is believed that scattered adjacent cones existed on Moua Pihaa as was true for Teahitia. Some are now partially covered by talus along the rift zones, while others forming isolated eruptive cones are located along the extension of the rift zones. Most volcanics consist of alkali basalt, basanite and tephrite (Hekinian et al. 1991).

5.2.3.4
Cyana

Cyana Seamount, is intermediate size (about 1 200 m high with a basal diameter of 16 km) and represents an ancient edifice with smooth surface morphology due to flow brecciation, landslides, and sedimentation (Table 5.1, Fig. 5.6c, dive TH22). The summit is at 1 450 m depth and shows another volcanic cone about 1 550 m deep (Fig. 5.6c,d). Lava flows are exposed along numerous scarps, giving rise to abundant talus. Hydrothermal deposits, pelagic sediment, and iron-manganese crusts blanket a large area (Fig. 5.8g). An increase in the size of flows forming giant lava tubes with thick glassy margins is noticed for rocks enriched in silica (trachyte and trachy-andesite). Two dives conducted on the northern (TH22) and southern (TH25) slope show giant pillow lava, giant tubes and blocky/tabular flows of silica-enriched lava (trachy-andesite and trachyte) (Fig. 5.6d).

5.2.3.5
Rocard

Rocard is an intermediate-sized (about 1 000 m high with a basal diameter of 12 km) edifice made up of at least four volcanic cones coalesced together (Fig. 5.7a). One of

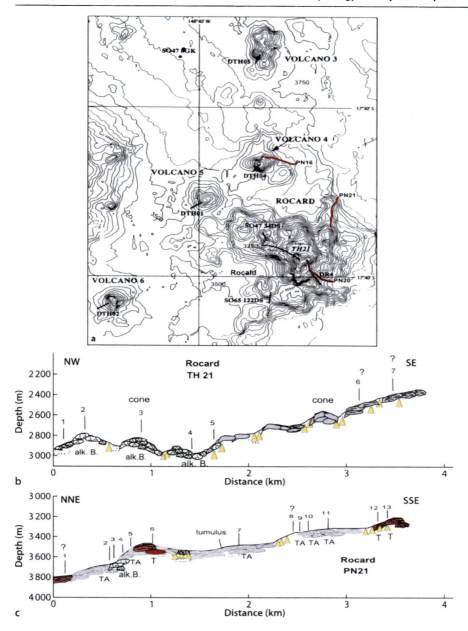

Fig. 5.7. a Detailed bathymetric map shows the locations of the major volcanic edifices in the vicinities of the Rocard Volcano (Society hotspot); sample locations: *SO...* = F.S. SONNE 1987, 1989; *DTH...* = dredge, N.O SUROIT 1989; and *PN* = *Nautile*/N.O. L'ATALANTE, Polynaut cruise 1999 (Binard et al. 1991); **b, c** profiles conducted on the Rocard composite volcano of the Society and Austral hotspot regions; **b** profile of dive TH21 along the southern slope of a volcanic cone shows pillow lava, flattened flow and giant tubes; steep scarps representing flow front are observed; **c** dive (PN21) along the northern slope of the volcanic cone composed essentially of trachy-andesite; the lithology is the same as in Fig. 5.6b

these shows a truncated cone and will be discussed later (see Sect. 5.5.2.2). Also, Rocard is morphologically similar to the Teahitia Seamount, with distinct rift zones despite its smaller size (1200 m in height) and its wide-spread rift zones. The rift zones on both Teahitia and Rocard are poorly developed compared to the larger edifices such as Moua Pihaa, Mehetia, and Macdonald. However, some lateral volcanic vents form ridges, whose orientations correlate to the main structural directions of the oceanic crust (Fig. 5.7a,b, TH21). Rocard differs from Teahitia in the abundance of more blocky and tabular silicic flows (trachyte, trachy-andesite) as well as giant tubes (Figs. 5.7b,c and 5.8i,j). The fragile nature of the flows has generated abundant talus, giving the edifice a highly brecciated and dislocated appearance. Giant lava tunnels, tabular/ blocky, and bulbous morphologies are found on the top and flank of the distinct eruptive cones. Also, fresh thick glass fragments (popping rocks, Hekinian et al. 1973), which degassed while reaching the surface, were recovered.

5.2.3.6
Volcanoes 16 and 17

Volcanoes 16 and 17 are intermediate in size (700–1000 m high, with 2–3 km of basal diameter) (Fig. 2a,b). The summits of the two volcanoes lie at a depth of 3200 m (Fig. 2a,b). *Volcano 16* has no rift zone, and it has a circular crater. Silica-enriched flows (trachyte and trachy-andesite) were recovered on these edifices (Binard et al. 1992).

5.2.3.7
The Small Edifices

Small edifices that are less than 500 m high and 2 km in basal diameter are commonly circular in shape (Figs. 5.2b and 5.7). They are found as adventive cones at the base (<3–5 km) and in the near vicinities (5–25 km) of the larger edifices (Figs. 5.2d, 5.5 and 5.7). There are no preferential orientations at the level of the individual edifices, but a regional pattern is inferred from the elongated shape of the edifices showing the same general orientation as that encountered on the ancient oceanic floor (horst and graben) of the Society hotspot region (Binard et al. 1991). *Volcano 3* shows structural orientations parallel to that of the main sea-floor direction (0–10° N) of the ancient oceanic crust (Binard et al. 1991). The circular, conical edifice *Volcano 15* observed by a deep-towed camera located 18.7 km from the main volcanic axis of Moua Pihaa is made up of bulbous and tubular pillows similar to the lava morphology forming the lateral vents of the large edifices. *Volcano 15* is interpreted as being an individual edifice located along the strike of the eastern rift zone of Moua Pihaa and oriented along the same linear crustal discontinuity. Abundant talus, sediment, and thin (<1 mm) manganese crusts covering the slopes and lava flows suggest the lack of recent volcanic activity. This small volcano was not sampled, but *volcanoes 3, 4, 5,* and *6* consist of evolved alkali-enriched flows ($K_2O > 5\%$), having similar compositions to those of the volcanics from the larger edifices. The *seismic volcanoes 1 and 2* that are located north of Teahitia near 17°25' S, 148°50' W are probably ancient, pre-hotspot edifices (see Sect. 5.5.2.2).

5.2.3.8
Mehetia Island

The island rises 435 m above sea level at Fareura Peak (Mottay 1976), and its summit is part of a large seamount (>4035 m in height) occupying the southeastern boundary of the hotspot region (Figs. 5.2b and 5.9). It sits on a 65 Ma old oceanic crust (Herron 1972). Despite the obviously young volcanic landscapes of Mehetia (well-preserved crater and surface features of lava flows), there is no historical record of volcanic activity. According to Polynesian legends, the island has been inhabited for several centuries; man-made stone artifacts can be found on top of the most recent volcanic units. The island is made up of two distinct groups of volcanic formations separated by an erosion event during which a small coral reef was formed (Fig. 5.9). Older formations contain hydromagmatic deposits (reworked palagonatized ash and "cauliflower bombs") lower lava flows (lava flow interbedded with poorly sorted breccia), strombolian ejecta (sand-sized pyroclastic debris) and summit flows (tabular flows), which constituted the main part of the island 74 000 years (±3 800 yr) ago (Steiger and Jager 1977). The summit flows form a central peak (Hurai Peak) bounded by vertical, eroded cliffs, which comprise the main summit of the island. The phreatomagmatic deposits are associated with the coral reef. More recent (25 000 ±9 800 yr, Steiger and Jager 1977) volcanic activity led to the construction of another volcanic cone and a southern volcanic unit. The young cone rises 400 m above the sea surface, on the northeastern flank of the central peak. It is made up of strombolian lava fountain ejecta interbedded with thin lava flows. The southern volcanic unit consists of a succession of lava flows with layered strombolian deposits (Fig. 5.9). Mehitia is tectonically active, and

▶

Fig. 5.8. (*See next page*); Bottom photograph taken by the submersibles *Cyana* and *Nautile* in the Society, Austral and Pitcairn hotspots; **a** dive TH07, radial fissure of 1 m wide strike to N200° direction on the northern flank of Teahitia at 2906 m depth; **b** dive TH19, shows "hornitos", which are volcanic glassy constructions rising from an adventive cone on the southern flank of Teahitia at 2694 m depth; **c** sheet flow (TH19) at 2674 m depth south flank of Teahitia; **d** "egg-shaped" frozen lava flow spilling out from bulbous pillow (TH19) at 2670 m depth; **e** tubular lava flow dropping down from a flow front along a rift zone on the eastern flank of Teahitia (TH09) at 2279 m depth; **f** active hydrothermal vent from the top of Teahitia (TH12-04) at 1453 m depth is associated with Fe-Mn-Si-oxihydroxide deposits; the particles are spewing out at a temperature of about 30 °C; **g** polygonal-shaped flat-lying crust of hyaloclastite from the summit of a cone located on the flank of the Cyana Volcano (TH22) at 1637 m depth; **h** blocky-layered flow covered with "pumice-like" glassy surface (TH03) at 2247 m depth from the summit of Turoi Volcano; **i** pillow lava with scoria-like surface at 2825 m depth (Rocard Volcano); **j** blocky with flattened surface of trachy-andesite on a gentle slope of the Rocald Volcano at 2530 m depth; **k** intrusive flows (dyke) at the edge of a slope near the summit (105 m depth) of Macdonald Seamount (dive TH29) in the Austral hotspot region; the intrusives (<1 m thick) are injected through volcanic ejecta; **l** fall of volcanic ejecta and lapilli partially burying sponge stems on the flank of the Macdonald crater at 190 m depth (deep-towed image, OFOS, R.V. SONNE Leg SO65, 1989); **m** release of volcanic gas from a crater of the Macdonald Seamount; an explosive event released important amount of CO_2 during the *Cyana* dive (TH30) in 1989 (Cheminée et al. 1991); **n** volcanic ejecta made up of sand-sized debris covered by finer grain volcanic ash is exposed by a small fault on the flank of Macdonald Seamount (TH28) at 1385 m depth; **o** adventive cone on the Bounty Volcano (Pitcairn hotspot): Flattened lobes of tabular flows (trachy-andesite, PN13-02) taken at 2974 m depth located on the flank, near the base of a small volcano; **p** Bounty Volcano: Giant lava tubes of picritic basalt (PN3-07) with twisted smooth-surfaced lava flowing down slope at 2280 m of the Bounty Volcano

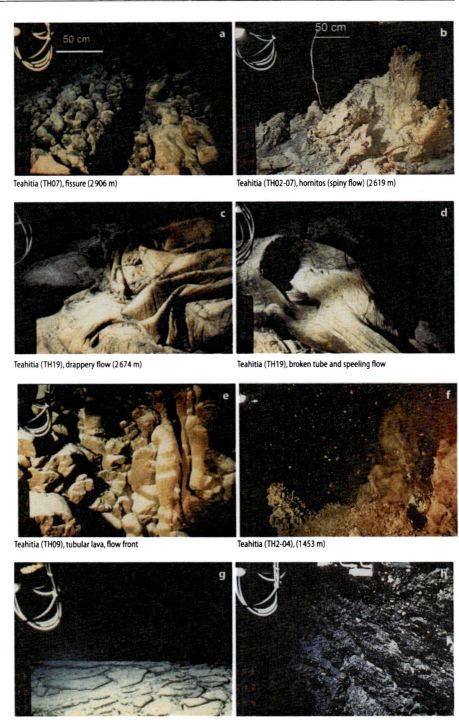

Chapter 5 · **South Pacific Intraplate Volcanism: Structure, Morphology and Style of Eruption**   173

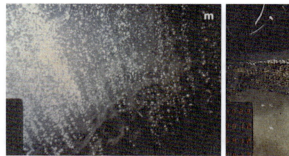

Rocard (PN21-02), blocky and tabular flow, trachy-andesite (3 664 m)   Rocard (TH21), blocky and flattened lobe of trachy-andesite (2 530 m)

Macdonald (TH29), intrusive (105 m)   Macdonald, ash, scoria-ejecta and sponge stems (190 m)

Macdonald crater (TH30), submarine eruption (156 m)   Macdonald Smt. (TH28), pyroclast (1 385 m)

PN03-03, haystack (2 501 m)   PN03-07, giant tubes (alk.B.) (2 280 m)

seismic swarms were detected by Talandier and Okal (1984a,b) along the southern flank of the edifice. From March to December 1981, the Polynesian Seismic Network recorded more than 3500 earthquakes (see Sects. 2.3.4.1 and 2.3.4.2). This crisis, correlated with the bathymetry, was divided into two episodes. The first was purely volcanic, with a sudden increase of activity and numerous small earthquakes during the first two months and was probably linked to underwater eruptions at a depth of about 1600 m on the southeast flank of the island. The second episode, mostly tectonic in character, consisted of a smaller number of higher-magnitude earthquakes (up to 4.3 ML), which were geographically dispersed and probably associated with underground tectonic readjustments (see Sect. 2.3.7.1). Such readjustments could be due to stress release resulting from lithospheric flexure caused by the gravity load of the volcanic edifice.

A submersible dive (TH10) took place during the Teahitia Cruise (December 1988 to January 1989) on the southern flank of Mehetia Island and detected numerous fissures (gjà), hyaloclastites, and yellow hydrothermal deposits occurring on the southern rift zone (1900-2300 m depth), which were attributed to recent volcanic activity. The large amount of dark hyaloclastites on the southeastern submarine flanks of Mehetia is due to volcanic ash derived from a Surtsean type of eruption when the Mehetia Island rose above the sea surface and/or during local submarine explosive events.

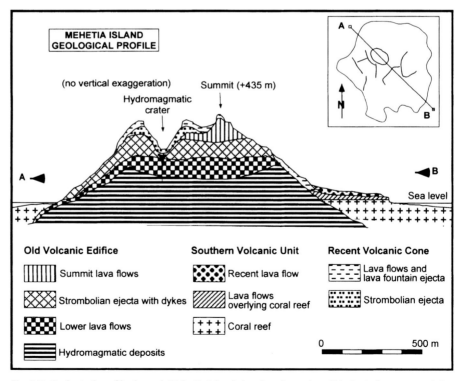

Fig. 5.9. Geological profile through Mehetia Island showing the various lithological sequences (after Binard et al. 1993)

5.3
Austral Hotspot

The Austral archipelago is located in the south-central Pacific Ocean where it forms the southeastern continuation of the Cook-Austral Island chain. The Austral region extends over more than 1 500 km in a ESE-WNW direction, from about 140° W, 29° S to about 155° W, 22° S, almost parallel to the Society and Tuamotu Island chains (Figs. 5.1 and 5.10a). The Austral archipelago is composed of seven main islands and several seamounts, guyots and shoals. With the exception of the currently active Macdonald Volcano discovered in 1967 (Norris and Johnson 1969), all the islands forming the Austral chain are extinct volcanoes rising from a sea floor of Paleocene age (Mayes et al. 1990).

A fracture zone running from ENE to WSW crosses the Austral chain between the islands of Raivavae and Tubuai (Mammerickx et al. 1975). The K/Ar dating of the Austral Islands (Duncan and McDougall 1976; Turner and Jarrard 1982) show an overall age progression from the Macdonald Seamount located at the southeastern end of the chain to Rimatara located to the northwest (Fig. 5.10a). The latter island has a minimum age in excess of 21 Ma (Turner and Jarrard 1982). On the basis of these ages, a hotspot origin for these islands was proposed. However, this age progression is far from regular, especially at Rurutu, where several groups of ages (1.1–12 Ma) have been obtained (Duncan and McDougall 1976), and at Tubuai where volcanic activity appears to span several million years (Bellon et al. 1974). At least two lines of hotspots aligned along the direction of the Pacific Plate motion are inferred in order to explain the observed jumps in the ages of these islands. The age pattern could also be explained by the intermittent activities in that region. Recently, based on the K/Ar age (230 ka) of a pillow lava sample, Bonneville et al. (2002) have suggested the existence of a new active edifice called the Arago Seamount. This seamount is located 130 km southeast of Rurutu Island and about 1 200 km west of Macdonald. However, most intense volcanic activity has been reported to be located within the southeastern extension of the Austral Island chains (Cheminée et al. 1989; Stoffers et al. 1989; Hekinian et al. 1991; Binard et al. 1991, 1992a) and is associated with the Macdonald Volcano.

5.3.1
The Submarine Edifices of the Austral Hotspot

The submarine volcanic edifices surveyed comprise two large edifices, the Macdonald and the Rà Volcanoes, and a smaller one (Macdocald) south of Macdonald Seamount (Fig. 5.10b).

5.3.1.1
Macdonald

Macdonald Seamount is a large edifice with identical volcanic morphology on its deep flanks to that observed on the other larger edifices of the Society and Pitcairn hotspots (Figs. 5.10b and 5.11). This is the most active submarine edifice so far encountered on the Pacific sea floor (Norris and Johnson 1969; Talandier and Okal 1982, 1984b). At least two volcanic events were identified and sampled since 1987, when hydromagmatic eruptions ejected lapilli and lithic debris (Stoffers et al. 1989; Hekinian et al. 1991). Erup-

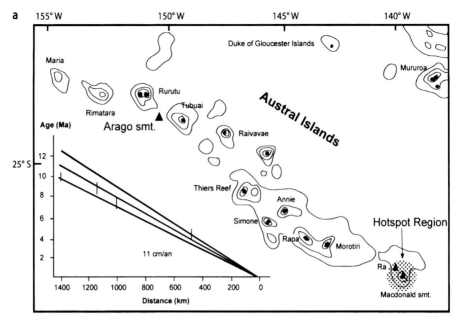

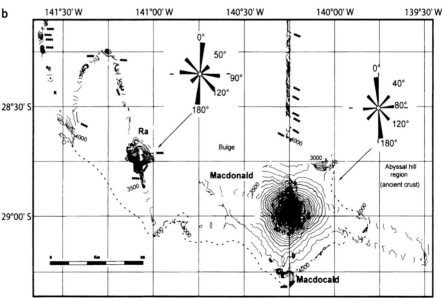

Fig. 5.10. a General location of the Austral volcanic chain about 1500 km long. The present-day volcanic activity is focused on the Macdonald Seamount. The K-Ar ages of the volcanic islands (Duncan et al. 1976) are plotted against their distances from Macdonald. Young age flow (230 ka) was found on the Arago Seamount (Bonneville et al. 2002). **b** Bathymetric chart of the Austral hotspot and orientations of the rift zones of Macdonald and Rà Seamounts, located in southeastern extension of the Austral Island volcanic chain are shown (after Binard et al. 1991). Also, a smaller edifice called Macdocald is located south of the Macdonald. The structural orientations are the same as those observed in the Society hotspot region

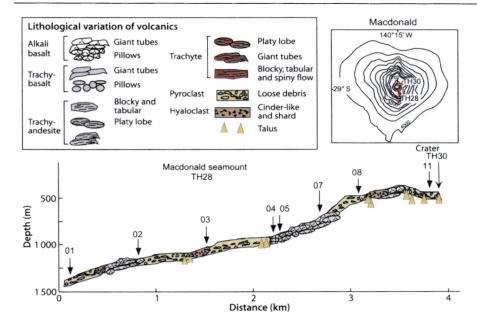

Fig. 5.11. Geological profile (CY1 011 = TH28) along the southern slope of the Macdonald Seamount. A large amount of volcanic ash and ejecta, from 1300 m depth up to the summit, through which intrusive outcrops occur at 500–900 m depth. The crater from which explosive activity was witnessed in 1989 (Cyana dive TH30, N.O. SUROIT) is located near the summit at about 200 m depth (Fig. 5.8m)

tions occurring on the summit plateau (at about 200 m depth) were simultaneously observed by submersible and on the sea surface by the ship's personnel (Cheminée et al. 1991) (Fig. 5.8m). The volcanic activity consisted of gas releases producing explosions and ash. Fine-grained ash (>30 cm thick) covers the edifice's surface down to 2000 m depth. The summit of the edifice is characterized by intrusives cutting through unconsolidated volcanic debris (ash and pyroclasts) (Fig. 5.8k). Volcanoclastic debris such as ash and accidental blocks of gabbro and dolerite were recovered from the flank of the volcano at 400–1500 m depth (see Sect. 2.4.2) (Fig. 5.8l and 5.8n). Most of the volcanics consist of alkali basalt, basanite and mugerite (Hekinian et al. 1991). Also, they are analogous to the flow morphology of the larger active Hawaiian hotspot such as the Kilauea submarine rift zone (Lonsdale 1989), Kealakekua Bay (Fornari et al. 1980), and the Loihi Seamount (Fornari et al. 1988; Karl et al. 1988).

5.3.1.2
Rà

The Rà Seamount located about 100 km from Macdonald Seamount is a large (3000 m high) edifice with a steep conical morphology discovered during the 1987 cruise of the F.S. SONNE (Stoffers et al. 1989). A deep-towed camera survey carried out near the top of this edifice revealed the presence of bulbous and giant tubular flows (Fig. 5.10b). Layered deposits of hyaloclastites coated by MnOx crust were seen. The recovered volcanics consist of alkali basalt, and nepheline tephrites (Hekinian et al. 1991).

5.3.1.3
Macdocald

Macdocald is an intermediate size (850 m high) edifice culminating at 3150 m depth and located at the southern base of the Macdonald Seamount within the 4000 m contour line. Its general morpho-structural appearance is comparable to that of the *seismic volcanoes 1* and *2* of the Society hotspot (Table 5.1, Fig. 5.10b). Altered E-MORBs (Ba < 30 ppm, Rb = 10 ppm, Sr = 160 ppm and K/Ti < 0.5) were dredged (Binard 1991).

In summary, the multibeam bathymetric data that have been collected suggest that the Austral hotspot is similar to the Society hotspot. There is a dramatic change of the sea-floor fabric at depths shallower than 4200 m (Binard et al. 1991; Hekinian et al. 1991) and a flat "bulged" sea floor paved with young volcanism at 3500 m depth on which the volcanic edifices were built.

5.4
Pitcairn Hotspot

The Pitcairn hotspot region built on a 30 Ma old oceanic crust, between the magnetic anomalies 7 and 9 (Herron 1972), is located 1500 km away from the East Pacific Rise near 129°20' W, 25°20' S, and extends up to about 100 km southeast of Pitcairn Island (Fig. 5.12a,b) (Hekinian et al. 2003). The basal sediment cored during the Deep Sea Drilling Project Leg 8 (site 75, 134°16' W, 12°31' S) at about 1500 km

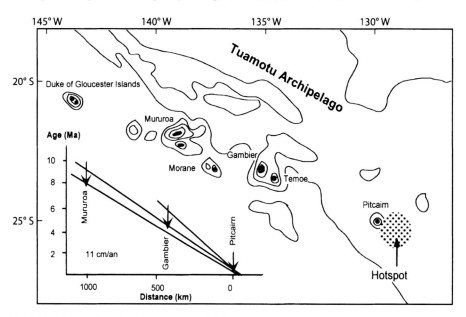

Fig. 5.12a. General location of the Pitcairn-Gambier volcanic chain, extending about 1000 km up to Mururoa is shown. The K-Ar ages of the volcanic islands (Duncan et al. 1976) are plotted against their distances from Pitcairn

Chapter 5 · South Pacific Intraplate Volcanism: Structure, Morphology and Style of Eruption

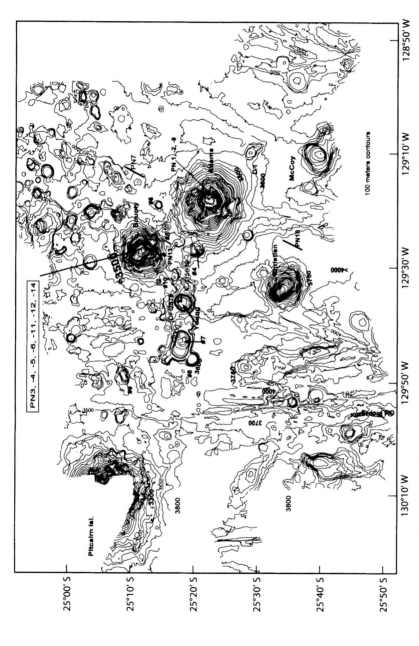

Fig. 5.12b. Bathymetric chart (100 m contour lines) obtained from multibeam Simrad EM12D system by the N.O. L'ATALANTE during the 1999 Polynaut cruise in the Pitcairn region in the South Pacific (Hekinian et al. 2003). The larger volcanic edifices are named and others are numbered (*Volcanoes #4, #5, #6, #7,* and *#8;* Binard et al. 1992b). The *Nautile*'s 1999 dives (*PN-*) (N.O. L'ATALANTE, Polynaut cruise 1999) are shown

north and on the same anomaly as that of Pitcairn Island, has been given an age of lower Oligocene (32 and 36 Ma) (Tracey et al. 1971). Pitcairn Island is located on the southeastern extension of the Duke of Gloucester-Mururoa-Gambier-Pitcairn volcanic alignment, with a N110° orientation (Duncan et al. 1974; Jarrard and Clague 1977). In the immediate vicinity of the Mururoa-Gambier-Pitcairn volcanic chain, two other parallel alignments are represented by the Oeno-Tureia and Oeno-Ducie Islands. According to Seasat (Satellite Remote Sensing) gravity anomaly data, it was suggested (Okal and Cazenave 1985) that two fracture zones (FZ1 and FZ2) forming a bend at the level of Oeno Island (FZ2) cut across these two later volcanic alignments. Okal and Cazenave (1985) have inferred that these two FZs are responsible for the hotspot direction changes (N095°). The activity in the Pitcairn hotspot is believed to have a separate origin from these two nearly parallel volcanic chains.

After recent (1999) bathymetric studies covering an area of about 9 500 km^2, it was calculated that the volcanic cones produced by intraplate volcanism cover an area of about 7022 km^2 (Hekinian et al. 2003) (Fig. 5.12b). The most prominent feature of the Pitcairn hotspot is the high density of the volcanic cones. The total number of volcanic cones is about ninety, and most of them are less than 500 m in height. The sea floor within the hotspot and on the surrounding abyssal hill regions lies at 3 750 m depth, and no apparent topographic swell (or "bulge") comparable to that observed in the Society and Austral regions was identified. The obvious trace of spreading a ridge propagator is also observed from the bathymetry (Fig. 5.12b). Recent volcanic activity within the Pitcairn hotspot region is limited to an area above the 3 500 m bathymetric contour line, where the ancient crustal orientations are covered by recent lava flows with large (>1 000 m high) and small (<500 m high) volcanic edifices.

In the Pitcairn hotspot, the sea-floor depth forming the ancient Farallon plate is estimated to be at 4 500 m (Parsons and Sclater 1977; Crough 1978). The observed mean depth of the surrounding sea floor of the Pitcairn hotspot region is 3 750 m (Mammerickx et al. 1975) (Fig. 5.12b). This is shallower than the 4181 m depth determined at deep-sea drilling site 75 (Leg 8) (Tracey et al. 1971) in the region. Considering a theoretical difference of about 750 m, (3 750 m instead of 4 500 m) for the Pitcairn hotspot, the discrepancy is equivalent to that measured (4 200 m instead of 5 000 m) in the Society and Austral hotspots (Binard et al. 1991).

5.4.1
Volcanic Edifices of the Pitcairn Hotspot

The Pitcairn region includes two large edifices, which are located in the southeastern prolongation at about 90–100 km from Pitcairn Island. These large seamounts are called Adams, after one of the mutineers, and Bounty, after the vessel H.M. Bounty. They are approximately 3 500 m in height, rising respectively to 55 m and 450 m below sea level and have erupted a volume of at least 525 km^3 and 275 km^3 of lava, respectively (Table 5.2, Fig. 5.12b). These two large edifices with a mean slope of about 15° are similar to the Macdonald (Austral) and Moua Pihaa (Society region) Seamounts (Binard et al. 1991).

Table 5.2. Morphology of submarine volcanic edifices from Pitcairn hotspot region

| | Latitude | Longitude | C Minimum depth (m) | D Basal depth (edifice) | E Height (m) | F Basal surface (km²) | G Basal diameter (km) | H Total volume (km³) | Summit plateau | | K Flatness | L Calculated edifice slope (°) |
									I Surface (km²)	J Diameter (km)		
Pitcairn Isl.	25°04'S	130°06'W	–347	3500	3847	–	–	–	–	–	–	–
Flatness <0.25												
Adams	25°23'S	129°15'W	60	3400	3340	525	26	460	5.5	2.6	0.10	15.9°
Bounty	25°11'S	129°24'W	450	3400	2950	275	19	245	2.2	1.7	0.08	18.8°
Volcano 5	25°15'S	129°20'W	2700	3100	400	8.3	3.3	1.3	0.25	0.6	0.17	16.3°
Volcano 6	25°13'S	129°17'W	2950	3250	300	8.3	3.2	1	0.4	0.7	0.21	19.4°
Volcano 7b	25°18'S	129°40'W	2200	2600	400	13	4	1.5	0.3	0.6	0.15	15.4°
Volcano 9	25°02'S	129°45'W	3150	3600	450	27	5.9	4.2	1.5	1.38	0.23	12.1°
Volcano 10	25°16'S	129°29'W	3200	2900	300	9.3	3.4	1.1	0.5	0.8	0.23	13.0°
Flatness >0.25												
Young	25°17'S	129°33'W	2200	3250	1050	47	7.7	21	5.5	2.6	0.35	22.4°
Volcano 4	25°20'S	129°24'W	2750	3300	550	24	5.5	7,4	5.2	2.6	0.47	20.8°
Volcano 7a	25°18'S	129°40'W	2600	3500	900	65	9	44	20	5	0.55	24.2°
Volcano 8	25°19'S	129°43'W	3000	3500	500	23	5.4	6.8	4.5	2.4	0.44	18.4°

[C]: minimum depth (m) (Pitcairn Island: height above sea level), [D]: basement depth of edifices (m), [E]: height (m) ([D] – [C]), [F]: basal surface (km²), [G]: basal diameter (km), [H]: total volume (km³), [I]: summit surface (km²), [J]: summit diameter (km), [K]: flatness ([J] / [G]), [L]: inferred slope of the edifice (Arctg(2[E] / ([G] – [J]))).
Pitcairn 7a and 7b correspond to the same Volcano #7 with two stage of constructions, 7a = ancient inheridted edifice (truncated type) from the Farallon plate and 7b is a conical shaped edifice (see Sect. 5.4).

5.4.1.1
Pitcairn Island

The island is located at 130°06' W, 25°04' S in the Pacific. The last eruptive event for this island was dated to be around 0.45 million years ago (Duncan et al. 1974). The first geological data from Pitcairn Island (Lacroix 1936) indicated the occurrence of alkali-enriched lava. A geological map of the island made by Carter (1967) shows the presence of a caldera (about 2 km in diameter) formed by trachyte, trachy-andesite and tuffaceous material (Fig. 5.13). Sr, Pb and Nb isotopic studies on samples collected from the island suggested that the two distinct magma types observed reflect a mixing of ancient subducted oceanic crust and sedimentary material beneath Pitcairn Island (Woodhead and McCulloch 1989). The volcanic formations were classified (Carter 1967) into four groups (Fig. 5.13): Tedside Volcanics (sheet flow interbedded with tuff and pyroclasts), Adamstown Volcanics (basalt intruded by trachyte and strombolian ejecta interbedded lava flows), Christian's Cave Formation (traces of base surges), and Pulawana Volcanics (trachy-

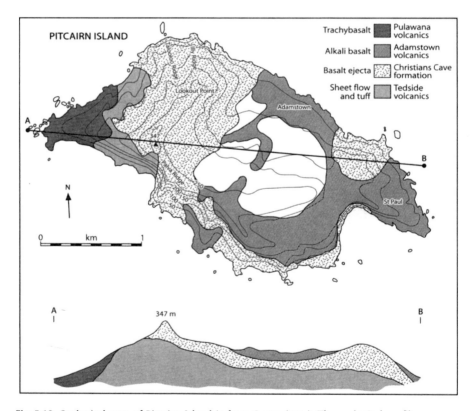

Fig. 5.13. Geological map of Pitcairn Island is from Carter (1967). The geological profile was constructed after the geological map of Carter (1967) and from independent field trip observations carried out in 1989 (Binard et al. 1992b). The major part of the island is made up of lava flows (basalt, trachyte), pyroclastites and tuffaceous material (see text). The general morphology of the island is formed by a semicircular tuff-ring (Christians Cave Formation), which is open in the northeast as a result of a large-scale hydromagmatic eruption

Chapter 5 · South Pacific Intraplate Volcanism: Structure, Morphology and Style of Eruption 183

basalt and alkali olivine basalt). The most prominent feature noticed at the Christian's Cave Formation is the result of hydromagmatic activity, which gave rise to a large semicircular tuff-ring open to the northeast. This tuff-ring feature was constructed during the hydromagmatic phase of eruption. The homogeneity of the volcanic ash produced suggests that the cone was constructed during a single eruptive event. The tuffaceous material is 300 m thick at Lookout Point. At Palva Valley Point, its thickness is inferred to be less than 50–70 m, above the Tedside Volcanics. It is believed that the semicircular ring morphology of Pitcairn Island is due to hydromagmatic deposits, which occurred on ancient inherited volcanic structures supporting the fragile ash formation. The trachytic composition of the lava linked to the formation is likely to be the most important factor in the creation of fine-grained and well-sorted pyroclastic rocks forming the Christian's Cave Formation. The last volcanic activity, found on top of the previous cone, is marked by a new basaltic magma supply, which has filled and overflowed the tuff-ring morphology toward the north (Adamstown) and the east (St. Paul) (Fig. 5.13).

5.4.1.2
Bounty

The Bounty Seamount has three summital cones and radial rift zones. A K/Ar age determination on the volcanics from Bounty gave an age of about 344 ±32 Ma (Guillou et al. 1997). Submersible and deep-towed video camera tracks, running from 500 to 3 000 m depth along the flank of the seamount show fresh pillows (giant tubular and bulbous) and abundant hyaloclastites and pyroclasts comparable to those observed on the flanks of Teahetia in the Society hotspot (Fig. 5.8 o and p). A composite geological profile along the western slope of the Bounty Volcano in the Pitcairn region shows a large variety of lava morphology and lava composition (Fig. 5.14a). At a depth of 800 m, the lava flow is highly vesiculated and consists of scoriaceous fragments. Some lobated flows and lava tunnels truncated by steep scarps with a relief of more than 50 m high occur (Fig. 5.14a). As on Adams Volcano, a scoriaceous surface of the flows, comparable to a subaerial "aa" flow, was also noticed and this could be the result of a rapid outpouring of fluid lava which then ran down the slope. This type of deposit was found on the slope, at 450–1 400 m depth, associated with small yellow hydrothermal chimneys only a few centimeters in height. A cross section through the exposed deposits at the summit shows alternating layers of ash and reddish yellow hydrothermal material (see Sect. 13.3), which represent repetitive successions of eruptive events. A sill formation at 650 m depth, interbedded with volcanic ejecta, was observed on the summit of Bounty. A small parasite cone built on the southern flank of Bounty at 2 900 m depth is essentially formed by silica-enriched flows such as trachy-andesite and trachyte, which were observed and sampled (Figs. 5.14b and 5.8 Part o).

5.4.1.3
Adams

Adams Seamount is morphologically similar to Bounty and has a summit formed by two cones less than 100 m in height with a roughly north-south orientation. The summit is heavily covered by coral sand, volcanic ejecta, dark hyaloclastites, blocky lava flows and scoriaceous material emplaced during hydromagmatic eruptions (Figs. 5.12b and 5.14c). During submersible and deep-towed camera observations on the flanks of this volcano, a large quantity of lapilli and some *cauliflower* bombs were identified. Hydrothermal deposits are intermixed with the summital ejecta, and no signs of recent activity were observed (see Sect. 13.2). Abundant dark patches of hyaloclastites occur on the slope, associated with fresh lava flows. The few scattered outcrops observed between 400 and 600 m depth show the presence of giant tubular and bulbous pillows, as well as occasional large flat and empty lava tunnels comparable to the lobate flow from the EPR. Broken up surfaces of shattered slabs of fluidal lava with a brecciated appearance (scoriaceous lava) similar to subaerial "aa" flows are seen. From 600 to 1400 m depth, abundant lapilli and scoriaceous material were mixed and interbedded on the slope along with red and yellow hydrothermal products. The angular fragments found on the slope are either flow breccia in the upper part of the edifice, or products expelled during hydromagmatic explosions. Trachytic lavas, made up of massive, light gray material and small, highly-vesicular pumice occur at 500–600 m depth on the northern side of the Adams Volcano (Fig. 5.14c). Also, the top (<500 m depth) of this seamount is covered by a coralline platform (Hekinian et al. 2003).

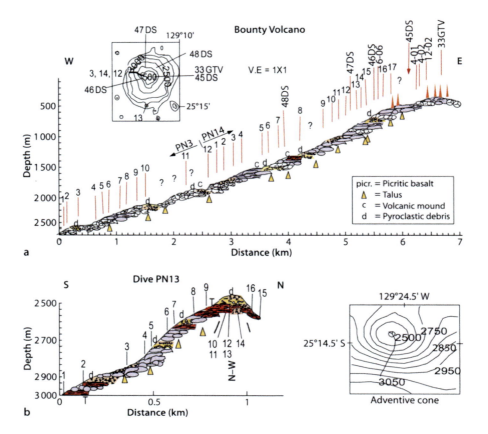

Fig. 5.14a,b. a Composite geological profiles were constructed along the western slope of the Bounty Volcano in the Pitcairn region; **b** geological profile on a small adventive cone located on the southern slope of the Bounty. *T* = trachyte

Chapter 5 · **South Pacific Intraplate Volcanism: Structure, Morphology and Style of Eruption** 185

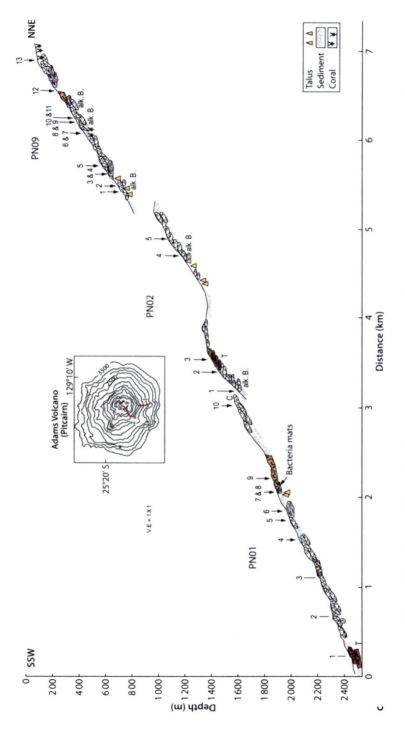

Fig. 5.14c. Composite dive profile (PN01, -02 and -09) shows the various lithologies encountered along the southern slope of the Adams Volcano. The legend is shown as an *inset*. The *vertical arrows* indicate the sample locations. The lithology is the same as in Fig. 5.11

5.4.1.4
Young

Young Seamount (named after another mutineer) is an intermediate-sized (2 km high) edifice similar to a few other edifices (*Volcano #4, 7 and 8*) (Fig. 5.12b). The Young Volcano (at 2 250 m depth) was investigated by a deep-towed camera (OFOS 78) run and shows bulbous pillows and lava tubes less than 1 m in diameter piled up on a steep slope (>40°) (Binard et al. 1992b). Numerous scarps a few meters high (interpreted as being flow front brecciation) cut these flows. These scarps have produced talus piles at their feet where thin (a few centimeters thick), pelagic sediment has dusted the large flow fragments and the breccia. Dark, flat slabs (5 cm thick) of manganese crusts that are sometimes partially buried by sediment, lava flows, and rock debris are also observed. As discussed later, this flat-topped seamount represents the remains of an old edifice built during spreading ridge activity on the ancient the Farallon plate.

5.4.1.5
Volcano #4

Volcano #4 is an intermediate-sized edifice with a truncated cone (3 000 m depth) having a summital depression (crater of 2.6 km in diameter) similar to those described on off-axis seamounts and interpreted as having a caldera structure (Hollister et al. 1978; Lonsdale and Spiess 1979; Batiza et al. 1984; Fornari et al. 1984). A television camera profile made on the summit within the inner part of the crater shows old pillow and tubular flows covered by dark manganese crust. Around the crater rim, most scarps (5–10 m high) are made up of pillow lava flow, and contain abundant slumped and weathered blocks. Heavily covered sediment coated with manganese (about 5–7 cm thick) occurs on the slopes formed by hollow pillows. This type of feature is similar to the old truncated vent of *Seismic Volcanoes 1* and *2* in the Society hotspot (Cheminée et al. 1989), thought to represent an ancient volcano inherited from the abyssal hill region. The ancient vents could be either conical or truncated, their morphology being determined during a caldera-building stage.

5.4.1.6
Volcano #7

Volcano #7 is an irregularly shaped, intermediate-sized edifice with slopes that vary with depth (2 250 m depth) at about 24.2° from 3 500 m to 2 600 m depth and 15.4° at depths shallower than 2 600 m. This slope variation is the result of two distinct stages of volcano growth (Table 5.2, Fig. 5.12b):

1. The first stage (*Volcano #7a*, Table 5.2) gave rise to a volcanic cone approximately 1 500 m high, with a polygonal to roughly square shape feature and having north-south and east-west oriented flanks. Similar types of polygonal structures studied by Fornari et al. (1987a) on off-axis volcanoes at 12°40' N near the EPR were interpreted as being the result of the magma supply during the tectonics of accretion. A summital collapsed feature creating a caldera structure and a truncated cone explains the change of slopes;
2. A later stage (*Volcano #7b*, Table 5.2) of growth gave rise to the summital cone (from 2 600 m to 2 200 m) during an extrusion of new lava flows.

5.4.1.7
The Small Edifices

The small edifices differ from mounds and are about 2 km in diameter and less than <0.5 km high. They also are the most abundant (about 90), and all of them together erupted a total of about 188 km³ of volcanics (Hekinian et al. 2003) (Table 5.2). The small edifices found at a distance of 5–35 km from the base of the larger volcanoes are generally aligned along two main directions, N-S (N160°) and NW 55°. Other small edifices can also be found as adventive cones on the flanks and in the immediate surroundings less than 5 km away from the base of the large volcanoes. *Volcanoes #5, 6, 9* and *10* are examples of small edifices with a circular cone generally made up of giant tubes of silica-enriched flows (>5 km³) (Table 5.2, Fig. 5.12b) (Hekinian et al. 2003). However, most of the small edifices are located at some distance (5–35 km) from the base of larger volcanoes within the 3 700 m bathymetric contour lines, and these structures consist essentially of silicic lava (trachyte and trachy-andesite (Fig. 5.12b). Hence, a major difference between the different-sizes of edifices is related to their range in compositional variability and their geological setting. The relationship of these small edifices with the larger ones and their origin are discussed in the last section of this chapter (see Sect. 5.7). They seem to have been supplied by feeder channels during puntiform (localized) eruptions.

5.4.2
The Distribution and Extent of Hotspot Volcanism

The extent and distribution of the hotspot activity was inferred from the back-scattered acoustic signal (reflectivity). This back-scatter reflectivity (Simrad EM12D) depends on the nature of the material extruded and is used to interpret the type of landscape. The various types of sea-floor morphologies encountered are presented in Hekinian et al. (2003) and summarized in Fig. 5.15a,b.

The structures with light reflectivity (LR) covering about 26% of the surveyed area are inherited from the ancient crust of the Farallon plate, now almost totally covered by sediment and ash – e.g., spreading ridge and small axial and off-axial volcanoes (Fig. 5.15a,b). The next darker area has lower amounts (20–40%) of sediment cover with volcanoclastic debris and a hilly topography. These features consist of pillows, lava tubes and blocky-tabular flows, loose talus debris, hyaloclastites and Fe-Si-Mn oxyhydroxide-enriched hydrothermal sediment and crust (e.g., station OFOS 78, Volcano #7 and Young) (Figs. 5.12b and 5.15a,b). This area is called "patchy" (P) because of the appearance of light-gray streaks and a hummocky surface reflectivity. This type of reflectivity comprising about 24% of the surveyed region is found essentially surrounding the large and the intermediate-sized volcanoes (e.g., *Volcanoes #7, #8*, Christian and Young) (Figs. 5.12b and 5.15a,b). Similarly, side scan sonar investigations from north EPR off-axis seamounts with hyaloclastites and pyroclasts show distinct reflectivity (Batiza et al. 1989). A middle-level reflectivity (MR) is seen in about 26% of the explored area and eventually includes the deeper (>3 500 m depth) volcanic landscape. The MR comprises areas having a smooth surface, a large wavelength topography and small constructional edifices (<300 m high) (Fig. 5.15a,b). The darker gray area, having a middle-high reflectivity (MHR) comprising about 4% of the surveyed zone, is mainly on the flanks of the most recent edifices such as the Bounty and Adams

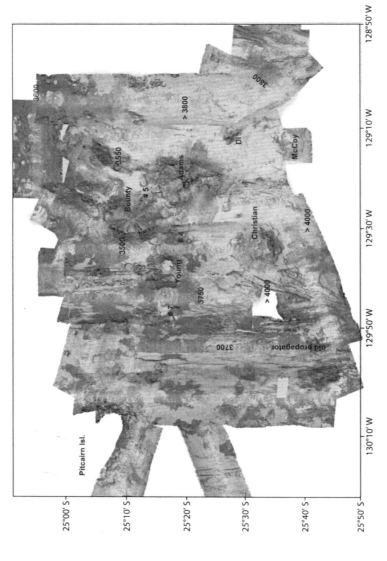

Fig. 5.15a. Back-scatter sonar image obtained from the multichannel Simrad EM12D system (N.O. L'ATALANTE) during the Polynaut cruise (1999) in the Pitcairn region (Hekinian et al. 2003). The reflectivity of the sea-floor is a function of the back-scattered signal. The *dark gray* represents the youngest terrain, which has the highest reflectivity (*HR*). The middle-high reflectivity (*MHR*) is comprised of pillows, giant tubes and blocky/tabular flows with a small amount (10–20%) of interconnecting sediment as well as volcanic ejecta. The middle-low reflectivity (*MR*) is comprised of small cones and low relief topography and moderate sediment cover (20–30%). The hummocky and patchy (*P*) area represents middle-low reflectivity with loose debris (pyroclastic material, slumped material, hyaloclastite and sediment (25–40%)). The lightest area with the lowest reflectivity (*LR*) consists of abundant sediment (>50% sediment) and cinder-like debris

Chapter 5 · South Pacific Intraplate Volcanism: Structure, Morphology and Style of Eruption

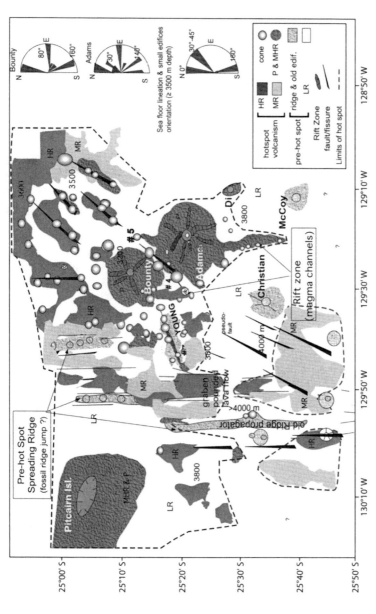

Fig. 5.15b. Schematic summary interpretation of the major structural directions and volcanic constructions of the Pitcairn hotspot reconstructed from back-scattered signal (Simrad EM12D). In addition, field observations using deep-towed camera, submersible and sampling stations were used (Hekinian et al. 2003). The hotspot activity includes the high (*HR*), middle-high (*MHR*), middle (*MR*) and patchy (*P*) (hummocky) type of sea-floor reflectivity. The inferred boundary of the hotspot has been indicated by a *heavy dashed contour line*. The older terrain has low reflectivity (*LR*) such as on the Christian and the fossil-spreading ridges, which were probably formed during pre-hotspot activity on the Farallon plate. The area around #7 and Young Volcanoes showing both low (*LR*) and high (*HR*) sea-floor reflectivity could have been a site of hotspot rejuvenation (see Sect. 5.4.2)

Volcanoes. These are essentially made of bulbous pillows, giant tubular and blocky/tabular flows, hyaloclastite crust and pyroclastic material. Their structures differ from those in other areas with MR and P reflectivity by their relatively low sediment distribution (<20%). The darkest area corresponds to the zone of highest reflectivity (HR), comprised of fresh volcanics with less than 5% sediment of hydrothermal origin. The HR area covers about 20% of the explored hotspot region, comprising small volcanic cones and star-shaped ridges (rift zones) with sharp near vertical breaks representing flow fronts (Figs. 5.12b and 5.15a,b). Deep (3 600–4 000 m depth) depressions associated with fault scarps including a graben-like structure also showing HR occur mainly in the western surveyed area near the ridge propagator and faulted terrain (Fig. 5.15a,b). These topographic lows show little relief and a flatter sea floor containing a few small volcanic cones. It is conceivable that such topographic lows are the sites of relatively young fissural eruptions giving rise to lava ponds.

The area of the Pitcairn region showing hotspot activity is patchy and more diffused than the sea floor surrounding the Macdonald Volcano and the larger edifices of the Society hotspot. It was observed that most of the deepest area (3 800–4 200 m depth) is heavily sedimented and that the more recent volcanic constructions having high back-scattered reflectivity (HR, MHR and most P and MR) occur at shallower depths than 3 800 m (Figs. 5.12b and 5.15b). The structures with LR (>3 800 m depth) are inherited from the ancient crust of the Farallon plate, now almost totally covered by sediment and ash. It is assumed that the farthest eastern limit of the hotspot is now located near the Bounty and Adams Volcano, approximately 110 km east of Pitcairn Island. From the interpretation of bottom reflectivity and multichannel bathymetry (Simrad EM12) studies in the Pitcairn region, it is thought that submarine hotspot activity extends over an area of about 7 022 km^2 (Hekinian et al. 2003).

The volume and extent of hotspot activity was evaluated with the assumption that the base of the ancient oceanic crust constructed on the Farallon plate lies near the 3 800 m contour line. This excludes the intermediate-sized edifices (e.g., Christian, Young and *Volcano #7*), which are presumed to have been formed near and/or on spreading ridges (see Sect. 5.5.2.2). Thus, it is found that the volume of the submarine hotspot volcanism covering an area shallower than 3 800 m depth with high (HR), patchy (P), middle (MR) and middle-high (MH) reflectivity includes about 5 906 km^3 (Hekinian et al. 2003) (Table 5.2).

5.5
Hotspot Versus Non-Hotspot Volcanoes

Not all the volcanoes within hotspot regions have been created during intraplate activity. In intraplate regions and especially within a hotspot area, it is often difficult to distinguish between structures caused by hotspot volcanism from those formed by crustal accretion at spreading ridges. As was seen in the Society and Pitcairn regions, some of the structures are inherited from the ancient spreading center of the Farallon plate; they show old, abundantly altered surface features covered by manganese crust and have a MORB-type composition (Cheminée et al. 1989; Hekinian et al. 1991). The criteria used to differentiate between ancient and newly-formed crust is primarily based on surface morphology (lava and shape of edifice), sea-floor structural lineation obtained from back-scatter reflectivity, and when available, the composition of the erupted lavas.

5.5.1
Sea-Floor Lineation and Seamount Distribution

Sea-floor lineation is more prominent on the larger edifices and on the surrounding sea floor (>3500 m depth) forming the Society, Austral and Pitcairn hotspots (Figs. 5.10b, 5.15b and 5.16). These lineations are expressed as fissures or as ridges forming rift zones and extending from near the summits of the edifices in a "starfish" shape going down slope along their flanks. The rift zones, such as those recognized in the Society and Pitcairn hotspots, vary in width from 2–50 m wide and are often associated with volcanic mounds, cones and flow front marked by sudden breaks in continuity (Figs. 5.6b,c and 5.14a). The maximum length of the rift zones is equal to the edifice's radius, if we assume that the main volcanic activity of a seamount remains within the area defined by its base contour lines. Therefore, the rift zone average length for the larger edifices is approximately 0.7 times the radius of each volcano (Binard et al. 1992).

The fact that several rift zones of large edifices and some smaller conical edifices are rigorously oriented in a preferential direction suggests that zones of weaknesses in the crust have became pathways for magma channeling and/or upwelling. This sea-floor lineation inherited from the ancient oceanic lithosphere of pre-hotspot origin and/or from directions created during plate tectonic reorientation is observed. Also, it is observed that the volcanic constructions are more prominent at depths shallower than 3750 m, where sea-floor structural lineations are more scattered and where small edifices are more frequently encountered (Figs. 5.16 and 5.17). This implies that the morpho-structural development of the seamounts is likely to be controlled by the pre-existing structural fabric of the ocean floor. The sea-floor fabric of the Society and Pitcairn hotspots is oriented in the preferential directions described below.

The N150°–180° orientations are observed at all the stages of volcano formation and correspond to structural discontinuities (fracture zone and spreading centers) due to the spreading activity of the ancient Farallon plate, which has been inactive for the past 12.5 Ma (Klitgord and Mammerickx 1982; Mammerickx and Klitgord 1982). This N150° direction is also seen in the rift zones of the Hawaiian chain (Smoot 1982) and especially on the Loihi Seamount (Fornari et al. 1979, 1988; Karl et al. 1988; Lonsdale 1989) and in the Austral chain with the Macdonald hotspot (Fig. 5.13a,b). In fact, the N-S directions (170–180°) on the large edifices as well as on the sea floor (>3500 m depth) prevail in all the hotspots and in the abyssal hill (>3500–4000 m depth) provinces. An alternative explanation for the dominant N-S direction of rift zones could be due to a preferential rift zone growth transverse to the hotspot chain, as observed along many of the guyots and seamounts of the Hawaiian-Emperor Bend (Smoot 1985). The N170°–180° direction deeper than the 3500 m contour line is parallel to the magnetic anomalies and represents the direction of the spreading axis of the ancient Farallon ridge. This is delineated by horst and graben (linear ridges and faults) features with scarps of 50–200 m high in the Society and the Pitcairn hotspots.

The N030–050° oblique orientation with respect to the main north-south direction is observed on spreading ridge off-axis seamounts and has been attributed by Fornari et al. (1987a) to a Riedel shear. These northeasterly directions correspond to the inferred orientations of the regional stress field caused by earthquake focal mechanisms detected in the south-central Pacific (Okal et al. 1980). In the present case, this hypothesis is the most plausible for explaining the orientations included between N180° and N080°, which

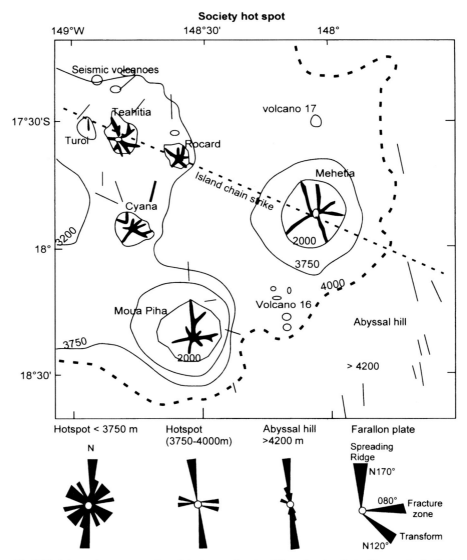

Fig. 5.16. Schematic representation of the large volcanic edifices showing their rift zones in the Society hotspot. The starfish type of rift zone is well developed on the mature edifices such as Moua Pihaa and Mehetia. The *dashed line* shows the N115° strike of the Society Island chain. The limit of the Society hotspot region is represented by the 4000 m contour line (limit of "bulge"). The *short continuous lines* oriented to N170° direction emphasize the main orientation of the small hills and volcanoes inside and outside the limit of the Society hotspot. The general morphostructural orientations of the hotspot are compared to that of the ancient crust (abyssal hill and Farallon plate regions)

▶

Fig. 5.17. Schematic reconstruction from bathymetry shows the structural lineation along the flanks of the individual seamounts from the Society hotspot (after Binard et al. 1991). The radial arrangement (starfish shape) distribution indicates magmatic channeling through rift zones. A statistical analysis of the rift zone directions for each seamount is defined according to four main orientations: N160–180°, N040°, N080°, N120°

Chapter 5 · **South Pacific Intraplate Volcanism: Structure, Morphology and Style of Eruption** 193

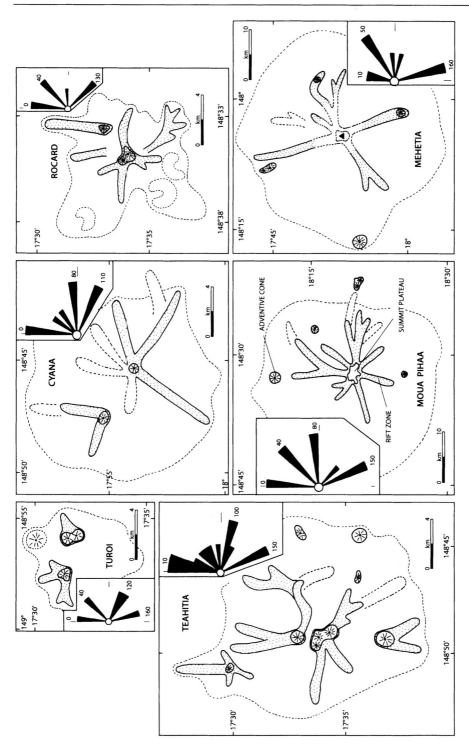

also correspond to the directions of the major fracture zones. The N045° direction in the Pitcairn hotspot corresponds to the pseudofault lineations associated with the ridge propagators inherited from a spreading ridge jump and/or from small-scale discontinuities due to spreading (Fig. 5.12b). In addition, most small edifices (<500 m depth) in the Pitcairn hotspot are oriented in the N050° direction (Figs. 5.12b and 5.15b).

The N080° orientation found mainly on the large hotspot edifices is likely to be associated with traces of ancient N80° transform faults (McNutt et al. 1989) such as the Tuamotu and Austral transform faults in the vicinity of the Society volcanic chain hotspot region (Herron 1972). This is the second type of major structural discontinuity observed to be parallel to the Pacific fracture zones (Mammerickx et al. 1975) associated with the ancient ridge system. (Figs. 5.10b, 5.12b, 5.15b and 5.16).

The N110–130° direction is mainly associated with the rift zones of the larger edifices forming the hotspots (Figs. 5.10b, 5.15b and 5.16). The most probable hypothesis for this orientation is a Reidel shear conjugated with the N030–040° direction according to the regional stress field (Okal et al. 1980). This coincides with the direction of displacement of the Pacific Plate as also shown by the orientation of the major volcanic island chains. On the Pacific Plate between the Clipperton and Galapagos fracture zones, the presence of en échelon ridges with N095°-oriented cross-grain fabric were attributed to N-S lithospheric tensile stress (Wintered and Sandwell 1987). For example, the presence of a similar stress under the Society hotspot region could enhance the development of the N110–130° lineations.

In summary, even if the preferential orientations of the larger edifices remain in a roughly north-south direction; other directions such as the N030–050° and N110–130°, which are not initially well developed will become prominent as the volcanic edifices grow. Assuming that the N030–045° and N110–130° directions represent an oblique shear system oriented perpendicularly to the major discontinuities of the crust, the N030–040° direction is probably related to a regional stress related to plate readjustment during spreading.

5.5.2
Morphological Classification of Intraplate Volcanoes

Using the flatness factor (summit diameter/basal diameter), Smith and Jordan (1987) and Smith (1988) have reported two types of shapes for Pacific seamounts: the conical (flatness < 0.25) and the truncated edifices (flatness > 0.25) (Tables 5.1 and 5.2). In this study, a similar approach is used for the volcanic edifices (small and large) from the Society and the Austral hotspot regions, and a comparison of morphology has been made (Tables 5.1 and 5.2, Fig. 5.18).

5.5.2.1
Conical Edifices

The conical shape corresponds to edifices with variable heights from 500 m (small vents) to 4 035 m (Mehetia). Interestingly, this shape is also that of some small ancient edifices inherited from the East Pacific Rise volcanic activity and not affected by a renewed hotspot volcanism. However, conical seamounts more than 500 m high found near the East Pacific Rise axis are less common than truncated ones. Most geomorphologic stud-

Chapter 5 · South Pacific Intraplate Volcanism: Structure, Morphology and Style of Eruption

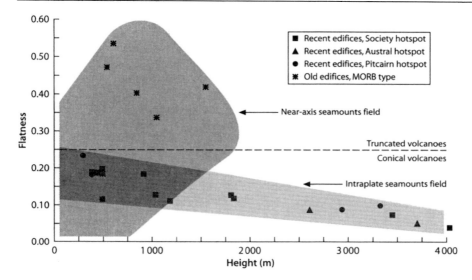

Fig. 5.18. The variation in seamount morphologies is related to the rate of flatness (summital diameter/basal diameter ratio); truncated (flatness > 0.25) and conical (flatness < 0.25) volcanoes are grouped inside two distinct fields (Tables 5.1 and 5.2). The fields of off-axis volcanoes are reported (Lonsdale and Spiess 1979; Fornari et al. 1984, 1987; Hekinian and Fouquet 1985; Hekinian et al. 1985; Batiza 1980; Lonsdale and Batiza 1980)

ies (Batiza 1977, 1982; Batiza and Vanko 1983, 1984; Batiza et al. 1984; Fornari et al. 1987b) show that seamount edification results in a juxtaposition of small volcanoes, less than 500 m tall, which are often elongated in shape. Also, the conical morphology is obtained by a pile of bulbous and tubular pillows derived from a centrally located eruptive source as illustrated by the numerous vents along the rift zones and those scattered on the hotspot-affected sea floor (3 250–3 750 m depth). As these volcanoes grow and mature, the presence of rift zones creates a starfish shape, but the edifices continue to retain their roughly conical profiles.

5.5.2.2
Truncated Edifices

Five truncated edifices, *Seismic Volcano 1* (Society hotspot), Macdocald (Austral), Mn-Seamount (abyssal hill region of the Society) and *Volcano #4 and #7* (Pitcairn hotspot), were recognized as being ancient seamounts (Figs. 5.2b, 5.12b and 5.15b). The truncated morphology of the off-axis volcanoes commonly observed (Batiza and Vanko 1983; Fornari et al. 1984) was interpreted by Lonsdale and Spiess (1979) as resulting from the draining and collapsing of a magmatic reservoir. This was a consequence of the volcano's drifting away from the East Pacific Rise axis where volcanoes are likely to be fed by cone sheets rather than from a simple conduit source (Simkin 1972). Compared to the off-axis volcanoes near the East Pacific Rise, the truncation of an intraplate edifice such as Rocald (Society hotspot) is interpreted as being due to a lateral migration of the magma supply located underneath the smaller edifice toward the larger one, Rocard (Fig. 5.7a). This will give rise to a summit collapse of the small edifice (Rocald). But if this tall edifice is derived from hotspot magmatism, its general morphology suggests similar collapse processes to those observed near/or on the East Pacific Rise. This is the submarine version of the caldera formation process related to the lateral transport of magma through dykes during flank eruptions (Ryan et al. 1981, 1983). However, the truncated edifice of Rocald could also represent an ancient off-axis volcano on which recent hotspot types of flows have risen through the summital circular fractures.

The ancient small edifice *Seismic Volcano 1* in the Society hotspot is characterized by a central caldera (about 300 m deep) (Fig. 5.2b). Pillow lavas were identified on the slopes and on the scarps forming the steep walls around the centrally collapsed crater. The surface of the volcanic edifice is heavily covered by pelagic sediment, rock fragments, and dark manganese crusts coating the lava flows (about 8–10 cm thick). Also, *Seismic Volcano 1* (SV1) has two different types of volcanics: (1) ankaramite (enriched in K_2O) and (2) extremely altered MORB-type rocks (depleted in K_2O) (Cheminée et al. 1989). The same type of MORB was also collected from the truncated Macdonald edifice (Austral region), which is made up of altered pillows covered by 4–6 cm thick manganese crusts; this MORB was also recovered from the small conical *Volcano #7* (Pitcairn), and from Christian and Young (Pitcairn) (Fig. 12b). This indicates that numerous conical and truncated edifices (<500 m in height) within the Society, Austral and Pitcairn hotspot regions are inherited from older volcanoes formed near or on the East Pacific Rise. Furthermore, some of the truncated structures (SV1, Rocald) having a hotspot type of lava (alkali enriched) may have been reactivated during hotspot volcanism as reported by Hekinian et al. (1991). In the Pitcairn region, bulbous and tubular flows partially buried by sediment (about 20–25%) are found in the area of Young and *Volcano #7*, both of which are volcanoes showing flat tops and collapsed craters from which Fe-Si-Mn oxyhydroxide crusts were dredged (Fig. 5.12b). The presence of MORBs on Young Volcano and on a small cone on the flank of Christian Seamount is believed to be due to pre-hotspot volcanism (Hekinian et al. 2003).

The degree of flatness indicates that truncated shapes are preferentially related to ancient drifting volcanoes inherited from an ancient spreading center. All these edifices show higher degrees of flatness with more gentle slopes (18–25°) (Fig. 5.18), while the most pronounced conical edifices having a flatness of less than 0.25 with slopes of 10–20° are associated with recent intraplate volcanic activity (Fig. 5.18). The intraplate seamounts are morphologically comparable to large subaerial shield volcanoes (Peterson and Moore 1987). Pillow lava and sheet flows have built the main part of the edifices, from their bases to about 500 m depth, leading to a high structural stability that prevents the occurrence of large magmatic reservoirs and collapsed caldera morphologies. Magmatic segregation is related to small reservoirs probably derived from the main, centrally-located conduit of intraplate volcanoes.

In summary, two types of morphologies are present in the Society, Austral and Pitcairn hotspot regions: (1) a few of the small and intermediate-sized structures (<1500 m in height) have truncated edifices with a flat top, similar in morphology and lava composition (MORBs) to seamounts inherited from EPR off-axis volcanic activities; and (2) most small (<500 m high) and large (up to 3700 m in height) conical edifices are formed during hotspot magmatism. This agrees with previous observations reported on the distribution of Pacific seamounts (Hollister et al. 1978; Smith and Jordan 1987; Smith 1988), where morphology (truncated to conical types) correlates to the height of the volcanic constructions (Tables 5.1 and 5.2, Fig. 5.18).

5.6
Style of Eruption and Formation of Hotspot Edifices

Different types of flow morphologies are encountered on the large edifices of the Society, Austral and Pitcairn hotspots (Fig. 5.8a–p). The hotspot volcanics differ from normal MORBs by their higher degree of vesicularity (>15 up to 80% vesicles) as well as in their form. For example, blocky/tabular flows and hornitos made up of silica-enriched lava are rare in spreading ridge environments. These submarine volcanic features from hotspot areas are comparable to those found on the active Hawaiian hotspot, such as the Loihi Seamount (Fornari et al. 1988; Karl et al. 1988; Lonsdale 1989). The morphological appearance of the flows including their shape, degree of fragmentation, surface textures (corrugation) and the size of the flows, are a good indication of their mode of emplacement and volcano construction. The types of volcanics forming an edifice depend on the style of eruption, which correlates to the physical properties and the composition of the lava. Many of the physical parameters controlling the style of eruption such the cooling rate, the size of the eruptive conduit, the degree of rock viscosity and the degree of crystallinity are discussed by Bonatti and Harrison (1988) and Perfit and Chadwick (1998).

5.6.1
Types of Eruption

A *Quiet* type of eruption indicates the absence of volatiles or the ease of gas escape where the volcanic landscape is characterized by the occurrence of low viscosity lava such as sheet flows and pillow lava that predominate at spreading ridges and subaerial shield volcanoes. The rate of effusion that increases from pillow to sheet flows (lobate and flat flow) controls their shape. Thus, in both intraplate and spreading ridge areas, the presence of sheet flows results from a high magma supply and a rapid flow rate.

An *Explosive* type of activity is related to magma enriched in volatiles, which is able to increase its gas pressure above the lithostatic pressure, and this results in violent eruptions. The explosive eruptions are associated with accidental debris (pyroclasts) and hyloclastites. Volatile enrichment is primarily due to the concentration of CO_2 and H_2O. These compounds are found in the mantle (about 1%) and are the essential contributors to rock vesicularity and gas concentration during magmatic solidification. When correlating the degree of rock vesicularity with the volatile content, (CO_2 and H_2O), it was shown that the degrees of vesicularity and pressure of the volatile phases are mainly due to the early exsolution of CO_2 from an alkali melt (Hekinian et al. 2000; Pineau et al. 2003). The exsolution of significant amounts of dissolved water from the alkali melt could also contribute to the expansion of bubbles accumulating in the upper part of the conduit and trigger explosions at water depths of 1 000–3 000 m. When the volatile-enriched melt rises and starts to form bubbles (mainly CO_2), the amount of vesicles increases, and the decoupling of the magma-bubble system will accelerate. This will increase the vesicle ascent and the accumulation of CO_2-rich gas pockets towards the top of the magmatic column. Consequently, the gas pressure exerted on the overlying volcanic construction will rise until it overcomes the load pressure. Gas-magma decoupling during magmatic ascent could explain the formation of hyaloclastites and pyroclasts. The explosion is initiated at the top of the magmatic column by a breaking up of the gas vesicles or gas pockets that produced the hyaloclastites at the same time or immediately after the pyroclasts (Hekinian et al. 2000; Pineau et al. 2003). Similarly, intraplate volcanism has generated abundantly vesicular flows that are often associated with pyroclastic and hyaloclastite deposits commonly found together on the large edifices at various depths (200–3 000 m depth) (Fig. 5.8n).

5.6.2
The Formation of a Volcanic Edifice

A compilation of data on the lava morphologies with respect to their setting obtained from on site (submersible, deep-towed camera and side scan sonar) observations in the Society, Austral and Pitcairn hotspot regions is used to reconstruct a "typical volcanic edifice" shown in Fig. 5.19 and described below (Binard et al. 1992a).

The Edifice Base shows broad-sheet flows made up of non-vesicular lava (vesicles <5% in volume). This type of flow morphology is comparable to that observed on the Mid-Oceanic Ridge axis (Arcyana 1978; Ballard and Moore 1977), where the lava flows are extruded at a high discharge rate and are confined within tectonic depressions (rift valleys and/or grabens). Since the Society hotspot does not have such tectonic depressions (Binard et al. 1991), a high rate of fluidal lava discharge is needed in order to give rise to the broad-sheet flows observed on the ancient sea floor (Fig. 5.8c). The sheet flow morphology correlates to a high degree of fluidity, which is mainly controlled by the temperature, the crystal/liquid ratio, and the chemical composition of the magma. Since variations in the crystal/liquid ratios or lava compositions at different depths were not observed for rocks sampled from the Society and the Austral hotspot regions, it is believed that temperature and magma supply could be the most important factor influ-

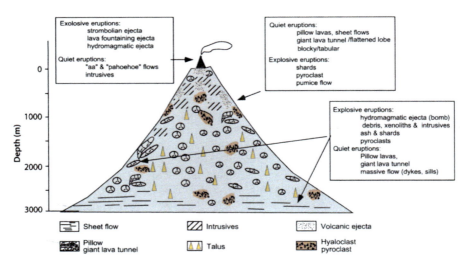

Fig. 5.19. Schematic representation of 'typical' volcano's eruptive styles occurring during the construction of an intraplate volcano. The reconstruction was made from field observations mainly suggested from the Society, Austral and Pitcairn hotspot regions (after Binard et al. 1992a). Pillow, shards, volcanoclast, and talus material are found from the base to the top of seamounts. Volcanic ejecta (hyaloclast and pyroclast) are abundantly located at the summit, but they are also found on small adventive cones on the flank of large edifices. The intrusive flows are difficult to see at the middle and base of the edifice where they are masked by erupted flows

Chapter 5 · South Pacific Intraplate Volcanism: Structure, Morphology and Style of Eruption 199

encing the viscosity and the eruption rate. Hence, rapid eruption could reduce the cooling rate once a chilled crust is formed. The high temperature is compatible to a flow discharge for a short distance from the magma reservoir (Bonatti and Harrison 1988).

The Edifice Flank is characterized by bulbous and giant tubular pillows and lava tunnels encountered on the lateral vents making up the rift zones (Fig. 5.8e). Pillows and lava tube formations likely to have originated from quiet eruptions were reported below 500–1 000 m depth (Cheminée et al. 1989; Fornari et al. 1979, 1988; Lonsdale 1989). The lava erupting onto the edifice flanks is still highly fluid as shown by the presence of folded surfaces, elongated and narrow lava tubes (<10 m length, 30 cm in diameter), stalactites, and hornitos (Fig. 5.8b). The change in flow morphology (pillows) at the level of rift zones compared to the deeper sheet flows could be due to a lower rate of discharge induced by scattered eruptive vents. In the Society, Pitcairn and the Austral hotspot regions, the shards and pyroclasts are mainly concentrated between the bulbous flows (pillows and lava tunnels) and on flat sedimented surfaces (Fig. 5.8d and 5.8i). Hyaloclastites are found as flat lying layers on top of lava flows, and are essentially encountered on seamounts that may be in both a shallow and deep-seated (>2 000 m depth) environment (Figs. 5.8g and 5.19). Their lithological settings and their composition suggest that the hyaloclastites were formed locally during the emplacement of basaltic flows having a thick glassy crust (<2 cm thick). Instead, the thicker and extended hyaloclastite layers mixed with hydrothermal deposits and pelagic sediment forming semi-indurate products covering lava flows are believed to have been formed during explosive eruptions. They are often associated with pyroclastic deposits.

The summit of shallow edifices (<500 m depth) comprises porous reservoirs, where magmatic gases are concentrated and where seawater changes into steam in contact with magma (Figs. 5.8m and 5.19). This involves hydromagmatic explosions, which give rise to circular craters. The submarine association of volcanic ejecta and lava flows is similar to the eruptive sequences observed during subaerial strombolian or lava fountain activities. The summit volcanic formations made up of easily reworked clastic material might be conducive for the occurrence of small magmatic reservoirs.

A shallower summit depth and reduced hydrostatic pressure could lead to the release of gases, which give rise to conical depressions within the volcanic ash deposits. The eruptive processes produced the pumice flow covering the summit of the shallow (<300 m depth) Turoi, Moua Pihaa and Macdonald Seamounts. The flow brecciation, the pumiceous character of the lava, and the "bread crust" surfaces of the fragments suggest the predominant role of magmatic gases during eruption. The evolved nature of the rock ($SiO_2 = 61\%$) and the low volume of eruption are compatible to the occurrence of magma and volatile segregation inside a reservoir. Similar pumice-type rocks encountered on subaerial volcanoes commonly result from volcanic eruptions when magmatic gases concentrated underneath the top of the edifices cause the sudden extrusion of volcanic bombs. It has been suggested that the pumice flow observed on the Turoi Seamount was formed from a similar (but submarine) volcanic eruption where explosive activities were prevented by the hydrostatic pressure (Fig. 5.8h). Shallow explosive events are probably more common than deeper ones and give rise to a more extensive distance of debris. For example, on the Macdonald Seamount, the debris due to a hydromagmatic explosion was found to extend from near the surface (50–100 m) down to about 1 000 m on the flank of the edifice. However, it is also known that reactive and explosive events also occur on the summit of volcanic edifices at depths greater than 1 000 m (Hekinian et al. 2000).

5.6.3
Relationship Between Hotspot Volcanic Edifices

Hotspot volcanism is expressed either by the formation of a single edifice (i.e., Macdonald) or a cluster of different sized edifices (i.e., Society and Pitcairn hotspots) within an area of 120–200 km in diameter (Figs. 5.2b, 5.10b and 5.12b). The interrelationship between volcanic edifices could be enhanced by subsurface magma channeling through dyke propagation. The dynamics of crack propagation in an elastic medium (Lister and Kerr 1991; Turcotte 1982) are important for enhancing magma circulation in the lithosphere. Evidence of dyke propagation comes essentially from ground deformation and eruptions observed in Iceland, Hawaii (Sigurdson 1987; Rubin and Pollard 1987) and in spreading centers (Chadwick et al. 1995).

The construction of large- and intermediate-sized edifices (such as Adams, Bounty, and Moua Pihaa in the Society hotspot and Volcano #4 from Pitcairn hotspot) is believed to be the result of a sustained magma supply and magma chamber replenishment during hotspot activity. A diapir melting front underlying the volcanoes is probably the source of the magma injection (Figs. 5.12b and 5.20). The replenishment of magma chambers underneath these larger edifices is in agreement with the observed diversities in isotopic composition (Woodhead and Devey 1993; Devey et al. 2003), suggesting several parental melts and/or mantle sources for the origin of the volcanics.

The small cones on the flanks (adventive), at the foot (<5 km from the base) and in the vicinity (<50 km away) of the larger edifices are believed to have formed during

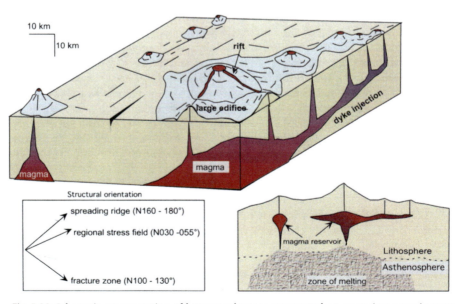

Fig. 5.20. Schematic representation of hotspot volcanoes constructed on an ancient oceanic crust, showing different structural orientations such as fracture zones and spreading-ridge directions. The relationship between the various-sized edifices is inferred through rift zone formation and sub-crustal magma channeling. The inset shows a broader area of the upper mantle in a partially molten zone from which magma reservoirs are supplied underneath the larger hotspot edifices. The main orientations of the sea-floor structural fabric are indicated (see Sect. 5.5.1)

localized (puntiform) eruptions supplied through magma channels along preferential tectonic directions. The adventive cones located on the flank and at the foot of the larger edifices are often interconnected with rift-zone channeling magma from the main edifice conduits. As a volcano grows, its mass will increase and raise the lithostatic pressure on the magma reservoir. As the height of the volcano increases, probably less magma will erupt from the summit and more from lateral vents, rift zones and parasitic cones. In order for the extrusive flow (P_{ext} = overpressure of eruption) to continue to overcome the lithostatic pressure (P_{lithos}; $P_{ext} > P_{lithos}$), it is necessary for the magma reservoir to become more evolved and/or be refilled with new melt. Evolved lavas (e.g., trachyte, andesite and rhyolite) such as those erupted in these small cones are enriched in volatiles. They are less dense and less viscous, so they are more likely to overcome the load pressure. Also, the pressure of silicic magma stored in a reservoir might increase as a result of the additional magma injection from greater depths. For shallow reservoirs, the depth inferred from petrological and geophysical studies is less than 5–10 km beneath active volcanoes (Davidson and de Silva 2000). As pressure in the chamber increases, lateral dikes, sills and other fractures propagate from the chamber to the surface (Wilson et al. 1980; Bower and Woods 1997). Fracture propagation is also facilitated by the bursting of gas pockets trapped in the conduit during the passage of new melt. The model of Bower and Woods (1997) applied to several historical eruptions shows the dependence of an eruption to the depth, volume and degree of crystallization of lava in a magma reservoir.

The other small cones located at some distance (<50 km) from the larger edifices are likely to be fed by deeper-seated subsurface magma channels than the adventive cones at the flank and/or foot of the larger volcanoes. These small edifices could be supplied from the replenished and differentiated magma reservoirs located underneath the larger edifices when the magma flows through pre-existing conduits or channels within a fractured lithosphere. For example in the Pitcairn hotspot, the evidence for subsurface feeder channels and eruptions is inferred from the distribution of volcanic cones along preferential directions that are aligned along elongated topographic highs emanating from the foot of the larger edifices (e.g., Adams and Bounty, Figs. 5.14a, 5.15a and 5.15b). The most obvious volcanic channels building edifices having a relief of at least 200 m with respect to the surrounding sea floor are observed to extend in a south-southwesterly (25°15' S, 120°30' W) and in a southern (25°30' S, 129°19' W) direction from the Bounty and Adams Volcanoes, respectively. We observed elongated and bulging structures up to about 20 km in length and 1–4 km wide with scattered small volcanic edifices erected along their length (Figs. 5.12b, 5.15a and 5.15b). Hence, when considering the distribution of small volcanic cones along a preferential direction (e.g., rift zone, fault, and other ancient tectonic feature), it is inferred that melt could laterally flow for a long distance (tens of kilometers) from its source (magma reservoir) (Figs. 5.15b, 5.16, 5.17 and 5.20).

5.7
Summary and Conclusions

The formation of hotspot seamounts is facilitated by existing crustal weaknesses such as fracture zones, fissures, or horst-graben structures inherited from ancient structures as well as recent intraplate tectonic motions.

The development of individual rift zones, such as the "star fish" branching features, is also linked to the size of the volcanoes. The orientations correlate to the major structural directions (N160–180°, N040°, N080°, N100–130°) observed in much larger edifices in the Pacific Ocean and correspond to those of spreading ridges (N160–180°), the ancient Farallon plate, fracture zones (N80°) and transform faults (N100–130°). In addition, the N030–050° orientations of rift zones on the large and small edifices is comparable to the structural directions inherited from spreading ridge jumps and/or ridge discontinuities (pseudofaults) and the regional stress field of the Farallon plate. The maximum length of the rift zones is equal to the edifice's radius however the larger edifices could also be interconnected with smaller ones through subsurface magma channeling. Evidence for subsurface feeder channels giving rise to small edifices is inferred from the distribution of volcanic cones along preferential directions, which also represent the same tectonic lineation mentioned above.

The larger edifices from the Society, Pitcairn and Austral hotspots are submarine shield volcanoes, whose inferred slopes (11–17°) are steeper than those of subaerial shield volcanoes (6–12°). Steeper slopes in a submarine environment might be due to the relatively faster cooling rate and higher viscosity of the flowing lava than what occurs on land-based eruptions. This is also illustrated by most of the volcanic adventive cones of the large edifices, which have slopes of more than 50–60°.

The structure of the seamounts is formed by a network of dyke swarms, joining together in a main magmatic conduit along an axis, which is located vertically underneath the summit. This main conduit plays the role of a magmatic reservoir, whose limited size was unable to provoke a caldera collapse in the hotspot volcanoes. The conical-shaped edifices are most likely to have originated during hotspot magmatism, while most of the edifices with truncated summits are believed to have been formed during spreading ridge (axial and/or off axial) activities.

The volume of the volcanics, which have built the various edifices, varies considerably. The smallest edifices have erupted a total volume of 1–4 km^3 of material, while the tallest one has a volume close to 1 000 km^3. From the interpretation of bottom reflectivity and multichannel bathymetry studies in the Pitcairn region, it is shown that submarine hotspot activity extends over an area of about 7 022 km^2. The largest edifices have erupted about 200 to 1 000 km^3 of volcanics.

The submarine volcanoes of the Society, Pitcairn and Austral hotspot regions have undergone similar eruptive events as indicated from their structural setting, lava flow morphologies and stage of edifice growth. The following model for the construction of a 'typical' intraplate volcano is proposed:

1. A quiet eruptive event with high discharge rate of fluid lava as found on modern East Pacific Ridge segments gives rise to pillows and extended flat flows on a smooth topography associated with hotspot activity.
2. On the flanks of the larger edifices, the quiet eruptions located within linear rift zones show a succession of volcanic cones and mounds made up of bulbous and tubular pillows related to steep slopes. Also, giant tubes and lava tunnels associated with the gentle slopes on top of lateral vents occur. Adventive cones found on the flank of these large edifices giving rise to both silicic lava and pyroclast-hyaloclast association suggest explosive events at 1 000–3 000 m depth. Alkali-enriched lavas are enriched in volatiles and produce vesicular (>15 to 70% vesicles) flows. The presence of silica rich lavas (trachyte

Chapter 5 · South Pacific Intraplate Volcanism: Structure, Morphology and Style of Eruption

and trachy-andesite) forming hornitos and massive blocky flow with abundant gas cavities and scoria-like surface suggest forceful magma injection of viscous flows. The succession of lava flows and pyroclastite/hyaloclastite units encountered along the slopes of the various edifices implies rhythmic variations in quiet and explosive events.

3. When the edifice summits are shallower than 500 m depth, pillow and intrusive flows derived from quiet eruptive events are interbedded with volcanic ejecta produced by hydromagmatic eruptions.

The subaerial volcanic constructions such as the islands of Mehitia and Pitcairn are shield-types of volcanoes showing a succession of pyroclastite/hyaloclastite layers and lava flows. In both examples, the abundance of fragmented debris of an explosive nature is mainly due to the magma/water interaction and to magmatic gas channeled and accumulated at the top of the volcano. Nevertheless, their earliest stage of seamount formation was also related to hotspot volcanism where quiet eruptions gave rise to sheet flows (i.e., Tedside Volcanics of Pitcairn Island) (see Sects. 5.4.1.1 and 5.6.2).

Acknowledgements

We are indebted to the captains, the officers and the crew of the N.O. J. CHARCOT, Le SUROIT, L'ATALANTE and the submersible (*Cyana* and *Nautile*) teams of GENAVIR for their efficiency, expertise and patience during the many dives that were done in the course of this ongoing study. We also express our gratitude to the captains and the crew during Legs 47 and 65 of the F.S. SONNE for their expertise. These cruises were supported by BMFT grant AZ.03R397A5 to P. Stoffers. The Polynaut cruise was directed and sponsored by INSU (France) and the Bundesministerium für Bildung und Forschung (BMBF, Germany). This work was initiated at IFREMER (Department of Marine Geosciences), and the authors are particularly indebted to the continued support of several members of the Department in Marine Geosciences. We are grateful to Dr. R. Batiza for reviewing the manuscript. The bottom photographs taken by submersibles are courtesy of IFREMER (Département de la Communication, Brest, Paris).

References

Arcyana (1978) Famous, Atlas Photographique. Gauthiers-Villars, Paris, pp 128
Ballard RD, Moore JG (1977) Photographic atlas of the Mid-Atlantic Ridge Rift Valley. Springer-Verlag, New York Heidelberg Berlin, pp 114
Batiza R (1977) Age, volume, compositional and spatial relations of small isolated oceanic central volcanoes. Mar Geol 24:169–183
Batiza R (1980) Origin and petrology of young oceanic central volcanoes: Are most tholeiitic rather than alkalic? Geology 8:477–482
Batiza R (1982) Abundances, distribution and sizes of volcanoes in the Pacific Ocean and implications for the origin of non-hotspot volcanoes. Earth Planet Sci Let 60:195–206
Batiza R (1989) Seamounts and seamount chains of the eastern Pacific. In: Winterer EL, Hussong DM, Decker RW (eds) The Eastern Pacific Ocean and Hawaii. Geological Society of America, pp 289–306
Batiza R, Vanko D (1983) Volcanic development of small oceanic central volcanoes on the flanks of the East Pacific rise inferred from narrow-beam echo-sounder surveys. Mar Geol 54:53–90
Batiza R, Vanko D (1984) Petrology of young Pacific seamounts. J Geophys Res 89:11235–11260
Batiza, R, Fornari DJ, Vanko D, Lonsdale P (1984) Craters, calderas and hyaloclastites: Common features of young Pacific seamounts. J Geophys Res 89:8371–8390
Bellon H (1974) Histoire géochronométrique des îles Gambier. Cahiers du Pacifique, 18:159–244

Binard N, Hekinian R, Cheminée JL, Searle RC, and Stoffers P (1991) Morphological and structural studies of the Society and Austral hot spot regions in the South Pacific. Tectonophysics 186:293–312

Binard N, Hekinian R, Cheminée JL, Searle RC, Stoffers P (1992a) Style of eruptive activity on intraplate volcanoes in the Society and Austral hot spot regions: Bathymetry, petrology, and submersible observations. J Geophys Res 97:13999–14015

Binard N, Hekinian R, Stoffers P (1992b) Morphostructural study and type of submarine volcanoes over the Pitcairn hot spot in the South Pacific. Tectonophysics 206:245–264

Binard N, Maury RC, Guille G, Talandier J, Gillot PY, Cotten J (1993) Mehetia Island, South Pacific: Geology and petrology of the emerged part of the Society hot spot. J Volcanol Geotherm Res 55:239–260

Bonatti E., Harrison CGA (1988) Eruption styles of basalt in oceanic spreading ridges and seamounts: Effect of magma temperature and viscosity. J Geophy Res 93:2967–2980

Bonneville A, Suavé R, Audin L, Clouard V, Dosso L, Gillot PY, Janney P, Jordahl K, Maamaatuaiahutapu K (2002) Arago Seamount: The missing hotspot found in the Austral Islands. Geol Soc Amer 30:1023–1026

Bower SM, Woods AW (1997) Control on magma volatile content and chamber depth on the mass erupted during explosive volcanic eruptions. J Geophys Res 102:10273–10290

Carter RM (1967) The geology of Pitcairn Island, South Pacific Ocean. Bernice Pauhi Bishop Museum Bull 231:1–38

Chadwick WW Jr., Embley RW, Fox CG (1995) SeaBeam depth changes associated with recent flows, CoAxial segment, Juan de Fuca Ridge: Evidence for multiple eruptions between 1981–1993. Geophys Res Letters 22:167–170

Cheminée JL, Hekinian R, Talandier J, Albarède F, Devey CW, Francheteau J, Lancelot Y (1989) Geology of an active hot spot: Teahitia-Mehetia region in the south central Pacific. Mar Geophys Res 11:27–50

Cheminée JL, Stoffers P, McMurtry G, Richnow H, Puteanus D, Sedwick P (1991) Gas-rich submarine exhalations during the 1989 eruption of Macdonald seamount. Earth Planet Sci Lett 107:318–327

Choukroune P, Francheteau J, Hekinian R (1984) Tectonics of the East Pacific Rise near 12°50'N: A submersible study. Earth Planet Sci Lett 68:115–127

Clague, DA, Dalrymple GB (1987) The Hawaiian-Emperor volcanic chain, Part I. In: Decker RW, Wright TL, Stauffers PH (eds) Volcanism in Hawaii. U.S. Geological Survey pp 5–54

Crandell DR, Miller CD, Glicken HX, Christiansen RL, Newhall CG (1984) Catastrophic debris avalanche from ancestral Mount Shasta volcano, California. Geology 12:143–146

Crough ST (1978) Thermal origin of mid-plate, hot-spot swells. Geophys J Roy Astronom Soc 55: 451–469

Crough ST, Jurdy DM (1980) Subducted lithosphere, hotspots, and the geoid Earth. Planet Sci Lett 48:15–22

Davidson J, de Silva S (2000) Composite volcanoes. In: Sigurdsson H (ed) Encyclopedia of volcanoes. Academic Press London, pp 663–681

Detrick RS, Crough ST (1978) Island subsidence, hotspots and lithosphere thinning. J Geophys Res 83:1236–1244

Devey CW, Albarede F, Cheminée JL, Michard A, Muhe R, Stoffers P (1990) Active submarine volcanism on the Society hotspot swell (West Pacific): A geochemical study. J Geophys Res 95:5049–5067

Devey CW, Mertz DF, Bourdon B, Cheminée JL, Dubois J, Guivel C, Hekinian R, Stoffers P (2003) Giving birth to hotspot volcanoes: Distribution and composition of young seamounts from seafloor near Tahiti and Pitcairn Islands. Geology 31:395–398

Duncan RA, Mcdougall I (1974) Migration of volcanism with time in the Marquesas Islands, French Polynesia. Earth Planet Sci Lett 21:414–420

Duncan RA, Mcdougall I (1976) Linear volcanism in French Polynesia. J Volcanol Geother Res 1: 197–227

Duncan RA, Mcdougall I, Carter RM, Coombs DS (1974) Pitcairn Island – another Pacific hot spot? Nature 251:679–682

Fisher RL, Norris RM (1960) Bathymetry and geology of Sala y Gómez, Southeast Pacific. Geol Soc Amer Bull 71:497–502

Fornari DJ, Peterson DW, Lockwood JP, Malahoff A, Heezen BC (1979) Submarine extension of the southwest rift zone of Mauna Loa Volcano, Hawaii: Visual observations from U.S. Navy Deep Submergence Vehicle DSV Sea Cliff. Geol Soc Amer Bull 90:435–443

Fornari DJ, Lockwood JP, Lipman PW, Rawson M, Malahoff A (1980) Submarine volcanic features west of Kealakekua Bay, Hawaii. J Volcanol Geotherm Res 7:323–337

Fornari DJ, Ryan WBF, Fox PJ (1984) The evolution of craters and calderas on young seamounts: Insights from Sea Mark I and Sea Beam sonar surveys of a small seamount group near the axis of the East Pacific Rise at 10° N. J Geophys Res 89:11069–11083

Chapter 5 · South Pacific Intraplate Volcanism: Structure, Morphology and Style of Eruption

Fornari DJ, Batiza R, Allan JF (1987a) Irregularly shaped seamounts near the East Pacific Rise: Implications for seamount origin and rise axis processes. In: Keating BH, Fryer P, Batiza R, Boehlert GW (eds) Seamounts, islands, and atolls. Geophys Monograph 43, Amer Geophys Union, Washington D.C., pp 35–47

Fornari DJ, Batiza R, Luckman MA (1987b) Seamount abundances and distribution near the East Pacific Rise 0°–24° N based on seabeam data. In: Keating BH, Fryer P Batiza R, Boehlert GW (eds) Seamounts, islands, and atolls. Geophys Monograph 43, Amer Geophysical Union, Washington D.C., pp 13–21

Fornari DJ, Garcia MO, Tyce R, Gallo DG (1988) Morphology and structure of Loihi Seamount based on seabeam sonar mapping. J Geophys Res 93:15227–15238

Guillou H, Garcia MO, Turpin L (1997) Unspiked K-Ar dating of young volcanic rocks from Loihi and Pitcairn hot spot seamounts. J Volvanol Geotherm Res 78:239–249

Hekinian R, Fouquet Y (1985) Volcanism and metallogenesis of axial and off-axial structures on the East Pacific Rise near 13° N. Economic Geology 80:221–249

Hekinian R, Chaigneau M, Cheminée JL (1973) Popping rocks and lava tubes from the Mid-Atlantic rift valley at 36° N. Nature 245:371–373

Hekinian R, Francheteau J, Ballard RD (1985) Morphology and evolution of hydrothermal deposits at the axis of the East Pacific Rise. Oceanologica Acta 8(2):147–155

Hekinian R, Bideau D, Stoffers P, Cheminée JL, Muhe R, Puteanus D, Binard N (1991) Submarine intraplate volcanism in the South Pacific: Geological setting and petrology of the Society and the Austral regions. J Geophys Res 96:2109–2138

Hekinian R, Pineau F, .Shilobreeva S, Bideau D, Gracia E, Javoy M (2000) Deep sea explosive activity on the Mid-Atlantic Ridge near 34°50' N: Magma composition, vesicularity and volatile content. J Volcanol Geotherm Res 98:49–77

Hekinian R, Cheminée JL, Dubois J, Stoffers P, Scott S, Guivel C, Garbe-Schönberg D, Devey CW, Bourdon B, Lackschewitz K, McMurtry G, Le Drezen E (2003) The Pitcairn hotspot in the South Pacific: Distribution and composition of submarine volcanic sequences. J Volcanol Geotherm Res 121:219–245

Hémond C, Devey CW, Chauvel C, (1994), Source compositions and melting processes in the Society and Austral plumes (South Pacific Ocean): Element and isotope (Sr, Nd, Pb, Th) geochemistry. Chem Geol 115:7–45

Herron EM (1972) Sea-floor spreading and the Cenozoic history of the East-Central Pacific. Geol Soc Amer Bull 83:1671–1692

Hoffert M, Cheminée JL, Person A, Larque P (1987) Dépôt hydrothermal associé au volcanisme sousmarin intraplaque. Prélèvement effectué avec Cyana sur le volcan actif de Teahitia (Polynésie Française). Comptes Rendus Académie Sci Paris (II) 304(14):829–832

Hollister CD, Glenn MF, Lonsdale PF (1978) Morphology of seamounts in the Western Pacific and Philippine Basin from multi-beam sonar data. Earth Planet Sci Lett 41:405–418

Jarrard RD, Clague DA (1977) Implications of Pacific island and seamount ages for the origin of volcanic chains. Reviews of Geophysics Space Physics 15:57–76

Johnson RH (1970) Active submarine volcanism in the Austral islands. Science 167:977–979

Johnson RH (1984) Exploration of three submarine volcanoes in the South Pacific. National Geographic Soc Res Report 16:405–419

Karl DM, McMurtry GM, Malahoff A, Garcia MO (1988) Loihi Seamount, Hawaii: A mid plate volcano with a distinctive hydrothermal system. Nature 335:532–535

Klitgord KD, Mammerickx J (1982) Northern East Pacific Rise: Magnetic anomaly and bathymetric framework. J Geophys Res 87:6725–6750

Lacroix A (1936) Les roches volcaniques de l'île de Pitcairn (Océan Pacifique Austral). Comptes Rendus Académie Sci Paris 202:788–79

Lister JR, Kerr RC (1991) Fluid-mechanical model of cracks propagation and their applications to magma transport in dykes. J Geophys Res 96:10049–10077

Lonsdale P (1988) Geography and history of Louisville hotspot chain in the south west Pacific. J Geophys Res 93:3078–3104

Lonsdale P (1989) A geomorphological reconnaissance of the submarine part of the East Rift Zone of Kilauea Volcano, Hawaii. Bull Volcanol 51:123–144

Lonsdale P, Batiza R (1980) Hyaloclastite and lava flows on young seamounts examined with a submersible. Geol Soc Amer Bull 91:545–554

Lonsdale P, Spiess FN (1979) A pair of young cratered volcanoes on the East Pacific Rise. J Geology 87:157–173

Macdonald GA, Abbott AT (1970) Volcanoes in the sea: The geology of Hawaii. The University Press of Hawaii, Honolulu, pp 441

Mammerickx J, Klitgord KD (1982) Northern East Pacific Rise: Evolution from 25 m.y. B.P. to the Present. J Geophys Res 87:6751–6759

Mammerickx J, Anderson RN, Menard HW, Smith SM (1975) Morphology and tectonic evolution of the East-Central Pacific. Geol Soc Amer Bull 86:111–118

Mayes CL, Lawver LA, Sandwell DT (1990) Tectonic history and new isochron chart of the South Pacific. J Geophys Res 95(B6):8543–8567

McNutt MK, Fischer K, Kruse S, Natland J (1989) The origin of the Marquesas fracture zone ridge and its implications for the nature of hot spots. Earth Planet Sci Lett 91:381–393

Menard HW (1956) Archipelagic aprons. Amer Assoc Petroleum Geologists Bull 40:2195–2210

Michard A, Michard G, Javoy M, Cheminée JL, Binard N (1989) Chemistry of submarine springs from the Teahitia Seamount. EOS Transaction Amerer Geophys Union 70:495

Minster JB, Jordan TG (1978) Present day plate motions. J Geophys Res 83: 5331–5354

Moore JG, Clague DA, Holcomb RT, Lipman PW, Normark WR, Torresan ME (1989) Prodigious submarine landslides on the Hawaiian ridge. J Geophys Res 94:17465–17484

Morgan WJ (1972) Plate motion and deep mantle convection. Geol Soc America Memoirs 132:7–22

Mottay G (1976) Contribution à l'étude géologique de la Polynésie Française: Archipel des Australes-Mehetia (Archipel de la Société). Doctoral thesis, Université de Paris Sud, 4 place Jussieu, Paris

Norris RA, Johnson RH (1969) Submarine volcanic eruptions recently located in the Pacific by sofar hydrophones. J Geophys Res 74:650–664

Okal EA, Cazenave A (1985) A model for the plate tectonic evolution of the east-central Pacific based on SEASAT investigations. Earth and Planetary Science Letters 72:99–116

Okal EA, Talandier J, Sverdrup KA, Jordan TH (1980) Seismicity and tectonic stress in the South-Central Pacific. J Geophys Res 85:6479–6495

Parsons B, Sclater JG (1977) An Analysis of the variation of ocean floor bathymetry and heat flow with age. J Geophys Res 82:803–827

Perfit M, Chadwick WW Jr (1998) Magmatism at mid-ocean ridges: Constraints from volcanological and geochemical investigations. In: Buck RW, Delaney PT, Karson JA, Lagabrielle Y (eds) Faulting and magmatism at mid-ocean ridges. Geophysical Monograph Amer Geophys Union, Washington, D.C., 106:59–115

Peterson DW, Moore RB (1987) Geological history and evolution of geologic concepts, island of Hawaii. In: Decker RW, Wright TL, Stauffers PH (eds) Volcanism in Hawaii. U.S. Geological Survey, Denver, pp 149–189

Pinneau F, Shilobreeva S, Bideau D, Javoy M, Hekinian R (2003) Deep sea explosive activity on the Mid-Atlantic Ridge near 34°50' N: Stable isotope (C, H, O) of lava and their volatile content. J Volcanol Geotherm Res (submitted)

Phipps Morgan J, Morgan WJ, Zhang Y-S, Smith WHF (1995) Observational hints for a plume-fed suboceanic asthenosphere and its role in mantle convection. J Geophys Res 100:12753–12768

Puteanus D, Glasby GP, Stoffers P, Kunzendorf H (1991) Hydrothermal iron-rich deposits from the Teahitia-Mehetia and Macdonald hot spot areas, Southwest Pacific. Mar Geology 98:389–409

Rubin AM, Pollard DD (1987) Origin of blade-like dikes in volcanic rift zone. In: Decker RD, Wright TL, Stauffer PH (eds) Volcanism in Hawaii. US Geol Survey Professional Papers 1350:1449–1470

Ryan MP, Koyanagi RY, Fiske R (1981) Modeling the three-dimensional structure of macroscopic magma transport systems: Application to Kilauea, Hawaii. J Geophys Res 86:7111–7129

Ryan MP, Blevins JYK, Okamura AT, Koyanagi RY (1983) Magma reservoir subsidence mechanics: Theorical summary and application to Kilauea volcano, Hawaii. J Geophys Res 88:4147–4181

Schlanger SO, Garcia MO, Keating BH, Naughton JJ, Sager WW, Haggerty JA, Philpotts JA, Duncan RA (1984) Geology and geochronology of the Line Islands. J Geophys Res 89:11261–11272

Searle RC, Francheteau J, Cornaglia B (1995) New observations on mid-plate volcanism and the tectonic history of the Pacific plate, Tahiti to Easter microplate. Earth Planet Sci Lett 131:395–421

Sigurdsson H (1987) Dyke injection in Iceland: A review. In: Halls HC, Fahrig WF (eds) Mafic dyke swarms. Geol Assoc Canada Special papers 34:55–64

Simkin T (1972) Origin of some flat-topped volcanoes and guyots. Geol Soc Amer Memoirs 132:183–193

Smith DK, Jordan TH (1987) The size and distribution of Pacific seamounts. Geophys Res Lett 14:1119–1122

Smith DK, Jordan TH (1988) Seamount statistics in the Pacific Ocean. J Geophys Res 93(B4):2899–2918

Smith WHF, Sandwell DT (1997) Global sea floor topography from satellite altimetry and ship ship depth soundings. Science 277:1956–1962

Smoot NC (1982) Guyots of the Mid-Emperor chain, mapped with multibeam sonar. Mar Geology 47:153–163

Steiger RH, Jäger E (1977) Subcommission on geochronology: Convention on the use of decay constants in geo- and cosmochronology. Earth Planet Sci Lett 36:359–362

Stoffers P, Botz R, Cheminée JL, Devey CW, Froger V, Glasby G, Hartmann M, Hekinian R, Kogler F, Laschek D, Larque P, Michaelis W, Muhe R, Putanus D, Richnow HH (1989) Geology of Macdonald seamount: Recent submarine eruption in the South Pacific. Mar Geophys Res 11:101–112

Stoffers P, Scientific Party (1990) Active Pitcairn hotspot found. Marine Geol 95:51–55

Talandier J, Okal EA (1982) Crises sismiques du volcan Macdonald (Océan Calme Sud). Comptes Rendus de l'Académie des Sciences de Paris (II) 295:195–200

Talandier J, Okal EA (1984a) The volcanoseismic swarms of 1981–1983 in the Tahiti-Mehetia area, French Polynesia. J Geophys Res 89:11216–11234

Talandier J, Okal EA (1984b) New surveys of Macdonald Seamount, Southcentral Pacific, following volcanoseismic activity, 1977–1983. Geophys Res Lett 1:813–816

Tracey JI, Sutton GH,.Nesteroff WD, Galehouse J, Vonderborch CC, Moore T, Lipps J, Bilal ul Haq UZ, Beckmann JP (1971) Leg 8 Site 75 summary. In: Initial Reports of the Deep Sea Drilling Project, U.S. Government Printing Office, Washington D.C., pp 17–42

Turcotte DT (1982) Magma migration. Ann Rev Earth Planet Sci 10:397–408

Turner DL, Jarrard RD (1982) K-Ar dating of the Cook-Austral Island chain: A test of the hot-spot hypothesis. J Volcanol Geotherm Res 12:187–220

Vogt PR, Smoot NC (1984) The Geisha Guyots: Multibeam bathymetry and morphologic interpretation. J Geophys Res 89:11085–11107

Wilson L, Sparks RSJ, Walker GPL (1980) explosive eruptions; IV: The control of magma properties and conduit geometry on eruption column behavior. Geophys J Roy Astrnom Soc 63:117–148

Winterer EL, Sandwell DT (1987) Evidence from en-echelon cross-grain ridges for tensional cracks in the Pacific plate. Nature 329:534–537

Woodhead JD, Devey CW (1993) Geochemistry of the Pitcairn seamounts; I: source character and temporal trends. Earth Planet Lett 116:81–99

Woodhead JD, Mcculloch MT (1989) Ancient seafloor signals in Pitcairn Island lavas and evidence for large amplitude, small length-scale mantle heterogeneities. Earth Planet Lett 94:257–273

Chapter 6

Submarine Landslides in French Polynesia

V. Clouard · A. Bonneville

6.1
Introduction

Landslides are common features of oceanic islands and play a key role in their evolution. Caused by caldera collapse or flank collapses, they can be classified into three types: (1) rock falls, (2) slumps or (3) debris avalanches (Moore et al. 1989). Rock falls, or superficial landslides, are mainly related to erosion processes of the subaerial parts of the island. The pieces of debris are less than 1 m in size, and their surface is rippled. Flank collapses generally produce giant submarine landslides, with a horseshoe-shaped feature at their head (Moore et al. 1989). The landslides due to a deep listric fault are cataclysmic events producing fast moving debris avalanches. Deposits can extend over several hundred kilometers away from an island and are characterized by thicknesses less than 2 km, with a hummocky terrain at their lower part. Side-slip over deep fault is termed slump (Fig. 6.1). Slumps are slow-moving slope instabilities. The thickness of the deposits can be as much as 10 km, since the primitive volcano flank is less shattered and disrupted than in the case of a debris avalanche. The causes of major lateral collapses are still a matter of debate, but in most cases they are thought to be related to magma intrusion in the rift zones (Denlinger and Okubo 1995; Keating and McGuire 2000).

Oceanic landslides were first noticed on the Hawaiian Ridge (Moore 1964), and detailed studies have been conducted on the Hawaiian Islands (Fornari et al. 1979; Moore et al. 1989, 1995; Smith et al. 1999), the Canary Islands (Watts and Masson 1995; Holcomb and Searle 1991; Stillman 1999), on Réunion Island (Lenat et al. 1989; Gillot et al. 1994), and on the Cabo Verde Islands (Day et al. 1999).

In this chapter, we will present evidence of forty submarine landslides in the Society and Austral Archipelagos of French Polynesia. These landslides were found during a systematic study of submarine island slopes. The detailed bathymetry of the sea floor around the Society and Austral Islands is the result of a complete synthesis of multibeam data collected in these areas since the 1980s (ZEPOLYF cruise 2003; see Sect. 1.2). Most of the data were acquired during a survey conducted in July 1999 with the R/V L'ATALANTE (ZEPOLYF cruise 1999). During this study, the multibeam echo sounder also provided acoustic bottom imagery (Simrad EM12) used to interpret and describe the landslides. Finally, the inland origin of the landslides is deduced from a geomorphological analysis of the subaerial topography of the volcanic edifices. In the discussion section, a classification for these landslides is proposed.

Fig. 6.1. Example of lateral mass movement presumed to occur in Hawaii; **a** superficial landslide; **b** movement on deep-reaching listric fault; **c** side-slip on deep "decollement" (From Walker 1988)

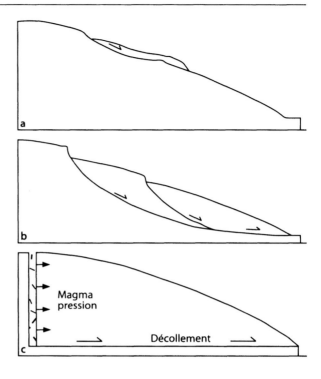

6.2
Geological Setting

The Society and Austral Islands are located on the south-central part of the Pacific Plate (Fig. 6.2). The average velocity of Pacific Plate motion is 11 cm yr^{-1} (Herron 1972; Minster and Jordan 1978). The region is crossed by the Austral and Marquesas Fracture Zones (FZ) with a N70° strike. These FZs are the main tectonic features of the sea floor and are related to the Pacific Farallon Ridge, which is now extinct. The age of the oceanic crust ranges between 50 and 85 Ma (Herron 1972; Mayes et al. 1990; Munschy et al. 1996). The Society and Austral Island Chains trend N120°, which is the same direction as the present Pacific Plate motion, and these island chains are associated with hotspot volcanism.

The Society Island chain (Fig. 6.3) extends over 750 km from the present hotspot location under Mehetia to three atolls in the west. It is composed of five atolls and nine islands. Ages increase from the southeast to the northwest, based on radiometric dating of volcanism from the present at Mehetia (0 to 0.3 Ma, White and Duncan 1996) to about 4.3 Ma at Maupiti (Duncan and McDougall 1976). So far, no ages have been determined for the atolls. The geochronology, geochemistry, volcanology, and geology of the Society Islands have been reviewed in some detail by Diraison et al. (1991a) and more recently by White and Duncan (1996). The aerial activity of these volcanoes can be described in terms of three main sequential stages: (1) the edification of a shield volcano, (2) the formation of a caldera, and (3) the occurrence of post-caldera volcanism. The submarine volcanic activity in the hotspot area is described in Sect. 5.2. The Society Islands exhibit every transition stage of oceanic islands in low latitudes, from an elevated island with a growing coral reef to the atoll stage.

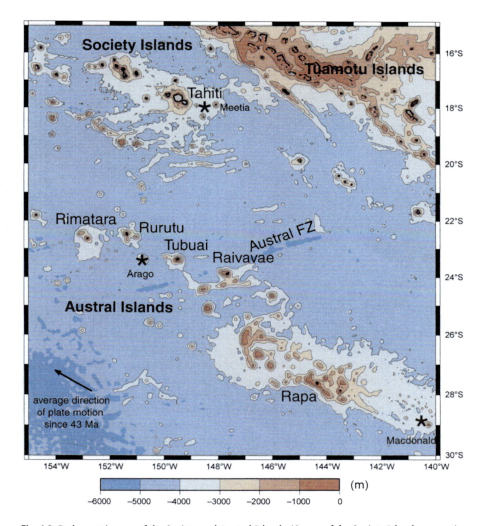

Fig. 6.2. Bathymetric map of the Society and Austral Islands. Names of the Society Islands appear in Fig. 6.3. *FZ* stands for fracture zone. *Black stars* represent the active volcanic areas of the Society, Rurutu and Macdonald hotspots

The Austral Island chain extends northwest for >1500 km from the active submarine Macdonald Volcano (Norris and Johnson 1969; Johnson and Mahaloff 1971; Talandier and Okal 1984) to the atoll of Maria (Fig. 6.4). The chain is composed of five small islands, one atoll, and numerous seamounts. The Austral fracture zone crosses the Austral alignment between the islands of Tubuai and Raivavae. The age of the oceanic crust ranges from 35 Ma at the southeast to 80 Ma in the northwest (Mayes et al. 1990). Because the ages obtained by Duncan and McDougall (1976) on Rurutu (12 Ma for the first volcanic stage), Tubuai (9.3 Ma), Raivavae (6.5 Ma) and Rapa (5.1 Ma) are compatible with a hotspot origin, these authors suggested that the Austral Chain was related to the Macdonald hotspot. However, a more recent reconstruction of the Macdonald hotspot

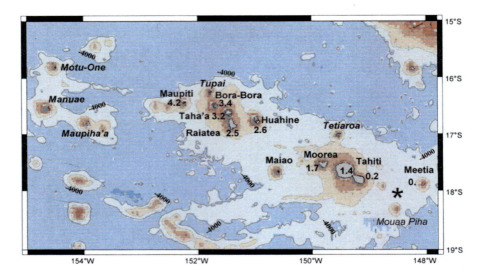

Fig. 6.3. Bathymetric map of the Society Islands (isobaths every 1000 m). Radiometric ages in Ma are in *bold, black numbers*. A *star* represents the Society hotspot. *Bold letter* names are used for islands, *bold-italic print* for the atolls, and *normal italic print* for the seamounts

track and newly calculated seamount ages (Bonneville et al. 2002) suggest that the Macdonald hotspot was only responsible for building Rapa and for the seamounts located south of the Austral alignment, while the islands of Rurutu, Tubuai and Raivavae correspond to another hotspot. A second and recent volcanic stage exists at Rurutu for which Turner and Jarrard (1982) also postulated the existence of another hotspot, which could be responsible for building some of the Cook Islands as well. The existence of this hotspot is confirmed by a 0.2 Ma age (Bonneville et al. 2002) on the Arago Seamount located 120 km to the southeast of Rurutu Island. In addition to these three recent hotspot alignments, older submarine seamount chains also exist in the area. To the north of the Macdonald Volcano, there are two seamount chains, with ages ranging from 40 to 23 Ma (McNutt et al. 1997). Furthermore, old guyots covering most of the Austral area were mapped during the ZEPOLYF2 (2001) cruise. These successive volcanic episodes suggest the existence of a complex sea-floor pattern in the Austral alignment, and this structure might have partially weakened the lithosphere.

6.2.1
Data

Submarine landslides were seen on detailed bathymetric maps around the Society and Austral Islands. These maps have been drawn using 125 m × 125 m grids extracted from a synthesis of multibeam and single-beam data collected in the zone since the 1980s (ZEPOLYF 2003; see Sect. 1.2). We also used acoustic imagery gathered by Simrad EM12D echo sounder on board R/V L'ATALANTE (ZEPOLYF1 and ZEPOLYF2 cruises). Acoustic imagery data, corresponding to the backscattering strength of the echo-sounder beams, are precise enough to pick out landslide deposits on the sea floor. The inland origin of the landslides is deduced from the geomorphological analysis of the aerial

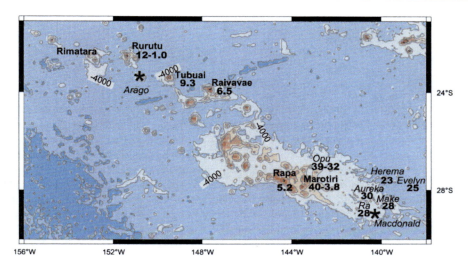

Fig. 6.4. Bathymetric map of the Austral Islands (isobaths every 1000 m). Radiometric ages in Ma are in *bold, black numbers* (see references in text). The active areas of the Macdonald and Rurutu hotspots are represented by *stars*. *Bold letters* are used to name the islands, *bold-italic print* for the atolls and normal italic print for the seamounts

part of the edifices. The elevation data on the islands were obtained by digitizing topographic maps of French Polynesia (Service de l'Urbanisme 1992; ORSTOM 1993). For the island of Tahiti, we used a digital terrain model derived from two stereoscopic SPOT images. Caldera locations reported on the island figures are those usually recognized in geological studies and have been confirmed by gravity analysis (Clouard et al. 2000).

6.2.2
Landslide Characterization

A localization of the landslides has been possible through the combined analyses of bathymetric data and acoustic imagery. First, the bathymetry provides information on the submarine slopes of the volcanoes. Landslide channels as well as landslide deposits perturb the otherwise undisturbed volcanic slopes, which are deeper and smoother in the case of the headwall or the slide, and higher and more disorganized in the case of deposits. Slopes are also irregular along the volcano rift zones, and in this case, the acoustic imagery can be relatively precise. Rift zones present almost dark uniform facies due to the lack of sediment, whereas landslide deposits have a dotted and speckled pattern due to the mixing of fragmented material of various sizes. Secondly, the bathymetry reveals large hummocks or mega-blocks, whose sizes are several hundred meters wide. Large hummocks are present in the case of giant lateral collapses such as slumps and debris avalanches. To complete our analyses, acoustic imagery provides information on the limits of the landslide deposits, even when the relief is too small to be seen using just the bathymetric data.

Descriptions of the landslides are presented below for each surveyed island, from the southeast to the northwest, i.e., for islands that are increasingly older, first in the Society Archipelago, then in the Austral Islands.

6.3
Landslides of the Society Islands

6.3.1
Mehetia

Mehetia Island is the emerged portion of the Society hotspot, located on the southeastern boundary of the Society hotspot area. Underwater eruptions on the southern flank were recorded as having occurred in 1981 (see Sect. 2.3.5.1). However, the two main volcanic building stages and major volcanic activity took place between 74 000 and 25 000 years ago (Steiger and Jaeger 1977), with a shift in activity to the south of the island (see Sect. 5.2.3.7). A mass wasting that presents a homogenous radial distribution marks all the flanks of the island. In the upper part of the submarine edifice, the scars of several landslides are present (Fig. 6.5) and can be related to the deposits of the lower slopes. Basically, these deposits correspond to a superficial landslide of fragmented material, but large rocky blocks also exist on the northeastern and on the western flank. Only on the southern flank does one landslide look like a debris avalanche. Recent volcanic activity on the southern flank of the island might be the cause of the observed debris avalanche.

6.3.2
Moua Pihaa Seamount

Moua Pihaa is the shallowest (143 m) active seamount of the Society hotspot (Cheminée et al. 1989; Hekinian et al. 1991), with its volcanic activity concentrated in the summit crater (see Sect. 5.2.3.5). This activity is located to the southwest of Mehetia (Fig. 6.3). The upper submarine slopes of Moua Pihaa are hollowed out by three indentations, each

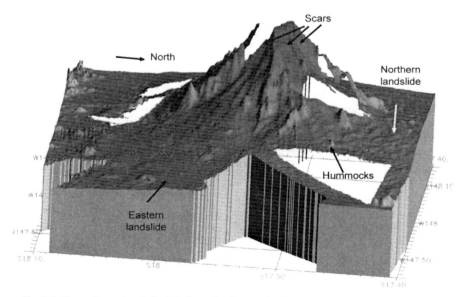

Fig. 6.5. Three-dimensional shaded view of Mehetia Island

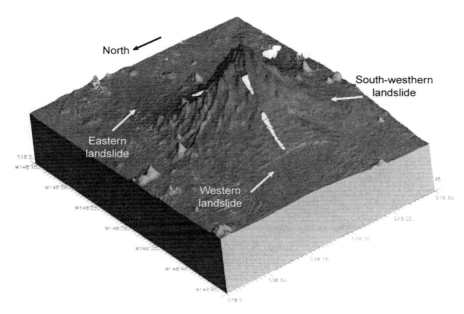

Fig. 6.6. Three-dimensional shaded view of Moua Pihaa Seamount

of which corresponds to the headwall of a landslide (Fig. 6.6). A ripple surface typical of a debris avalanche characterizes the lower slopes. The deposits are made up of small-sized fragmented material and do not show the presence of hummocks larger than 100 m, which corresponds to the spatial resolution of the multibeam data, except for the northwestern landslide where a few large blocks are present. The scar-heads range in depths between 500 m for the western landslide, to 1 500 m for the northwestern slide.

6.3.3
Tahiti

Tahiti Island is the main island of the Society Island chain. The island is composed of two coalescent volcanoes, Tahiti Nui to the northwest and Tahiti Iti to the southeast. Tahiti Nui is a shield volcano whose subaerial volcanism is dated between 1.4 and 0.2 Ma (Duncan and McDougall 1976; Diraison et al. 1991; Duncan et al. 1994; Leroy 1994). The southern landslide of Tahiti (Fig. 6.7) has been described by Clouard et al. (2001). It corresponds to a debris avalanche due to a subaerial listric fault. The top of the listric fault is the main rift zone of the island. The landslide deposits extend up to 60 km away from the island's shore. The volume of the deposits has been estimated to be 1 150 km^3, whereas the landslide scar represents a volume of only 299 km^3. Successive landslides along the same listric fault whose scar was then refilled by later lava flows explain this difference in volume. The origin of these landslides is related to repeated intrusions of magma inside the main rift zone, the main event occurring at the beginning of the late shield stage. To the north, the bathymetry of the Tahiti-Nui flank reveals the existence of a major catastrophic landslide event that occurred during its eruptive history (Hildenbrand et al., to be published). The related debris avalanche has a volume estimated at about 800 km^3,

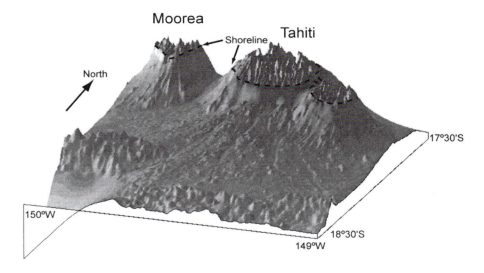

Fig. 6.7. Three-dimensional shaded view of Moorea and Tahiti's southern area. Tahiti's giant landslide deposits cover all the southern flank of the island

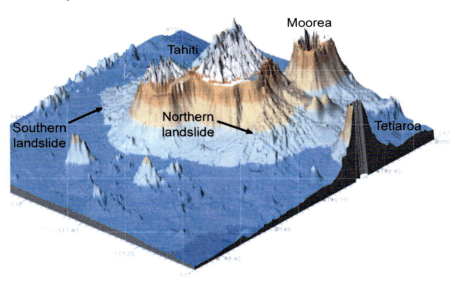

Fig. 6.8. Three-dimensional shaded view of Tahiti Island's northern area. Tahiti's giant landslide deposits cover all the northern flank of the island

and extends as far as 50 km to the north (Fig. 6.8). This avalanche is covered by subsequent volcanic activity, as evidenced by the presence of a smooth surface at a mean depth of 1 500 m. The on-land and off-shore analyses suggest a listric fault geometry for the landslide scar, allowing an estimation of the slide volume to be about 460 km^3. Compared to the southern landslide, the collapse is apparently very compact, indicating a relatively low mobility, which may be due to the slide's submarine triggering.

6.3.4
Moorea

Moorea Island is a triangular island whose age ranges from 2.2 to 1.5 Ma (Diraison et al. 1991). A central collapsed feature exists on the island topography. It presents a horseshoe shape that is open towards the sea. To explain the particular shape of the northern shore of Moorea Island, the hypothesis of a giant landslide along an east-west fault has been proposed (Blanchard et al. 1981). An alternative explanation based on the measurement of horizontal and vertical throw (1 000 m and almost zero, respectively) of the northern part of the island is in favor of a decollement at shallow depth with no subsequent collapse (Ledez et al. 1998). Bathymetry to the north of the island (Fig. 6.8) shows no trace of landslide deposits off-shore. The hypothesis of a giant landslide is therefore unlikely to explain Moorea's particular shape. The central depression of Moorea is merely related to classic caldera subsidence, as revealed by a gravity study (Clouard et al. 2000) coupled with the horizontal displacement of the northern part of the aerial edifice.

6.3.5
Huahine

Huahine is composed of two separate islands, Huahine Nui and Huahine Iti, located in the same lagoon. These two islands are related to a single volcano separated by a large submerged caldera (Deneufbourg 1965; Clouard et al. 2000). The period of the main volcanic activity ranges from 2 to 3.2 Ma (Diraison et al. 1991; Duncan and McDougall 1976). There is no evidence for any landslide on the flanks of Huahine (Fig. 6.9), even though a lack of bathymetric coverage prevents us from drawing any conclusions for the east flank.

6.3.6
Raiatea-Tahaa

As was true for Huahine, the two islands of Raiatea and Tahaa (Fig. 6.10) are enclosed in the same lagoon, but in this case, the islands correspond to two distinct volcanoes. However, since it is a difficult to distinguish between them for the origin of submarine landslides, the two islands have been analyzed together. Raiatea's age ranges from 2.4 to 3.4 Ma and Tahaa's from 2.8 to 4.2 Ma (Diraison et al. 1991; Duncan and McDougall 1976). A southern central depression was recognized at Raiatea and attributed to caldera collapse (Brousse and Berger 1985), but a more recent interpretation indicates that this collapse is related to successive landslides (Blais et al. 1997). Figure 6.10 clearly shows landslide deposits on the eastern submarine slope of Raiatea, and their origin corresponds to the southern central depression. The complicated shape of the limit of the debris could confirm the occurrence of at least two landslides. The presence of large hummocks and the pattern of the debris flow on the acoustic imagery are compatible with the presence of a debris avalanche. Figure 6.9 also reveals debris-avalanche deposits originating from the south of Raiatea. Traces of two landslides are present on the submarine slopes of Tahaa (Fig. 6.9). The eastern slide apparently originates in the caldera area. The outline of the displaced mass is narrow and does not extend very

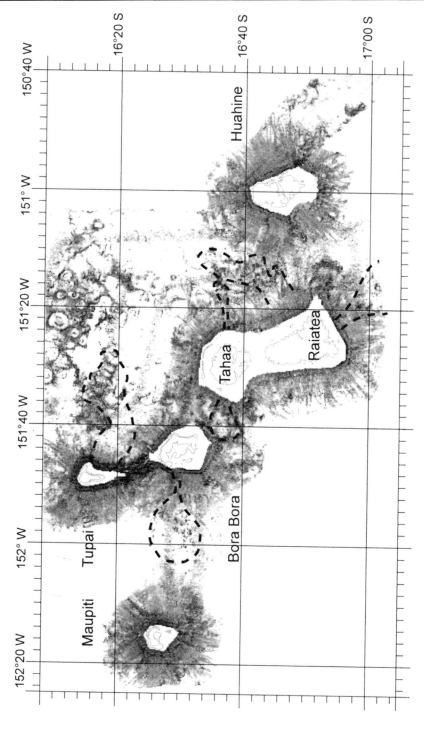

Fig. 6.9. Acoustic bottom imagery around the western Society Islands acquired by EM12D echo sounder (N.O. L'ATALANTE, ZEPOLYF cruise 2003). High reflectivity is in *dark gray*, low reflectivity in *light gray*. Speckled pattern of the acoustic imagery is typical of the rugged morphology of down slope deposits. Limits of the submarine landslides' extent are indicated by *thick dashed lines*

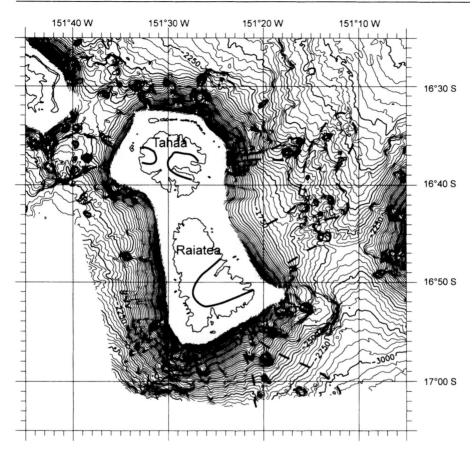

Fig. 6.10. Bathymetry around Raiatea, Tahaa Islands. Aerial collapse locations are drawn in *black*. Limits of the landslide deposits are drawn using *thick dashed lines*

far from the island shore (Fig. 6.10). The volume of the debris is smaller than the volume of material that should have come from the caldera. The lower surface of this landslide probably corresponds to only the eastern wall of the caldera, and the rest of the missing debris volume probably corresponds to classic caldera subsidence. The western landslide of Tahaa is larger. Bathymetry shows giant blocks in the channel between Tahaa and Bora Bora, which might correspond to a lateral collapse of the subaerial western flank (Fig. 6.10).

6.3.7
Bora Bora

Age determinations of Bora Bora range from 3.1 to 4.4 Ma (Diraison et al. 1991; Duncan and McDougall 1976). This island is characterized by a well-preserved caldera and a tilting of the volcanic structure towards the southwest (Blais et al. 2000). A giant landslide is present on the western submarine flank (Fig. 6.11). The pattern of this landslide corresponds to a debris avalanche as revealed by hummocky terrain in the lower

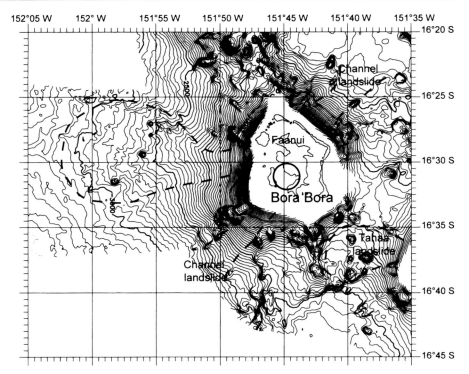

Fig. 6.11. Bathymetry around Bora Bora Island. The location of the caldera is represented by a *circle*. The aerial western collapse location is drawn in *black*. The *dashed line* indicates the limits of landslide deposits. Two other landslides probably exist to the north and to the south

part. The hummocks reach a height of 200 m for a width of nearly 1 km. They are spread over smaller-sized material, which also suggests the characteristic speckled pattern of debris avalanches. The upper limit of the landslide is facing the only opening in the coral reef. Two locations are candidates for the aerial origin of the slope failure: the caldera and Faanui Valley. The hypothesis of a vertical subsidence mechanism for the caldera is well constrained by geomorphological observations and by the location of the ancient magma chamber interpreted on the basis of gravity analysis (Clouard et al. 2000). Moreover, the location of the caldera has shifted southward from the origin of the landslide. Thus, we propose that the origin of Bora Bora's western landslide is the Faanui Valley, at the northwest of the island. Two other landslides exist to the north and to the south of the island (Fig. 6.11). To the south, only the landslide channel is mapped by our bathymetric data, and to the south, the deposits overlap the Tupai landslide deposits (see Sect. 6.3.8) and thus, the specific characteristics of the Bora Bora slide cannot be determined.

6.3.8
Tupai

Tupai is a small atoll north of Bora Bora (see location in Fig. 6.9). No age has been determined for this atoll, but it is assumed to be the result of Society hotspot activity

Chapter 6 · Submarine Landslides in French Polynesia   221

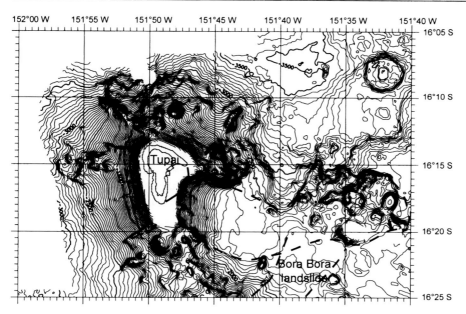

Fig. 6.12. Bathymetry around Tupai Atoll. Limits of the landslide deposits are drawn with a *dashed line*. This landslide was diverted to the north by Bora Bora's slopes. The proposed origin of the slide is drawn in *black*

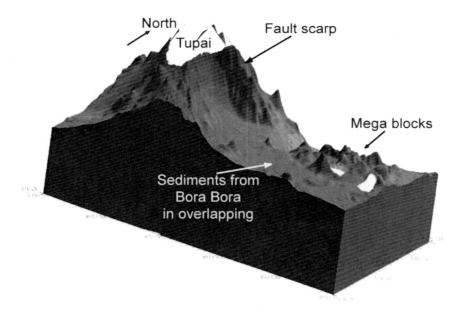

Fig. 6.13. Three-dimensional shaded view of Tupai Atoll. The average height of the scarp is 700 m, and it is 7 km wide. The largest blocks are 3 km × 2.5 km. In the south, the deposits are partially covered by sediments coming from Bora Bora

(Fig. 6.9). Traces of a large landslide exist on the eastern side of the edifice (Fig. 6.12). However, there do not appear to be any hummocks spread along the landslide flow. All the debris is concentrated in the lower part of the landslide, presents a steep toe, and the width of greatest debris is several kilometers (3 km × 2.5 km for the largest). These are the characteristics of a debris avalanche, which heads out in a giant scarp, reaching up to 1 000 m in height and 7 km in length (Fig. 6.13). This might correspond to the scar of the mega blocks. We propose that this landslide is the result of a collapse where the whole subaerial and submarine eastern flank of the island slid down into the sea. This explains the non-circular shape of the present-day coral reef. The sediment cover of the deposits gives an indication of the date of this landslide. Figure 6.13 indicates that the Bora Bora sediment drifts overlap the southern hummocks. On the other hand, the displacement of the landslide was not linear but was diverted to the north, probably by the submarine volcanic slope of Bora Bora. Therefore, we can assume that this collapse occurred after the end of Bora Bora's building stage.

6.4
Austral Island Landslides

6.4.1
Macdonald

The Macdonald Seamount is the active expression of the Austral hotspot (Norris and Johnson 1969). It is located 430 km to the southeast of Rapa (Fig. 6.4), and its summit reaches a height of 26 m below sea level (Talandier and Okal 1984). The Macdonald Seamount presents a star-shaped morphology. Between the rift zones, its slopes are covered by landslide deposits characterized by small-sized granular deposits that produced a smooth pattern on the edifice's lower slopes (Fig. 6.14). The morphology of the upper part of the volcano reveals scars that correspond to the heads of the various collapses. To the east and to the west, sediment waves or ripple marks are present. These undulations are believed to have been caused by turbidity currents (e.g., Wynn et al. 2000). To the south, a small relief could in fact be a large rocky block that fell from the top of the volcano. Collapse scars are visible on the northwest, the northeast, and on the south of this seamount.

6.4.2
Rapa

Rapa is the southernmost island of French Polynesia. It was generated by the Macdonald hotspot 5 Myr ago (Duncan and McDougall 1976). On the island, the central collapse opened to the west is related to caldera subsidence (Clouard et al. 2000). If a landslide is associated with the caldera collapse, as suggested by Rapa's topography, our data are unable to prove this. The bathymetric data only partially cover the southern landslide track previously noticed by Jordahl (1999), but not the lower part where most of the deposits are usually concentrated (Fig. 6.15); therefore, we cannot infer the classification type of this landslide, whose track is about the same width as the island itself.

Chapter 6 · **Submarine Landslides in French Polynesia** 223

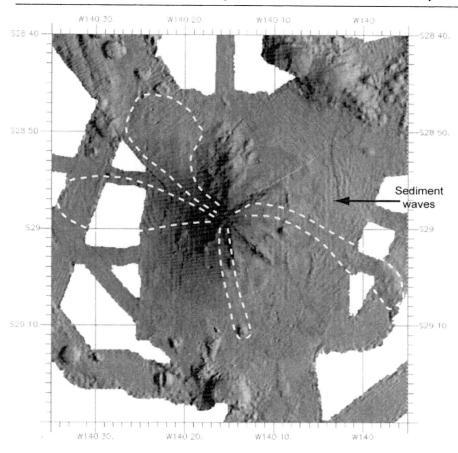

Fig. 6.14. Macdonald Seamount. Limits of the submarine landslides' extent are indicated by *white dashed lines*. Sediment waves are present on the western and the eastern flanks

Fig. 6.15. Bathymetric map around Rapa. *Thick dashed lines* are the outline of the landslide channel. The lower deposits are not mapped

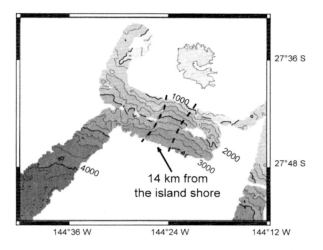

6.4.3
Raivavae

Raivavae Island is located to the south of the Austral fracture zone (Fig. 6.4). Its age (6.5 Ma, Duncan and McDougall 1976) is compatible with its being formed by the same hotspot that generated Tubuai and Rurutu. The aerial part of the island is enclosed in a lagoon. To the south of the edifice, a coalescent seamount or a large rift zone is the extension of the aerial part of Raivavae (Fig. 6.16). The lack of bathymetric data on the top of this rift zone is due to its shallow depth, which prevented us from doing a survey for security reasons during the ZEPOLYF2 cruise. All the submarine landslides existing around Raivavae originate from this extension. To the southwest and to the south-southeast we can observe debris-avalanche landslides as evidenced by hummocks (Figs. 6.16 and 6.17). The scar of the southwestern landslide can be seen along the first 1 500 m of the submarine slope. The southeastern landslide corresponds to a superficial landslide. The occurrence of two giant landslides only in the southern part of the island is probably related to a shift of the volcanic activity to the southern extension of the island.

6.4.4
Tubuai

Tubuai Island was formed 9.3 Ma ago (Duncan and McDougall 1976) by the same hotspot responsible for Rurutu's older stage of activity. It is also enclosed in a 4 km wide lagoon. Three landslides blanket the submarine slope of the island (Fig. 6.18). Their type

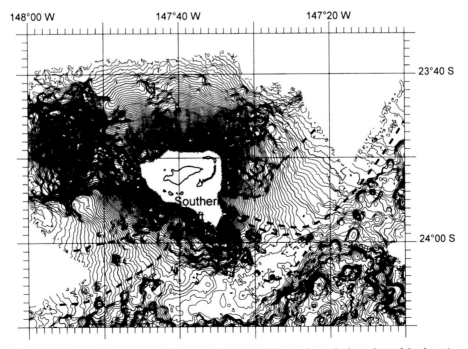

Fig. 6.16. Bathymetric map around Raivavae. *Thick dashed lines* indicate the boundary of the deposits

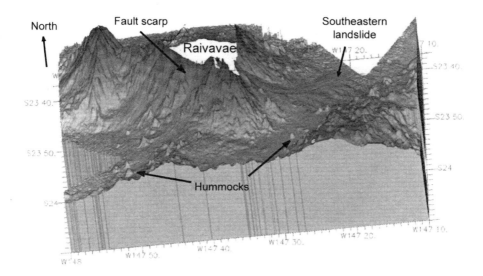

Fig. 6.17. Three-dimensional shaded view of Raivavae Island. The two southernmost landslides are characterized by hummocky terrane; the southeastern landslide is represented by small size debris. The scar of the southwestern landslide is visible on the upper slope

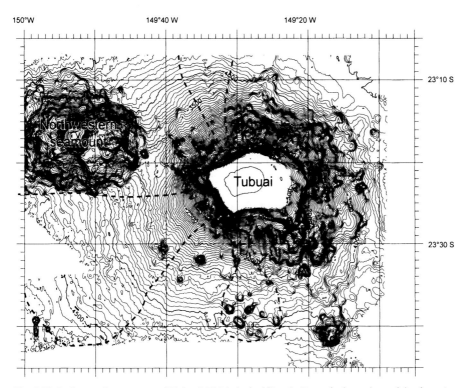

Fig. 6.18. Bathymetric map around Tubuai. *Thick dashed lines* indicate the boundary of the deposits. The western landslide corresponds to a superficial landslide, the southern one to a debris avalanche

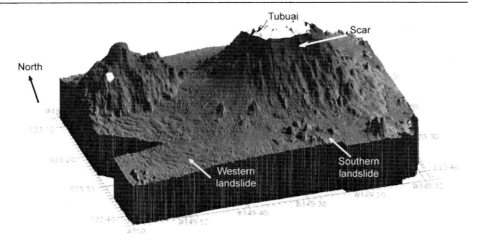

Fig. 6.19. Three-dimensional shaded view of Tubuai Island. Undulations on the western landslide indicate that it is a superficial type. Hummocks on the southern slide suggest that it is a debris-avalanche type

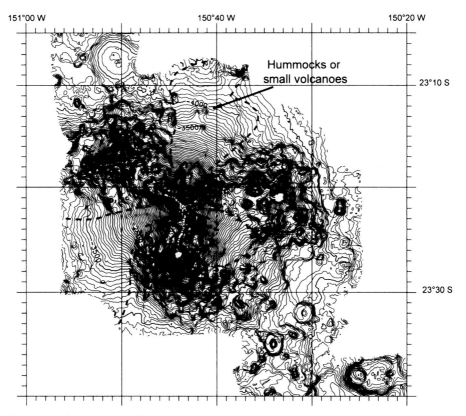

Fig. 6.20. Bathymetric map of Arago Seamount. *Thick dashed lines* show the boundary of the deposits. The western landslide is a superficial one, and the northern slide corresponds to a debris avalanche

is determined on the basis of a three-dimensional view (Fig. 6.19). The western landslide is a superficial landslide originating from the western subaerial part of the island and presents sediment waves in the extremity of the deposits with a mean wavelength of 2 000 m, and a mean height of 40 m. This landslide is posterior to the formation of the northwestern seamount, as the deposits overlap the lower slope of this seamount. The southern landslide is characterized by large hummocks located among smaller deposits in its lower part. It is the result of a debris avalanche. A scar exists in the upper submarine slope (Fig. 6.19). When extended to the aerial part of the edifice, it could explain the presence of a large sedimentary plain between the western and the eastern volcanic hills of the island. To the north, the third landslide cannot be well defined, as our data do not cover the lower part of the deposits; however, the lack of large rocky blocks along the track suggests that it is another superficial landslide.

6.4.5
Arago

The Arago Seamount is located 120 km to the southeast of Rurutu. Radiometric dating on a dredged basalt sample provides an age of 0.2 Ma for this seamount, which is therefore considered the location of the active hotspot that corresponds to the second stage of Rurutu (Bonneville et al. 2002). Arago is composed of three shallow coalescent volcanoes, the depth of the shallowest reaching 26 m. below sea level. These three volcanoes are linked by rift zones, and the depth of the center is about 150 m. Each flank of Arago is covered by submarine landslide deposits (Figs. 6.20 and 6.21). The

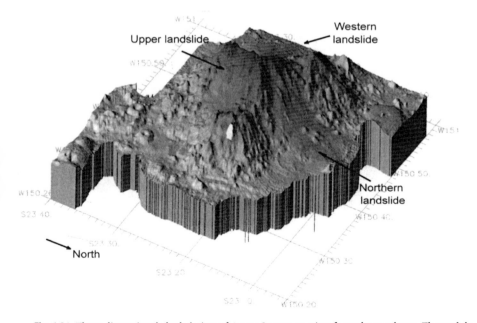

Fig. 6.21. Three-dimensional shaded view of Arago Seamount, view from the northeast. The undulations on the western landslide indicate that it is a superficial type, hummocks on the northern one suggest that it is a debris-avalanche type. The scarp of the upper landslide is visible, but not the deposits

western mass wasting, which spreads over the largest surface, corresponds to a superficial landslide, with sediment waves present at the extremity of the lobate tongue (Fig. 6.21). Smooth slopes and a few hummocks characterize the northern landslide. Its characteristics are in between those of a superficial landslide of fragmented material and those of a debris avalanche. A third landslide exists on the upper southeastern flank of the volcano. Its aspect is the same as the northern one. The size of the deposits is small and gives a smooth aspect to the bathymetry. These two landslides are similar to the Moua Pihaa Seamount's mass wasting, and seem to characterize the landslides that occur on submarine active volcanoes.

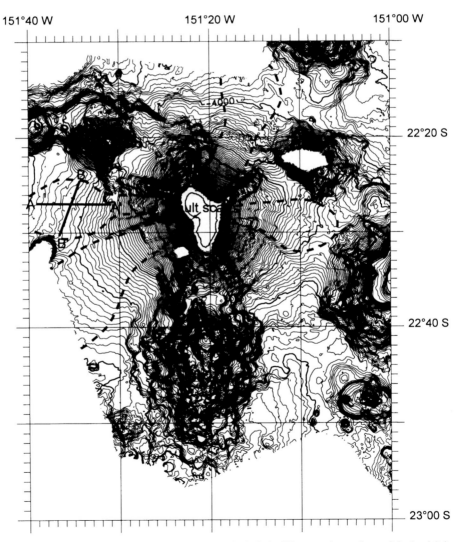

Fig. 6.22. Bathymetry around Rurutu Island. The *thick dashed lines* are the outlines of the landslide flow. Locations of the bathymetric profiles AA', BB' (Figs. 6.24 and 6.25) are used to characterize the deposits that are reported

6.4.6
Rurutu

Rurutu Island is characterized by two periods of volcanic activity separated by 11 Ma and by the uplift of carbonate plateaus. The first volcanic stage occurred between 10 and 12 Ma and the second one between 0.65 and 1.65 Ma (Duncan and McDougall 1976; Turner and Jarrard 1982; Matsuda et al. 1984). This second stage corresponds to hotspot activity, whose location is now the Arago Seamount (Bonneville et al. 2002). Uplifted carbonate plateaus have been explained by a thermal rejuvenation of the lithosphere due to the

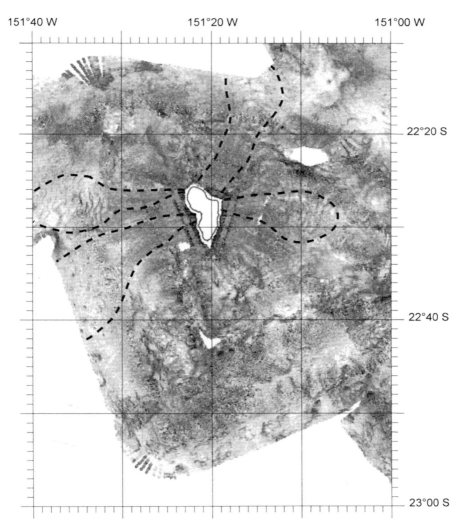

Fig. 6.23. Acoustic imagery around Rurutu Island. The *thick dashed lines* are the boundary of the deposits. High reflectivity is in *dark gray* and low reflectivity in *light gray*. Only a few blocks are present for the western landslide, but the large majority of the blocks are not seen

second hotspot's activity (Calmant and Cazenave 1986). On Rurutu Island, two different types of landslide exist (Fig. 6.22). The western and the eastern landslides are characterized by sediment waves, which are visible on the acoustic imagery (Fig. 6.23) and indicative of small size deposits, corresponding to superficial landslides. The geometric characteristics of the western deposits are deduced from two bathymetric profiles (Figs. 6.24 and 6.25). The mean wavelength of the sediment waves is 1 200 m, their mean height is 35 m, and the thickness of the deposits is about 50 m. This type of flank destabilization results in the accumulation of fragmented volcanic material due to erosion

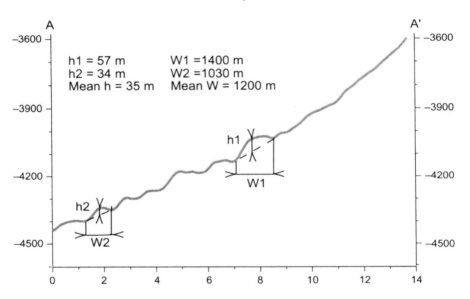

Fig. 6.24. Bathymetric profile through the western Rurutu superficial landslide along AA' direction (see profile location in Fig. 6.22). The mean wavelength (W), and height (H) of the sediment waves are obtained for six undulations of the deposits

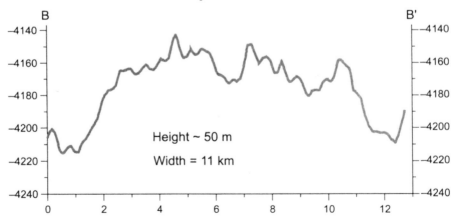

Fig. 6.25. Bathymetric profile through the western Rurutu landslide along BB' profile (see profile location in Fig. 6.22)

processes (Moore and Fiske 1969; Fornari et al. 1979). The southwestern landslide is characterized by hummocks in its lower part, and a regular slope from the island shore to the distant part of the deposits. Unfortunately, we don't have bathymetric data for the lower part of the deposits. The on-land origin of the landslide corresponds to the present-day western bay of Rurutu. Between the sea's surface and a depth of 1 500 m, there is a scarp higher than 700 m and parallel to the failed mass channel that could be the scar of the collapse. Therefore, by analogy with Tupai's shape, this landslide could correspond to a debris avalanche where the entire subaerial and submarine southwestern part of the island disappeared into the sea. The fourth landslide on the northern slope of the island is not completely covered by our survey and thus cannot be classified.

6.4.7
Rimatara

Rimatara is the easternmost and smallest island of the Austral Archipelago (ca. 5 km of diameter) without a coral reef. No reliable radiometric dating has been done so far (Duncan and McDougall 1976). Two small landslides exist, one on the southwest flank and the other on the southeast flank (Fig. 6.26). The largest slide trends directly down the southeast shore (Fig. 6.27). A small north-south ridge separates the lobate flow in

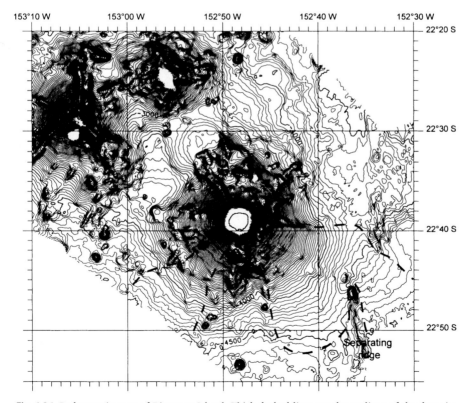

Fig. 6.26. Bathymetric map of Rimatara Island. *Thick dashed lines* are the outlines of the deposits. The southeastern landslide corresponds to a superficial landslide

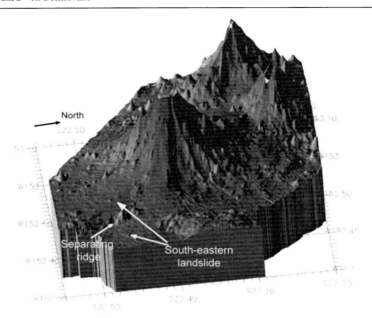

Fig. 6.27. Three-dimensional shaded view of Rimatara Island, view from the west. Undulations of the southeastern landslide deposits reveal a superficial landslide type

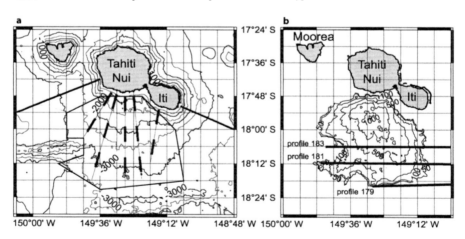

Fig. 6.28. a True bathymetry (ZEPOLYF1 cruise, 1996) of the southern flank of Tahiti (isobaths every 500 m), and location of the profiles used to produce the theoretical bathymetry; **b** map of the thickness of the landslide deposits deduced from the differences between the true and the theoretical bathymetry (isobaths every 250 m). The *black lines* correspond to the seismic profiles presented Fig. 6.29

two in its lower part. The thickness of the deposits estimated from the bathymetric contour is about 50 m. This erosion feature is characterized by small ripple marks in the lower part of the deposits. These characteristics are those of a superficial landslide. The second landslide, on the southwestern flank, cannot be completely classified, as our data do not cover the lower part of the deposits.

6.5
Classification of the Society and Austral Landslides

6.5.1
Geometric Characteristics

For each landslide for which we are able to determine the extent, we have measured the runout length (L) and the elevation difference (h) between the top of the scar to the lower part of the deposits. Except when the headwall is clearly marked on the submarine slopes, the origin of the landslide is considered to be at sea level. The outline of the deposits and thus the runout length (L) corresponds to the limit where the deposit topography vanishes. Some debris might have escaped detection, but we consider that this amount is not significant. The ratio h/L, or coefficient of Heim (1932), represents the apparent coefficient of friction, with no water interaction. It provides an indication on the mobility of the landslide.

Table 6.1.
Characteristics of Society, Austral landslides. N = north, E = east, W = west, S = south $D.A.$ = debris avalanche, $S.L.$ = superficial landslide, $bl.$ = block/s

Island	Location	Runout length (km)	Height (m)	h/L	Type
Mehetia	W	25	3 700	0.148	S.L.
	NW	20	2 700	0.135	S.L.
	N	21	2 600	0.123	S.L.
	NE	27	3 800	0.141	S.L., bl.
	E	32	3 900	0.122	S.L.
	S	19	3 700	0.195	D.A.
Mouaa Piha	W	25	2 800	0.112	S.L.
	W	24	3 050	0.127	S.L.
	NE	22	2 300	0.105	S.L., bl.
Tahiti	S	60	3 500	0.058	D.A.
	N	50	3 750	0.075	D.A.
Raiatea	W	35	3 200	0.091	D.A.
	S	>23	>3 100		D.A.
Tahaa	W	15	2 650	0.177	D.A.
	E	15	2 000	0.133	D.A
Bora Bora	W	25	3 000	0.120	D.A.
	S	>15	>3 000		?
	N	25	3 000	0.120	?
Tupai	E	31	2 750	0.110	D.A.
Macdonald	W, E	30	3 800	0.127	S.L.
	S	21	3 800	0.181	1 bl.
	SE	34	4 000	0.118	S.L.
	NW	29	3 800	0.131	S.L.
Rapa	S	>12	>3 000		
Raivavae	SW	>36	>4 000		D.A.
	SSE	19	3 200	0.168	D.A.
	SE	>49	>4 200		S.L.
Tubuai	W	55	>4 700		S.L.
	S	30	4 000	0.133	D.A.
	N	>25	>4 300		S.L.
Arago	W	>26	>4 000		S.L.
	N	24	4 500	0.187	S.L., bl.
	SE	11	1 800	0.164	?
Rurutu	W	31	4 500	0.145	S.L.
	N	>28	>4 250		
	SW	>28	>4 300		D.A.
	E	20	3 000	0.150	S.L.
Rimatara	SE	33	5 000	0.152	S.L.
	SW	>20	>4 500		?

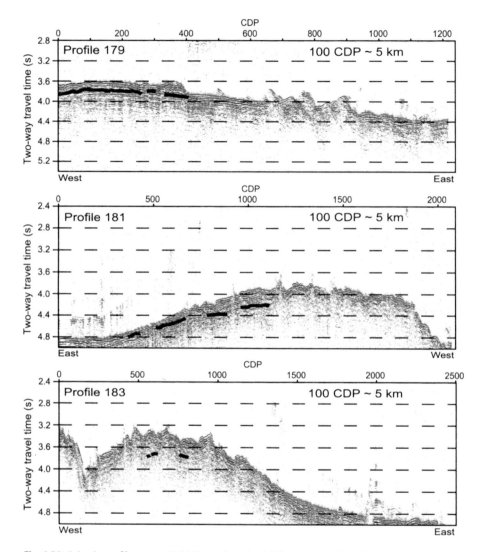

Fig. 6.29. Seismic profiles across Tahiti's southern landslide. Location of profiles is given in Fig. 6.28. The thicker portions of the lines represent a strong reflector assumed to be the basement interface. *CDP* stands for Common Depth Point

Table 6.2.
Seismic velocity in Tahiti landslide deposits

Profile	Deposit thickness		seismic velocity (m s^{-1})		
	measured on the				
	bathymetry	seismic profiles	2000	3000	4000
PR179	200 m	0.18 s	180 m	270 m	360 m
PR181	400 m	0.27 s	270 m	405 m	540 m
PR183	600 m	0.38 s	380 m	570 m	760 m

We find higher h/L ratios, from 0.06 to 0.19 (Table 6.1), than those given in the review of Hampton et al. (1996) where these ratios are around 0.05. This is a sign of the poor mobility of the deposits. The other noticeable point is that no distinction appears between the values of the apparent coefficient of friction regardless of what the landslide type may be. In other words, the mobility remains constant and doesn't depend on the size of the sliding material.

6.5.2
Seismic Velocity

To determine the seismic velocity in debris-avalanche deposits, we used seismic profiles across Tahiti's southern landslide. The thickness of the deposit can be estimated from bathymetric data. We assumed the hypothesis that the island slopes have similar profiles from sea level to the sea floor at the foot of the edifice all around the island, except in the landslide deposit area, where the slope was partially filled by the deposits. The true bathymetry in the landslide area is removed and replaced by bathymetric profiles extracted from non-landslide areas, and a theoretical bathymetric grid was prepared. The difference between this theoretical grid and the real bathymetry provides a map of the thickness of the deposits (Fig. 6.28). Their thickness reaches up to about 1 000 m close to the island. Three seismic profiles were collected cross the landslide area (Fig. 6.28). They are perpendicular to the direction of the slide. Data were binned to a common depth point after performing a frequency filtering. For the three profiles, the seismic reflectors were discontinuous and weak (Fig. 6.29). Numerous diffraction hyperboles corresponded to large debris like the hummocks. However, a reflector exists on the three profiles and is assumed to correspond to the basement interface, i.e., the bottom of the landslide deposits. The depth in time of this reflector is used to calculate the deposit thickness with different velocity values (Table 6.2). The results are compared to the thickness directly measured on the thickness map above the seismic profile.

The best results were obtained for a velocity of 3 000 m s^{-1} (Table 6.2). This is greater than the 2 000 m s^{-1} used for pelagic sediments in other regional studies (Ouassaa 1996) but is still within the proposed range of 2 400 to 4 000 m s^{-1} estimated for sediment layer over the oceanic crust around Oahu Island in the Hawaiian Chain (Watts et al. 1985).

6.6
Evolution of the Mass Wasting with the Age of the Edifices

6.6.1
Landslide Related to Submarine Active Volcanoes

Moua Pihaa, the shallowest active seamount of the Society Chain, and the two active seamounts of the Austral Islands, Arago and Macdonald, present the same landslide characteristics, showing only fine grained debris-avalanche deposits, with very few isolated blocks. Hence, relatively fine-grained debris avalanches seem to characterize the landslides that occur on active submarine volcanoes. Blocks forming hummocks are virtually absent. The morphology of the deposits suggests that the flows were laid down smoothly, and there is no sign of any cataclysmic events. These early landslides are usually explained by gravitational processes. The flanks of active seamounts are too steep and have not yet reached their gravitational stability, as was shown for the Canary Islands (Carracedo 1999).

6.6.2
Landslide Related to Young Oceanic Islands (<4 Ma)

Debris avalanches characterize young island landslides. Mehetia, Tahiti Island, Raiatea, Tahaa and Bora Bora all present evidence of debris avalanches. In the entire Society Island chain except for Mehetia, there are no traces of fine-grained deposits. They could have been covered by subsequent debris-avalanche deposits or sediments that settled on the same areas, or they were removed by submarine currents. This is in good agreement with the Hawaiian landslide study of Moore et al. (1989), who proposed that the main cataclysmic events happen at the end of the shield-building phase.

6.6.3
Landslide Related to Older Oceanic Islands (>4 Ma)

The slopes of the oldest islands of the Austral Archipelago, (such as Rapa (5 Ma), Tubuai (9 Ma), Raivavae (6.5 Ma) and Rimatara) present evidence of both fine-grained and blocky debris avalanches. If, as the study of the Society landslides suggests, the traces of fine-grained debris avalanches rapidly disappear, then the Austral slides are recent. They could correspond to erosion processes of old volcanic material between rift zones that lead to the star-shaped characteristics of the atolls and guyots in the area (Mitchell 2001).

6.6.4
Landslide Related to Tectonic Events

Finally, there are also the major landslides induced by vertical motions of the oceanic crust such as what was observed on Tupai, in the Society Chain, and Rurutu Island, in the Austral Chain.

6.7
Conclusion

Our study has revealed forty major landslides that have covered the submarine slopes of the Society and Austral Islands. These submarine landslides can be classified into three types that are related to three stages of edifice building: (1) Active submarine volcanoes are subject to numerous mass wasting events due to gravitational flank instabilities. These landslides are characterized by small volumes and small-sized debris in the deposits, visible scars on the upper slopes of the edifices and a radial distribution of the landslides. (2) During the subaerial volcanic stage, the islands are subjected to giant collapses associated with debris avalanches. This corresponds to isolated events, where a large volume of subaerial material is moved. (3) The end of the volcanic activity is marked by the end of giant lateral collapses; however, mass wasting can continue, but the events are smaller. Erosion processes have occurred between the rift zones and produced the star-shaped morphology observed on old islands and on Pacific guyots. In addition, two islands that were submitted to significant vertical tectonic motions have also been the sites of giant flank landslides.

Acknowledgements

The data used in this study were mainly collected during the ZEPOLYF1 and ZEPOLYF2 cruises performed with the R/V L'ATALANTE. We are grateful to the crews (GENAVIR) and to the IFREMER staff for the quality of their work. These cruises were sponsored by the French Government and by the local government of French Polynesia as a part of the ZEPOLYF program. We wish to thank Christine Deplus for constructive comments on an earlier version of the manuscript, Roger Hekinian for his careful review and Ginny Hekinian for her remarks.

References

Blais S, Guille G, Maury R, Guillou H, Miau H, Cotten J (1997) Géologie et pétrologie de l'île de Raiatea (Société, Polynésie Française). C R Acad Sci Paris, série IIa 324:435–442

Blais S, Guille G, Guillou H, Chauvel C, Maury RC, Caroff M (2000) Géologie, géochimie et géochronologie de l'île de Bora Bora (Société, Polynésie Française), C R Acad Sci Paris, série IIa 331:579–585

Blanchard F, Liotard JM, Brousse R (1981) Origine mantellique des benmoréites de Moorea (îles de la Société, Pacifique), Bull Volc 44:691–710

Bonneville A, Le Suavé R, Audin L, Clouard V, Dosso L, Gillot P-Y, Janney P, Jordahl K, Maamaatuaiahutapu K (2002) Arago Seamount, the missing hotspot found in the Austral Islands. Geology 30:1023–1026

Brousse R, Berger ET (1985) Raiatea dans l'archipel de la Société (Polynésie française). C R Acad Sci Paris, série IIa 301:115–118

Calmant S, Cazenave A (1986) The effective elastic lithosphere under the Cook Austral and Society Islands. Earth Planet Sci Lett 77:187–202

Carracedo JC (1999) Growth, structure, instability and collapse of Canarian volcanoes and comparisons with Hawaiian volcanoes. J Volcanol Geotherm Res 94:1–19

Cheminée J-L, Hekinian R, Talandier J, Albarède F, Devey CW, Francheteau J, Lancelot Y (1989) Geology of an active hotspot: Teahitia-Mehetia region in the south Central Pacific. Mar Geol Res 11:27–50

Clouard V, Bonneville A, Barsczus HG (2000) Size and depth of frozen magma chambers under atolls and islands of French Polynesia using detailed gravity studies. J Geophys Res 105:8173–8192

Clouard V, Bonneville A, Gillot P-Y (2001) A giant landslide on the southern flank of Tahiti Island, French Polynesia. Geophys Res Lett 28:2253–2256

Day SJ, Heleno da Silva SIN, Fonseca JFBD (1999) A past giant lateral collapse and present-day flank instability of Fogo, Cape Verde Islands. J Volcanol Geotherm Res 94:191–218

Deneufbourg G (1965) Carte géologique, notice explicative sur la feuille Huahine, Bur Rech Géol Minier

Denlinger RP, Okubo P (1995) Structure of the mobile south flank of Kilauea Volcano, Hawaii. J Geophys Res 100:24499–24507

Diraison C, Bellon H, Léotot C, Brousse R, Barsczus HG (1991) L'alignement de la Société (Polynésie française): volcanologie, géochronologie, proposition d'un modèle de point chaud. Bull Soc Geol Fran. 162:479–496

Duncan RA, McDougall I (1976) Linear volcanism in French Polynesia. J Volcanol Geotherm Res 1:197–227

Duncan RA, Fisk MR, White WM, Nielsen RL (1994) Tahiti: Geochemical evolution of a French Polynesian volcano. J Geophys Res 99:24341–24357

Fornari DJ, Moore JG, Calk L (1979) A large submarine sand-rubble flow on the Kilauea Volcano, Hawaii. J Volcanol Geotherm Res 5:239–256

Gillot P-Y, Lefèvre J-C, Nativel P-E (1994) Model for the structural evolution of the volcanoes of Réunion Island. Earth Planet Sci Lett 122:291–302

Hampton MA, Lee HJ, Locat J (1996) Submarine landslides. Rev Geophys 34:33–59

Hekinian R, Bideau D, Stoffers P, Cheminée J-L, Muhe R, Puteanus G, Binard N (1991) Submarine intraplate volcanism in the South Pacific; Geological setting and petrology of the Society and the Austral regions. J Geophys Res 96:2109–2138

Herron EM (1972) Sea-floor spreading and the Cenozoic history of the east-central Pacific. Geol Soc Am Bull 83:1671–1692

Hildenbrand A, Bonneville A, Gillot P-Y (2003) Off shore evidence for a landslide on the northern flank of Tahiti-Nui, (French Polynesia). Geophys Res Lett, *submitted*

Holcomb RT, Searle RC (1991) Large landslides from oceanic volcanoes. Mar Geotechnol 10:19–32

Johnson RH, Mahaloff A (1971) Relation of Macdonald Volcano to migration of volcanism along the Austral Chain. J Geophys Res 76:3282–3290

Jordahl KA (1999) Tectonic evolution and midplate volcanism in the south Pacific. PhD thesis, MIT

Keating BH, McGuire WJ (2000) Island edifice failures and associated tsunami hazards. Pure Appl Geophys 157:899–55

Le Dez A, Maury RC, Guillou H, Cotten J, Blais S, Guille G (1998) L'île de Moorea (Société): Édification rapide d'un volcan-bouclier polynésien. Geol France 3:51–64

Lénat J-F, Vincent P, Bachelery P (1989) The off-shore continuation of an active basaltic volcano: Piton de la Fournaise (Reunion Island, Indian Ocean); Structural and geomorphological interpretation from seabeam mapping. J Volcanol Geotherm Res 36:1–36

Leroy I (1994) Evolution des volcans en système de point chaud: île de Tahiti, archipel de la Société (Polynésie française). PhD thesis, Univ. Paris XI, Orsay

Matsuda J, Notsu K, Okano J, Yaskawa K, Chungue L (1984) Geochemical implications from Sr isotopes and K-Ar age determinations for the Cook Austral Islands chain. Tectonophysics 104:145–154

Mayes CL, Lawver LA, Sandwell DT (1990) Tectonic history and new isochron chart of the South Pacific. J Geophys Res 95:8543–8567

McNutt MK, Caress DW, Reynolds J, Jordahl KA, Duncan RA (1997) Failure of plume theory to explain midplate volcanism in the Southern Austral Islands. Nature 389:479–482

Minster JB, Jordan TG (1978) Present-day plate motions. J Geophys Res 83:5331–5354

Mitchell NC (2001) The transition from circular to stellate forms of submarine volcanoes. J Geophys Res 106:1987–2003

Moore JG (1964) Giant submarine landslides on Hawaiian Ridge. US Geol Surv Prof Pap 501-D:D95–D98

Moore JG, Fiske RS (1969) Volcanic substructure inferred from dredge samples and ocean-bottom photographs. Hawaii Geol Soc Am Bull 80:1191–1202

Moore JG, Clague DA, Holcomb RT, Lipman PW, Normark WR, Torresan MT (1989) Prodigeous submarine landslides on Hawaiian Ridge. J Geophys Res 94:17645–17484

Moore JG, Bryan WB, Besson MH, Normark WR (1995) Giant blocks in the South Kona landslide, Hawaii. Geophysics 23:125–128

Munschy M, Antoine C, Gachon A (1996) Evolution tectonique de la région des Tuamotu, océan Pacifique Central. C R Acad Sci Paris, série IIa 323:941–948

Norris A, Johnson RH (1969) Submarine volcanic eruptions recently located in the Pacific by SOFAR hydrophones. J Geophys Res 74:650–664

ORSTOM (1993) Atlas de la Polynésie française. Ed de l'Orstom, Paris

Ouassaa K (1996) Etude de la structure sismique de la croûte océanique dans la partie active du point chaud de Tahiti. Traitement et interprétation des données sismiques des campagnes Midplate 2 et Teahitita 4. PhD thesis, Univ. Bretagne Occidentale, Brest

Service de l'Urbanisme (1992) Carte topographique 1:20 000, Papeete, Tahiti, French Polynesia

Smith JR, Malahoff A, Shor AN (1999) Submarine geology of the Hilina Slump and morpho-structural evolution of Kilauea Volcano, Hawaii. J Volcanol Geotherm Res 94:59–88

Steiger RH, Jaeger E (1977) Subcomission on geochronology: Convention on the use of decay constants in geo-, cosmo-chronology. Earth Planet Sci Lett 36:359–362

Stillman CJ (1999) Giant Miocene landslides and the evolution of Fuerteventura, Canary Islands. J Volcanol Geotherm Res 94:89–104

Talandier J (2004) Seismicity of the Society and Austral hotspot in the South pacific: Seismic direction, monitoring and interpretations of underwater volcanism. In: Hekinian R (ed) Oceanic hotspots. Springer-Verlag, Berlin Heidelberg New York, this volume

Talandier J, Okal EA (1984) New surveys of Macdonald Seamount, south Central Pacific, following volcanoseismic activity, 1977–1983. Geophys Res Lett 11:813–816

Turner DL, Jarrard RD (1982) K-Ar dating of the Cook-Austral island chain: A test of the hot-spot hypothesis. J Volcanol Geotherm Res 12:187–220

Walker GPL (1988) Three Hawaiian calderas: An origin through loading by shallow intrusions? J Geophys Res 93:14773–14784

Watts AB, Masson DG (1995) A giant landslide on the north flank of Tenerife, Canary Islands. J Geophys Res 100:24487–24498

Watts AB, ten Brink US, Buhl P, Brocher TM (1985) A multichannel seismic study of the lithosphere flexure across the Hawaiian-Emperor seamount chain. Nature 315:105–111

White WM, Duncan RA (1996) Geochemistry and geochronology of the Society Islands: New evidence for deep mantle recycling. In: Basu A, Hart S (eds) Earth processes: Reading the isotopic code. Am Geophys Union Geophys Monograph. 95:183–206

Wynn RB, Masson DG, Stow DAV, Weaver PPE (2000) Turbidity current sediment waves on the submarine slopes of the western Canary Islands. Mar Geol 163:185–198

ZEPOLYF (2003) Bathymetry of French Polynesia: Digital bathymetric model based on a multi-beam and single-beam data compilation. Univ. Polynésie française, Tahiti, French Polynesia

ZEPOLYF1 (1996) Rapport de mission de la campagne ZEPOLYF1. Univ. Polynésie française, IFREMER, SHOM, Tahiti, French Polynesia

ZEPOLYF2 (2001) Rapport de campagne, documents et travaux ZEPOLYF. Univ. Polynésie française, Tahiti, French Polynesia

Chapter 7

Mantle Plumes are NOT From Ancient Oceanic Crust

Y. Niu · M. J. O'Hara

7.1
Introduction

Basaltic volcanism mainly occurs in three tectonic settings on the Earth. Volcanism along sea-floor spreading centers produces Mid-Ocean Ridge basalts (MORB) that are depleted in incompatible elements. Volcanism above intra-oceanic subduction zones produces island arc basalts (IAB) that are enriched in water-soluble incompatible elements (e.g., Ba, Rb, Cs, Th, U, K, Pb, Sr), but depleted in water-insoluble incompatible elements (e.g., Nb, Ta, Zr, Hf, Ti). MORB and IAB are products of plate tectonics, and their geochemical differences result from differences in their respective sources and physical mechanisms through which they form. MORB are formed by plate-separation-induced passive mantle upwelling and decompression melting, thus sampling the uppermost mantle that is depleted in incompatible elements. Depletion of the MORB mantle is widely accepted as resulting from the extraction of incompatible element-enriched continental crust during the Earth's early history (Armstrong 1968; Gast 1968; O'Nions and Hamilton 1979; Jacobsen and Wasserburg 1979; DePaolo 1980; Allègre et al. 1983; Hofmann 1998). IAB are widely accepted as resulting from subducting slab-dehydration-induced melting of mantle wedge peridotites, giving rise to the characteristic geochemical signatures of slab "component", which is rich in water and water-soluble elements (e.g., Gill 1981; Tatsumi et al. 1986; McCulloch and Gamble 1991; Stolper and Newman 1994; Hawkins 1995; Pearce and Peate 1995; Davidson 1996).

Basalts are also produced by intraplate volcanism away from plate boundaries. These basalts include flood basalts erupted on land and those erupted/erupting on many oceanic islands. In contrast to MORB and IAB, these basalts are enriched in all incompatible elements and more enriched in the more incompatible elements. If the enriched characteristics in continental flood basalts were caused by continental crust contamination, then the enriched oceanic equivalent, termed ocean-island basalts (OIB), must reflect a mantle source that is enriched in incompatible elements. The observation that the shallow mantle for MORB is depleted in incompatible elements suggests that OIB must be derived from regions deeper than the MORB mantle. As the inferred OIB sources differ from undifferentiated "primitive mantle", it has thus been speculated that OIB source materials must have been previously processed and brought to the upper mantle melting regions by mantle plumes. Among the many contributions endeavoring to understand the origin of mantle plumes and OIB sources in the context of plate tectonics is the classic paper titled *"Mantle plumes from ancient oceanic crust"* by Hofmann and White (1982). These authors proposed *"oceanic crust is returned to the [lower] mantle during subduction … … Eventually, it becomes unstable as a consequence of internal heating, and the resulting diapirs [at the core-mantle boundary] become the source plumes of oceanic island basalts (OIB) and hot-spot volcanism."*

In this chapter, we show with evidence that there is no genetic link between ancient subducted oceanic crust and the source materials of OIB. Our arguments are based on well-understood petrology, geochemistry, and experimental data on mineral physics.

7.2
Petrological Arguments

7.2.1
Melting of Oceanic Crust Cannot Produce the High Magnesian Melts Parental to Many OIB Suites

Christensen and Hofmann (1994) explored physical scenarios about how subducted oceanic crust can isolate itself from the attached lithospheric mantle during mantle convection so as to form deep-rooted plumes to rise and feed hotspot volcanism in the upper mantle. It should be understood that the bulk oceanic crust is picritic/basaltic in composition (Niu 1997) and cannot, by melting, produce the high magnesian lavas seen in many OIB suites. It has been well established for many years that basaltic melts are derived from more magnesian picritic melts produced by partial melting of mantle peridotites (O'Hara 1968a,b; Stolper 1980; Falloon et al. 1988; Herzberg and O'Hara 1998, 2002; O'Hara and Herzberg 2002). Partial melting of recycled oceanic crust, which is compositionally basaltic/picritic (e.g., Niu 1997) and petrologically eclogitic (O'Hara and Yoder 1967; O'Hara and Herzberg 2002), will not produce basaltic/picritic melts, but melts of more silicic composition (Green and Ringwood 1968; Wyllie 1970). If total melting had occurred, the melts would be basaltic/picritic in composition but would still differ in both major and trace element systematics from those of average OIB. In fact, primitive OIB melts are much more magnesian than the most primitive MORB and likely to be more magnesian than bulk oceanic crust (Herzberg and O'Hara 1998, 2002; Clague et al. 1991; Norman and Garcia 1999). Therefore, petrologically ancient recycled oceanic crusts cannot become sources of mantle plumes feeding hotspot volcanisms and OIB.

7.3
Geochemical Arguments

7.3.1
Melting of Subduction-Zone Dehydrated Residual Oceanic Crusts Cannot Yield the Trace Element Systematics in OIB

The oceanic crust is altered during its accretion at ocean ridges and subsequently pervasively weathered/hydrated on the sea floor. This crust that is atop the subducting slab endures the greatest extent of dehydration in subduction zones. It is widely accepted that the fluids released from this dehydration lowers the solidus of the overlying mantle wedge that melts to produce arc lavas (Gill 1981; Tatsumi et al. 1986; McCulloch and Gamble 1991; Pearce and Peate 1995; Davidson 1996; Tatsumi and Kogiso 1997) as reflected in IAB geochemistry (Fig. 7.1). If this interpretation of IAB genesis

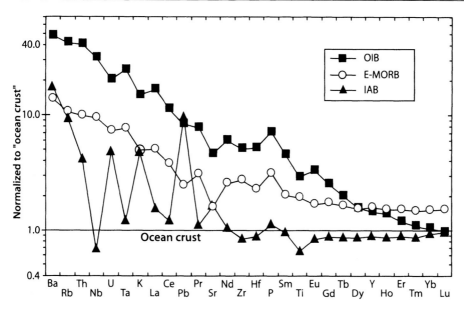

Fig. 7.1. Multi-element "spider-diagrams" of average ocean-island basalts – *OIB* (Sun and McDonough 1989), enriched E-MORB (Niu et al. 2002a), average island arc tholeiites – *IAB* (Ewart et al. 1988; Y. Niu, unpublished data for Tonga and Mariana arc tholeiites) normalized to present-day mean composition of oceanic crust (see Table 7.1 for data)

is indeed correct, then the residual subducted crust that has passed through subduction-zone dehydration reactions will have geochemical signatures that are complementary to the signatures of arc lavas (McDonough 1991; Niu et al. 1999, 2002a). In other words, this residual crust should be relatively enriched in water-insoluble incompatible elements (e.g., Nb, Ta, Zr, Hf and Ti) but highly depleted in water-soluble incompatible elements (e.g., Ba, Rb, Cs, Th, U, K, Sr, Pb etc.) (Fig. 7.2b). It follows logically that if the recycled oceanic crust were geochemically responsible for OIB, then OIB would be highly depleted in these water-soluble incompatible elements. This is not observed. In fact, OIB are enriched in these water-soluble incompatible elements as well as water-insoluble elements (Figs. 7.1 and 7.2a,b) in spite of super-chondritic Nb/Th and Ta/U ratios (Sun and McDonough 1989; Niu and Batiza 1997; Niu et al. 1999). In summary, melting or partially melting residual oceanic crust that has passed through subduction-zone dehydration reactions with the geochemical signatures shown in Fig. 7.2b will neither produce OIB nor any volcanic rocks ever sampled on the Earth's surface. Recycled terrigenous sediments would be enriched in water-soluble elements, but they also dehydrate or even melt in subduction zones, contributing to arc volcanism (Plank and Langmuir 1998; Elliot et al. 1997). If the terrigenous sediments are neither dehydrated nor melted, they still fail to explain the elevated Ce/Pb and Nb/U ratios in most OIB (Hofmann et al. 1986; Niu et al. 1999). Furthermore, terrigenous sediments with detrital zircon crystals will lead to unpredicted Zr-Hf fractionation from REE, thus giving bizarre Hf isotopes (Patchett et al. 1984; White et al. 1986), which is not observed in OIB (see below).

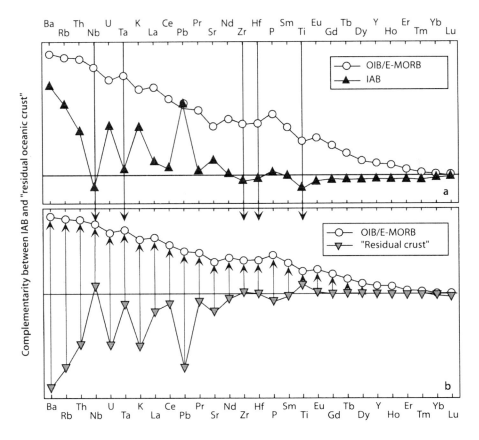

Fig. 7.2. a Schematic representation of Fig. 7.1; **b** schematic illustration of trace element systematics a mean oceanic crust would have after passing through subduction-zone dehydration reactions. The point is that such subduction-zone filtered *residual* oceanic crust is depleted in water-soluble incompatible elements like Ba, Rb, Th, U, K, Pb and Sr while relatively enriched in water-insoluble incompatible elements like Nb, Ta, Zr, Hf, and Ti. It is materially impossible to melt such a *residual* crust to produce magmas with OIB geochemical signatures, unless some form of refertilisation in the deep mantle took place as indicated by the *vertical arrows*. The latter "refertilisation" is entirely *ad hoc* with neither evidence nor physical mechanisms. *The horizontal lines* in *Part a* and *Part b* represent present-day composition of oceanic crust as shown in Fig. 7.1 and Table 7.1)

7.3.2
OIB Sr-Nd-Hf Isotopes Record no Subduction-Zone Dehydration Signatures

If ancient subducted oceanic crusts had indeed played a role in the petrogenesis of OIB, then the isotopic signatures of subduction-zone dehydration, which is non-magmatic, should be preserved in the OIB. This is, however, not observed in Sr-Nd-Hf isotopic systems. We choose these simple isotopic systems instead of complex Pb isotopes or poorly understood systems such as Os and noble gases. Such a choice is logical, because it can avoid ambiguous conclusions. Also, if the hypothesis fails for simple isotopic systems, interpretations based on complex isotopic systems in favor of that hypothesis will collapse accordingly.

Chapter 7 · Mantle Plumes are NOT From Ancient Oceanic Crust

Figure 7.3a plots averages of 40 OIB suites in ε_{Sr}–ε_{Nd} space (data from Albarède 1995). Figure 7.3b plots a number of OIB suites in ε_{Hf}–ε_{Nd} space (data from Salters and White 1998). Except for the so-called EM2 and HIMU OIB suites (Zindler and Hart 1986 for the acronyms) in Fig. 7.3a, all other 36 OIB suites define a scattered, yet statistically significant (at >99.9% confidence levels) inverse linear trend. Except for the HIMU-

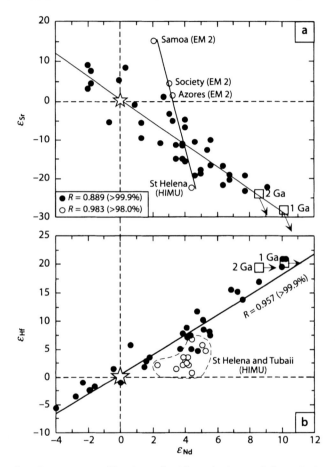

Fig. 7.3. a Plot of OIB in ε_{Sr}–ε_{Nd} space (data from Albarède 1995), where each data point is an average of an ocean island group; **b** plot of OIB in Hf-Nd space (data from Salters and White 1998), where the data represent a number of ocean islands. Note that except for three EM2 OIB suites and one HIMU suite in Part a and HIMU suites in Part b, all the rest of the OIB data define significant linear trends in these two spaces with recommended chondrite uniform reservoir (CHUR, *open stars*) values lying within the trends. These suggest that the ultimate process/processes that have led to the linear trends must be simple and magmatic because of the very similar effective bulk distribution coefficients of Sr, Nd and Hf (Table 7.1). Note also that the three EM2 and one HIMU OIB suites in Part a define a significant linear trend, suggesting that EM2 and HIMU OIB suites may be genetically related. The ε-notations are calculated (use εNd as an example) as εNd = (^{143}Nd/^{144}Nd$_{[sample]}$ / ^{143}Nd/^{144}Nd$_{(CHUR)}$ – 1) × 10 000, using the recommended (e.g., Faure 1986; Dickin 1997) present-day CHUR values: ^{87}Sr/^{86}Sr = 0.70450; ^{143}Nd/^{144}Nd = 0.512638 and ^{176}Hf/^{177}Hf = 0.282818. The latter is similar to 0.282772 ±29 of Blichert-Toft and Albarède (1997). The *open squares* are present-day isotopic compositions the 2 Ga and 1 Ga old oceanic crusts would have. The ancient oceanic crusts were calculated by 20% batch melting from depleted MORB mantle, whose mean age is assumed to be 2.5 Ga. The *arrows* point to the effect of subduction-zone dehydration

Table 7.1a. Compositions of various oceanic rocks, mean crust, and effective bulk distribution coefficients

	Average Bulk Ocean Crust[a]	Average OIB[b]	Average IAB[c]	Average Effective D[d]
Ba	7.384	350	125.6	0.0063
Rb	0.747	31.0	6.800	0.0208
Th	0.100	4.00	0.405	0.0547
Nb	1.570	48.0	1.041	0.0651
U	0.051	1.02	0.245	0.0877
Ta	0.112	2.70	0.135	0.0922
K	822.7	12000	3811	0.1293
La	2.250	37.0	3.407	0.1546
Ce	7.161	80.0	8.546	0.1985
Pb	0.390	3.20	3.696	0.2328
Pr	1.270	9.70	1.384	0.2270
Sr	146.3	660	237.3	0.2500
Nd	6.519	38.5	6.700	0.2533
Zr	55.34	280	45.24	0.2587
Hf	1.537	7.80	1.333	0.2753
P	387.6	2700	432.5	0.2737
Sm	2.242	10.0	2.112	0.2814
Ti	6034	17200	3858	0.2932
Eu	0.925	3.00	0.761	0.2938
Gd	3.064	7.62	2.625	0.2925
Tb	0.536	1.05	0.452	0.2966
Dy	3.610	5.60	3.045	0.2987
Y	20.28	29.0	17.59	0.2979
Ho	0.776	1.06	0.662	0.2987
Er	2.226	2.62	1.938	0.2983
Tm	0.324	0.35	0.276	0.2994
Yb	2.102	2.16	1.908	0.2990
Lu	0.315	0.30	0.295	0.2981

[a] Average composition of "bulk ocean crust" (BOC) is calculated by combining 40% N-MORB (average of 132 N-MORB glass samples analyzed using ICP-MS by Y. Niu, see Niu et al. (2002a); assumed to represent erupted lavas and unerupted feeding dikes of upper ocean crust, equivalent to seismic layers 2a and 2b, respectively) with 60% oceanic gabbros (average of 87 whole-rock gabbroic samples of ODP Hole 735B analyzed using ICP-MS by Y. Niu, see Niu et al. (2002b); assumed to be lower ocean crust, equivalent seismic layer 3).

[b] Average ocean island basalts (OIB) composition of Sun and McDonough (1989).

[c] Average island arc tholeiitic basalts (IAB) composition derived from Ewart et al. (1998) for Tonga arc, and unpublished data by Y. Niu for Tonga and Mariana arcs.

[d] Effective bulk distribution coefficients for the tabulated elements are determined by relative variability (defined as $RDS\% = 1\sigma/mean \times 100$) of MORB data (Niu and Batiza 1997; Niu et al. 1999, 2002a), which is proportional to relative incompatibility, by assuming bulk D for Ba being close to zero while bulk D for heavy rare earth elements being about 0.2997 determined from various published Kd data and polybaric melting relation of Niu (1997).

Table 7.1b. Normalized to immobile Nb

	Average Bulk Ocean Crust	Average IAB	IAB/BOC[a]
Hf/Nb	0.979	1.280	1.31
Lu/Nb	0.2003	0.2831	1.41
Sm/Nb	1.428	2.028	1.42
Nd/Nb	4.151	6.433	1.55
Sr/Nb	93.16	227.9	2.45
K/Nb	523.9	3659	6.98
U/Nb	0.033	0.235	7.19
Rb/Nb	0.476	6.529	13.7
Pb/Nb	0.248	3.549	14.3
Ba/Nb	4.701	120.579	25.6

[a] By normalizing mobile element abundances with respect to the abundances of immobile elements such as Nb, and by comparing IAB with average bulk ocean crust (BOC), we can take the ratios as reflecting and proportional to the mobility of these elements during subduction-zone dehydration reactions.

like OIB, the data define a statistically significant (>99.9% confidence levels) positive linear trend (Fig. 7.3b). Given the relatively minor occurrences of EM2 and HIMU OIB suites on a global scale, we will first focus our discussion on the majority of OIB suites here and discuss the implications of EM2 and HIMU OIB suites later.

Because of the large differences in relative mobility of Rb > Sr > Nd > Sm > Lu > Hf during subduction dehydration inferred from observations (Table 7.1) and determined experimentally (Kogiso et al. 1997), the significant correlations in Fig. 7.3a,b would not exist or would have been destroyed if sources of these OIB had been involved in, or actually part of, ancient oceanic crusts passing through subduction-zone dehydration reactions. The significant linear correlations thus suggest that (1) the elements Sr, Nd and Hf have behaved similarly in the respective sources of these OIB suites in the past >1 Ga; (2) the similar behavior would be unlikely if these OIB sources had experienced subduction-zone dehydration, but is to be expected if the process or processes these OIB sources had experienced were magmatic because of the similar effective bulk D's of these elements (Table 7.1); and (3) coupled correlations of a radioactive parent over radiogenic daughter (P/D) for ratios such as Rb/Sr, Sm/Nd and Lu/Hf in these OIB sources must also have existed without having been disturbed in the last >1 Ga.

The open squares in Fig. 7.3a,b represent the present-day isotopic compositions that ancient oceanic crusts would have if they were produced 2 Ga and 1 Ga ago, respectively, from the depleted MORB mantle (DMM). The calculation assumes a mean age of 2.5 Ga for the DMM corresponding to the mean age of continental crust (Jacobsen and Wasserburg 1979; Taylor and McLennan 1985) because of the agreement that DMM resulted from continental crust extraction in the Earth's early history (Armstrong 1968; O'Nions and Hamilton 1979; Jacobsen and Wasserburg 1979; DePaolo 1980; Allègre et al. 1983). Regardless of model details, we cannot avoid the conclusion that the ancient subducted oceanic crusts are isotopically too depleted (too unradiogenic Sr, and too radiogenic Nd and Hf) to meet the required isotopic values of present-day OIB. Therefore, ancient recycled oceanic crusts cannot be mantle plume sources feeding intraplate volcanism and OIB. The arrows next to the open squares point to the effect of subduction-zone dehydration. Obviously, ancient oceanic crusts passing through subduction-zone dehydration would be isotopically even more depleted, therefore even more unlikely to be sources of OIB.

Although relatively minor in occurrence, the deviation of EM2 and HIMU OIB suites from the main linear OIB trends needs attention (Fig. 7.3a,b). In particular, the three EM2 OIB suites (Samoa, Society and Azores) and the HIMU OIB suite (St. Helena) define a simple but statistically significant (at >98% confidence level) linear trend (Fig. 7.3a). This suggests that their origin may be somehow related. It is generally thought that EM2 OIB reflect a source component of recycled terrigenous sediments (Weaver 1991; Hofmann 1997), whereas HIMU OIB reflect a source component of recycled oceanic crust (Hofmann 1997). Surface or near-surface processes are likely to have caused P/D (e.g., Rb/Sr, Sm/Nd, Lu/Hf, U/Pb, Th/Pb etc.) fractionation. Therefore, these near-surface, processed materials with fractionated P/D ratios, with time and when returned to mantle source regions of oceanic basalts, would produce peculiar isotopic signatures such as EM2, HIMU etc. in some OIB. However, it is imperative to note that neither EM2 nor HIMU OIB suites show trace element systematics (Weaver 1991) that are consistent with having experienced subduction-zone dehydration reactions (see Fig. 7.2b). The only physical scenario in which terrigenous sediments and "oceanic crust" could be introduced into the mantle without possibly experiencing subduction-zone dehydration is where the subduction zone begins to initiate along passive continental margins (Niu et al. 2003). Many parts of passive margins are characterized by thick sequences of volcanics and intrusives of mantle plume origin during continental break-up (Eldholm and Coffin 2000). Such mantle plume generated magmatic constructions are unlikely to have been heavily altered and hydrated (vs. normal oceanic crust) and thus should not have experienced significant dehydration when subducted as metamorphosed dense eclogites, along with the loaded terrigenous sediments, into the mantle. In this case, however, (1) the "basaltic crust" subducted is not normal oceanic crust, and (2) this subducted "crust" does not go down to the lower mantle; otherwise, it will never come back to the source regions of oceanic basalts in the upper mantle (see below). The significance of sediment and crustal subduction during subduction initiation at passive margins as a speculative hypothesis (Niu et al. 2003) requires further evaluation.

7.4
Mineral Physics Arguments

Recent mantle tomographic studies have reached the consensus that subducting oceanic lithosphere can penetrate the 660 km seismic discontinuity (660-D) into the lower mantle (van der Hist et al. 1997; Grand et al. 1997). This supports the whole mantle convection model and the proposal that mantle plumes originate from the core-mantle boundary (Griffiths and Campbell 1990; Davies and Richards 1992). This also lends support to the model by Hofmann and White (1982) that ancient oceanic crusts could be heated and segregated from the ambient mantle at the core-mantle boundary (Christensen and Hofmann 1994), feeding mantle plumes and hotspot volcanism. To balance the downward flow of subduction, upward mass transfer from the lower mantle to the upper mantle is required. Plume flux from the deep mantle may be the most important upward flow that feeds the upper mantle (Phipps Morgan et al. 1995; Niu et al. 1999), but it is also feasible that the upward mass transfer takes place in the form of a regional "swell" across the 660-D. In either scenario, the fundamental question is whether or not the subducted crust can return to the upper mantle source regions of oceanic basalts. Recent mineral physics studies indicate that this is physically unlikely.

7.4.1
Subducted Oceanic Crusts are too Dense to Rise to the Upper Mantle

Ono et al. (2001) have shown that subducted basaltic oceanic crust turns into an assemblage of stishovite (~24 vol.%), Mg-perovskite (~33%), Ca-perovskite (~23%) and Ca-ferrite (~20%) at shallow upper mantle conditions. This assemblage is significantly denser than the ambient peridotitic mantle. Figure 7.4 compares the bulk density of oceanic crust with that of ambient mantle peridotite under shallow lower mantle conditions as a function of depth. Assuming a whole-mantle convection scenario, and considering a depth of 780 km, the temperature of 2 000 K at this depth is reasonable. There, the subducted oceanic crust is >2.3% denser than the ambient peridotite mantle. Such huge negative buoyancy will impede the rise of the subducted oceanic crust into the upper mantle. If the crustal portion of the subducted lithosphere was segregated at greater depths as proposed (Christensen and Hofmann 1994), then this crust would only rise to the level of neutral buoyancy, which is at depths of about 1 600 km (Kesson et al. 1998; Ono et al. 2002). If the observed seismological heterogeneity at this and deeper depths (Kaneshima and Helffrich 1999; Kellogg et al. 1999; van der Hilst and Kárason 1999) is controlled by the level of neutral buoyancy, then it is unlikely that oceanic crust subducted to the deep mantle at the core-mantle boundary will rise to the upper mantle source regions of oceanic basalts.

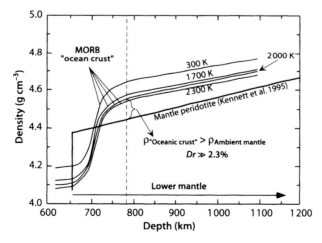

Fig. 7.4. Modified from Ono et al. (2001) to show that oceanic crust subducted into the lower mantle will be transformed to a high-pressure mineral assemblage whose bulk-rock density is significantly greater than that of the ambient peridotite mantle (Kennett et al. 1995). For a whole-mantle convection scenario, the mantle temperature would be about ~2 000 K at ~780 km. In this case, the subducted oceanic crust would be >2.3% denser than the ambient mantle. Such huge negative buoyancy impedes the rise of subducted crust into the upper mantle source regions of oceanic basalts

7.4.2
Basaltic Melts in the Lower Mantle Conditions are Denser than Ambient Solid Peridotites

Melts may exist in the seismic D" region near the core-mantle boundary (Williams and Garnero 1996). As oceanic crusts likely have lower solidus temperatures relative to the ambient mantle, it is possible that subducted oceanic crusts may have contributed to the partial or total melting. However, basaltic melts are again too dense in comparison to the ambient mantle to rise (Suzuki et al. 1998; Ohtani and Maeda 2001; Agee 2001). Figure 7.5 demonstrates that a basaltic melt (compositionally equivalent to oceanic crust) becomes denser and progressively more so than solid mantle minerals (perovskite and magnesiowuestite) and bulk mantle peridotites at depths of >1 400 km. At depths close to the core-mantle boundary (or D" region), this basaltic melt is >~15% denser than the bulk peridotitic solid mantle. The negative buoyancy of the basaltic melt is so large that it is physically difficult to rise at all, let alone arrive in the source regions of oceanic basalts in the upper mantle. Because of the classic geochemical interpretation of "mantle plumes from ancient oceanic crust" (Hofmann and White 1982), Ohtani and Maeda (2001) had to invoke slab-derived water to lower the basaltic melt density in order to be consistent with that geochemical interpretation. It is practically and physically impossible to overcome the 15% negative buoyancy by adding finite water in the basaltic melt. Furthermore, the subducted crust, which is atop the subducting lithosphere, experiences the greatest extent of dehydration in subduction zones, and is thus water poor relative to the serpentinized peridotites atop the lithospheric mantle beneath the crust (Dick 1989; Niu and Hekinian 1997) and the metasomatized deep portions of oceanic lithosphere (Niu et al. 2002a; Niu and O'Hara, 2003).

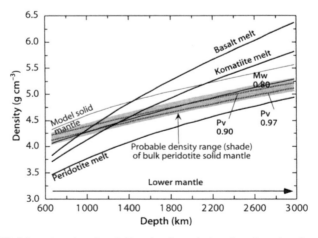

Fig. 7.5. Modified from Agee (1998) and Ohtani and Maeda (2001) to show that if oceanic crust subducted into the lower mantle melts, this melt of basaltic composition will be denser and progressively more so with depth than solid minerals of proper compositions (Mw: magnesiowuestite, Pv: perovskite) and the bulk mantle peridotites at depths in excess of 1 400 km. At depths approaching the core-mantle boundary, the basaltic melt is >15% denser than the bulk solid mantle. Therefore, subducted oceanic crust in the lower mantle cannot return to the upper mantle in the form of melt

Figure 7.5 also shows that peridotite melts are less dense than basaltic melts, komatiitic melts, and most importantly, than the solid mantle peridotites in the lower mantle conditions. We emphasize that peridotites and peridotite melts are better candidates for mantle plumes feeding intraplate volcanism and OIB. Physically, peridotites and peridotite melts are the least dense in the lower mantle conditions. Chemically, peridotites and peridotite melts can produce high magnesian basaltic/picritic melts as required by primitive OIB, which is impossible by melting recycled oceanic crusts of basaltic composition.

7.5
Summary

We have shown in terms of straightforward petrology, geochemistry and mineral physics that ancient subducted oceanic crusts cannot be source materials of mantle plumes feeding intraplate volcanism and OIB. Melting of oceanic crusts cannot produce high magnesian OIB lavas. Oceanic crusts produced from depleted mantle >1 Ga ago are isotopically too depleted to meet the required values of most OIB. Subducted oceanic crusts that have passed through subduction-zone dehydration must be depleted in water-soluble incompatible elements such as Ba, Rb, Cs, Th, U, K, Sr, Pb but relatively enriched in water-insoluble incompatible elements such as Nb, Ta, Zr, Hf, Ti. Melting of residual crusts with such trace element composition cannot produce OIB or any volcanic rocks sampled on the Earth. Oceanic crusts subducted into the lower mantle will be >2% denser than the ambient mantle at shallow lower-mantle depths. Such huge negative buoyancy will impede the subducted oceanic crusts from rising into the upper mantle. If subducted oceanic crusts melt at depths near the core-mantle boundary, such melts are even denser, up to ~15%, than the ambient peridotitic mantle. Therefore, subducted bulk oceanic crusts can neither in the solid state nor in the melt form rise into the upper mantle source regions of oceanic basalts. However, we cannot rule out the possibility that minor elements of subducted oceanic crusts might be carried into the upper mantle along with buoyant ascending plumes of peridotite compositions. Models invoking recycled oceanic crust to explain the geochemistry of OIB must be able to demonstrate how such crust can, by melting, produce the high magnesian lavas observed in many OIB suites. They must also be able to explain the lack of subduction-zone dehydration signatures in OIB. Models that require ancient subducted oceanic crusts become plume sources derived from the lower mantle must also explain the physical mechanisms needed to overcome the huge negative buoyancy of the subducted crusts in both solid state and melt form.

We suggest, following Niu et al. (2002a), that deep portions of recycled oceanic lithosphere are the best candidates for mantle plume sources. These deep portions of oceanic lithosphere are filled with dykes or veins enriched in volatiles, alkalis, and all incompatible elements as a result of low-degree melt metasomatism at the interface between the low velocity zone and the cooling and thickening oceanic lithosphere. These metasomatized lithospheric materials are peridotitic in bulk composition and can, by partial or locally total melting, produce the high magnesian melts required for primitive OIB. Such peridotite-dominated material will necessarily develop positive thermal buoyancy upon heating in the deep mantle, with or without the presence of a melt phase, making it possible for the material to ascend as plumes.

Acknowledgements

Both authors thank Cardiff University for support. Y. Niu acknowledges the support of UK NERC for a Senior Research Fellowship. Discussions with Rodey Batiza, Marcel Regelous, Jack Casey, and Roger Hekinian were very useful. Y. Niu also acknowledges the support from University of Houston for completing the research.

References

Agee CB (1998) Crystal-liquid density inversions in terrestrial and lunar magmas. Phys Earth Planet Inter 107:63–74

Albarède F (1996) Introduction to geochemical modeling. Cambridge University Press Cambridge, pp 543

Allègre CJ, Hart SR, Minster J-F (1983) Chemical structure and evolution of the mantle and continents determined by inversion of Nd and Sr isotopic data; I. Theoretical methods. Earth Planet Sci Lett 66:177–190

Armstrong RL (1968) A model for the evolution of strontium and lead isotopes in a dynamic earth. Rev Geophys Space Phys 6:175–200

Blichert-Toft J, Albarède F (1997) The Lu-Hf isotope geochemistry of chondrites and evolution of the mantle-crust system. Earth Planet Sci Lett 148:243–258

Christensen UR, Hofmann AW (1994) Segregation of subducted oceanic crust in the convecting mantle. J Geophys Res 99:19867–19884

Clague D, Weber WS, Dixon JE (1991) Picrite glasses from Hawaii. Nature 353:553–556

Davidson JP (1996) Deciphering mantle and crustal signatures in subduction zone magmatism. In: Bebout GE, Scholl DW, Kirby SH, Platt JP (eds) Subduction – top to bottom. Geophys Monogr 96:251–264

Davies GF, Richards MA (1992) Mantle convection. J Geol 100:151–206

DePaolo DJ (1980) Crustal growth and mantel evolution: Inferences from models of element transport and Nd and Sr isotopes. Geochim Cosmochim Acta 44:1185–1196

Dick HJB (1989) Abyssal peridotites, very slow spreading ridges and ocean ridge magmatism. Geol Soc Spec Publ 42:71–105

Dickin AP (1997) Radiogenic isotope geology. Cambridge University Press, Cambridge

Eldholm O, Coffin MF (1998) Large igneous provinces and plate tectonics. Geophys Monogr 121: 309–326

Elliott T, Plank T, Zindler A, White W, Bourdon B (1997) Element transport from slab to volcanic front at the Mariana Ar. J Geophys Res 102:14991–15019

Ewart A, Collerson KD, Regelous M, Wendt JI, Niu Y (1988) Geochemical evolution within the Tonga-Kermadec-Lau Arc-Backarc system: The role of varying mantle wedge composition in space and time. J Petrol 39:331–368

Falloon TJ, Green DH, Hatton CJ, Harris KL (1988) Anhydrous partial melting of a fertile and depleted peridotite from 2 to 30 kb and application to basalt petrogenesis. J Petrol 29:1257–1282

Faure G (1986) Principles of Isotope Geology. John Wiley and Sons, Inc., New York

Gast PW (1968) Trace element fractionation and the origin of tholeiitic and alkaline magma types. Geochim Cosmochim Acta 32:1055–1086

Gill JB (1981) Orogenic andesites and plate tectonics. Springer-Verlag, Berlin

Grand SP, van der Hilst RD, Widiyantoro S (1997) Global seismic tomography: A snapshot of convection in the Earth. GSA Today 7:1–7

Green TH, Ringwood AE (1968) Genesis of the calc-alkaline igneous rock suite. Contrib Mineral Petrol 18:105–162

Griffiths RW, Campbell IH (1990) Stirring and structure in mantle starting plumes. Earth Planet Sci Lett 99:66–78

Hawkins JW (1995) Evolution of the Lau Basin – insights from ODP Leg 135. Geophys Monogr 88:125–174

Herzberg C, O'Hara MJ (1998) Phase equilibrium constraints on the origin of basalts, picrites, and komatiites. Earth Sci Rev 44:39–79

Herzberg C, O'Hara MJ (2002) Plume-associated ultramafic magmas of Phanerzoic age. J. Petrol 43:1857–1883

Hofmann AW (1988) Chemical differentiation of the Earth: The relationship between mantle, continental crust, and oceanic crust. Earth Planet Sci Lett 90:297–314

Hofmann AW (1997) Mantle geochemistry: The message from oceanic volcanism. Nature 385:219–229

Chapter 7 · Mantle Plumes are NOT From Ancient Oceanic Crust

Hofmann AW, Jochum KP (1996) Source characteristics derived from very incompatible trace elements in Mauna Loa and Mauna Kea basalts, Hawaii Scientific Drilling Project. J Geophys Res 101: 11831–11839

Hofmann AW, White WM (1982) Mantle plumes from ancient oceanic crust. Earth Planet Sci Lett 57: 421–436

Jacobsen SB, Wasserburg GJ (1979) The mean age of mantle and crustal reservoirs. J Geophys Res 84:7411–7427

Kaneshima S, Helffrich G (1999) Dipping low-velocity layer in the mid-lower mantle: Evidence for geochemical heterogeneity. Science 283:1888–1891

Kellogg LH, Hager BH, Van der Hilst RD (1999) Compositional stratification in the deep mantle. Science 283:1881–1884

Kennett BLN, Engdahl ER, Buland R (1995) Constraints on seismic velocities in the Earth from travel times. Geophys J Int. 122:108–124

Kesson SE, Fitz Gerald JD, Shelley JM (1998) Mineralogy and dynamics of a pyrolite lower mantle. Nature 393:252–255

Kogiso T, Tatsumi Y, Nakano S (1997) Trace element transport during dehydration processes in the subducted oceanic crust: 1. Experiments and implications for the origin of ocean island basalts. Earth Planet Sci Lett 148:193–205

McCulloch MT, Gamble JA (1991) Geochemical and geodynamical constraints on subduction zone magmatism. Earth Planet Sci Lett 102:358–374

McDonough WF (1991) Partial melting of subducted oceanic crust and isolation of its residual eclogitic lithology. Phil Trans R Soc Lond A33:407–418

Niu Y (1997) Mantle melting and melt extraction processes beneath ocean ridges: Evidence from abyssal peridotites. J Petrol 38:1047–1074

Niu Y, Batiza R (1997) Trace element evidence from seamounts for recycled oceanic crust in the eastern Pacific mantle. Earth Planet Sci Lett 148:471–483

Niu Y, Hekinian R (1997) Basaltic liquids and harzburgitic residues in the Garrett transform: A case study at fast-spreading ridges Earth Planet Sci Lett 146:243–258

Niu Y, O'Hara MJ (2003) The origin of ocean island basalts: A new perspective from petrology, geochemistry and mineral physics considerations. J Geophys Res 108:10.1029/2002JB002048

Niu Y, Collerson KD, Batiza R, Wendt JI, Regelous M (1999) The origin of E-Type MORB at ridges far from mantle plumes: The East Pacific Rise at 11°20'. J Geophys Res 104:7067–7087

Niu Y, Regelous M, Wendt JI, Batiza R, O'Hara JM (2002a) Geochemistry of near-EPR seamounts: Importance of source vs. process and the origin of enriched mantle component. Earth Planet Sci Lett 199:329–348

Niu Y, Gilmore T, Mackie S, Greig A, Bach W (2002b) Mineral chemistry, whole-rock compositions and petrogenesis of ODP Leg 176 gabbros: Data and discussion. Proc ODP Sci Results 176:1–60 (on line)

Niu Y, O'Hara MJ, Pearce JA (2003) Initiation of subduction zones as a consequence of lateral compositional buoyancy contrast within the lithosphere: A petrologic perspective. J Petrol 44:851–866

Norman MD, Garcia MO (1999) Primitive magmas and source characteristics of the Hawaiian plume: petrology and geochemistry of shield picrites. Earth Planet Sci Lett 168: 27–44

O'Hara MJ (1968a) The bearing of phase equilibria studies in synthetic and natural systems on the origin and evolution of basic and ultrabasic rocks. Earth Sci Rev 4:69–133

O'Hara MJ (1968b) Are ocean floor basalts primary magmas? Nature 220:683–686

O'Hara MJ, Herzberg C (2002) Interpretation of trace element and isotope features of basalts: Relevance of field relations, petrology, major element data, phase equilibria, and magma chamber modeling in basalt petrogenesis. Geochim Cosmochim Acta 66:2167–2191

O'Hara MJ, Yoder Jr HS (1967) Formation and fractionation of basic magmas at high pressures. Scott J Geol 3:67–117

O'Nions RK, Evensen NM, Hamilton PJ (1979) Geochemical modeling of mantle differentiation and crustal growth. J Geophys Res 84:6091–6101

Ohtani E, Maeda M (2001) Density of basaltic melt at high pressure and stability of the melt at the base of the lower mantle. Earth Planet Sci Lett 193:69–75

Ono S, Ito E, Katsura T (2001) Mineralogy of subducted basaltic crust (MORB) from 25 to 37 GPa, and chemical heterogeneity of the lower mantle. Earth Planet Sci Lett 190:57–63

Patchet PJ, White WM, Feldmann H, Kielinczuk S, Hofmann AW (1984) Hafnium/rare earth element fractionation in the sedimentary system and crustal recycling into the Earth's mantle. Earth Planet Sci Lett 69:365–378

Pearce JA, Peate DW (1995) Tectonic implications of the composition of volcanic arc magmas. Ann Rev Earth Planet Sci 23:251–285

Phipps Morgan J, Morgan WJ, Zhang Y-S, Smith WHF (1995) Observational hints for a plume-fed, suboceanic asthenosphere and its role in mantle convection. J. Geophys Res 100:12753–12767

Plank T, Langmuir CH (1998) The chemical compositions of subducting sediments and its consequences for the crust and mantle. Chem Geol 145:325–394

Salters VJM, White WM (1998) Hf isotope constraints on mantle evolution. Chem Geol 145:447–460

Stolper E (1980) A phase diagram for mid-ocean ridge basalts: Preliminary results and implications for petrogenesis. Contrib Mineral Petrol 74:13–27

Stolper E, Newman S (1994) The role of water in the petrogenesis of Mariana trough magmas. Earth Planet Sci Lett 121:293–325

Sun S-S, McDonough WF (1989) Chemical and isotopic systematics of ocean basalt: Implications for mantle composition and processes. Geol Soc Spec Publ 42:323–345

Suzuki A, Ohtani E, Kato T (1998) Density and thermal expansion of a peridotite melt at high pressure. Phys Earth Planet Inter 107:53–61

Tatsumi Y, Kogiso T (1997) Trace element transport during dehydration processes in the subducted oceanic crust: 2. Origin of chemical and physical characteristics in arc magmatism. Earth Planet Sci Lett 148:207–221

Tatsumi Y, Hamilton DL, Nesbitt RW (1986) Chemical characteristics of fluid phase from a subducted lithosphere and origin of arc magmas: Evidence from high-pressure experiments and natural rocks. J Volcanol Geotherm Res 29:293–309

Taylor SR, McLennan SM (1985) The continental crust: Its composition and evolution. Oxford University Press, New York

van der Hilst RD, Kárason H (1999) Compositional heterogeneity in the bottom 1000 kilometers of Earth's mantle: Toward a hybrid convection model. Science 283:1885–1888

van der Hilst RD, Widiyantoro S, Engdahl ER (1997) Evidence for deep mantle circulation from global tomography. Nature 386:578–584

Weaver BL (1991) The origin of ocean island basalt end-member compositions: Trace element and isotopic constraints. Earth Planet Sci Lett 104:381–397

White WM, Patchett PJ, BenOthman D (1986) Hf isotope ratios of marine sediments and Mn nodules: evidence for a mantle source of Hf in seawater. Earth Planet Sci Lett 79:46–54

Williams Q, Garnero EJ (1996) Seismic evidence of partial melt at the base of Earth's mantle. Science 273:1528–1530

Wyllie PJ (1970) Ultramafic rocks and upper mantle. Mineral Soc Am Spec Paper 3:3–32

Zindler A, Hart SR (1986) Chemical geodynamics. Annu Rev Earth Planet Sci 14:493–571

Chapter 8

The Sources for Hotspot Volcanism in the South Pacific Ocean

C. W. Devey · K. M. Haase

8.1
Introduction

The South Pacific is characterized by a large number of active hotspots, many of which have been active for long periods of time (possibly as long as 120 Ma, Staudigel et al. 1991) producing extensive island and/or seamount chains (see Introduction, Fig. 0.1). The hotspots are presently located either beneath relatively old lithosphere (e.g., Society, Pitcairn, Australs, Marquesas, Juan Fernandez) or lie closer to the spreading axis (Foundation, Easter) (see Sect. 5.1). During the last fifteen years, the German-French initiative to study these hotspots has resulted in a vast amount of petrological and geochemical data being collected on fresh, mainly submarine volcanics.

Oceanic basalts potentially provide information about the chemical composition and evolution of the Earth's mantle (see Sect. 7.2). It has long been known that ocean-island basalt (OIB) compositions are by far more variable in incompatible trace element and radiogenic isotope compositions than basalts from mid-ocean ridges (MORB) (Gast et al. 1964; Sun 1980). The large amount of data has also shown that the global variation of oceanic basalt compositions can be explained in terms of mixing between four end-member compositions, which have been termed depleted MORB mantle (DMM), high μ ($= {}^{238}U/{}^{204}Pb$, HIMU), enriched mantle 1 (EM1), and enriched mantle 2 (EM2) (Zindler and Hart 1986). The volcanic chains of the South East Pacific erupt magmas that cover the whole globally known range of isotopic and trace element compositions. Some of the lavas from the South East Pacific volcanoes represent the extreme end-members of radiogenic isotope compositions on Earth: for example, Pitcairn hotspot lavas are classic EM1 samples (Woodhead and McCulloch 1989) and Mangaia yields lavas which define the isotopic nature of the HIMU end-member (White and Hofmann 1982; Zindler and Hart 1986; Woodhead 1996). However, most volcanoes are built from lavas spanning a range of compositions, indicating that mixing processes between different magma sources occur beneath the volcanoes. These mixing trends imply relatively small-scale mantle heterogeneity; a variety of models has been suggested to account for this heterogeneity.

McNutt and Fischer (1987) noted in the late 1980s that the sea floor in the South Pacific is unusually shallow (see Sect. 1.3). They attributed this to an anomalously hot mantle domain in this region. Whether the anomalously hot mantle was the cause or the result of the high concentration of hotspots in this region has since been, and is still being, intensely debated. Deep mantle seismic tomography (e.g., Gu et al. 2001; Tanaka 2002) shows that this region of the Pacific is underlain by an anomalously slow (and hence probably hot) region as far down as the core-mantle boundary. Experimental results (Davaille 1999) have led to the suggestion that the Superswell is associated with a whole-mantle hot upwelling and that the individual hotspots may be generated from the upper surface of this large-scale thermal upwelling. Work on the temperature anomaly in the mantle associated with the Society hotspot (Niu et al. 2002) suggests that this hotspot at least is underlain by a distinct thermal anomaly less than 500 km wide that covers at least the 410 km and 660 km seismic discontinuities. Geochemical work on volcanics from the Puka Puka Ridge (Janney et al. 2000), erupted away from hotspots but on the Superswell have been interpreted as showing that the upper mantle in the Superswell region is geochemically anomalous, and perhaps also evidence for large-scale mantle upwelling in the region.

This chapter deals with the nature and possible origin of magma sources for the South East Pacific intraplate volcanoes. The study of the magma sources gives insight into the chemical evolution of the Earth's mantle and the mixing processes occurring in the mantle. Furthermore, it allows us to place constraints on processes occurring both during magma genesis and magma ascent. Particularly interesting in this respect is the possibility of using the compositions of the hotspot magmas to constrain geophysical models of hotspot origin over the South Pacific Superswell (McNutt and Judge 1990).

8.2
The Hotspot Chains of the South East Pacific

Before we can begin to look for common features in the magmatic chemistry of the South East Pacific hotspots, we need to examine the compositional variations present in the individual chains. In doing this, we need to take into account both the stratigraphic variations within single volcanoes as well as the along-chain variations between the different edifices. A complete separation of these two effects is only possible with detailed knowledge of the stratigraphy of all the volcanoes being studied, knowledge which, for the most part, is not available. As a consequence, we are forced to make a hybrid study, attempting wherever possible to compare volcanoes at similar stratigraphic levels, always being aware, however, that non-systematic stratigraphy on some volcanoes could produce spurious results. The clear evidence from structural studies of ocean-island volcanoes that these edifices are subject to extensive erosion mainly through mass-wasting (Clouard and Bonneville, s. Sect. 6.2 and 6.3) long after volcanic activity has ceased suggests that comparing similar stratigraphic levels will not be an easy task.

The evolution of hotspot volcanoes worldwide is typically compared to that of the Hawaiian volcanoes, probably the best-studied examples (Macdonald and Katsura 1964; Clague and Dalrymple 1987). Most Hawaiian volcanoes are built, after an initial submarine stage, of a large lava shield comprising more than 90% of the volume of the volcano. Later stages erupt only minor volumes of material often after a period of quiescence and erosion of the main volcanic edifice. Accordingly, these late phases are called post-shield or post-erosional stages of volcanism.

The hotspot volcanoes of the South East Pacific region generally show periods of activity ranging from several hundred thousand to a few million years. For several volcanoes, a detailed stratigraphy has been determined that generally shows the volcanoes following a pattern of evolution from shield to post-shield lavas. These studies have shown that the magma sources as well as the processes of partial melting in the mantle have varied significantly with time. In most cases (e.g., Tahiti, Marquesas), the South

East Pacific volcanoes erupt more Si-undersaturated and more incompatible element-enriched lavas with time (Fig. 8.1a,b), indicating lower degrees of partial melting, possibly at greater depths during the post-shield rather than during the shield stage (Le Dez et al. 1996). The time-scales over which the compositional variation of the lavas occurs ranges from less than one million years such as in the case of Tahiti (Fig. 8.2a) to breaks between phases of volcanic activity of up to 6 to 10 million years on the Austral Islands of Aitutaki and Rurutu (Turner and Jarrard 1982; Duncan et al. 1994) (Fig. 8.1a,b and 8.2c). Interestingly, even the near-ridge hotspots such as Easter Island show a variation of the lava compositions where the early stages are tholeiitic while the later lavas become more enriched and slightly alkaline (Haase et al. 1997). The short time-scales of compositional variation in many hotspot volcanoes can be explained by the model of the lithospheric plate drifting across a mantle plume (Morgan 1972; Watson and McKenzie 1991), although the breaks of several million years in the Cook-Austral volcano chain appear to require other models (see Sect. 8.2.1).

We will now examine the chains individually and then compare them all to look for common features.

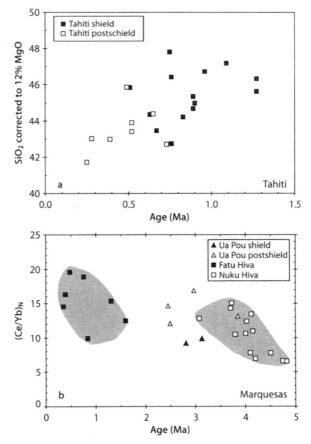

Fig. 8.1. a SiO$_2$ corrected to 12% MgO versus age for Tahiti; b (Ce/Yb)$_N$ versus age for three Marquesas volcanoes that show decreasing SiO$_2$ contents and increasing incompatible element enrichment with time. *Data sources:* Duncan et al. (1986, 1994); Woodhead (1992); Cheng et al. (1993); Desonie et al. (1993); Le Dez et al. (1996)

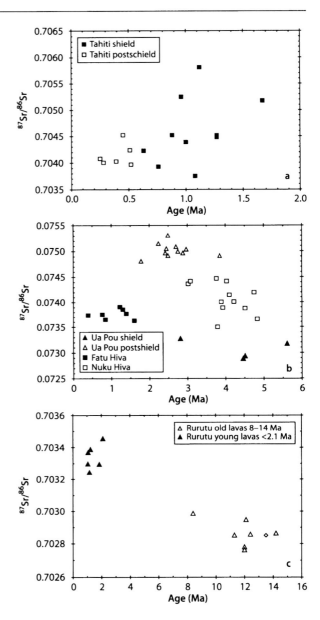

Fig. 8.2. Sr isotopes versus age for; **a** Tahiti; **b** the Marquesas; and **c** the Austral Islands. *Data sources:* McNutt and Fisher (1987); Chauvel et al. (1997)

8.2.1
Cook-Australs

This is perhaps one of the best-studied Polynesian chains in terms of along-chain variation but also probably one of the most complicated! Modern work on Austral geochemistry began in the 1980s at the time when the influence of subduction on mantle geochemistry in general and hotspot compositions in particular (e.g., Hofmann and White 1982; White and Hofmann 1982) was beginning to be appreciated. Dupuy et al. (1988) used trace element signatures in Austral magmas to demonstrate their complementary relationship to island-arc tholeiites, supporting the conclusions based on early isotopic work (Vidal et al. 1984) that magmas in this area of French Polynesia were derived from a "megalithic source containing variable amounts of former oceanic crust" (Dupuy et al. 1988). Vidal et al. (1984) showed that the Australs, isotopically speaking, contained some of the most extreme magmas on Earth. This and other studies in the 1980s and early 1990s based on limited, stratigraphically mostly uncontrolled sample sets (e.g., Palacz and Saunders 1986; Nakamura and Tatsumoto 1988) did not provide the solution to Austral petrogenesis but ended in many cases instead in what could be called the "Multi-component Mantle Mixing Model Maze" from which there was no dignified escape! The trace element data on Austral Island magmas available at the present day (Fig. 8.3) still support Dupuy's original conclusion – the Austral magmas show trace element signatures characteristic of subduction residues (depletions in Pb, K Cs, Rb, maximum enrichments in Nb, Ta and Th) as were found for many other ocean islands (e.g., Weaver et al. 1987; Devey et al. 2000). Recent isotopic modeling (Chauvel et al. 1992, 1995; Dostal et al. 1998) has allowed this picture to be refined markedly, showing that the Cook-Austral and the Pitcairn-Gambier volcanics are produced from a source with a strong HIMU (recycled oceanic crust) component mixed with variable amounts of subducted pelagic sediment. Both of these components have undergone chemical changes during the subduction process associated with dehydration and prograde metamorphism.

Geophysical (e.g., McNutt et al. 1997) and geochronological (e.g., Turner and Jarrard 1982) studies have provided evidence for polyphase activity of the individual Austral

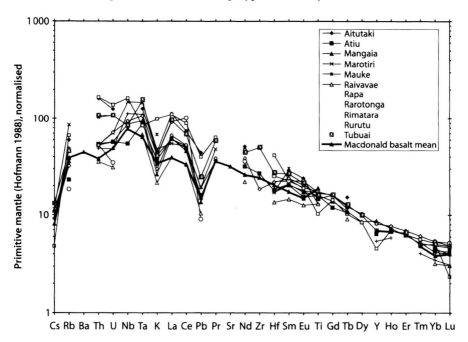

Fig. 8.3. Average trace element patterns for Austral Islands and Macdonald Seamount (data from the Georoc compilation, Max-Planck-Institut für Chemie Mainz, November 2002)

Fig. 8.4. A question of taste – two possible interpretations of age-distance information from the Cook-Austral Islands; a,b redrawn after Chauvel et al. (1997); c after Woodhead (1996)

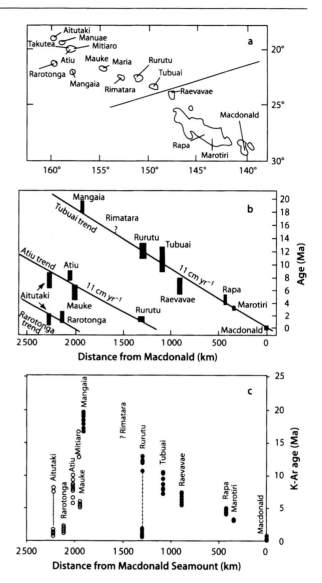

edifices covering time spans of several tens of Ma. The problem has been graphically illustrated by Chauvel et al. (1997, their Figure 1) in a figure reproduced here as Fig. 8.4a,b. They show that geochronological data from the Cook-Austral Islands may need up to three age-distance trends to explain them. Their observations of two distinct volcanic episodes with distinct geochemical signatures on the Island of Rurutu, for example, led these authors to propose the involvement of two plumes for the generation of the island. The recent finding of a 0.2 Ma old volcano southeast of Rurutu (Arago Seamount) with lavas of similar Pb isotopic composition supports the model of at least two and possibly three plumes being responsible for the formation of the Cook-Austral volcanic chain (Bonneville et al. 2002). Woodhead (1996), using the same geochronological data as Chauvel et al. (Fig. 8.4c), comes to a somewhat different conclusion: he suggests that, with the exception of Rurutu, all volcanoes between Mangaia and Macdonald define an age-progressive hotspot trace, while the volcanoes northwest of Mangaia (called by him the Southern Cooks) do not define any age-progressive trend at all. Compositional information shows that the older Rurutu lavas (13–10.8 Ma) are derived from a HIMU source, which Chauvel et al. (1997) linked to the source presently feeding the active Macdonald seamount, located at the southeast end of the Austral Chain (Figs. 8.4a–c). Macdonald, although not at present erupting lavas with the high $^{206}Pb/^{204}Pb$ isotope ratios characteristic of HIMU volcanoes, has been shown (Hémond et al. 1994) to be fed by a source with strong HIMU trace element characteristics. The isotopic characteristics of the putative source for the older Rurutu volcanics appear (Chauvel et al. 1997) to have varied in time. Thus the isotopic compositions of the islands on the "Tubuai Trend" (Fig. 8.4a) are not constant but appear to vary non-systematically along the chain. A similar conclusion has been reached by Woodhead (1996). The younger Rurutu volcanics (1.8–1.1 Ma) also appear to contain a component derived from the older, HIMU-like "Tubuai Trend" source mixed with a contribution from a more EM-like source, proposed by Chauvel et al. (1997) to be the plume responsible for the younger volcanism. Thus, these authors propose a variation of the model of plume interaction with a lithospheric source for Rurutu in which the older lavas resemble HIMU basalts formed in a mantle plume, while the younger lavas have less radiogenic Pb isotopes and were generated by mixing of material of a second plume with metasomatized lithosphere that had been enriched by carbonatitic melts from the first plume.

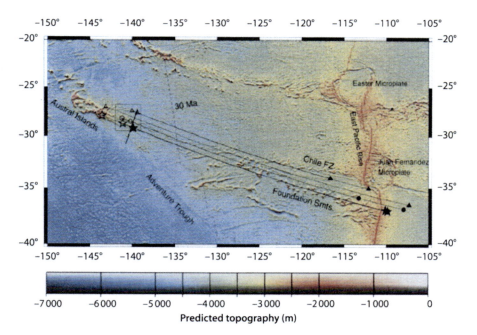

Fig. 8.5. Satellite altimetry of the South Pacific Ocean from Smith and Sandwell (1997) and the Austral Chain in the larger picture from McNutt et al. (1997)

Appealing to such multiple plumes aligned along the plate rotation vector does not suit all workers in the Australs. McNutt et al. (1997) state, based on their analysis of effective elastic thickness and age of the lithosphere and with reference to the multiple plume or rejuvenation hypotheses, *"Our data suggest that such minor adjustments to plume theory are inadequate"*. For further thoughts on this problem, the reader is referred to Clouard and Bonneville (Chap. 6). They use their data to bring the whole idea of plumes beneath the Australs, and indeed all of French Polynesia (McNutt 1998) into question. They propose instead that either stresses within the lithosphere associated with loading by older seamount chains (the Ngatemato Chain, in the case of the Australs) or lithospheric weaknesses inherited from the accretion process control where melts from the incipiently hot Pacific Superswell mantle reach the surface (Fig. 8.5). Although this is an apparently plausible model from the geophysical viewpoint and perhaps applicable in the specific case of the Australs, it fails to provide a causative reason for the first order observation that the present-day hotspot in all Polynesian chains, e.g., Pitcairn (Cheminée et al. 1989; Stoffers and SO-65 1990), Foundation (Cheminée et al. 1989; Devey et al. 1997; O'Connor et al. 1998), Society (Talandier and Okal 1984; Cheminée et al. 1989), Australs (Johnson 1970; Clague and Dalrymple 1987; Stoffers et al. 1989), Marquesas (Clague and Dalrymple 1987; Jordahl et al. 1995), and Easter (Haase and Devey 1996) lies at the southeast end of the chain.

8.2.2
Society Islands

The age-progressive nature of volcanism in the Society Islands was demonstrated by Duncan and McDougall (1976). They collected samples on all the main islands and showed that the data were compatible with an 11 cm yr^{-1} rate of plate movement over a stationary melting anomaly in the mantle (Fig. 8.6). The melting anomaly itself has been shown by geophysical methods to be associated with a warming of the mantle beneath Tahiti (Niu et al. 2002).

Geochemical studies of the Society Islands have been mostly confined to studies of individual edifices (see Sect. 5.2.3) – there has not been an attempt to make an along-chain comparison of the compositions of the volcanoes and their variations with time. Fig-

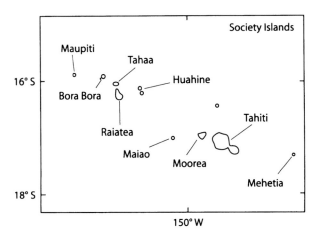

Fig. 8.6. The location of the islands in the Society Islands chain, modified after Duncan and McDougall (1976, p 203)

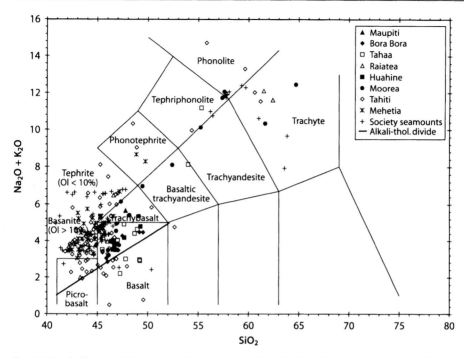

Fig. 8.7. Total-alkali vs. silica diagram for the Society Islands. Data from the Georoc compilation, Max-Planck-Institut für Chemie Mainz, November 2002

ure 8.7 gives an overview of the rock types found on the islands. Characteristic for the Society Islands are highly alkaline lavas ranging from basanites to phonolites. Notable in Fig. 8.7 is that the younger volcanoes (Tahiti, Mehetia and the Society Seamounts) seem to have on average much more alkalic compositions, making up most of the basanites and tephrites for example (although Tahiti also has some less undersaturated magmas). Duncan et al. (1994) have shown for the island of Tahiti that there is a clear change of composition of magmas with time, from more silica-saturated shield-building lavas to much more alkaline post-shield magmas. There is no evidence of this on the youngest Society island, Mehetia, since the lavas recovered there (Binard et al. 1993; Cheng et al. 1993; Hémond et al. 1994) all fall in the strongly alkaline field typical of the seamount magmas. By analogy with the evolution of Hawaiian volcanoes (e.g., Clague and Dalrymple 1987), it seems quite possible that Mehetia and the Society Seamounts are all in the more alkaline preshield phase; the main Tahitian volcano, Tahiti Nui, has already passed this stage and the subsequent later shield phase and has now entered (and possibly even finished) the post-shield phase. The less-undersaturated compositions seen in samples from the other Society islands suggest that erosion, combined with subsidence as the volcanoes are carried away from the active hotspot, has led to only the voluminous shield-building volcanic phases being exposed at the present-day surface.

A Sr-Nd isotope plot (Fig. 8.8) does not show any clear indication of systematic change in isotopic signature of the magmas along the chain that might in any way mimic the stratigraphic changes in isotope ratios seen by Duncan et al. (1994) on Tahiti. If we try to quantify the degree of alkalinity of the samples by calculating the dif-

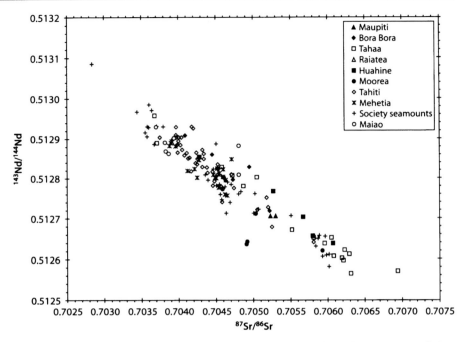

Fig. 8.8. Sr-Nd isotope diagram for the Society volcanoes. Data taken from the Georoc compilation, Max-Planck-Institut für Chemie Mainz, November 2002

ference between their total alkali content and the line defining the tholeiite/alkali basalt divide for Hawaii (from Macdonald and Katsura 1964) (equivalent to the vertical distance between the sample and the alkalic/tholiitic divide in Fig. 8.7), we find that there is, however, a correlation between alkalinity and isotopic ratio (Fig. 8.9). This plot is somewhat limited by the fact that major and trace elements and isotope ratios have seldom been determined on the same subaerial samples; the available data define, however, an unmistakable trend. This trend can be interpreted in terms of more alkaline magmas being derived from a source with lower time-integrated incompatible element enrichment. This is, on a multi-volcano scale, similar to the observation made by Duncan et al. (1994) for Tahiti itself. It is also in accord with recent observations on the initial sea-floor stages of Society volcanism, which imply that the source initially tapped by magmatism has the highest time-integrated incompatible element enrichment (Devey et al. 2001; Devey et al. 2003).

The whole Society Chain has long been a typical example for volcanoes with EM2 (Zindler and Hart 1986) source compositions. Numerous authors have presented evidence (both isotopic and trace-elemental) to show that this source is influenced by subducted continental material (e.g., White and Hofmann 1982; Devey 1990; Duncan et al. 1994; Hémond et al. 1994). The trace element patterns presented in Fig. 8.10 show that in trace element terms, the islands are very similar. Relative to the Austral Chain represented by the average Macdonald basalt, the Society basic magmas have slightly steeper patterns overall and do not show a maximum enrichment of Nb and Ta. There is also some indication that the pronounced negative K-anomaly seen in the Austral magmas is not as strongly developed in the Society's samples.

Chapter 8 · The Sources for Hotspot Volcanism in the South Pacific Ocean

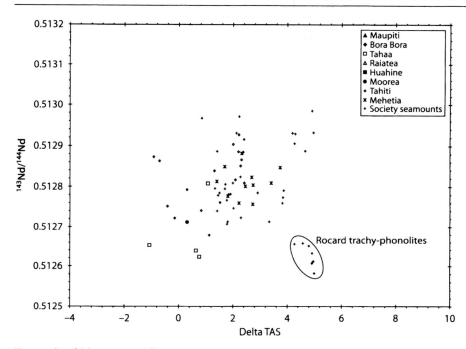

Fig. 8.9. Plot of delta-TAS (the difference in total alkali content between the sample and the line defining the alkali/tholeiitic divide on Hawaii (Macdonald and Katsura 1964) defined as total alkalies = 0.3509 SiO$_2$ – 13.316 vs. Nd isotopic ratio. When the highly fractionated Rocard trachyphonolites are excluded, it appears that the most alkaline magmas (highest delta-TAS) are derived from a source with the highest Nd isotopic ratios

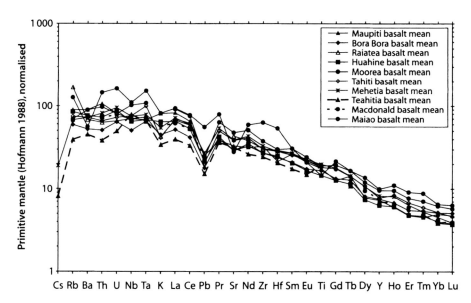

Fig. 8.10. Incompatible trace element patterns for the average basaltic magmas from the Society Islands, the main submarine basaltic edifice Teahitia, and Macdonald (Austral Chain, see Fig. 8.3). Averages compiled from data from the Georoc database, Max-Planck-Institut für Chemie, Mainz, November 2002

8.2.3
Pitcairn-Gambier Chain

The Pitcairn-Gambier Chain was shown to be the product of age-progressive volcanism in 1974 (Duncan et al. 1974). The chain stretches from the young volcanic island of Pitcairn to the atolls of Mururoa and Fangataufa and may even reach as far as the Line Islands (Garcia et al. 1993). Dating of samples from submarine edifices southeast of Pitcairn Island itself (Guillou et al. 1997) has shown them to be the youngest volcanoes in the chain. Most geochemical work has concentrated on Pitcairn Island and the adjacent seamounts (e.g., Woodhead and McCulloch 1989; Woodhead and Devey 1993; Woodhead et al. 1993; Eiler et al. 1995; Eisele et al. 2002), although some work on the older islands in the chain has been published (Dupuy et al. 1993). The work on Pitcairn Island (Woodhead and McCulloch 1989) has shown a volcanic stratigraphy in which the magmas become more MORB-like in their isotopic values with time. The Pitcairn Seamounts cover a similar isotopic range and also appear to show some compositional variations with the height of the individual edifices. The subaerial shield and the submarine lavas show the most extreme EM1 compositions with very low Nd and Pb isotope ratios. The EM1 mantle signature has been attributed to the presence of subducted pelagic sediments (Weaver 1991; Chauvel et al. 1992; Eisele et al. 2002), delaminated metasomatized subcontinental lithospheric mantle (McKenzie and O'Nions 1983; Mahoney et al. 1991), and recycled oceanic plateaus (Gasperini et al. 2000).

Dupuy et al. (1993) presented both trace element and isotopic data along the chain. They found no systematic changes in degrees of trace element enrichment with dis-

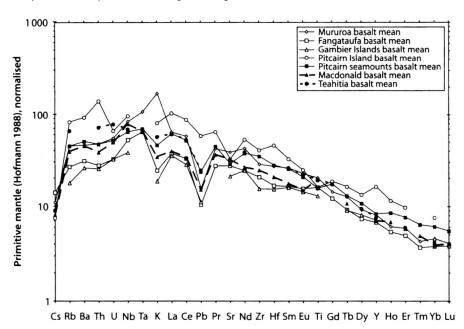

Fig. 8.11. Trace element patterns of basalt averages from the Pitcairn-Gambier Chain (averages calculated from data in the Georoc database, Max-Planck-Institut für Chemie, Mainz). Also shown are average values for Macdonald (Australs) and Teahitia (Society) Seamounts (see Fig. 8.10)

tance from Pitcairn, a feature confirmed by Fig. 8.11. What they did find, however, is that various isotopic parameters, especially ^{143}Nd/^{144}Nd vs. ^{206}Pb/^{204}Pb, varied systematically with age of the lithosphere (Fig. 8.12). They attributed this feature to contamination of the lithosphere below the Pitcairn-Gambier Chain by material from the Easter plume at the time of creation of this part of the Pacific Plate. The absence of a Pitcairn geochemical signature in the older edifices was ascribed by Dupuy et al. (1993) to the fact that samples from these extinct volcanoes probably represent the last eruptive products formed, a stage at which observations from other Pacific hotspots have shown the magma source to be most strongly influenced by melting in the lithosphere. The work on the geochemistry of the up-to-90 Ma Line Island basalts (Garcia et al. 1993) has also shown them to have a clear Easter/Sala y Gómez chemical signature, implying that the Easter hotspot may have been active for at least this time. Easter Island lies along a flow-line to Pitcairn, and as such this hypothesis certainly appears viable.

8.2.4
Marquesas Islands

The Marquesas Islands (Fig. 8.13) form one of the most enigmatic island chains in the Pacific – it shows signs of age-progressive volcanism (Duncan and McDougall 1974) but the plate motion vector defined (towards the NW at 74 ±6 mm yr^{-1}) is 30° more northerly and 30 mm yr^{-1} slower than that defined by other Polynesian chains (Desonie et al. 1993). McNutt et al. (1989) explained this in terms of the influence of the Marquesas Fracture Zone. They postulated that this fracture zone and associated spreading axis fabric in the crust

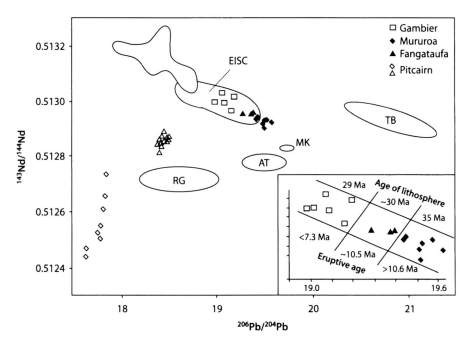

Fig. 8.12. ^{143}Nd/^{144}Nd vs. ^{206}Pb/^{204}Pb isotope diagram for the Pitcairn-Gambier Chain redrawn after Dupuy et al. (1993, their Fig. 7)

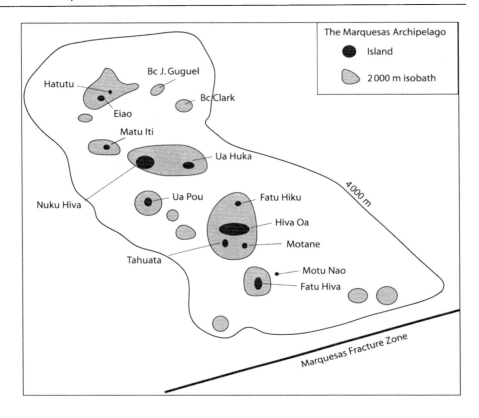

Fig. 8.13. The islands of the Marquesas Chain (redrawn after Desonie et al. 1993)

controlled the location of the eruptive centers for a very weak Marquesas plume. Desonie et al. (1993), on the other hand, explain the aberrant Marquesas plate velocity information in terms of a displacement of the Marquesas plume by mantle flow.

The Marquesas Chain is also somewhat unusual in having been studied by several authors in terms of the temporal evolution of volcanism along the chain (e.g., Vidal et al. 1984; Dupuy et al. 1987; Woodhead 1992; Desonie et al. 1993). The individual volcanoes have also been the subject of detailed study (e.g., Liotard and Barsczus 1983a,b, 1984), in some cases going as far as borehole studies (Caroff et al. 1995, 1999). Volcanic activity on many of the islands has been found to span a considerable time range, in particular on Ua Pou (Duncan et al. 1986). All volcanoes appear to show a temporal evolution of magma compositions from a tholeiitic shield stage to a more alkaline late stage. Figure 8.14 shows, however, that the degree of undersaturation (in terms of the delta-TAS quantity introduced earlier) and the degree of incompatible-element enrichment (using La/Yb as a proxy) vary from island to island. With the exception of the observation that the alkaline stages generally have higher trace element contents than the tholeiitic stages, there is little sign of any clear systematics in the trace element enrichment of the different volcano stages on a primitive-mantle-normalized incompatible-element diagram (Fig. 8.15). Relative to the other Polynesian hotspots discussed so far, the Marquesan tholeiites show a similar REE pattern slope to the other hotspots; the highly

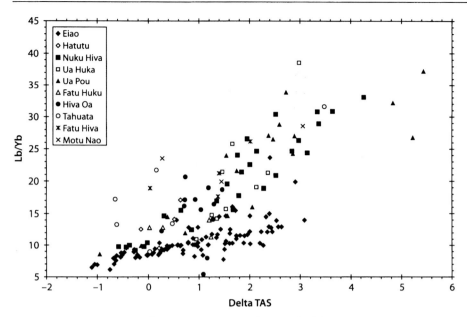

Fig. 8.14. Delta-TAS vs. La/Yb for the Marquesas Islands. Note the different slopes of the trends for the different islands; the group Hiva Oa + Tahuata + Fatu Huku (which probably, together with Motane (no data), all form parts of the same volcano) shows no correlation

incompatible element (Nb-Rb) slope is generally steeper in the Marquesan tholeiite than in the other hotspots, perhaps indicating a more depleted source.

In contrast to other island chains where the magma source appears to change from plume-dominated to lithosphere-dominated with time (e.g., Pitcairn-Gambiers; see above), the Marquesas volcanoes appear to become isotopically less lithosphere-like progressively during their evolution. This feature has taxed the imagination of several workers on the islands. Woodhead (1992) showed that the extent of the isotopic change between shield and later alkaline stage was related to the time gap between these two periods of volcanism, although the significance of this relationship has been challenged by Desonie et al. (1993). The shield building stage of volcanism on all Marquesas Islands was shown by Woodhead to be isotopically similar, with $^{87}Sr/^{86}Sr$ 0.7030–0.7040 and $^{143}Nd/^{144}Nd$ of 0.5129–0.5130. This depleted nature is in agreement with the observation from the trace element patterns that the Marquesan source appears to be more strongly depleted in highly incompatible elements than other Polynesian hotspots discussed so far (Fig. 8.15). The trace-element ratios of the volcanic shields remain relatively constant along-chain (Woodhead 1992) (Fig. 8.15). The late stage volcanics are somewhat more heterogeneous in nature; Woodhead (1992) suggests that this may be linked to variable residence times and magma chamber sizes in the lithosphere. Desonie et al. (1993) favor an explanation for the similar time evolution of all Marquesan volcanoes, which involves mixing of plume and asthenospheric material as a result of a weak Marquesas plume diapir traversing a horizontally flowing asthenosphere. With reference to the experimental work of Richards and Griffiths (1989) and Griffiths and Campbell (1991), they show that such a diapir could develop a core dominated by asthenospheric material wrapped in a sheath of predominantly

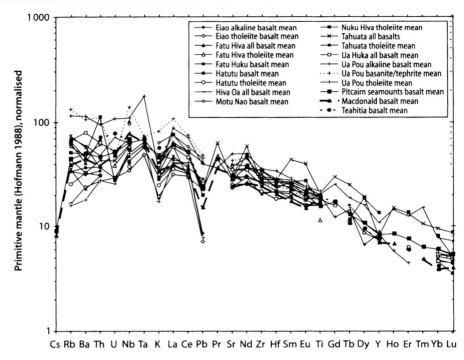

Fig. 8.15. Trace element patterns for the Marquesas Islands together with averages from the Society, Austral and Pitcairn hotspots. Averages calculated from data in the Georoc database, Max-Planck-Institut für Chemie, Mainz

plume-derived material. Desonie et al. (1993) also note an apparent zonation of hotspot products in terms of isotopic and trace element ratios perpendicular to the plume trace (i.e., from southwest to northeast). A similar feature has been noted on the active submarine Society hotspot (Devey 1990). Neither of these authors provides a clear explanation for this observation.

All workers on the along-chain variations in composition of Marquesas volcanics (Dupuy et al. 1987; Woodhead 1992; Desonie et al. 1993) have noted that their isotopic compositions lie between DMM, EM2 and HIMU (using the nomenclature of Zindler and Hart 1986). Desonie et al. (1993) have shown that these compositions can be modeled both in terms of isotopes and trace elements by simultaneous variation of source enrichment and degree of melting (Fig. 8.16).

8.2.5
Juan Fernandez Chain

The Juan Fernandez Chain is an east-west trending chain on the Nazca plate, east of the East Pacific Rise (see Introduction, Fig. 0.1). It consists of two islands (Robinson Crusoe and Alexander Selkirk) associated with several seamounts both east and west of the islands. Dating of the islands (Baker et al. 1987) has yielded results consistent with age-progressive volcanism: the more easterly island Robinson Crusoe (also known as Mas a Tierra, "landward" in the sense of nearer to Chile) appears to have been active

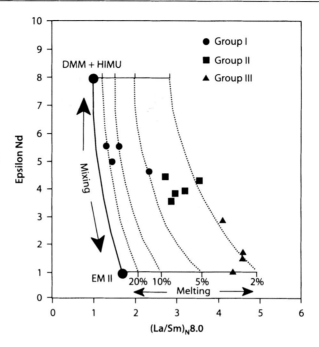

Fig. 8.16. Modeling of the isotopic and trace element variations in a suite of Marquesan basalts redrawn after Desonie et al. (1993). The observed magmatic values can be explained by covariation in the degree of trace element depletion of the source (vertical variation) and degree of melting (horizontal variation). The sample groups were defined on major and trace element grounds. Group I contains relatively trace element depleted, unradiogenic tholeiites, Group III, at the other extreme, are trace element enriched, radiogenic and alkalic lavas. DMM = depleted MORB mantle, HIMU = high μ (= $^{238}U/^{204}Pb$), EM I = enriched mantle 1, EM II = enriched mantle 2

around 4 Ma whilst Alexander Selkirk (also known as Mas Afuera, "further away" in the sense of further from Chile) has been dated at ca. 1 Ma. Baker et al. (1987) describe the volcanic stratigraphy of Robinson Crusoe as consisting of a tholeiitic shield stage followed by low-volume alkali basalt and basanites; the difference in age between the tholeiitic and alkaline groups is, however, not known. Alexander Selkirk is composed apparently solely of tholeiites and some highly differentiated trachytes. Both Baker et al. (1987) and more recently Farley et al. (1993) have shown that the Juan Fernandez Islands are characterized by very homogeneous isotopic signatures, consistent with derivation from a source with little geochemical variation. Gerlach et al. (1986) showed that this source has a composition very similar to that found as a component in many other hotspot magmas around the globe. Farley et al. (1993) showed, however, using helium isotopes that there are differences in the sources between the shield and post-shield (alkaline) phases of Robinson Crusoe volcanism. They found that the shield stage is derived from a more Loihi-like (plume) source with $^3He/^4He$ = 14.5–18 R_A, whereas the source for the alkaline magmatism appears more MORB-like (11.2–13.6 R_A). In this respect, the evolution appears similar to the majority of hotspot volcanoes in the Pacific (with the clear exception of the Marquesas as noted above), i.e., from plume-dominated to more strongly lithosphere-influenced with time. Alexander Selkirk shows even lower $^3He/^4He$ ratios of 8.3 R_A, which is surprising, considering the similarity between the volcanoes in terms of other trace elements and isotopes.

Work on the young submarine volcanoes "Friday" and "Domingo" to the west of Alexander Selkirk (Devey et al. 2000) (Fig. 8.17) has confirmed the general isotopic homogeneity of Juan Fernandez magmas and allowed, due to the extreme freshness of the submarine glasses, details of the plume-MORB interaction to be determined. All magmas from these volcanoes are extremely alkaline. Devey et al. (2000) show that subtle variations of the

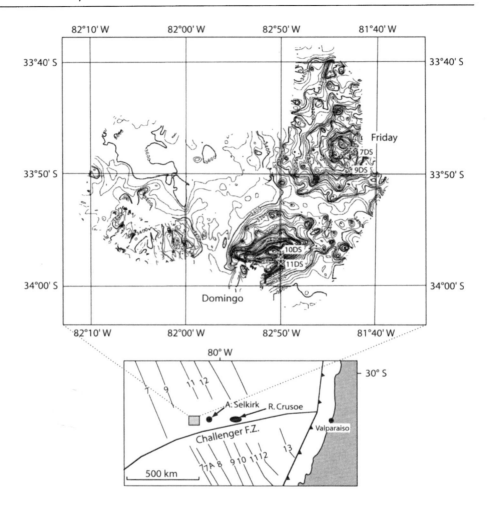

Fig. 8.17. Location map for the Juan Fernandez Chain, from Devey et al. (2000)

isotopic and trace element signatures of the seamount magmas can best be explained as the result of progressive metasomatic reactions between carbonated plume magmas and the lithospheric mantle through which they pass. The progressive "armoring" of the conduits through which the magmas pass leads to a gradual decrease in the influence of the lithospheric component on the magma compositions during growth of the submarine volcanoes (Fig. 8.18). How this can be linked to the helium isotope observations on Robinson Crusoe outlined above is somewhat unclear – it is possible that the Juan Fernandez volcanoes undergo a multistage evolution in source characteristics from *(a)* early submarine, lithosphere-influenced highly alkaline magmas to *(b)* more plume-influenced submarine alkaline magmas to *(c)* plume-dominated shield stage tholeiites and finally to *(d)* more lithosphere-influenced late alkaline magmas. The plume producing the magmas appears to be very similar to that which has been proposed as a common component in many hotspot systems (PREvalent MAntle or PREMA) (Zindler and Hart 1986).

Chapter 8 · **The Sources for Hotspot Volcanism in the South Pacific Ocean**   271

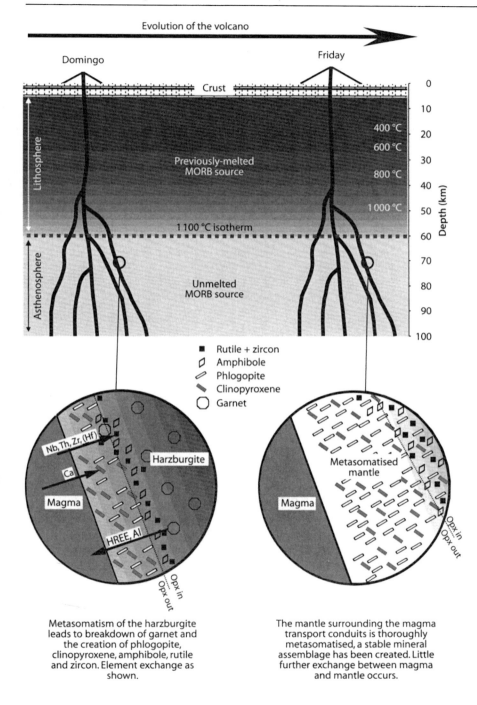

Fig. 8.18. Cartoon of the proposed reactions between plume-derived melts and the harzburgitic lithospheric mantle beneath the Juan Fernandez Seamounts, after Devey et al. (2000)

8.2.6
Foundation Seamounts

The Foundation Seamounts were only discovered in 1992 (Mammerickx 1992) but have since been the subject of several studies (Devey et al. 1997; Hekinian et al. 1997, 1999; O'Connor et al. 1998, 2001; Maia et al. 2000, 2001). The chain is clearly the product of age-progressive volcanism with individual volcanic edifices becoming progressively younger towards the southeast from 22–2 Ma. In the last ca. 6 Ma, the age progression has become more complicated due to enhanced plume-ridge interaction as the spreading axis migrates towards the hotspot, leading to the formation of coeval volcanic elongated ridges (O'Connor et al. 2001; see Sect. 11.3). Studies of the relative age of hotspot volcanism and underlying lithosphere along the chain show a complex history of interplay between the hotspot and the spreading axis (O'Connor et al. 1998; Maia et al. 2000). However, in our opinion, this interplay is unlikely to be as complex as that envisaged by Sleep (2002), who would like the Foundation plume to have also created the Austral Chain. He proposes that the seamount-free gap between the Austral and Foundation Chains (see Sect. 5.1, Fig. 5.1) is the result of a channeling of plume material to the spreading axis when the axis was situated close to the plume. This idea doesn't appear to be reconcilable with present-day observations: the plume and spreading axis are close to one another, and yet large volcanic edifices are being constructed on the Pacific Plate (Devey et al. 1997; Maia et al. 2000; O'Connor et al. 2001).

Very little geochemical work on the Foundation Chain has been published up to the present; the data that exist in the literature are mainly concentrated on the area on and close to the Pacific-Antarctic spreading axis (Hekinian et al. 1997, 1999). Hekinian et al. (1999) show a broad variation in magmatic petrogenesis along the chain (Fig. 8.19), from magmas strongly affected by plume-ridge interaction in the west (2 000–1 300 km from the present spreading axis) via relatively pure plume-like magmas in the central portion (1 300–300 km from the present axis) to plume-ridge interaction in the easternmost part of the chain (300–0 km from the present axis). This type of interaction sequence is also consistent with geophysical models for the lithospheric flexure associated with the volcanic loading (Maia et al. 2000). The sudden change to apparently wholly plume-dominated compositions at 1 300 km from the present axis is related to an eastward jump of the spreading axis at that time; since then, the plume and spread-

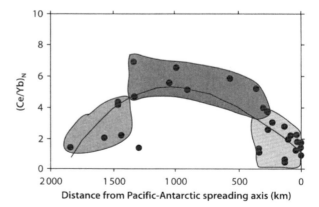

Fig. 8.19. Ce/Yb variation along the Foundation Chain (expressed as distance to the west of the present Pacific-Antarctic Ridge (PAR)), redrawn after Hekinian et al. (1999)

Chapter 8 · The Sources for Hotspot Volcanism in the South Pacific Ocean 273

ing axis have been gradually approaching one another yet again. Hémond and Devey (1996) have shown that this evolution is mirrored by the isotopic ratios of the magmas and that the pure plume component has a strong HIMU signature.

8.2.7
Easter/Sala y Gómez-Nazca Chain

The Easter/Sala y Gómez Seamount chain has received a lot of attention in the past ten years because of one American and one joint German-French cruise. One of the major issues was to resolve the question whether the seamount chain formed above a hot line (Bonatti and Harrison 1976; Bonatti et al. 1977) or a hotspot or mantle plume (Wilson 1973). Young (<0.5 Ma) subaerial volcanism was long known to exist on Easter Island (Clark and Dymond 1977; Kaneoka and Katsui 1985), but the recent cruises found much more abundant young submarine volcanism (Hagen et al. 1990; Stoffers et al. 1994; Rappaport et al. 1997; Pan and Batiza 1998). While most authors agree now that the Easter/Sala y Gómez Chain formed above a mantle plume, the distribution of young lavas and new age dates have led to a discussion of the location of this mantle plume. On the one hand, some authors argue that the plume must be in the area where the most enriched young (<1.5 Ma old) lavas occur, i.e., close to Sala y Gómez (Hanan and Schilling 1989; O'Connor et al. 1995), while another group of authors suggest that the largest volume of young volcanism is close to Easter Island, reflecting the presence of the plume beneath this region (Hagen et al. 1990; Haase et al. 1996). A third possible location west of the Easter Microplate East Rift was suggested to account for the bathymetry of the Tuamotu Islands (Okal and Cazenave 1985; Clouard and Bonneville 2001).

The lavas of the Easter Seamount chain consist of tholeiitic to mildly alkaline rocks, ranging from basalts to rhyolites and trachytes on Easter Island. In terms of highly incompatible elements, both depleted and enriched lavas occur. The depleted lavas are mainly known by the well-studied volcanic fields west of Easter Island (Fretzdorff et al. 1996; Haase and Devey 1996), i.e., closest to the East Rift spreading center of the Easter Microplate. These depleted tholeiites are comparable to depleted lavas from the Galapagos Hotspot (White et al. 1993). Most lavas found in the Easter Hotspot region and along the Easter Seamount chain are slightly enriched in their incompatible-elements with a K/Ti ratio of about 0.2 to 0.3 (Fig. 8.20). More alkaline and enriched lavas are also found and may represent the final stages of the volcanism in each region (Baker et al. 1974; Haase et al. 1997). The most enriched lavas of the Easter Seamount chain generally show the highest Sr and $^{206}Pb/^{204}Pb$ isotope ratios ranging to about 0.7032 and 20.1, respectively, but one older lava has been recovered with a higher $^{206}Pb/^{204}Pb$ of 20.4 (Kingsley and Schilling 1998). In terms of incompatible element composition, the Easter Seamount chain lavas resemble HIMU type magmas with large positive peaks of Nb and Ta and relatively low concentrations of Rb, Ba, and U. Isotopically, the plume source may either represent a mixture of HIMU and EM1 types of mantle (Cheng et al. 1999) or a mixture between HIMU and FOZO/C mantle types (Dixon et al. 2002). While the HIMU type source may consist largely of recycled MORB, the FOZO type source of the Easter Seamount chain plume may also contain relatively primitive but metasomatized mantle material that leads to high $^{3}He/^{4}He$ and volatile contents (Dixon et al. 2002; Kingsley et al. 2002; Simons et al. 2002). For a discussion of these different mantle components, see the next section.

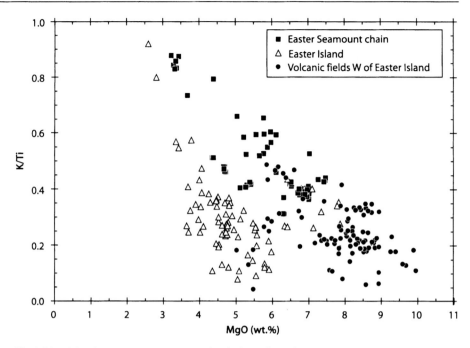

Fig. 8.20. K/Ti ratios versus MgO contents for the lavas from the Easter Seamount chain, Easter Island and the submarine volcanic fields west of Easter Island. *Data sources:* Baker et al. (1974); Fretzdorff et al. (1996); Haase and Devey (1996); Haase et al. (1997); Pan and Batiza (1998); Cheng et al. (1999)

8.3
Discussion: Petrogenesis of South East Pacific Hotspots

8.3.1
Location of Magma Sources: Plume, Asthenosphere or Lithosphere?

In terms of the magma source evolution, the shield stage lavas of South East Pacific hotspots are in several cases more radiogenic in Sr and Pb isotope ratios than the post-shield magmas, while the $^{143}Nd/^{144}Nd$ are lower. Examples for this case are the Society hotspot volcanoes and Pitcairn. This type of source evolution has been called the "Hawaiian type" by Woodhead (1992). Other volcanoes evolve from a shield stage with relatively high $^{143}Nd/^{144}Nd$ and low Sr and Pb isotope ratios toward a post-shield stage with more radiogenic Sr and Pb isotopic lava compositions. Woodhead (1992) referred to this type of evolution as the "Marquesan type". In many cases, the variation between enriched and depleted sources has been interpreted as being due to mixing between material from a deep mantle plume and the lithosphere. For example, the variation of the Marquesas magma composition has been suggested to reflect mixing between an EM2 deep mantle plume and DMM-HIMU lithospheric mantle sources (e.g., Duncan et al. 1986; Woodhead 1992; Desonie et al. 1993; White and Duncan 1996). Other authors have proposed that it is the lithosphere that carries the DMM composition while the EM and HIMU sources reside in the plume (Caroff et al. 1995; Le Dez et al. 1996). According to Woodhead (1992), the type of source variation with time depends on the

Chapter 8 · The Sources for Hotspot Volcanism in the South Pacific Ocean   275

strength of the plume. Weak plumes cannot penetrate or significantly melt the lithosphere, and thus the main magmatic shield stage is formed by magmas from mixtures of enriched plume and depleted lithospheric sources, while the late stage basalts represent the pure plume material (Marquesan type). In contrast, the shield stage of strong plumes (Hawaiian type) is formed by melts from the plume only while the late-stage magmas are generated by mixing between plume and lithospheric mantle. While these models imply mixing at deep levels in the mantle, Caroff et al. (1995) suggest that the DMM material is assimilated in the oceanic crust during the ascent of the magmas.

Several authors have presented the view that the different magma sources, even the depleted DMM source, all reside in the mantle plume and that the incompatible element-depleted material is entrained in the mantle during the ascent of the plume. Such a model has been put forward for Tahiti (Duncan et al. 1994; White and Duncan 1996). Another model suggests that the deep mantle plume entrains material from the lower mantle with an intermediate isotopic composition between EM, HIMU, and MORB (called FOZO: FOcal ZOne, the region in multi-isotopic space toward which many isotopic arrays for oceanic magmatism seem to point) and that this entrained material surrounds the plume and forms the source of the post-shield and post-erosional lavas (Hauri et al. 1994). A series of authors (Zindler and Hart 1986; Farley and Craig 1992; Farley et al. 1992; Hanan and Graham 1996) have suggested mantle reservoirs with comparable compositions to FOZO that are called PREMA ("PREvalent MAntle"), PHEM ("Primitive HElium Mantle"), and C ("Common"). The FOZO-type mantle is the source of many hotspot magmas that form close to spreading axes. For example, the Easter and Galapagos Hotspots erupt lavas that have FOZO-composition reflecting generation from a mantle with relatively low $^{87}Sr/^{86}Sr$ but high $^{143}Nd/^{144}Nd$ and Pb isotope ratios as well as high $^3He/^4He$ (Hauri et al. 1994; Hanan and Graham 1996). It is not clear whether this is due to the fact that increased mixing of the plume source with upper mantle material is taking place close to the spreading axis or whether increased degrees of melting lead to a homogenization of the melts from various plume components producing magmas with an average FOZO composition.

8.3.2
Superswell – How Geochemically Different is It?

It is clear from maps of present-day distributions of hotspots (e.g., Crough and Jurdy 1980) that the South Pacific Superswell, together perhaps with the Northeast Atlantic, contains an unusually high concentration of hotspots. McNutt (1998) has calculated that on the Superswell, 14% of the present-day hotspots occur in an area representing <5% of the global surface. It is also clear from the geophysical work on the Superswell outlined previously that the mantle there is anomalously warm. Is this somehow reflected in the compositions of the magmas? Are they different from those of other hotspots? McNutt (1998) has approached this problem to some extent, but we would like to try and examine it further here.

If the excess Superswell temperature comes from enhanced levels of radioactive decay in the mantle, we might expect the Superswell hotspots to show higher concentrations of the heat-producing elements than hotspots elsewhere. Table 8.1 and Fig. 8.21 show average magma compositions taken in a narrow (4–6 wt.%) MgO range from a selection of Superswell and Atlantic hotspots. The MgO filter was used to minimize

Table 8.1a. Average compositions of magmas lying in the range MgO = 4–6% from various Superswell and Atlantic hotspots: major and minor oxides (Wt%) (data from the compilation Georoc, Max-Planck-Institut für Chemie, Mainz)

Hotspot	SiO₂	TiO₂	Al₂O₃	Fe₂O₃	FeOT	FeO	CaO	MgO	MnO	K₂O	Na₂O	P₂O₅	LOI
Pitcairn-Gambier 4–6% MgO	46.28	3.58	15.42	8.90	11.37	8.46	9.24	5.03	0.16	1.40	3.03	0.66	3.13
Society Islands 4–6% MgO	45.94	3.70	15.79	4.70	11.36	7.17	9.51	4.91	0.14	2.02	3.71	0.88	1.98
Marquesas 4–6% MgO	46.96	3.78	15.10	5.82	11.44	7.17	9.18	5.11	0.17	1.45	3.32	0.60	1.71
Austral-Cook 4–6% MgO	44.16	3.22	15.20	8.96	12.51	8.50	10.73	5.25	0.20	1.25	3.50	0.66	2.10
Tristan 4–6% MgO	46.61	3.36	16.84	3.60	10.52	7.10	9.34	4.77	0.18	2.77	4.02	0.79	0.86
Canaries 4–6% MgO	45.98	3.48	15.89	5.07	11.39	5.52	9.59	5.07	0.19	1.77	3.99	0.88	1.05

Table 8.1b. Average compositions of magmas lying in the range MgO = 4–6% from various Superswell and Atlantic hotspots: Rare Earths (ppm) (data from the compilation Georoc, Max-Planck-Institut für Chemie, Mainz)

Hotspot	La	Ce	Pr	Nd	Sm	Eu	Gd	Tb	Dy	Ho	Er	Tm	Yb	Lu
Pitcairn-Gambier 4–6% MgO	34.5	78.2	10.7	42.0	8.96	2.98	8.33	1.12	6.15	0.91	2.59	0.28	1.97	0.27
Society Islands 4–6% MgO	53.3	109.9	10.8	69.6	11.62	3.85	9.47	1.64	7.45	1.24	3.12	0.41	2.47	0.37
Marquesas 4–6% MgO	36.6	82.6	15.1	49.7	11.69	3.51	13.01	1.66	8.17	2.77	3.78	1.42	2.83	0.41
Austral-Cook 4–6% MgO	53.6	113.9	11.5	53.6	9.75	3.11	8.66	1.24	6.85	1.13	3.06	0.38	2.34	0.34
Tristan 4–6% MgO	108.4	140.5	12.2	73.8	9.13	3.33	7.77	1.13	5.70	0.97	2.80	–	2.00	0.27
Canaries 4–6% MgO	65.8	133.4	16.0	66.6	12.79	4.02	11.18	1.55	7.51	1.32	3.33	0.41	2.75	0.43

Table 8.1c. Average compositions of magmas lying in the range MgO = 4–6% from various Superswell and Atlantic hotspots: trace elements (ppm) (data from the compilation Georoc, Max-Planck-Institut für Chemie, Mainz)

Hotspot	Sc	V	Cr	Co	Ni	Cu	Zn	Ga	Rb	Sr	Y	Zr	Nb	Cs	Ba	Hf	Ta	Pb	Th	U
Pitcairn-Gambier 4–6% MgO	26	242	65	42	50	26	121	–	29	664	37	284	51	0.45	300	6.5	3.6	3.8	3.6	1.0
Society Islands 4–6% MgO	137	289	60	37	62	65	145	–	56	780	38	351	57	1.00	577	8.1	3.7	2.8	6.2	1.6
Marquesas 4–6% MgO	20	290	68	38	56	49	131	25	34	696	40	351	48	–	358	7.8	2.6	7.9	4.5	1.1
Austral-Cook 4–6% MgO	18	233	65	41	55	67	124	25	30	827	31	278	69	0.43	388	6.4	4.6	3.9	6.5	1.7
Tristan 4–6% MgO	14	242	34	32	27	31	102	27	104	1127	35	314	90	0.77	782	7.5	6.2	12.1	7.3	1.6
Canaries 4–6% MgO	22	280	65	33	44	68	112	22	34	1031	36	356	83	0.46	513	7.6	5.6	4.5	7.2	2.1

Table 8.1d. Average compositions of magmas lying in the range MgO = 4–6% from various Superswell and Atlantic hotspots: isotopes (data from the compilation Georoc, Max-Planck-Institut für Chemie, Mainz)

Hotspot	$^{87}Sr/^{86}Sr$	$^{143}Nd/^{144}Nd$	$^{206}Pb/^{204}Pb$	$^{207}Pb/^{204}Pb$	$^{208}Pb/^{204}Pb$
Pitcairn-Gambier 4–6% MgO average	0.703639	0.512827	18.42	15.51	38.93
Society Islands 4–6% MgO average	0.704443	0.512821	19.11	15.58	38.78
Marquesas 4–6% MgO average	0.703835	0.512909	19.23	15.57	38.99
Austral-Cook 4–6% MgO average	0.703297	0.512866	20.47	15.72	39.85
Tristan 4–6% MgO average	0.704864	0.512574	18.60	15.54	39.03
Canaries 4–6% MgO average	0.703204	0.512896	19.70	15.62	39.51

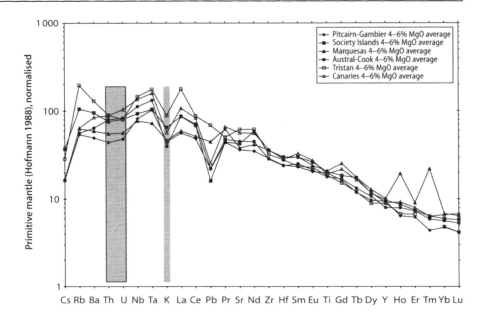

Fig. 8.21. Incompatible trace element patterns for averages of magma compositions from various Superswell and Atlantic hotspots (all data compiled from the Georoc database, Max-Planck-Institut für Chemie, Mainz). The magmas used to build the averages fall in a limited MgO range of 4–6 wt.%. The main radioactive heat-producing elements are marked; note that the concentrations of these elements are not significantly higher in the Superswell hotspots

the effects of trace element enrichment due to crystal fractionation and thus compare element concentrations between hotspots on the same basis. The table and figure show that if anything, the Superswell hotspots appear to have lower concentrations of heat-producing elements at a particular MgO than the Tristan or Canaries hotspot, taken to represent non-Superswell, non-Pacific hotspots that have been relatively well-studied geochemically. This seems to rule out an origin for the excess Superswell mantle temperature through enrichment in radioactive elements, although an increased proportion of hotspot-source mantle generally in the Superswell mantle could lead to it becoming more heated.

So how might higher proportions of hotspot-source mantle arise? We have seen earlier that most of the mantle components identified in hotspots (EM1, EM2, HIMU) can be related to the residues of subducted oceanic crust. As more material will be subducted with time, it is possible that larger proportions of hotspot-source material in the mantle could be related to a very long-term collection and storage of the subduction residues. Here it is interesting to note (as has been previously remarked by McNutt 1998) that many of the most extreme compositions defining the isotopic endmembers globally come from the Superswell area. As the isotopic values are time-integrated reflections of the parent-daughter ratios in the respective decay chain, generally speaking the older the subducted residue is, the more isotopically extreme it will become. Figure 8.22 shows data from the most isotopically extreme hotspots worldwide. It is notable that both the HIMU and EM1 extremes are clearly marked by South Pacific Superswell hotspots (Fig. 8.22a–d). The most extreme EM2 isotopic composi-

Chapter 8 · **The Sources for Hotspot Volcanism in the South Pacific Ocean**

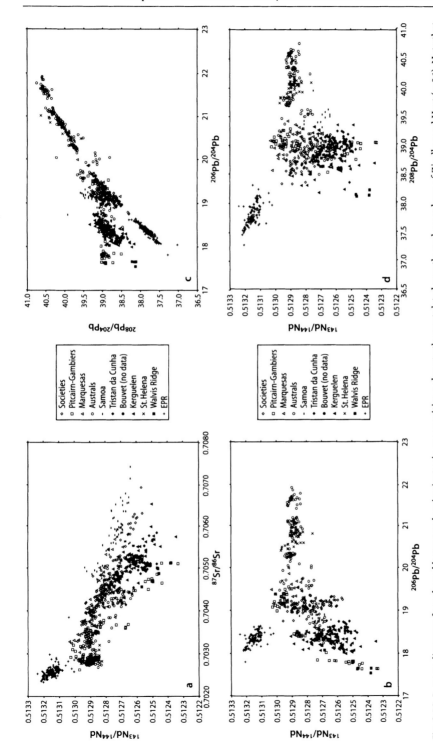

Fig. 8.22. Isotope diagrams for selected hotspots showing isotopic compositions close to the postulated mantle-end-members of Zindler and Hart (1986). Note that in most cases, the most extreme values are marked by hotspots from the South Pacific. Data from the Georoc (www.georoc.mpch-mainz.gwdg.de) and PetDB (petdb.ldeo.columbia.edu) databases

tions are found on Samoa, which is outside the area covered by the Superswell, although it lies in the South Pacific. Slightly less radiogenic than Samoa are then the Society magmas, part of the Superswell area.

The bathymetric Superswell anomaly is associated with some of the most extreme isotopic compositions known from hotspots worldwide. This in turn suggests that the mantle sources for these hotspots are relatively old. This is in accordance with a Superswell mantle that is proportionally over-enriched in subduction debris as a result of an extremely long-term accumulation of subducted material.

Acknowledgements

We would like to thank the initiators and leaders of the VIP initiative, Prof. P. Stoffers and Prof. J.-L. Cheminée † for their constant enthusiasm for Polynesian volcanism. Numerous research cruises to and publications about this part of the world with both of them were formative experiences for us both. The vast amounts of geochemical data on submarine volcanics generated during VIP could not have been produced without the expertise of many geochemists, especially Dieter Garbe-Schönberg in Kiel. Finally, we thank all our colleagues who have participated in German-French cruises and especially Roger Hekinian for fruitful discussions over the years.

References

Baker PE, Buckley F, Holland JG (1974) Petrology and geochemistry of Easter Island. Contrib Mineral Petrol 44:85–100

Baker PE, Gledhill A, Harvey PK, Hawkesworth CJ (1987) Geochemical evolution of the Juan Fernandez Islands, SE Pacific. J Geol Soc London 144:933–944

Binard N, Maury RC, Guille G, Talandier J, Gillot PY, Cotten J (1993) Mehetia Island, South Pacific: Geology and petrology of the emerged part of the Society hot spot. J Volcanol Geotherm res 55:239–260

Bonatti E, Harrison CGA (1976) Hot lines in the Earth's mantle. Nature 263:402–404

Bonatti E, Harrison CGA, Fisher DE, Honnorez J, Schilling J-G, Stipp JJ, Zentilli M (1977) Easter volcanic chain (Southeast Pacific): A mantle hot line. J Geophys Res 82:2457–2478

Bonneville A, Suavé RL, Audin L, Clouard V, Dosso L, Gillot PY, Janney P, Jordahl K, Maamaatuaiahutapu K (2002) Arago Seamount: The missing hotspot found in the Austral Islands. Geology 30:1023–1026

Caroff M, Maury RC, Vidal P, Guille G, Dupuy C, Cotten J, Guillou H, Gillot P-Y (1995) Rapid temporal changes in ocean island basalt composition: Evidence from an 800 m deep drill hole in Eiao Shield (Marquesas). J Petrol 36:1333–1363

Caroff M, Guillou H, Lamiaux M, Maury RC, Guille G, Cotten J (1999) Assimilation of ocean crust by hawaiitic and mugearitic magmas: An example from Eiao (Marquesas). Lithos 46:235–258

Chauvel C, Hofmann AW, Vidal P (1992) HIMU–EM: The French Polynesian connection. Earth Planet Sci Letts 110:99–119

Chauvel C, Goldstein SL, Hofmann AW (1995) Hydration and dehydration of oceanic crust controls Pb evolution in the mantle. Chem Geol 126:65–75

Chauvel C, McDonough W, Guille G, Maury R, Duncan RA (1997) Contrasting old and young volcanism in Rurutu Island, Austral Chain. Chem Geol 139:125–143

Cheminée JL, Hekinian R, Talandier J, Albarède F, Devey CW, Francheteau J, Lancelot Y (1989) Geology of an active hot spot: Teahitia-Mehetia region in the south-central Pacific. Mar Geophys Res 11:27–50

Cheng QC, Macdougall JD, Lugmair GW (1993) Geochemical studies of Tahiti, Teahitia and Mehetia, Society Island chain. J Volcanol Geotherm Res 55:155–184

Cheng QC, Macdougall JD, Zhu P (1999) Isotopic constraints on the Easter Seamount chain source. Contrib Mineral Petrol 135:225–233

Clague DA, Dalrymple GB (1987) The Hawaiian-Emperor volcanic chain: Part 1 geologic evolution. US Geol Surv Prof Paper 1350:5–54

Clark JG, Dymond J (1977) Geochronology and petrochemistry of Easter and Sala y Gómez Islands: Implications for the origin of the Sala y Gómez Ridge. J Volcanol Geotherm Res 2:29–48

Chapter 8 · The Sources for Hotspot Volcanism in the South Pacific Ocean   281

Clouard V, Bonneville A (2001) How many Pacific hotspots are fed by deep mantle plumes? Geology 29:695-698

Crough ST, Jurdy DM (1980) Subducted lithosphere, hotspots, and the geoid. Earth Planet Sci Lett 48:15-22

Davaille A (1999) Simultaneous generation of hotspots and superswells by convection in a heterogeneous planetary mantle. Nature 402:756-760

Desonie DL, Duncan RA, Natland JH (1993) Temoral and geochemical variability of volcanic products of the Marquesas Hotspot. J Geophys Res 98:17649-17665

Devey CW, Albarede F, Cheminee J-L, Michard A, Mühe R, Stoffers P, (1990) Active submarine volcanism on the Society hotspot swell (west Pacific): A geochemical study. J Geophys Res 95:5049-5066

Devey CW, Hekinian R, Ackermand D, Binard N, Francke B, Hémond C, Kapsimalis V, Lorenc S, Maia M, Möller H, Perrot K, Pracht J, Rogers T, Stattegger K, Steinke S, Victor P (1997) The Foundation Seamount chain: A first survey and sampling. Mar Geol 137:191-200

Devey CW, Hémond C, Stoffers P (2000) Metasomatic reactions between carbonated plume melts and mantle harzburgite: The evidence from Friday and Domingo Seamounts (Juan Fernandez Chain, SE Pacific). Contrib Mineral Petrol 139:68-84

Devey CW, Lackschewitz KS, Mertz DF, Bourdon B, Cheminée J-L, Dubois J, Guivel C, Hekinian R, Stoffers P (2001) Evidence for preferential melting of the enriched components in Polynesian plumes. Eos Trans AGU Fall Meet Suppl 82:F1397

Devey CW, Lackschewitz KS, Mertz DF, Bourdon B, Cheminée J-L, Dubois J, Guivel C, Hekinian R, Stoffers P (2003) Giving birth to hotspot volcanoes: Distribution and composition of young seamounts from the seafloor near Tahiti and Pitcairn Islands. Geology 31(5):395-398

Dixon JE, Leist L, Langmuir CH, Schilling J-G (2002) Recycled dehydrated lithosphere observed in plume-influenced mid-ocean-ridge basalts. Nature 420:385-389

Dostal J, Cousens B, Dupuy C (1998) The incompatible element characteristics of an ancient subducted sedimentary component in ocean island basalts from French Polynesia. J Petrol 39(5):937-952

Duncan RA, McDougall I (1976) Linear volcanism in French Polynesia. J Volcanol Geotherm Res 1:197-227

Duncan RA, McDougall I, Carter RM, Coombs DS (1974) Pitcairn Island - another Pacific hot spot. Nature 251:679-682

Duncan RA, McCulloch MT, Barsczus HG, Nelson DR (1986) Plume versus lithospheric sources for melts at Ua Pou, Marquesas Islands. Nature 322:534-538

Duncan RA, Fisk MR, White WM, Nielsen RL (1994) Tahiti: Geochemical evolution of a French Polynesian volcano. J Geophys Res 99:24341-24357

Dupuy C, Vidal P, Barsczus HG, Chauvel C (1987) Origin of basalts from the Marquesas Archipelago (south central Pacific Ocean): Isotope and trace element constraints. Earth Planet Sci Letts 82:145-152

Dupuy C, Barsczus HG, Liotard JM, Dostal J (1988) Trace element evidence for the origin of ocean island basalts: An example from the Austral Islands (French Polynesia). Contrib Mineral Petrol 98:293-302

Dupuy C, Vidal P, Maury RC, Guille G (1993) Basalts from Mururoa, Fangataufa and Gambier Islands (French Polynesia): Geochemical dependence on the age of the lithosphere. Earth Planet Sci Lett 117:89-110

Eiler JM, Farley KA, Valley JW, Stopler EM, Hauri EH, Craig H (1995) Oxygen isotope evidence against bulk recycled sediment in the mantle sources of Pitcairn Island lavas. Nature 377:138-141

Eisele J, Sharma M, Galer SJG, Blichert-Toft J, Devey CW, Hofmann AW (2002) The role of sediment recycling in EM-1 inferred from Os, Pb, Hf, Nd, Sr isotope and trace element systematics of the Pitcairn hotspot. Earth Planet Sci Lett 196:197-212

Farley KA, Craig H (1992) Mantle plumes and mantle sources. Science 258:821

Farley KA, Natland JH, Craig H (1992) Binary mixing of enriched and undegassed (primitive?) mantle components (He, Sr, Nd, Pb) in Samoan lavas. Earth Planet Sci Lett 111:183-199

Farley KA, Basu AR, Craig H (1993) He, Sr and Nd isotopic variations in lavas from the Juan Fernandez Archipelago, SE Pacific. Contrib Mineral Petrol 115:75-87

Fretzdorff S, Haase KM, Garbe-Schönberg C-D (1996) Petrogenesis of lavas from the Umu volcanic field in the young hotspot region west of Easter Island, southeastern Pacific. Lithos 38:23-40

Garcia MO, Park K-H, Davis GT, Staudigel H, Mattey DP (1993) Petrology and isotope geochemistry of lavas from the Line islands chain, Central Pacific Basin. In: Pringle MS, Sager WW, Sliter WV, Stein S (eds) The Mesozoic Pacific: Geology, tectonics and volcanism. AGU Geophys Monograph, Washington DC 77:217-231

Gasperini D, Blichert-Toft J, Bosch D, Del Moro A, Macera P, Télouk P, Albarède F (2000) Evidence from Sardinian basalt geochemistry for recycling of plume heads into the Earth's mantle. Nature 408:701-704

Gast PW, Tilton GR, Hedge C (1964) Isotopic composition of lead and strontium from Ascension and Gough Islands. Science 145:1181-1185

Gerlach DC, Hart SR, Morales VWJ, Palacios C (1986) Mantle heterogeneity beneath the Nazca plate: San Felix and Juan Fernandez islands. Nature 322:165–169

Griffiths RW, Campbell IH (1991) On the dynamics of long-lived plume conduits in the convecting mantle. Earth Planet Sci Lett 103:214–227

Gu YJ, Dziewonski AM, Su W, Ekström G (2001) Models of the mantle shear velocity and discontinuities in the pattern of lateral heterogeneities. J Geophys Res 106:11169–11199

Guillou H, Garcia MO, Turpin L (1997) Unspiked K-Ar dating of young volcanic rocks from Loihi and Pitcairn hot spot seamounts. J Volcanol Geotherm Res 78:239–249

Haase KM, Devey CW (1996) Geochemistry of lavas from the Ahu and Tupa volcanic fields, Easter Hotspot, southeast Pacific: Implications for intraplate magma genesis near a spreading axis. Earth Planet Sci Letts 137:129–143

Haase KM, Devey CW, Goldstein SL (1996) Two-way exchange between the Easter mantle plume and the Easter microplate spreading axis. Nature 382:344–346

Haase KM, Stoffers P, Garbe-Schönberg C-D (1997) The petrogenetic evolution of lavas from Easter Island and neighbouring seamounts, near-ridge hotspot volcanoes in the SE Pacific. J Petrol 38:785–813

Hagen RA, Baker NA, Naar DF, Hey RN (1990) A SeaMARC II survey of Recent submarine volcanism near Easter Island. Mar Geophys Res 12:297–315

Hanan BB, Graham DW (1996) Lead and helium isotope evidence from oceanic basalts for a common deep source of mantle plumes. Science 272:991–995

Hanan BB, Schilling J-G (1989) Easter Microplate evolution: Pb isotope evidence. J Geophys Res 94: 7432–7448

Hauri EH, Whitehead JA, Hart SR (1994) Fluid dynamic and geochemical aspects of entrainment in mantle plumes. J Geophys Res 99:24275–24300

Hekinian R, Stoffers P, Devey CW, Ackermand D, Hémond C, O'Connor J, Binard N, Maia M (1997) Intraplate versus ridge volcanism on the Pacific-Antarctic Ridge near 37° S–111° W. J Geophys Res 102:12265–12286

Hekinian R, Stoffers P, Ackermand D, Revillion S, Maia M, Bohn M (1999) Ridge-hotspot interaction: The Pacific-Antarctic Ridge and the Foundation seamounts. Mar Geol 160:199–233

Hémond C, Devey CW (1996) The Foundation Seamount chain, Southeast Pacific: First isotopic evidence of a newly discovered hotspot track. J Conf Abstr 1:255

Hémond C, Devey CW, Chauvel C (1994) Source compositions and melting processes in the Society and Austral plumes (South Pacific Ocean): Element and isotope (Sr, Nd, Pb, Th) geochemistry. Chem Geol 115:7–45

Hofmann AW, White WM (1982) Mantle plumes from ancient oceanic crust. Earth Planet Sci Letts 57: 421–436

Janney PE, Macdougall JD, Natland JH, Lynch MA (2000) Geochemical evidence from the Pukapuka volcanic ridge system for a shallow enriched mantle domain beneath the South Pacific Superswell. Earth Planet Sci Lett 181:47–60

Johnson RH (1970) Active submarine volcanism in the Austral Islands. Science 167:977–979

Jordahl KA, McNutt MK, Webb HF, Kruse SE, Kuykendall MG (1995) Why there are no earthquakes on the Marquesas Fracture Zone. J Geophys Res 100:24431–24447

Kaneoka I, Katsui Y (1985) K-Ar ages of volcanic rocks from Easter Island. Bull Volcanol Soc Japan 30: 33–36

Kingsley R, Schilling J-G (1998) Plume-ridge interaction in the Easter-Salas y Gómez Seamount chain-Easter Microplate system: Pb isotope evidence. J Geophys Res 103:24159–24177

Kingsley RH, Schilling J-G, Dixon JE, Swart P, Poreda R, Simons K (2002) D/H ratios in basalt glasses from the Salas y Gómez mantle plume interacting with the East Pacific Rise: Water from old D-rich recycled crust or primordial water from the lower mantle? Geochem Geophys Geosys 3

Le Dez A, Maury RC, Vidal P, Bellon H, Cotten J, Brousse R (1996) Geology and geochemistry of Nuku Hiva, Marquesas: Temporal trends in a large Polynesian shield volcano. Bull Soc Geol France 167: 197–209

Liotard J-M, Barsczus HG (1983a) Contribution à la connaissance pétrographique et géochemique de l'ile de Hatutu, Archipel des Marquises, Polynésie francaise (Océan Pacifique Centre – Sud). CR Acad Sc Paris 297:725–728

Liotard J-M, Barsczus HG (1983b) Contribution à la connaissance pétrographique et géochemique de l'ile de Fatu Huku, Archipel des Marquises, Polynésie francaise (Océan Pacifique Centre – Sud). CR Acad Sc Paris 297:509–512

Liotard J-M, Barsczus HG (1984) Contribution à la connaissance pétrographique et géochemique de l'ile d'Eiao, Archipel des Marquises, Polynésie francaise (Océan Pacifique Centre – Sud). CR Acad Sc Paris 298:347–349

Macdonald GA, Katsura T (1964) Chemical composition of Hawaiian lavas. J Petrol 5:82–133

Mahoney JJ, Nicollet C, Dupuy C (1991) Madagascar basalts: Tracking oceanic and continental sources. Earth Planet Sci Lett 104:350–363

Chapter 8 · The Sources for Hotspot Volcanism in the South Pacific Ocean

Maia M, Ackermand D, Dehghani GA, Gente P, Hekinian R, Naar D, O'Connor J, Perrot K, Phipps Morgan J, Ramillien G, Revillon S, Sabetian A, Sandwell D, Stoffers P (2000) The Pacific-Arctic Ridge Foundation hotspot interaction: A case study of a ridge approaching a hotspot. Mar Geol 167:61–84

Maia M, Hémond C, Gente P (2001) Contrasted interactions between plume, upper mantle, and lithosphere: Foundation Chain case. Geochem Geophys Geosys 2:101029/2000GC000117

Mammerickx J (1992) The Foundation Seamounts: Tectonic setting of a newly discovered seamount chain in the South Pacific. Earth Planet Sci Lett 113:293–306

McKenzie D, O'Nions RK (1983) Mantle reservoirs and ocean island basalts. Nature 301:229–231

McNutt MK (1998) Superswells. Rev Geophys 36(2):211–244

McNutt MK, Fischer KM (1987) The south Pacific superswell. In: Keating BH, Fryer P, Batiza R, Boehlert GW (eds) Seamounts, islands and atolls. Am Geophys Union geophys Monograph 43:25–34

McNutt MK, Judge AV (1990) The Superswell and mantle dynamics beneath the South Pacific. Science 248:969–975

McNutt MK, Fischer K, Kruse S, Natland J (1989) The origin of the Marquesas fracture zone ridge and its implications for the nature of hot spots. Earth Planet Sci Letts 91:381–393

McNutt MK, Caress DW, Reynolds J, Jordahl KA, Duncan RA (1997) Failure of plume theory to explain midplate volcanism in the southern Austral islands. Nature 389:479–482

Morgan WJ (1972) Deep mantle convection plumes and plate motions. American Association of Petroleum Geologists Memoir 56:203–213

Nakamura Y, Tatsumoto M (1988) Pb, Nd, and Sr isotopic evidence for a multicomponent source for rocks of Cook-Austral Islands and heterogeneities of mantle plumes. Geochim Cosmochim Acta 52:2909–2924

Niu F, Solomon SC, Silver PG, Suetsugu D, Inoue H (2002) Mantle transition-zone structure beneath the South Pacific superswell and evidence for a mantle plume underlying the Society hotspot. Earth Planet Sci Lett 198:371–380

O'Connor JM, Stoffers P, McWilliams MO (1995) Time-space mapping of Easter Chain volcanism. Earth Planet Sci Lett 136:197–212

O'Connor JM, Stoffers P, Wijbrans JR (1998) Migration rate of volcanism along the Foundation Chain, SE Pacific. Earth Planet Sci Lett 164:41–59

O'Connor JM, Stoffers P, Wijbrans JR (2001) En echelon volcanic elongate ridges connecting intraplate Foundation Chain volcanism to the Pacific-Antarctic spreading center. Earth Planet Sci Letts 192:633–648

Okal EA, Cazenave A (1985) A model for the plate tectonic evolution of the east-central Pacific based on SEASAT investigations. Earth Planet Sci Lett 72:99–116

Palacz ZA, Saunders AD (1986) Coupled trace element and isotope enrichment in the Cook-Austral-Samoa islands, southwest Pacific. Earth Planet Sci Letts 79:270–280

Pan Y, Batiza R (1998) Major element chemistry of volcanic glasses from the Easter Seamount chain: Constraints on melting conditions in the plume channel. J Geophys Res 103:5287–5304

Rappaport Y, Naar DF, Barton CC, Liu ZJ, Hey RN (1997) Morphology and distribution of seamounts surrounding Easter Island. J Geophys Res 102:24713–24728

Richards MA, Griffiths RW (1989) Thermal entrainment by deflected mantle plumes. Nature 342:900–902

Simons K, Dixon J, Schilling J, Kingsley R, Poreda R (2002) Volatiles in basaltic glasses from the Easter-Salas Gómez Seamount chain and Easter Microplate: Implications for geochemical cycling of volatile elements. Geochem Geophys Geosys 3(7):1–29

Sleep NH (2002) Ridge-crossing mantle plumes and gaps in tracks. Geochem Geophys Geosys 3:101029/2001GC000290

Smith W, Sandwell D (1997) Measured and estimated seafloor topography (version 42) World Data Center A for Marine Geology and Geophysics research publication RP-1

Staudigel H, Park K-H, Pringle MS, Rubenstone JL, Smith WHF, Zindler A (1991) The longevity of the South Pacific isotopic and thermal anomaly. Earth Planet Sci Letts 102:24–44

Stoffers P, Botz R, Cheminée J-L, Devey CW, Froger V, Glasby GP, Hartmann M, Hekinian R, Kögler F, Laschek D, Larqué P, Michaelis W, Mühe RK, Puteanus D, Richnow HH (1989) Geology of Macdonald Seamount region, Austral Islands: Recent hotspot volcanism in the south Pacific. Mar Geophys Res 11:101–112

Stoffers P, Hekinian R, Haase KM, Scientific Party (1994) Geology of young submarine volcanoes west of Easter Island, Southeast Pacific. Mar Geol 118:177–185

Stoffers P, SO-65 tsp (1990) Active Pitcairn hotspot found. Mar Geol 95:51–55

Sun S-S (1980) Lead isotopic study of young volcanic rocks from mid-ocean ridges, ocean islands and island arcs. Phil Trans R Soc Lond A 297:409–445

Talandier J, Okal EA (1984) The volcanoseismic swarms of 1981–1983 in the Tahiti-Mehetia area, French Polynesia. J Geophys Res 89:11216–11234

Tanaka S (2002) Very low shear wave velocity at the base of the mantle under the South Pacific superswell. Earth Planet Sci Letts 203:879–893

Turner DL, Jarrard RD (1982) K-Ar dating of the Cook-Austral island chain: A test of the hotspot hypothesis. J Volcanol Geotherm Res 12:187–220

Vidal P, Chauvel C, Brousse R (1984) Large mantle heterogeneity beneath French Polynesia. Nature 307:536–538

Watson S, McKenzie D (1991) Melt generation by plumes: A study of Hawaiian volcanism. J Petrol 32:501–537

Weaver BL (1991) The origin of ocean island basalt end-member compositions: Trace element and isotopic constraints. Earth Planet Sci Letts 104:381–397

Weaver BL, Wood DA, Tarney J, Joron JL (1987) Geochemistry of ocean island basalts from the South Atlantic: Ascension, Bouvet, St Helena, Gough and Tristan da Cunha alkaline igneous rocks. Geol Soc Spec Publ 30:253–267

White WM, Duncan RA (1996) Geochemistry and geochronology of the Society Islands: New evidence for deep mantle recycling. In: Basu A, Hart S (eds) Earth processes: Reading the isotopic code, vol 95. Am Geophys Union, Geophys Monogr, Washington, DC, pp 183–206

White WM, Hofmann AW (1982) Sr and Nd isotope geochemistry of oceanic basalts and mantle evolution. Nature 296:821–825

White WM, McBirney AR, Duncan RA (1993) Petrology and geochemistry of the Galapagos Islands: Portrait of a pathological mantle plume. J Geophys Res 98:19533–19563

Wilson JT (1973) Mantle plumes and plate motions. Tectonophys 19:149–164

Woodhead JD (1992) Temporal geochemical evolution in oceanic intra-plate volcanics: A case study from the Marquesas (French Polynesia) and comparison with other hotspots. Contrib Mineral Petrol 111:458–467

Woodhead JD (1996) Extreme HIMU in an oceanic setting: the geochemistry of Mangaia Island (Polynesia), and temporal evolution of the Cook-Austral hotspot. J Volcanol Geotherm Res 72:1–19

Woodhead JD, Devey CW (1993) Geochemistry of the Pitcairn seamounts, I: Source character and temporal trends. Earth Planet Sci Lett 116:81–99

Woodhead JD, McCulloch MT (1989) Ancient seafloor signals in Pitcairn Island lavas and evidence for large amplitude, small length-scale mantle heterogeneities. Earth Planet Sci Letts 94:257–273

Woodhead JD, Greenwood P, Harmon RS, Stoffers P (1993) Oxygen isotope evidence for recycled crust in the source of EM-type ocean island basalts. Nature 362:809–813

Zindler A, Hart S (1986) Chemical geodynamics. Ann Rev Earth Planet Sci 14:493–571

Chapter 9

Ridge Suction Drives Plume-Ridge Interactions

Y. Niu · R. Hekinian

9.1
Introduction

Geological processes are consequences of the Earth's thermal evolution. Plate tectonics, which explain geological phenomena along plate boundaries, elegantly illustrate this concept. For example, the origin of oceanic plates at ocean ridges, the movement and growth of these plates, and their ultimate consumption back into the Earth's interior through subduction zones provide an efficient mechanism to cool the Earth's mantle, leading to large-scale mantle convection. Mantle plumes, which explain another set of global geological phenomena, cool the Earth's deep interior (probably the Earth's core) and represent another mode of Earth's thermal convection (e.g., Davies and Richards 1992). Plate tectonics and plume tectonics are thus genetically independent from each other. However, when the rising plumes approach the lithospheric plates, interactions between the two inevitably result. Such interactions are most prominent near ocean ridges, where the lithosphere is thin and the effect of mantle plumes is best revealed. "Plume-ridge interaction" has been a hot topic in recent years, and much effort has been expended in this area aimed at understanding the geological, geochemical, and geodynamic consequences (Schilling et al. 1983, 1994, 1995, 1996, 1999; Schilling 1991; Feighner and Richards 1995; Ito and Lin 1995a,b; Ito et al. 1996; Kincaid et al. 1995, 1996; Ribe 1996; Sleep 1996; Haase and Devey 1996; Hekinian et al. 1996, 1997, 1999; Pan and Batiza 1998; Niu et al. 1999; Graham et al. 1999; Maia et al. 2000; Georgen et al. 2001; Haase 2002).

In this study, instead of reviewing details of existing models, we present our new perspectives on the geochemical and geological consequences of plume-ridge interactions in the form of schematic models. These models differ from most common perceptions, but are consistent with observations and comply with simple physics. We focus on first-order observations and stress the importance of several fundamental concepts and variables required to fully understand the "expression" and "intensity" of plume-ridge interactions. These include: (1) what mantle plumes are; (2) the nature and composition of plume sources; (3) the actual role of ocean ridges; (4) the effect of plate separation rate; and (5) plume-ridge distance. We illustrate these concepts/variables with representative examples.

9.2
Concepts

9.2.1
Mantle Plumes:
Deep-Rooted Hot Materials or Wet Shallow Mantle Melting Anomalies?

The theory of plate tectonics, by its definition, explains tectonic activities along plate boundaries between two adjacent plates. For example, volcanism at ocean ridges and overlaying subduction zones is the consequence of plate tectonics. However, volcanism occurring away from plate boundaries cannot be explained by plate tectonics theory. This "intraplate" volcanism is often interpreted as resulting from hotspots or mantle plumes. Volumetrically large and temporally long-lived intraplate volcanism represented by the islands of Hawaii in the Pacific, the Azores in the Atlantic and the Kerguelen plateau in the southern ocean may indeed be hotspots genetically associated with deep-rooted mantle plumes (Morgan 1971, 1981; Duncan and Richards 1991). However, other intraplate volcanism such as Cenozoic volcanic activities widespread in eastern Australia (e.g., Johnson ed. 1989) and eastern China (Deng et al. 1998; Zhang et al. 1998), the well-known Cameroon volcanic line (Halliday et al. 1988) and numerous seamounts scattered throughout much of the Earth's ocean floor (Batiza 1982) away from any plate boundaries are likely to be melting anomalies resulting from easily melted fertile mantle source materials (enriched in volatiles, alkalis and other incompatible elements) in the shallow asthenosphere. The latter may be more appropriately termed "wet spots" (e.g., Green and Falloon 1998). We must emphasize, however, that we do not yet at present have sufficient data to draw a clear distinction between "hot-spots" and "wet spots". It is possible that some or many of the so-called hotspots may prove to be wet spots when more relevant data become available. In the following discussion, we simply use the terms "hotspots" or "mantle plumes" as widely used in current literature for convenience. The key is that the greater the heat/mass flux a plume has, the greater the effect of the plume-ridge interaction will be for a given plume-ridge distance and a given plate separation rate (see below).

9.2.2
Nature of Plume Materials

Compared to most Mid-Ocean Ridge basalts (MORBs), ocean-island basalts (OIB) are highly enriched in incompatible elements. Hofmann and White (1982) proposed that mantle plumes are from ancient oceanic crust. Many subsequent studies of OIB and enriched MORB favor this proposal, although some form of mantle metasomatism is clearly needed (Sun and McDonough 1989; Halliday et al. 1995; Niu et al. 1996 1999). Following the suggestion by Niu et al. (2002), Niu and O'Hara (see Sects. 7.4.2 and 7.5) using knowledge based on well-understood petrology, geochemistry and mineral physics demonstrate that ancient recycled oceanic crust cannot be the source material supplying OIB. Instead, they show deep portions of oceanic lithosphere that are the best candidates for mantle plume sources (Niu et al. 2002). These deep portions of oceanic lithosphere are filled with dykes or veins enriched in volatiles, alkalis, and all the other incompatible elements as a result of low-degree melt metasomatism at the interface between the low velocity zone and the cooling and thickening oceanic lithosphere. The

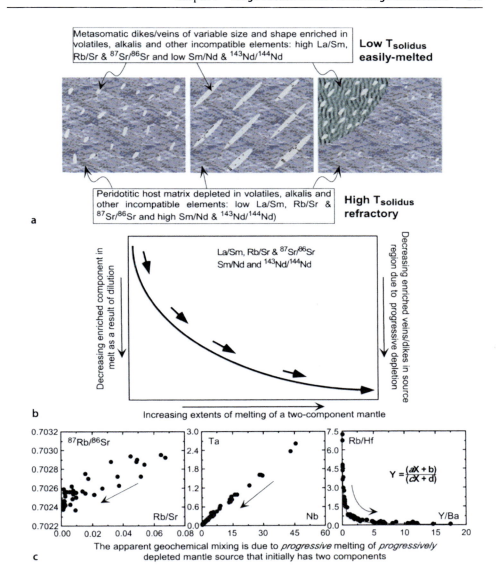

Fig. 9.1. a Schematic illustration of the concept of two-component fertile mantle sources, which include easily melted metasomatized dykes/veins of variable size and shape enriched in volatiles, alkalis and other incompatible elements dispersed in the refractory and predominantly depleted peridotite matrix. **b** Schematic illustration of the geochemical consequences of melting such a two-component mantle. The enriched dykes/veins have lower solidus temperatures and preferentially melt first. Consequently, the enriched component dominates the composition of the melt produced in the early stages and decreases with further melting as a result of dilution. Concurrently, the source region is progressively depleted in the enriched dykes/veins, and further melting of this depleted source material can only produce melts progressively depleted in volatiles and incompatible elements. **c** Three representative geochemical diagrams after Niu and Batiza (1997), showing apparent geochemical "mixing" as a consequence of the melting of a two-component mantle (or "melting-induced mixing") (Niu et al. 1999, 2002). But, this is often mistakenly interpreted as "mixing" relationships between two singular melts or solids in the source regions of melt generation. The *arrows* in *Part c* point to the direction of increasing extents of melting

key is that plume sources (1) are peridotitic in bulk composition and (2) contain incompatible element-enriched dykes/veins of metasomatic origin. In other words, fertile OIB sources are composite lithologies with two components: an easily melted component enriched in volatiles, alkalis and other incompatible elements dispersed in the predominantly more refractory peridotitic matrix (Fig. 9.1a).

The concept of two-component fertile mantle sources favors the two-stage melting model for OIB and MORB (Phipps Morgan and Morgan 1999) and is consistent with observations in MORB from the eastern Pacific (Batiza and Vanko 1984; Batiza and Niu 1992; Zindler et al. 1984; Langmuir et al. 1986; Hekinian et al. 1989, 1995; Sinton et al. 1991; Niu et al. 1996, 1999, 2002; Niu and Batiza 1997; Niu and Hekinian 1997a; Regelous et al. 1999; Wendt et al. 1999) and the Mid-Atlantic Ridge (Niu and Batiza 1994; Niu et al. 2001; Regelous et al. 2001). In other words, both OIB and MORB sources have two such components, but the enriched dykes/veins are far more abundant in source regions of OIB than beneath ocean ridges (Niu et al. 1999). Figure 9.1b shows schematically that melting such a two-component mantle will produce more enriched melt during the early stage of melting not just because of the commonly perceived low-degree melting of a uniform source, but largely because of the greater contributions of the enriched easily melted component. With continued melting, the source region is progressively depleted in this enriched component, leading to a progressively more depleted melt. This is important in understanding the spatial variation of lava composition in the context of plume-ridge interactions (see below). The geochemical consequence of melting such a two-component mantle is to produce melting-induced mixing relationships (not true mixing between two singular end-members) in geochemical diagrams (Niu et al. 1996, 1999, 2002; Niu and Batiza 1997) (Fig. 9.1c).

This two-component fertile plume source model differs from the popular plume dispersion model (e.g., Schilling 1991). The latter model (1) assumes compositional and isotopic distinction between plume material and fertile MORB source material and (2) interprets the geochemical data as mixing between the two singular materials in the form of melt or in the source regions prior to the major melting events (Schilling 1991; Schilling et al. 1994, 1995, 1999; Kingsley et al. 2002). In our opinion, a dispersion model that requires an invasion of the plume material into the MORB source is physically difficult.

9.2.3
Ocean Ridges:
Ridge Suction – The Active Driving Force for Plume-Ridge Interactions

Ocean ridges are mostly passive features in the sense that mantle upwelling is largely caused by plate separation (McKenzie and Bickle 1998). This passive upwelling brings hot material from depths to melt by decompression (Fig. 9.2a,b). Continued plate separation leads to continued mantle upwelling, decompression melting, and crust formation. Furthermore, the oceanic lithospheric mantle, which thickens by cooling, is accreted fastest near ridges, with >50% of the full thickness achieved within the first ~17–18 Myr (Fig. 9.3a). All these factors require a continued asthenospheric material supply towards ocean ridges to form the lithosphere (crust plus lithospheric mantle). The entirety of the process is continuous, because the newly formed lithosphere is continuously moved away from the ridge. In other words, the asthenospheric mantle beneath ocean ridges represents the regions of lowest pressure in the entire mantle,

Chapter 9 · Ridge Suction Drives Plume-Ridge Interactions

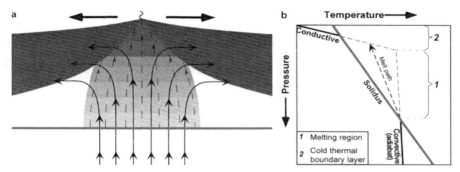

Fig. 9.2. a Cartoon showing plate separation-induced passive upwelling of the asthenospheric mantle and the resultant decompression melting beneath ocean ridges; **b** schematic illustration of the processes described in *Part a* in a P-T space. Passive upwelling brings hot mantle material to rise adiabatically and melt when encountering the solidus. Melting continues with upwelling until the upwelling mantle reaches the cold thermal boundary layer as a result of conductive heat loss to the surface (Niu 1997; Niu and Hekinian 1997b)

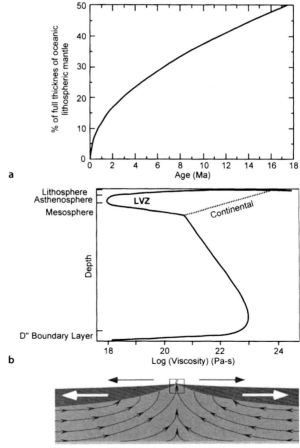

Fig. 9.3. a Assuming an oceanic plate reaches its full thickness (~95 km) after ~70 Myr (e.g., Stein and Stein 1997), ~50% of the lithospheric thickness is achieved in the first ~17.5 Myr (i.e., $t_{1/2} = (0.5 \times 70^{1/2})^2$). This, plus crust formation at ridges, requires ridgeward material supply. **b** As mantle viscosity is the lowest in the asthenosphere, in particular in the seismic low velocity zone (LVZ), at depths <~250 km, and increases exponentially with depth (Phipps Morgan et al. 1995; Lambeck and Johnston 1998), it is physically straightforward that the ridgeward mass flows must be mostly asthenospheric and horizontal. **c** Consequently, the regions of asthenosphere beneath ocean ridges have the lowest pressure that drives asthenospheric flows (i.e., ridge suction). This suggests that the spreading lithospheric plates are necessarily decoupled from the sublithospheric flow. The *square* at the ridge axis approximates the size of Fig. 9.2a

(1) Ridge suction. (2) Ridge-ward asthenospheric flow. (3) Necessary *decoupling* between the spreading lithospheric plates and the LVZ atop the asthenosphere.

which creates pressure gradients and sucks asthenospheric materials towards ocean ridges (Phipps Morgan et al. 1995; Niu et al. 1999). This argues that while ocean ridges are mostly passive features in the context of plate tectonics, they play an active role in dictating sublithospheric mantle flows.

The materials could either be supplied from below at great depths or transported laterally. As the mantle viscosity increases exponentially with depth beyond the low velocity zone (LVZ) (Fig. 9.3b after Phipps Morgan et al. 1995), it is physically straightforward that much of the material needed to create oceanic crust and oceanic lithosphere must be supplied by lateral sublithospheric flow through the LVZ (Niu et al. 1999). Mantle plume materials in the asthenosphere are generally hot (or "wet"), enriched in volatiles, and thus have the lowest viscosity. Consequently, mantle plume materials in the LVZ have the greatest tendency to flow and supply ridges because of the material needed beneath ocean ridges to form crust and lithospheric mantle (Niu et al. 1999). This is in fact apparent from mantle topographic studies (Zhang and Tanimoto 1993; Phipps Morgan et al. 1995). An inevitable conclusion of this analysis based on simple physics and mass conservation suggests that the lithospheric plate motion and the sublithospheric flow are necessarily decoupled (Fig. 9.3c), which is in fact seismically detected (Silver and Holt 2002).

9.2.4
Ridge Suction Increase with Increasing Spreading Rate

The amount of material required per time unit to form oceanic lithosphere (crust plus lithospheric mantle) near ocean ridges increases linearly with an increasing spreading rate. This linear relationship is better understood if we consider the oceanic lithosphere that has reached its full thickness of ~95 km after 70 Ma (Stein and Stein 1996). As the full thickness is independent of spreading rate, the volume of the lithosphere (L) formed per time unit per length unit parallel to the ridge would be the product of the half spreading rate ($R_{1/2}$) and the full thickness (T_{full}): $L = R_{1/2}T_{full}$. As T_{full} is constant, L is linearly proportional to $R_{1/2}$ with T_{full} being the simple proportionality. A less straightforward, but equally valid and more relevant scenario is what is taking place in the vicinity of ocean ridges where oceanic lithosphere is being created. Figure 9.4a shows that within the first 1 Myr, the lithospheric mantle reaches a thickness of 11 km due to conductive cooling ($L = 11 t^{0.5}$; see Fowler 1990). For a fast spreading ridge with $R_{1/2} = 60$ mm yr^{-1}, the 1 Myr isochron is 60 km away from the ridge axis, but for a slow ridge with $R_{1/2} = 10$ mm yr^{-1}, the 1 Myr isochron is 10 km from the ridge axis. The shaded areas are the lithospheric mantle thus far formed per unit length (per km along the ridge axis). For a half space, the volume of the lithospheric mantle formed per unit length in the first 1 Myr is 440 km^3 beneath the fast ridge, but is only 73.33 km^3 beneath the slow ridge. Again, a factor of 6 in volume, 440 / 73.33 = 6 is the same as the factor in spreading rate: 60 / 10 = 6. If the width of the active upwelling zone beneath ocean ridges is proportional to the spreading rate (Turcotte and Phipps Morgan 1992; Forsyth 1992) and is about the width equivalent to 1 Myr isochron, then the total material flux towards the ridge can be calculated. Figure 9.4b shows the total material flux (in volume, km^3) required to form the lithosphere on both sides of the ridge is linearly related to the spreading rate. Scenario A considers the material required to form the lithospheric mantle due only to conductive heat loss: Φ (km^3) = 14.667$R_{1/2}$ (km Myr^{-1}). Scenario B also includes a 5 km thick crust: Φ (km^3) = 24.667$R_{1/2}$ (km Myr^{-1}). Note that for simplicity it is assumed that the crust thickness is the same for both slow-

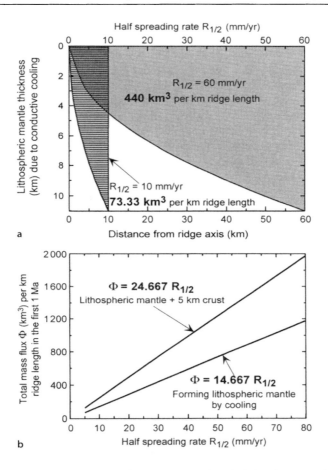

Fig. 9.4. a Comparison of materials (in km³) required to produce the lithospheric mantle on one side of the ridge in the first one million years between fast (e.g., $R_{1/2} = 60$ mm yr⁻¹; *shaded area* times per unit ridge parallel length in km) and slow (e.g., $R_{1/2} = 10$ mm yr⁻¹; *area with zebra pattern* times per unit length) spreading ridges. The volumes per unit length are calculated by simple integration of the *shaded area*: $c\int_a^b \sqrt{x}dx$, where x is the distance (km) from ridge axis $a = 0$ to b of interest (60 km, and 10 km respectively here) and $c = 11 b^{-1/2}$ (km$^{1/2}$). Note that the materialrequirement is linearly proportional to the spreading rate, i.e., 440 / 73.33 (volume ratio) = 60/10 (speed ratio) = 6. **b** Indeed, the mass fluxes required to form the lithosphere are linearly related to spreading rate. Plotted are the total mass (km³) requirements across the ridge in the first one million years against half spreading rate. The total mass required to form the lithospheric mantle as a result of conductive cooling is Φ (km³) = $15R_{1/2}$ (km Myr⁻¹), and it becomes Φ (km³) = $25R_{1/2}$ (km Myr⁻¹) if an average of 5 km thick crust is also considered. The important conclusion is that ridge suction force or ridgeward material flux is significant and increases with increasing spreading rate. Note that the calculations neglect the effect of density changes on the volume, but this effect is <2% and will not affect the conclusion here. Also note that the unit of the slope (~15 and 25 respectively) is km² Myr

and fast-spreading ridges here, which is incorrect in reality (Niu and Hekinian 1997b), but will not affect the conclusion here. This simple analysis indicates that the ridgeward mass flux, thus the ridge suction force, increases linearly with increasing plate-spreading rate. Φ (km³) = $25R_{1/2}$ (km Myr⁻¹) is a good approximation (assuming a ~5 km thick igneous crust) for the material flux to the ridge in the first 1 Myr.

9.2.5
The Effect of Plume-Ridge Distance

Mantle plumes, by definition, are columnar ascending flows derived from the lower mantle (Morgan 1971, 1981). The lower-mantle derived plume materials melt by decompression in the asthenosphere (Campbell and Griffiths 1990) while spreading laterally away from plume centers beneath the lithosphere (Hill et al. 1992). The actual distance of lateral flow or spread is unknown, but is inferred to be less than 2 000 km from the geochemistry of volcanics and topographic expression in the context of plume-ridge interaction studies (Schilling 1991; Ito and Lin 1995; Ribe 1996; Sleep 1996). This inference is not unreasonable if plume materials are compositionally distinct from the shallow mantle materials as reflected in MORB and if plume materials are indeed significantly hotter than the ambient asthenospheric mantle. However, any further conclusions derived from such an inference can be dangerous without understanding *(a)* the very nature of plume materials; *(b)* the driving force for lateral sublithospheric flow; and *(c)* the possible differentiation of plume materials during the lateral flow. It is in fact very likely that plume materials may flow as far as physically possible, perhaps in excess of several thousands of km (see Niu et al. 1999). The important point here is the relative roles of ridges and plumes. If the plume-ridge distance is short, the interaction would be intense, to which plumes would contribute significantly in both mass and heat. On the other hand, if the plume is far from the ridge, the interaction would be weak, but the ridge would play a more important role in creating the pressure gradient needed to drive the ridgeward flow (Figs. 9.3a–3c and 9.4a–b).

9.3
Examples

9.3.1
"Proximal" Versus "Distal" Plume-Ridge Interactions

Iceland and Hawaii are presently the largest volcanically active mantle plumes/hotspots detected on modern ocean floor. The Iceland hotspot is centered on the very axis of the Mid-Atlantic Ridge (MAR), whereas the Hawaiian hotspot is a typical intraplate hotspot on the Pacific Plate some ~5 000 km away from the East Pacific Rise. If there are some kinds of plume-ridge interactions in both cases, then the style, intensity and both geological and geophysical consequences must be different.

MORB along the Reykjanes Ridge show incompatible element enrichment with an amplitude increasing systematically from N-type MORBs at deep ridges to enriched MORBs and even OIB compositions at shallow ridges toward Iceland (Sun et al. 1975; Schilling et al. 1983; Taylor et al. 1997). The along-ridge depth and the geochemical variations, which are the thermal and chemical consequences of plume ridge interactions (Schilling 1991; Ito and Lin 1995; Ito et al. 1996; Ribe 1996; Sleep 1996), reflect the along-ridge asthenospheric flow of plume materials (Fig. 9.5a). According to the popular plume dispersion model (Schilling 1991), the along-ridge variation in lava compositions would indicate the spatial extent of plume-ridge interactions. For example, geochemical signals of Iceland plume material in Reykjanes Ridge axial lavas decline and approach the level of N-MORB some ~1 500 km south of Iceland, which is thought

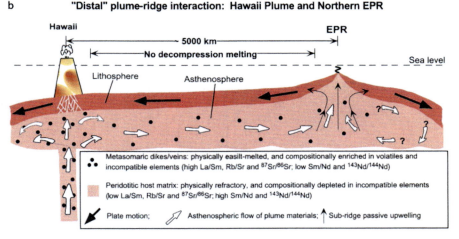

Fig. 9.5. Cartoons, modified from Niu et al. (1999), showing different consequences of plume-ridge interactions as a function of plume-ridge distance. If the plume is close to a ridge, for example, Iceland plume-Reykjanes Ridge, the buoyant upwelling and melting of the hot plume produces shallow ridge topography and enriched OIB-like basalt. The ambient sub-ridge mantle upwelling allows the plume material to flow laterally and to melt because of decompression, producing basalt whose extent of enrichment declines in the flow direction as the amount of the enriched component in the flowing plume mantle diminishes. The flow continues, but the enriched geochemical signals in the basalt do not. If the plume is far from a ridge, for example, Hawaiian plume-EPR, the low-viscosity Hawaiian plume material may flow laterally toward the EPR in response to the elevated ridge suction force beneath the fast-spreading EPR (Fig. 9.3a–3c and 9.4a,b). As the lateral flow is largely horizontal, decompression melting does not take place during flow, which allows the Hawaiian plume materials (both enriched and depleted) to survive long-distance transport to the EPR mantle

to be the maximum extent of the along-ridge effect of Iceland mantle plumes (e.g., Schilling 1991; Ito et al. 1996). This notion may be erroneous, since the depleted component of the plume material is likely to flow continuously. It is conceptually important to note that the declining geochemical enrichment in erupted lavas away from

Iceland is not the result of a simple dispersion or dilution effect, but the consequence of decompression melting during sub-ridge passive mantle upwelling, which preferentially melts and depletes the easily melted component within the remaining flowing plume material, leading to geochemically more depleted lavas erupted (Fig. 9.1b). The "residual" plume source materials depleted in enriched lithologies may continue to flow, but the "familiar" enriched geochemical signatures of plumes in erupted basalts diminish and are eventually replaced by "normal" depleted MORB.

Niu et al. (1999) termed this phenomenon as "flow differentiation" of an initially two-component plume mantle in a sub-ridge environment where the lithosphere is young and thin and decompression melting is possible because of the ambient passive mantle upwelling beneath the ridge (Fig. 9.5a). The increasing axial depth away from Iceland is the consequence of loss in both heat and mass (sub-ridge melting and cooling) of the plume material. The concept of a two-component plume source and "flow-differentiation" that we present here is physically straightforward and differs from the plume dispersion models (Schilling 1991; Ito et al. 1996; Schilling et al. 1999). The latter models, which require an "invasion" of enriched plume materials in the pre-existing depleted N-MORB mantle and a geochemical mixing between the two end-members, has space problems that are physically difficult to resolve.

Niu et al. (1999) showed that some enriched EPR MORBs define trends with the enriched end isotopically pointing to the field of Hawaiian lavas. This geochemical observation, plus mantle tomographic observation (Zhang and Tanimoto 1993; Phipps Morgan et al. 1995), led Niu et al. (1999) to suggest a sublithospheric flow of the Hawaiian mantle plume material towards the EPR (Fig. 9.5b). They termed such a physical scenario as "distal" plume-ridge interaction (versus the "proximal" interaction as in the case of Iceland-Reykjanes Ridge). This proposal would seem to be counter-intuitive because it requires some ~5 000 km sublithospheric flow in the LVZ against the vector of Pacific Plate motion. However, such lithosphere-asthenosphere decoupled flow in the LVZ is in fact physically straightforward (see above and Fig. 9.3a–c) because of ridge suction forces as a result of the material needs to form the crust and lithospheric mantle at ocean ridges. This is particularly the case beneath the fast-spreading EPR as the ridgeward mass flux increases with the increasing spreading rate (Fig. 9.4a,b). In contrast to the along-ridge flow of Iceland plume materials, which is focused along the ridge, the flow of Hawaiian mantle plume materials towards the EPR is more dispersed, which is unlikely to generate any significant thermal effect on the sea floor. However, as the asthenospheric flow is largely horizontal with essentially no decompression to cause melting, Hawaiian mantle plume materials (both enriched components and the depleted refractory peridotitic matrix) can survive a long distance travel to the EPR where they will melt by decompression and produce geochemically Hawaiian-like MORB (Niu et al. 1999).

9.3.2
Spreading Rate Directs Plume Flows

Figure 9.4 demonstrates that the ridge suction force or material needs/mass flux towards ridges increase with an increased spreading rate. As a result, plate-spreading rate plays a dictating role in both the direction and mass flux of plume material in the context of plume-ridge interactions. This explains many observations.

9.3.2.1
Cases of Ridge-Centered Hotspots

The along-ridge flow of Iceland plume material to both the north and south of Iceland (Mertz et al. 1991; Schilling et al. 1999) is asymmetrical. The apparent chemical and thermal effect of plume-ridge interaction extends more than ~1 500 km along the Reykjanes Ridge to the south (Schilling 1991; Ito et al. 1996), but it is no more than ~400 km along the South Kolbeinsey ridge north of Iceland (Schilling et al. 1999). The total spreading rate along the Reykjanes Ridge increases from >20 mm yr^{-1} southward, whereas the total spreading rate north of Iceland decreases from <20 mm yr^{-1} northward. Therefore, the Icelandic plume materials prefer to flow along the Reykjanes ridge to the south of Iceland rather than to the north (Fig. 9.6) because of the greater material demand beneath the faster-spreading Reykjanes ridge than the slower-spreading Kolbeinsey ridge north of Iceland (Fig. 9.4). This asymmetrical effect of the Iceland plume, which we believe results from spreading rate differences, has been overlooked in all models of plume-ridge interactions.

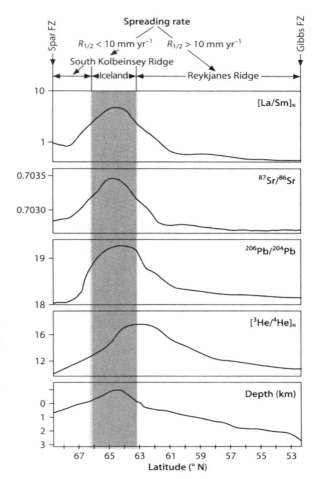

Fig. 9.6. Schematic representation of topography (*bottom panel*) and geochemical trends of MORB along the North Mid-Atlantic Ridge to illustrate the asymmetrical effects of Iceland plumes. Geochemically, the Iceland plume has greater influence on the Reykjanes Ridge (RR) to the south both in amplitude and distance than on the South Kolbeinsey Ridge (SKR) to the north. The large topographic gradient from Iceland down to the Gibbs transform reflects the large thermal effect of the Iceland plume. The apparent shallower depth along the SKR is likely due to the thermal effect of the Jan Mayen plume to the north. We interpret the asymmetrical Iceland plume effect results from the greater spreading rate (>20 mm yr^{-1} full spreading rate) along the RR than the SKP, where the full spreading rate is <20 mm yr^{-1}. Greater spreading rate requires greater mass supply to form the lithosphere (Fig. 9.4). The schematic lava compositional variation is based on data given by Schilling et al. (1983, 1999) and Hanan et al. (2000)

Another example is the Amsterdam/St. Paul hotspots and their interactions with the Southeast Indian Ridge (Graham et al. 1999). Away from the hotspots, the plume-influenced topographic expression extends >3 000 km along the Southeast Indian Ridge down to the Australian-Antarctic Discordance, where the full spreading rate is on average >70 mm yr^{-1}. However, the topographic expression is rather weak to the northwest, where the full spreading rate is <65 mm yr^{-1}. A systematic sampling and large geochemical database with good spatial coverage is needed to evaluate the geochemical consequences.

9.3.2.2
The East Pacific Rise-Mid-Atlantic Ridge "Paradox"

What is important in the context of plume-ridge interaction is the first-order differences between the fast-spreading EPR and the slow-spreading MAR. There are many topographically and geochemically identifiable mantle plumes/hotspots near the axis of the MAR (e.g., Iceland, Azores, 14° N anomalies, Ascension, Tristan, Gough, Shona, Bouvet and others), but the identifiable hotspots/plumes along the entire EPR are few, notably, the Easter hotspot at ~27° S on the Nazca plate. Such apparent unequal hotspot distribution would predict that enriched E-type MORB must be more abundant at the MAR than at the EPR. This is, however, not observed. In fact, along the MAR, away from those on- and near-axis hotspots/plumes and a few "wet" spots (e.g., Niu et al. 2001), the ridge segments are deep with depleted N-type MORB. In contrast, the EPR axis is generally shallow and enriched E-type MORBs occur along much of the axis where samples are available (Langmuir et al. 1986; Hekinian et al. 1989; Sinton et al. 1991; Batiza and Niu 1992; Reynolds et al. 1992; Perfit et al. 1994; Mahoney et al. 1994; Niu et al. 1996, 1999; Regelous et al. 1999). The EPR MORB and MAR MORB are compared in terms of average abundances of incompatible trace elements (Fig. 9.7a) and Sr, Nd and Pb isotopes (Fig. 9.7b) derived from the LDEO global MORB database (Lehnert et al. 2000). The statistically significant (>99.9% confidence level) linear correlation with close to unity slope (1.0229) (Fig. 9.7a) suggests that the mean abundance of incompatible trace elements in EPR and MAR MORB are quite similar. The highly incompatible elements such as Ba, Rb, Nb, Th, U, and Ta are plotted below the 1:1 line, suggesting that these elements are indeed slightly more abundant in MAR MORB than in EPR MORB while other elements are less so. However, as none of these elements deviate significantly from the 1:1 line, the overall abundance difference is insignificant. Note that the slope of 1.0229 suggests that on average EPR MORB have about 2% more incompatible elements as a whole than the MAR MORB, which is consistent with the greater extent of fractional crystallization in EPR MORB than in MAR MORB (Morel and Hekinian 1980; Natland 1980; Sinton and Detrick 1992). Figure 9.7b shows that isotopically the mean compositions of EPR and MAR MORB are identical. The relatively smaller variability (RSD% = 1σ / mean × 100) of EPR MORB below the 1:1 is also consistent with the EPR MORB being more uniform in composition as a result of a greater extent of homogenization during melt aggregation in the mantle and magma chamber processes in the crust (Batiza 1984; Niu et al. 1996, 1999, 2002) at the fast-spreading ridges (Fig. 9.7b inset).

The similar mean MORB composition of EPR to that of MAR suggests similar extents of mantle plume contributions to EPR MORB. We consider the apparent rarity

Fig. 9.7. Comparison of mean abundances of incompatible elements in **a** and Sr-Nd-Pb iso-topic ratios in **b** between MORB from the fast-spreading EPR (23° S to 23° N) and the slow-spreading MAR (55° S to 52° N) using the recently available global MORB database (Lehnert et al. 2000). Note the statistically significant correlation with a nearly unity (1.0229) slope in Fig. 9.7a, suggesting similar plume material contributions to the two ocean-ridge systems. In Fig. 9.7b, the mean Sr-Nd-Pb isotopic ratios are statistically identical, which reinforces that plume source contributions are identical at the EPR and MAR. The correlated smaller variability (RSD% = 1σ/ mean × 100) of EPR MORB isotopic ratios plotted in the *inset* reflects a well-known effect of greater extents of melt homogenization in EPR MORB. The RSD% for incompatible element abundance is not correlated, thus not shown, which is largely due to inhomogeneity in data quality found in the literature (analyzed by different means in different laboratories with variable precisions and accuracy). Nd-Sr-Pb isotopes are all determined by TIMS normalized to international standards in all laboratories. Note that logarithmic scales are used to show all the details

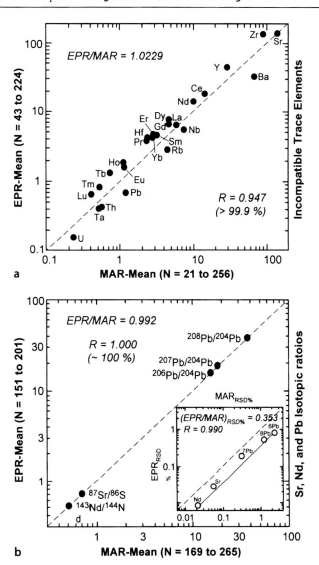

of near-ridge hotspots along the EPR results from fast-spreading. The fast spreading creates large suction forces (Figs. 9.3c and 9.4) that do not allow the development of surface expressions of mantle plumes as such, but draw plume materials to a broad zone of sub-ridge upwelling, giving rise to abundant E-type MORB and elevated and smooth axial topography. This scenario applies to the Louisville hotspot-ridge interaction at the Pacific-Antarctic Ridge in the southern Pacific, where the exact location of the current Louisville hotspot center is difficult to identify (Lonsdale 1988) because of the significant ridge suction effect. However, both the thermal effects (shoaling axis: <2 000 m below sea level) and geochemical consequences (abundant E-MORB and alkali basalts in near-ridges seamounts) of the plume-ridge interaction are conspicuous (Castillo et al. 1998).

The extensive EPR suction of plume materials is well expressed on the flanks of the southern EPR in the MELT region (Southern East Pacific Rise at 13° S–18° S; Forsyth et al. 1998), where both bathymetry and gravity display a conspicuous asymmetry across the axis with the Pacific Plate flank being much shallower, low in gravity, and importantly having many parallel volcanically active seamount chains perpendicular to the EPR axis. Some of these seamounts are volcanically active up to 150 km away from the EPR axis (Shen et al. 1995; Scheirer et al. 1996). We interpret these seamount chains as resulting from EPR suction of small mantle plumes (either hotspots or "wet" spots) beneath the Pacific Plate. These low viscosity plume materials must travel to the EPR because of the greater suction due to the high spreading rates in the area. The suction drives the flow to take the shortest path – perpendicular to the EPR axis. As the off-axis lithosphere is thin beneath the fastest spreading ridge and it is progressively thinner towards the axis, therefore the ridgeward flow of the plume materials has a component of decompression, which leads to decompression-melting and seamount formation. Predictably, (1) these seamounts must all be volcanically active, (2) most of these seamount lavas must be relatively enriched in volatiles, alkalis and other incompatible elements, and (3) the relative enrichment in lavas should decrease in seamounts closer to the axis because of progressive depletion of the easily melted components in the flowing plume material (Fig. 9.1b) as is the case along the Reykjanes Ridge (Fig. 9.5a). Petrological and geochemical data on samples with good spatial coverage are needed to test these predictions. In fact, the existing data along the Easter Seamount chain and the Foundation hotline elegantly illustrate this concept.

9.3.2.3
Geochemical Expression of Plume-Ridge Interactions Along the Easter Seamount Chain

The Easter Seamount chain (ESC) is located between Easter and Sala y Gómez islands and the East Rift of the Easter Microplate in the South East Pacific. This region has been well sampled and studied in recent years with the aim of understanding source compositions of the Easter mantle plume and the thermal and compositional linkage with the nearby EPR spreading center, which in this case is the East rift zone of the Easter microplate (Hanan and Schilling 1989; Fretzdorff et al. 1996; Haase and Devey 1996; Hekinian et al. 1996; Pan and Batiza 1998; Kingsley and Schilling 1998; Haase 2002; Kingsley et al. 2002). The ESC, also called Easter hotline, is composed of many volcanically active seamounts in a band with a width of ~150 km extending from the southern end of the East rift zone to the east for about ~900 km (Fig. 9.8a,b). Along the hotline are two conspicuous topographic anomalies marked by the Easter Island and the Sala y Gómez Islands, respectively about 400 km and 800 km away from the East Rift zone. The excess topography is proportional to the amount of melt delivered to the surface for volcanic constructions, and

▶

Fig. 9.8. a Lava geochemical systematics along the Easter Seamount chain as a function of distance to the East Rift of the Easter Microplate. While scattered, most data define systematic trends as highlighted by the *shaded bands*. **b** Cartoon illustrating that the observed lava geochemical variation is the consequence of progressively melting a two-component plume material. Ridge suction requires the hot (or "wet") plume material to flow towards the ridge with an upwelling component that causes decompression melting of the flowing plume material. The enriched dykes/veins with low solidus temperatures are progressively depleted during the ridgeward flow, thus leading to progressive melting of more depleted residual material and producing more depleted lavas towards the ridge (Fig. 9.1b,c). The geochemical data are from Pan and Batiza (1998), Kingsley et al. (1998, 2002), and the cartoon is modified from these authors

Chapter 9 · Ridge Suction Drives Plume-Ridge Interactions

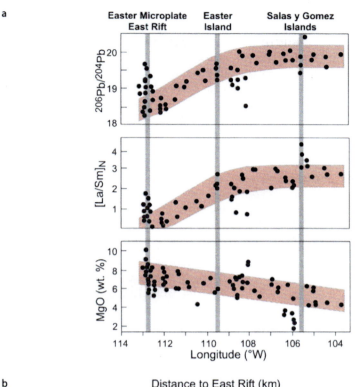

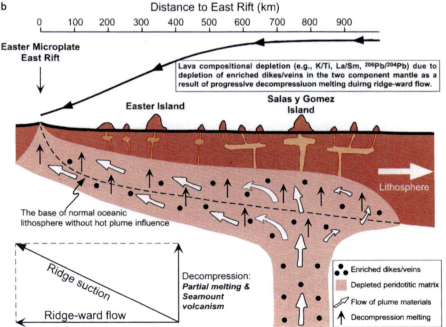

the amount of melt delivered to the surface is often considered to be the greatest over the hotspot center. Consequently, there has been a dispute about whether the present Easter hotspot is located beneath Easter Island (Haase and Devey 1996; Fretzdorff et al. 1996; Haase 2002) or is actually marked by the Sala y Gómez Islands (Hanan and Schilling 1989; Kingsley et al. 1998; Pan and Batiza 1998; Kingsley et al. 2002). Given the fact that ridge suction is the primary force that drives asthenospheric flow, it becomes physically straightforward to see that the ultimate hotspot source region must be located beneath the Sala y Gómez Islands or even to the east. If the hotspot source were beneath Easter Island, then an asthenospheric flow away from Easter Island to feed the volcanism of Sala y Gómez Islands is needed; however, there is no driving force for this flow direction. In fact, the systematic lava compositional variation (Pan and Batiza 1998; Kingsley et al. 1998, 2002) along the ESC as a function of distance to the East Rift zone supports the hypothesis that the actual hotspot is located beneath the Sala y Gómez Islands.

The scattered yet systematic declining in the enriched component towards the East Rift as indicated by geochemical parameters such as $^{206}Pb/^{204}Pb$ and $[La/Sm]_N$ (Fig. 9.8a; Also K/Ti, Na_2O, K_2O, and CaO/Al_2O_3, but not shown) is expectedly the consequence of "flow differentiation" discussed above (Figs. 9.1b and 9.5a). Note that a total spreading rate of this part of the EPR is about 160 mm yr^{-1}. As the Easter microplate rotates clockwise (Naar and Hey 1991), the East rift at the southeast corner (where the ESC meets) has a total spreading rate close to 160 mm yr^{-1}, thus having the greatest suction force for driving the asthenospheric flow of the Easter plume material. As the lithosphere is thinning towards the rift, this ridgeward asthenospheric flow also has a component of decompression (Fig. 9.8b). The latter inevitably leads to the partial melting of the flowing plume material beneath the lithosphere. The enriched dykes/veins of the plume source have lower solidus temperature, and will thus preferentially melt first, producing highly enriched lavas on the islands of Sala y Gómez. Decompression melting of the flowing plume material that is progressively depleted in the enriched and easily melted dykes/veins will produce melts that are progressively depleted in enriched components towards the East Rift as seen in erupted lavas (Fig. 9.8a). The physical mechanism is the same as described for lava geochemical signals along the Reykjanes Ridge – "flow differentiation" within the plume material that is flowing and melting. The ridgeward increase in MgO, which is proportional to eruption temperatures, results from composition-dependent melt evolution (Niu et al. 2002). Enriched melts are enriched in alkalis and volatiles that lower both the liquidus and solidus temperatures of silicate melts. Consequently, enriched melts cool to a lower liquidus temperature and crystallize to a greater extent (thus lower MgO) than depleted melts before solidification.

Our "flow-differentiation" model differs from the interpretations of a binary mixing between a distinctive (singular) plume source and the distinctive MORB source (Schilling 1991; Schilling et al. 1983, 1994, 1999; Kingsley et al. 1998, 2002). We emphasize that such ridgeward geochemical depletion in erupted lavas is an inevitable consequence of interaction between off-ridge mantle plumes and fast-spreading ridges. The same is also well expressed along the Foundation hotline volcanics (Hekinian et al. 1997, 1999) (see below). Note that the ridgeward flow of plume materials and the excess heat provided do not allow the normal development of the lithosphere by conductive cooling, thus leading to the thin lithosphere. This process is not the same as "thermal" erosion (Schilling 1991; Kincaid et al. 1995), which requires the pre-existence of a "perfect" lithosphere (as indicated by the *dashed line*) followed by erosion or channelization.

9.3.2.4
Geochemical Expression of Plume-Ridge Interactions Along the Foundation Hotline

The Foundation Seamount chain is a fairly recent discovery (e.g., Mammerickx 1992; Devey et al. 1996; Hekinian et al. 1997, 1999) that extends northwestward for >1700 km from the inferred present-day location of the Foundation hotspot at ~36° S, 114° W (Devey et al. 1996; Hekinian et al. 1997, 1999; O'Connor et al. 1998, 2001; Maia et al. 2000, 2001). The nature of the hotspot track is verified by the along-chain lava age progression (O'Connor et al. 1998, 2001) that is consistent with the notion that the Pacific Plate has spread northwestward at a speed of ~91±2 mm yr^{-1} for the last 21 Ma (O'Connor et al. 1998; see Sect. 11.5). The Foundation hotspot is inferred to be located in between ~35°45' S, 114° W and 37°35' S, 114°20' S with a radius of ~200 km (Maia et al. 2001), and is about 400 km from the Pacific-Antarctic Ridge axis at ~37°45' S, 111°7.5' W (Hekinian et al. 1999; Maia et al. 2001). The volcanically active seamounts between the PAR axis and the hotspot volcanoes form elongated volcanic ridges (Hekinian et al. 1997, 1999; O'Connor et al. 2001). These volcanically active ridges reveal a vivid expression of plume-ridge interaction between the Foundation hotspot and the Pacific-Antarctic Ridge. The systematic variation in the composition of these volcanics as a function of distance to the ridge axis shown in Fig. 9.9a is identical to that along the ESC (Fig. 9.8a) and is readily explained by the same process – ridge suction of the Foundation plume material that melts by decompression, and produces progressively more depleted lavas as a result of progressive depletion of easily melted dykes/veins in the ridgeward flowing plume material (Fig. 9.9b).

9.4
Summary and Conclusion

The main points of this contribution are listed below:

1. Plate tectonics and mantle plumes are genetically unrelated, but when the ascending mantle plumes approach lithospheric plates, interactions between the two occur. Such interactions are most prominent at or near ocean ridges, where the lithosphere is thin and the effect of mantle plumes is best revealed.
2. In order to fully understand plume-ridge interactions and their geophysical and geochemical expressions, some basic concepts must be understood and some variables must be considered. These include (a) the nature and composition of plume materials; (b) the actual role of ocean ridges; (c) the effect of plate spreading rate; and (d) plume-ridge distance.
3. Mantle plume materials are necessarily heterogeneous, and are likely to have two components: metasomatized, easily melted dykes/veins enriched in volatiles, alkalis and other incompatible elements dispersed in the more refractory, predominantly peridotitic matrix. Partial melting of such a two-component mantle produces geochemically enriched melts in its early stages, but progressive melting will produce progressively depleted melts because of progressive depletion of the enriched dykes/veins in the source regions. This is best manifested by systematic variations in lava composition (a) along the Reykjanes Ridge away from the Iceland, (b) along the Easter Seamount chain towards the East Rift, and (c) along the Foundation hotline volcanic ridges towards the Pacific-Antarctic Ridge.

4. While ocean ridges are mostly passive features in the context of plate tectonics, they play an active role in terms of plume-ridge interaction. This active role is manifested by the fact that the low velocity zone beneath ocean ridges represents regions of low pressure that allow asthenospheric flow (i.e., ridge suction). The ridge suction results from a need for material to form the oceanic crust and oceanic lithospheric mantle in the broad zone of upwelling beneath ocean ridges. The ridge suction or mass flux towards ocean ridges is linearly proportional to spreading rate.

5. Because the LVZ has the lowest viscosity that increases exponentially with depth, the mass flow towards ocean ridges is largely horizontal beneath the lithosphere. It follows that the spreading lithospheric plates must necessarily be decoupled from the sublithospheric flow. The degree of the decoupling increases with increasing plate spreading rate, because of the spreading-rate dependent material demand/flow towards the ridge.

6. The commonly interpreted geochemical "mixing" between compositionally distinct enriched plume material and depleted "MORB" source in the context of plume-ridge interaction is misleading. The apparent mixing relationship in geochemical diagrams is NOT a physical mixing either in the solid state or in the melt form, but the consequences of melting a two-component mantle. In other words, the so-called geochemical mixing is physically a differentiation of the two components from a composite lithology through partial melting.

7. The commonly perceived plume dispersion model, i.e., invasion of distinctively enriched plume materials into the depleted MORB mantle, has physical difficulties in making space available for the invasion. However, the "flow differentiation" model we present here in a sub-ridge or near-ridge environment is physically straightforward. Ridge suction requires that plume material flow towards the ridge. This ridgeward flow has a decompression component that induces partial melting of the flowing plume material, which in turn will generate volcanic activities between hotspots and the nearby ridges (e.g., the Easter Seamount chain and Foundation hotline volcanic ridges). The ridgeward flowing and melting plume material is progressively depleted in the enriched dykes/veins, resulting in the subsequent melts being progressively more depleted towards the ridge.

8. Because ridge suction increases with increasing spreading rate, ridge-centered mantle plumes, such as in Iceland, will have a greater thermal and compositional effect on the faster-spreading (>20 mm yr^{-1}) Reykjanes Ridge than the slower-spreading (<20 mm yr^{-1}) South Kolbeinsey Ridge to the north. The similar mean abundance of incompatible elements and identical mean Sr-Nd-Pb isotopic ratios between EPR MORB and MAR MORB suggest statistically similar plume material contributions to MORB melts in these two ocean-ridge systems. In contrast to the MAR, the rarity of near-ridge plumes along the EPR are due to its fast-spreading,

▶

Fig. 9.9. a Geochemical systematics along the Foundation hotline volcanic ridges as a result of interactions between the Foundation plume and the Pacific-Antarctic Ridge near 37° S, 111° W. The stippled columns mark the approximate location of Foundation hotspot seamounts. The data are from Hekinian et al. (1997, 1999) and Devey et al. (1997) with evolved samples (andesites, dacites and rhyolites having $SiO_2 > 52$ wt.%) excluded. The data systematics and interpretations are the same as for the Easter hotline in Fig. 9.8a. Note that, in contrast to the systematics-defined by major and trace element data shown here, Pb isotopic data by Maia et al. (2001) show no systematics at all, which also contrasts with Pb isotopic systematics along the Easter hotline (Fig. 9.8a). **b** Cartoon showing essentially the same as in Fig. 9.8b

Chapter 9 · Ridge Suction Drives Plume-Ridge Interactions   303

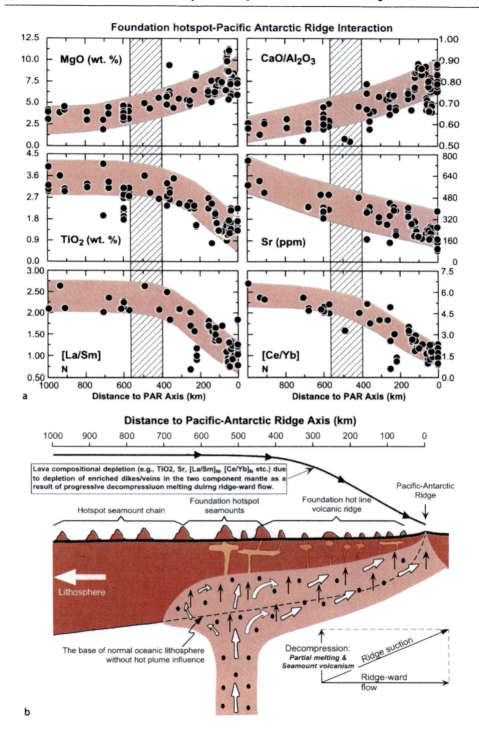

which results in extensive ridge suction forces or ridgeward mass flow and does not allow the development of surface expressions of small mantle plumes but incorporates them into the broad zone of mantle upwelling and fast oceanic crust accretion at the EPR.

Acknowledgements

R. Hekinian acknowledges the Alexandre von Humboldt award and the Department of Geology, University of Kiel (Germany) for their support. Y. Niu thanks Cardiff University for support and UK NERC for support through a Senior Research Fellowship. Y. Niu also acknowledges the support from University of Houston for completing the research.

References

Batiza R (1982) Abundances, distribution and sizes of volcanoes in the Pacific Ocean and implications for the origin of non-hotspot volcanoes. Earth Planet Sc Lett 60:195–206

Batiza R (1984) Inverse relationship between Sr isotope diversity and rate of oceanic volcanism has implications for mantle heterogeneity. Nature 309:440–441

Batiza R, Niu Y (1992) Petrology and magma chamber processes at the East Pacific Rise ~9°30′ N. J Geophys Res 97:6779–6797

Batiza R, Vanko DA (1984) Petrology of young Pacific seamounts. J Geophys Res 89:11235–11260

Campbell IH, Griffiths RW (1990) Implications of mantle plume structure for the evolution of flood basalts. Earth Planet Sc Lett 99:79–93

Castillo PR, Natland JH, Niu Y, Lonsdale P (1998) Sr, Nd, and Pb isotopic variation along the Pacific ridges from 53 to 56° S: Implications for mantle and crustal dynamic processes. Earth Planet Sc Lett 154:109–125

Davies GF, Richards MA (1992) Mantle convection. J Geol 100:151–206

Deng J, Zhao H, Luo, Guo Z, Mo X (1998) Mantle plumes and lithosphere motion in east Asia. Geodynam Ser 27:59–66

Devey CW, Hekinian R, Stoffers P, Ackermand D, Binard N, Drusch M, Francke B, Hémond C, Kapsimalis V, Lorenc S, Maia M, Möller H, O'Connor J, Perrot K, Pracht J, Ramm D, Rogers T, Stattegger K, Steinke S, Victor P (1997) A first survey and sampling of the Foundation Seamount Chain. Mar Geol 137:191–200

Duncan RA, Richards MA (1991) Hotspots, mantle plumes, flood basalts, and true polar wander. Rev Geophys 29:31–50

Feighner MA, Richards MA (1995) The dynamics of plume-ridge and plume plate interactions: An experimental investigation. Earth Planet Sc Lett 129:171–182

Forsyth DW (1998) Geophysical constraints on mantle flow and melt generation beneath mid-ocean ridges. AGU, Washington, D.C., Geophys Monogr Ser 71:1–66

Forsyth DW, The MELT Seismic Team (1998) Imaging the deep seismic structure beneath a mid-ocean ridge: The MELT experiment. Science 280:1215–1218

Fowler CMR (1990) The solid Earth: An introduction to global geophysics. Cambridge University Press Cambridge 472 pp

Fretzdorff S, Haase KM, Garbe-Schonberg C-D (1996) Petrogenesis of lavas from the Umu volcanic field in the young hotspot region west of Easter Island, SE Pacific. Lithos 38:23–40

Georgen JE, Lin J, Dick HJB (2001) Evidence from gravity anomalies for interactions of the Marion and Bouvet hotspots with the Southwest Indian Ridge: Effects of transform offsets. Earth Planet Sc Lett 187:283–300

Graham DW, Johnson KTM, Priebe LD, Lupton JE (1999) Hotspot-ridge interaction along the Southeast Indian Ridge near Amsterdam and St. Paul Islands: Helium isotope evidence. Earth Planet Sc Lett 167:297–310

Green DH, Falloon TJ (1998) Pyrolite: A Ringwood concept and its current expression. In: Jackson I (ed) The Earth's mantle - Composition, structure, and evolution. Cambridge University Press, Melbourne, pp 311–380

Hanan BB, Schilling J-G (1989) Easter microplate evolution: Pb isotope evidence. J Geophys Res 94: 7432–7448

Hanan BR, Blichert-Toft J, Kingsley R, Schilling J-G (2000) Depleted Iceland mantle plume geochemical signature: Artifact of multicomponent mixing? Geochem Geophy Geosy 1 (article) 1999GC000009

Haase KM (2002) Geochemical constraints on magma sources and mixing processes in Easter Microplate MORB (SE Pacific): a case study of plume-ridge interaction: Chem Geol 182:335-355

Haase KM, Devey CW (1996) Geochemistry of lavas from the Ahu and Tupa volcanic fields, Easter Hotspot, Southeast Pacific: Implications for intraplate magma genesis near a spreading axis. Earth Planet Sc Lett 137:129-143

Halliday AN, Dickin AP, Fallick AE, Fitton JG (1988) Mantle dynamics: A Nd, Sr, Pb and O isotope study of the Cameroon Line volcanic chain. J Petrol 29:181-211

Halliday AN, Lee D-C, Tommasini S, Davies GR, Paslick CR, Fitton JG, James DE (1995) Incompatible trace elements in OIB and MORB source enrichment in the sub-oceanic mantle. Earth Planet Sc Lett 133:379-395

Hanan BR, Blichert-Toft J, Kingsley R, Schilling J-G (2000) Depleted Iceland mantle plume geochemical signature: Artifact of multicomponent mixing? Geochem Geophy Geosy 1 (article) 1999GC000009

Hekinian R, Thompson G, Bideau D (1989) Axial and off-axial heterogeneity of basaltic rocks from the East Pacific Rise at 12°35' N-12°51' N and 11°26' N-11°30' N. J Geophys Res 94:17437-17463

Hekinian R, Bideau D, Herbért R, Niu Y (1995) Magmatic processes at upper mantle-crustal boundary zone: Garrett transform (EPR South). J Geophys Res 100:10163-10185

Hekinian R, Francheteau J, Armijo R, Cogne JP, Constantin M, Girardeau J, Hey RN, Naar DF, Searl R (1996) Petrology of the Easter microplate region in the south Pacific. J Volc Geotherm Res 72:259-289

Hekinian R., Stoffers P, Devey C, Ackermand D, Hémond C, O'Connor J, Binard N, Maia M (1997) Intraplate versus ridge volcanism on the Pacific Antarctic ridge near 37° S-111° W. J Geophys Res 94:12265-12286

Hekinian R, Stoffers P, Ackermand D, Revillon S, Maia M, Bohn M (1999) Ridge-hotspot interaction: the Pacific-Antarctic Ridge and the foundation seamounts. Mar Geol 160:199-223

Hill RI, Campbell IH, Davies GF, Griffiths RW (1992) Mantle plumes and continental tectonics. Science 256:186-193

Hofmann AW, White WM (1982) Mantle plumes from ancient oceanic crust. Earth Planet Sc Lett 57: 421-436

Ito G, Lin J (1995a) Oceanic spreading center-hotspot interactions: Constraints from along-isochron bathymetric and gravity anomalies. Geology 23:657-660

Ito G, Lin J (1995b) Mantle temperature anomalies along the present and paleoaxes of the Galapagos spreading center as inferred from gravity analyses. J Geophys Res 100:3733-3745

Ito G, Lin J, Gable CW (1996) Dynamics of mantle flow and melting at a ridge-centered hotspot: Iceland and the Mid-Atlantic Ridge. Earth Planet Sc Lett 144:53-74

Johnson RW (ed) (1989) Intraplate volcanism in eastern Australia and New Zealand. Cambridge University Press Melbourne, 408 pp

Kincaid C, Ito G, Gable C (1995) Laboratory investigations of the interaction of off-axis mantle plumes and spreading centres. Nature 376, 758-761

Kincaid C, Schilling J-G, Gable C (1996) The dynamics of off-axis plume-ridge interaction in the upper mantle. Earth Planet Sc Lett 137:29-43

Kingsley R, Schilling J-G (1998) Plume-ridge interaction in the Easter-Salas y Gómez seamount chain-Easter microplate system: Pb isotope evidence. J Geophys Res 103:24159-24177

Kingsley R, Schilling J-G, Dixon JE, Swart P, Poreda R, Simons K (2002) D/H ratios in basalt glasses from the Salas y Gómez mantle plume interacting with the East Pacific Rise: Water from old D-rich recycled crust or primordial water from the lower mantle? Geochem Geophy Geosy 1 (article) 2001GC000199

Lambeck K, Johnston P (1998) The viscosity of the mantle: Evidence from analyses of glacial-rebound phenomena. In: Jackson I (ed) The Earth's mantle - Composition, structure, and evolution. Cambridge University Press, Melbourne, pp 461-502

Langmuir CH, Bender JF, Batiza R (1986) Petrological and tectonic segmentation of the East Pacific Rise, 5°30'-14°30' N. Nature 332:422-429

Lehnert K, Su Y, Langmuir CH, Sarbas B, Nohl U (2000) A global geochemical database structure for rocks. Geochem Geophy Geosy 1 (technical brief) 1999GC000026

Lonsdale P (1988) Geography and history of the Louisville hotspot chain in the Southern Pacific. J Geophys Res 93:3078-3104

Mahoney JJ, Sinton JM, Kurz DM, Macdougall JD, Spencer KJ, Lugmair GW (1994) Isotope and trace element characteristics of a super-fast spreading ridge: East Pacific Rise, 13-23° S. Earth Planet Sc Lett 121:173-193

Maia M, Ackermand D, Dehghani GA, Gente P, Hekinian R, Naar D, O'Connor J, Perrot K, Phipps Morgan J, Ramillien G, Révillon S, Sabetian A, Sandwell D, Stoffers P (2000) The Pacific-Antarctic Ridge-Foundation hotspot interaction: A case study of a ridge approaching a hotspot. Mar Geol 167:61-84

Maia M, Hémond C, Gente P (2001) Contrasted interactions between plume and lithosphere: The Foundation chain case. Geochem Geophy Geosy 1 (article), 2000GC000117

McKenzie D, Bickle MJ (1988) The volume and composition of melt generated by extension of the lithosphere. J Petrol 29:625–679

Mertz DF, Devey CW, Todt W, Stoffers P, Hofmann AW (1991) Sr-Nd-Pb isotope evidence against plume-asthenosphere mixing north of Iceland. Earth Planet Sc Lett 107:243–255

Morel JM, Hekinian R (1980) Compositional variation of volcanics along segments of recent spreading ridges. Contrib Mineral Petrol 72:425–436

Morgan JW (1971) Convection plumes in the lower mantle. Nature 230:42–43

Morgan JW (1981) Hotspot tracks and opening of the Atlantic and Indian Oceans. In: Emiliani C (ed) The sea, vol. 7. Wiley, New York, pp 443–487

Naar DF, Hey RN (1991) Tectonic evolution of the Easter microplate. J Geophys Res 96:7961–7993

Natland JH (1980) Effect of axial magma chambers beneath spreading centers on the composition of basaltic rocks. Init Rep Deep Sea Drill Proj 54:833–850

Niu Y (1997) Mantle melting and melt extraction processes beneath ocean ridges: Evidence from abyssal peridotites. J Petrol 38:1047–1074

Niu Y, Batiza R (1994) Magmatic processes at the Mid-Atlantic ridge ~ 26° S. J Geophys Res 99:19719–19740

Niu Y, Batiza R (1997) Trace element evidence from seamounts for recycled oceanic crust in the eastern equatorial Pacific mantle. Earth Planet Sc Lett 148:471–484

Niu Y, Hekinian R (1997a) Basaltic liquids and harzburgitic residues in the Garrett transform: A case study at fast-spreading ridges. Earth Planet Sc Lett 146:243–258

Niu Y, Hekinian R (1997b) Spreading rate dependence of the extent of mantle melting beneath ocean ridges. Nature 385:326–329

Niu Y, Waggoner DG, Sinton JM, Mahoney JJ (1996) Mantle source heterogeneity and melting processes beneath seafloor spreading centers: The East Pacific Rise, 18°–19° S. J Geophys Res 101:27711–27733

Niu Y, Collerson KD, Batiza R, Wendt I, Regelous M (1999) The origin of E-type MORB at ridges far from mantle plumes: The East Pacific Rise at 11°20' N. J Geophys Res 104:7067–7087

Niu Y, Bideau D, Hekinian R, Batiza R (2001) Mantle compositional control on the extent of melting, crust production, gravity anomaly and ridge morphology: A case study at the Mid-Atlantic Ridge 33–35° N. Earth Planet Sc Lett 186:383–399

Niu Y, Regelous M, Wendt JI, Batiza R, O'Hara MJ (2002) Geochemistry of near-EPR seamounts: Importance of *source* vs. *process* and the origin of enriched mantle component. Earth Planet Sc Lett 199:329–348

O'Connor JM, Stoffers P, Wijbrans JR (1998) Migration rate of volcanism along the Foundation Chain, SE Pacific. Earth Planet Sc Lett 164:41–59

O'Connor JM, Stoffers P, Wijbrans JR (2001) En echelon volcanic elongate ridges connecting intraplate Foundation Chain volcanism to the Pacific-Antarctic spreading center. Earth Planet Sc Lett 189:93–102

Pan Y, Batiza R (1998) Major element chemistry of volcanic glasses from the Easter Seamount Chain: Constraints on melting conditions in the plume channel. J Geophys Res 103:5287–5304

Perfit MR, Fornari DJ, Smith MC, Bender JF, Langmuir CH, Haymon RM (1994) Small-scale spatial and temporal variations in mid-ocean ridge crest magmatic processes. Geology 22:375–379

Phipps Morgan J, Morgan JW (1998) Two-stage melting and the geochemical evolution of the mantle: A recipe for mantle plum-pudding. Earth Planet Sc Lett 170:215–239

Phipps Morgan J, Morgan JW, Zhang Y-S, Smith WHF (1995) Observational hints for a plume-fed, suboceanic asthenosphere and its role in mantle convection. J Geophys Res 100:12753–12767

Regelous M, Niu Y, Wendt JI, Batiza R, Greig A, Collerson KD (1999) An 800 ka record of the geochemistry of magmatism on the East Pacific Rise at 10°30' N: Insights into magma chamber processes beneath a fast-spreading ocean ridge. Earth Planet Sc Lett 168:45–63

Regelous M, Niu Y, Castillo P, Batiza R, Greig A (2001) Contrasting geochemistry of on- and off-axis magmatism, 26° S Mid-Atlantic Ridge. EOS Trans Am Geophys Union 82(47):F1275–1276

Reynolds JR, Langmuir CH, Bender JF, Kastens KA, Ryan WBF (1992) Spatial and temporal variability in the geochemistry of basalts from the East Pacific Rise. Nature 359:493–499

Ribe NM (1996) The dynamics of plume-ridge interaction, 2: Off-ridge plumes. J Geophys Res 101: 16195–16204

Scheirer DS, Macdonald KC, Forsyth DW, Shen Y (1996) Abundant seamounts of the Rano Rahi Seamount Field near the Southern East Pacific Rise, 15° S to 19° S. Mar Geophys Res 18:13–52

Schilling J-G (1991) Fluxes and excess temperatures of mantle plumes inferred from their interaction with migrating mid-ocean ridges. Nature 352:397–403

Schilling J-G, Zajac M, Evans R, Johnston T, White W, Devine JD, Kingsley R (1983) Petrological and geochemical variations along the Mid-Atlantic Ridge from 29° N to 73° N. Am J Sci 283:510–586

Schilling J-G, Hanan BB, McCully B, Kingsley RH, Fontignie D (1994) Influence of the Siera Leone mantle plume on the equatorial Mid-Atlantic Ridge: A Nd-Sr-Pb isotopic study. J Geophys Res 99:12005–12028

Schilling J-G, Ruppel C, Davis AN, McCully B, Tighe SA, Kingsley RH, Lin J (1995) Thermal structure of the mantle beneath Equatorial Mid-Atlantic Ridge: Inferences from spatial variations of dredged basalt glass composition. J Geophys Res 100:10057–10076

Schilling J-G, Kingsley R, Fontignie D, Poreda R, Xue S (1999) Dispersion of the Jan Mayen and Iceland mantle plumes in the Arctic: A He-Pb-Nd-Sr isotope tracer study of basalts from the Kolbeinsey, Mohns, and Knipovich ridges. J Geophys Res 104:10543–10569

Shen Y, Scheirer DS, Forsyth DW, Macdonald KC (1995) Trade-off in production between adjacent seamount chains near the East Pacific Rise. Nature 373:140–143

Sinton JM, Detrick RS (1992) Mid-ocean ridge magma chambers. J Geophys Res 97:197–216

Sinton JM, Smaglik SM, Mahoney JJ (1991) Magmatic processes at superfast spreading mid-ocean ridges: Glass compositional variations along the East Pacific Rise 13°–23° S. J Geophys Res 96:6133–6155

Silver PG, Holt WE (2002) The mantle flow field beneath western north America. Science 295:1054–1057

Sleep NH (1996) Lateral flow of hot plume material ponded at sublithospheric depths. J Geophys Res 101:28065–28083

Stein S, Stein CA, (1996) Thermo-mechanical evolution of oceanic lithosphere: Implications for the subduction processes and deep earthquake. AGU Geophys Monogr 96:1–17

Sun S-S, McDonough WF (1989) Chemical and isotopic systematics of ocean basalt: Implications for mantle composition and processes. Geol Soc Spec Publ 42:323–345

Sun S-S, Tatsumoto M, Schilling J-G (1975) Mantle plume mixing along the Reykjanes ridge axis: Lead isotope evidence. Science 190:143–147

Taylor RN, Thirwall MF, Morton JB, Hilton DR, Gee MAM (1997) Isotopic constraints on the influence of the Icelandic plume. Earth Planet Sc Lett 148:E1–E8

Turcotte D L, Morgan JP (1992) Magma migration and mantle flow beneath a mid-ocean ridge. AGU Geophys Monogr 71:155–182

Wendt JI, Regelous M, Niu Y, Hekinian R, Collerson KD (1999) Geochemistry of lavas from the Garrett transform fault: Insights into mantle heterogeneity beneath the eastern Pacific. Earth Planet Sc Lett 173:271–284

Zhang M, Zhou X-H, Zhang J-B (1998) Nature of the lithospheric mantle beneath NE China: Evidence from potash volcanic rocks and mantle xenoliths. In: Flower MFJ, Chung S-L, Lo C-H, Lee T-Y (eds) Mantle dynamics and plate interactions in East Asia. AGU Washington, D.C., Geodynam Ser 27:197–219

Zhang Y-S, Tanimoto T (1993) High-resolution global upper mantle structure and plate tectonics. J Geophys Res 98:9793–9823

Zindler A, Staudigel H, Batiza R (1984) Isotope and trace element geochemistry of young Pacific seamounts: Implications for the scale of upper mantle heterogeneity. Earth Planet Sc Lett 70:175–195

Chapter 10

Intraplate Gabbroic Rock Debris Ejected from the Magma Chamber of the Macdonald Seamount (Austral Hotspot): Comparison with Other Provinces

D. Bideau · R. Hekinian

10.1
Introduction

The Macdonald Seamount (Fig. 10.1a) is located at the tip of the Austral hotline (Johnson 1970, 1980; Talandier and Okal 1984). The activity of the Austral and Society hotspots has been closely monitored by the detection of seismic swarms recorded by the French Polynesian seismic network ("Réseau Sismique Polynésien", RSP). Since the Austral Islands are too far away from the receiving stations, only 'T' waves have been detected from the Macdonald Seamount (Talandier and Okal 1984; see Sects. 2.4.1 and 2.4.2). This seamount is one of the most active submarine volcanoes in the world (Cheminée et al. 1991) and was first noticed after a strong seismic swarm was detected by the hydrophones of the Hawaiian Institute of Geophysics Network (Norris and Johnson 1969). A multibeam bathymetric survey (NO JEAN CHARCOT, FS SONNE and NO L'ATALANTE) of the most recent seamounts of the Society and the Austral hotspots was undertaken in 1986 and 1987. The edifices were also sampled, and several dredge hauls were undertaken on top of the Macdonald Seamount (Fig. 10.1b). Related publications have mainly dealt with the morphology and the structure of the Society and Austral hotspots and the petrology of the volcanics (Stoffers et al. 1989; Sect. 5.3.1.1; Hekinian et al. 1991). Among the samples recovered from the Macdonald Seamount, highly vesicular pillow lavas, volcaniclastics and accidental rock debris were found. The gabbroic clasts were ejected during hydromagmatic explosive events nearly twenty years after the seamount was first discovered. Later, they were partially covered by basanite lapilli during further explosions (Sect. 5.3.1.1; Hekinian et al. 1991).

Based on the magnetic anomalies in the South Pacific, the age of the ocean floor at the base of the Society and Austral volcanic chains is about 45 to 65 Ma, and the spreading rate of the Pacific Plate is about 10 to 11 cm yr^{-1} in the northwest direction (Herron 1972; Duncan and McDougall 1976). The Austral hotspot corresponds to elongated clusters of volcanic edifices erected on top of a regional bulge (Stoffers et al. 1989), which is characterized by a topographic low, up to 100 km in diameter, separating the foot of the volcanoes (3 500–3 800 m deep) and the surrounding abyssal hill region (>4 000 m deep). This bulge exhibits smooth morphology and a thin, approximately 50 cm thick sediment cover. The hotspot swell rises about 300–500 m above the level of the neighboring deep ocean floor. Beyond this depth, the old oceanic crust is characterized by a sediment thickness of more than 20 m and by fault escarpments (Cheminée et al. 1989). The Austral volcanic chain extends approximately 1 300 km in the NW-SE direction (Fig. 10.1a) from the island of Maria (22° S, 155° W) to the islands of Marotiri (28° S, 143° W). The regional survey of its southeastern extension is centered on the Macdonald Seamount and extends about 70 km east and 170 km west towards the island of Marotiri

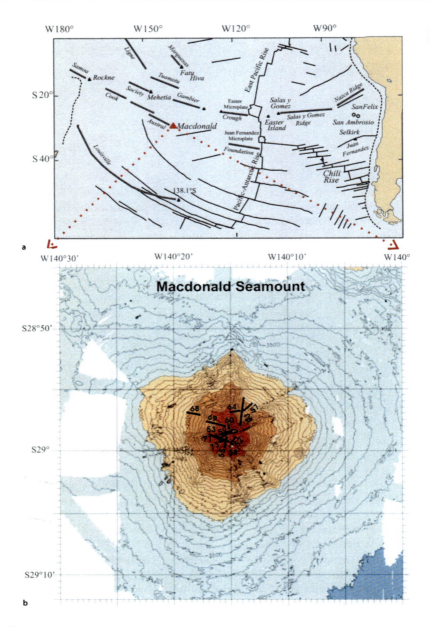

Fig. 10.1. a Main structural features of the South Pacific Ocean floor. The *black lines* indicate the location of the spreading centers, related transform faults and other fracture zones and microplate boundaries. The *lighter, gray lines* represent the intraplate hotlines. The *dashed lines* show the location of deep troughs (convergent plate boundaries). The *triangles* indicate the position of the main known hotspots (submarine volcanoes or islands), and the *circles* represent volcanically inactive islands. The location of the Macdonald Seamount at the tip of the Austral volcanic lineament is also indicated. b SeaBeam bathymetric map of the entire Macdonald Seamount edifice, with 100 m contour intervals (Bonneville, personal communication). The *dark lines* and *numbers* represent the dredge location during the 1987 cruise of the FS SONNE at the summit of the volcano, after Stoffers et al. (1989)

Chapter 10 · Intraplate Gabbroic Rock Debris Ejected from the Macdonald Seamount   311

(Stoffers et al. 1989). Along the 296° W trend of the volcanic chain, several seamounts were observed dominating a general ocean-floor level, which becomes shallower from the base of the Macdonald Seamount (3 700–3 950 m depth at 29° S, 139°30' W) towards the island of Marotiri (3 500–3 650 m depth at 28°30' S, 141°30' W). This depth decrease of the sea floor around the seamounts is also observed on a single channel bathymetric profile obtained in 1986 by the French Navy Vessel ESTAFETTE (Stoffers et al. 1989).

The southeastern end of the Macdonald neo-volcanic zone is located near 29° S, 139°50' W, about 20 km east of the base (3 850 m contour line) of the volcano. Further east, the deep ocean floor shows less than 300 m high westward-facing scarps, having NS fabric like the magnetic anomalies (Herron 1972). The thickness of the sediments estimated from the 3.5 kHz echo-sounder profiles exceeds 10 m in this region (Stoffers et al. 1989). The Macdonald Volcano (28.99° S, 140.15° W) is the only seamount surveyed in this area where volcanic and hydrothermal activity was observed on at least two sites (Stoffers et al. 1989; Hekinian et al. 1991, 1993b). The hydrothermal activity is probably related to the discharge of CH_4, CO_2 and SO_2 (Michaelis, personal communication; Sects. 13.4.1 and 13.4.2), which was observed near the summit (<200 m in depth) and in a small crater (2 000 m in depth) on the southeastern flank of the volcano. The Macdonald Seamount has a narrow summit at 39 m depth, which is made up of several volcanic pinnacles of scoriaceous agglutinated lapilli (Hekinian et al. 1991).

The present work is mainly focused on the gabbroic clasts that were scattered on the flanks of the Macdonald Seamount during hydromagmatic explosive events and partially covered by basanite lapilli during further explosions (see Sect. 5.3.1.1). The accidental rock debris, briefly described by Hekinian et al. (1991), consist of metabasalts, metadolerites, gabbros and metagabbros. The petrology of those types of rock debris has been generally neglected in other similar areas when compared with their ultramafic counterparts collected on volcanic islands, particularly among lava xenoliths. However, the magmatic and volcanic histories of these rocks are expected to offer some insights into the activity and the structure beneath the surface of the volcano. Deep-seated rock inclusions (xenoliths and xenocrysts) in volcanics are frequently found in continental and oceanic intraplate regions, in volcanic orogenic belts, in island arc systems, and more rarely on mid-ocean ridges. The study of ultramafic to felsic and metamorphic xenoliths from continental midplate areas and convergent plate boundary regions is important, because it bears information on the processes occurring in the deep lithosphere and on possible contamination during magma segregation and delivery to the surface. Ultramafic xenoliths are more commonly reported and studied than gabbroic inclusions in oceanic island lava, because they are potentially indicative of source processes in midplate mantle plumes. Since magmatic activity beneath mid-ocean ridge systems generally occurs in a thin lithosphere at relatively shallow depth, the xenoliths found are generally gabbroic in composition (Hekinian et al. 1985; Dixon et al. 1986; Davis and Clague 1990). However, they are rare and generally described as curious, anecdotic samples. The gabbroic material ejected from intraplate areas consists mainly of xenoliths included in lava, and very few volcanoes are known to have erupted accidental rock debris of gabbroic or ultramafic composition. One of them is the Piton de la Fournaise Volcano on Réunion Island, where wehrlitic and gabbroic cumulate clasts were recovered (Upton et al. 2000). These rocks are different from those studied here, because they are very fresh material containing small amounts of quenched vesicular glass in the interstices of the cumulate minerals. This provides no ambiguity on their origin from an active underlying magma chamber.

10.2
The Macdonald Seamount

10.2.1
Eruptive Activity

The Macdonald Seamount is currently considered to be the active expression of a hotspot having generated the southern alignment of the Austral Islands (see Sects. 1.3.2, 2.4.1 and 2.4.2). The edifice has been frequently visited (Johnson and Malahoff 1971; Johnson 1980; Talandier and Okal 1983, 1984; Sailor and Okal 1983), and two of its eruptions were visually observed on October 11, 1987 by the RV MELVILLE (New York Times, October 14, 1987) and in January 1989 by the NO LE SUROIT and the diving saucer *Cyana*. During one dive near the summit (200 m depth), gas release and eruption of pyroclastics were observed (Cheminée et al. 1991). The volcanic cone was about 4 200 m high and rose to within 40 m of the sea surface at this time. Between June 1987 and December 1988, the volcano was intermittently active, and after a brief pause of two weeks, became active again on January 19, 1989. At the last sounding, the summit was 27 m below sea level (see Sects. 2.4.1 and 2.4.2). A detailed time-event report of the Society and Austral hotspot activities over the last forty years of observation by the *RSP* is given by Talandier (Sect. 2.4.2). Since its discovery in 1967 until December 1977, no activity was detected. From 1977 to 1988, a total of twenty-nine seismic swarms were identified through detection of 'T' waves. According to Talandier (Sect. 2.4.2), the 'T' waves should be the result of degassing and a boiling of seawater on the sea floor during the extrusive phase of submarine eruptions. This mechanism is made possible when the source is located at shallow depths, because the resonance of bubbles is restricted to the first 800 m of the water column. In deeper environments (e.g., Mehetia and Teahitia near Tahiti Island), pressure has prevented the formation of a gas phase and 'T' waves are absent.

Figure 10.2 gives a short summary of the Macdonald Seamount's history according to Talandier (see Sects. 2.4.1 and 2.4.2). An explosive phase was suddenly recorded in December 1977, but with lower intensity than in 1967, and shorter events occurred in 1979. Then, a swarm similar to that of 1977 was detected in November 1980, with numerous short-duration explosive activities. Several swarms and explosive sequences occurred from 1980 to 1986. An explosive sequence characterized by three swarms of relatively short duration occurred in May and August 1986 and in June 1987. This was the beginning of a higher activity period of two years during which a total of 200 days of continuous noise separated by a few quiet sequences took place. The intensity and characteristics of the Macdonald Seamount's seismicity suggest an intense phase of volcanism over a few years, especially in 1986 and 1988. During this period of time, the expedition of the F.S. SONNE (January 1987) recovered samples of accidental rock debris lying on top of ash flows. Since 1989, no volcanic activity has been detected by the *RSP* network (Fig. 10.2). The observed extrusive events suggest superficial long-standing magmatic activity and the possible emergence of the Macdonald Seamount in the near future. The existence of this newly formed volcanic island above the sea surface will, however, probably be of short duration, depending on the conjugated effects of subsidence, collapse events and erosion with respect to the relative amount of volcanic supply. The absence of crisis during periods of time of about ten to twelve

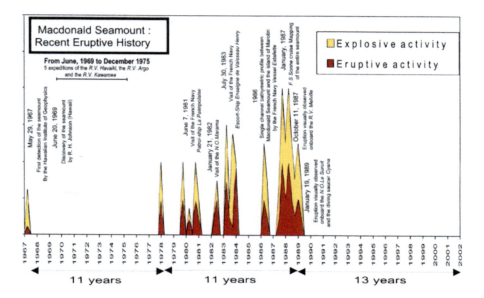

Fig. 10.2. Summary of the extrusive and explosive activity of the Macdonald Seamount since its discovery by R. H. Johnson (Hawaiian Institute of Geophysics, U.S.A.) in 1967 and during the past thirty-five years of monitoring by the French Polynesian Seismic Network (see Sect. 2.1). Also reported are the various visits to the volcano by navy and research vessels. See Sects. 2.4.1 and 2.4.2 for detailed compilation and discussion; the amplitudes are only indicative, and relative to Talandier's comments

years (1967–1977 and 1989–2002) and the fluctuations of the extrusive activity between 1977 and 1989 (eleven years) show that the past magmatic activity of the Macdonald Seamount was rather sporadic. In addition, the existence of fossil coral debris dredged at about 200 m below sea level suggests that the edifice had already emerged during the last period of glaciation and that its growth was rather slow until now.

10.2.2
Morphology and Structure

The Macdonald Seamount is a shallow submarine volcano, which could emerge at any time. Discussing the differences observed on top of the volcano by the various early expeditions, Talandier (Sect. 2.4.2) suggests that the pinnacle at 29 m depth did not exist at the time of Johnson's last survey in 1975. Nevertheless, this author also thinks that the summit of the volcano rose significantly between 1975 and 1982. In 1983, the French Navy Escort Ship E.V. HENRY, using both sounding and visual observations by scuba divers, confirmed the presence of a submittal plateau at an average depth of 40 m and a pinnacle at 27 m depth.

According to Stoffers et al. (1989), the Macdonald Seamount has a base diameter of about 45 km at a depth of 3 900 m, and a narrow summit less than 40 m in diameter. The slopes from 60 m down to about 600 m in depth are nearly vertical, and below this level the steepness gradually decreases towards the base of the edifice. Several volcanic aprons, corresponding to lava channels and/or lava flows, extend radially from the summit, with a preferential NS orientation (see Sect. 5.3.1.1). This fabric roughly corresponds to the general fault pattern and the magnetic anomalies on the ancient ocean floor in the area. These aprons are probably the result of eruptions along basement-controlled lateral rift zones (Binard et al. 1991), as proposed for the Geisha Seamounts (Vogt and Smoot 1984). Other isolated adventive cones about 300 to 400 m high occur on the Macdonald Seamount's flanks at depths greater than 100 m (Stoffers et al. 1989).

10.2.3
Sampling and Observations

Seventy-five days after the May 1983 seismic swarm, the Navy Ship E.V. HENRY reported a discoloration spot about 2 km east of the summit, suggesting the existence of some hydrothermal activity, which was confirmed by further observation and sampling (Stoffers et al. 1989; Cheminée et al. 1991; Hekinian et al. 1991, 1993b; see Sect. 5.3.1.1). The Macdonald Seamount was sampled by dredging (Fig. 10.1b) and visually observed (TV grab, OFOS) from the summit (39 m) down to 400 m, and from 800 m down to 2 020 m in depth in 1987 (Stoffers at al. 1989). From the top to about 100 m depth, a considerable amount of lapilli covers scoriaceous outcrops of dark volcanic flows. The thickness of the lapilli is estimated to be at least 50 cm. Several steep spatter cones made up entirely of very fresh scoriaceous material were found on the summit. Hydrothermal products were also recovered near the top at about 87 m depth, and further patches of powdery ochreous sediment and crusts were seen scattered along the flanks down to about 1 000 m below sea level (Hekinian et al. 1991, 1993b). A 2–3 m wide eruptive fissure was observed on the eastern flank of the volcano at about 127 m depth. Fresh lapilli partially covering concretions of scoriaceous lava blocks were scattered close to the fissure. At the edge of the fissure, a small spatter cone was observed. The shimmering, low visibility seawater surrounding the fissure probably results from the hydrothermal discharge and the emission of small gas bubbles. CH_4 concentrations up to six times higher than normal seawater were measured at two of the six water-sampling stations located at 30–100 m depth (Cheminée et al. 1991; see Sect. 13.4.1.1).

Several other scarps less than 100 m high were observed on the flanks between 150 and 470 m (Stoffers et al. 1989). They were described as flow fronts made up mainly of scoriaceous material and vesicular pillow lava. Between 620 and 1 000 m depths, other well-exposed flow fronts consist primarily of pillow lava and lobate flows with fewer vesicles than near the top of the volcano. Between the upper steep scarps (100–620 m depths), the gentler slopes are covered with lapilli. Here, coral communities are observed down slope and are partially buried by a recent fall of lapilli. About 90% of the slopes are covered by lapilli, whose thickness seems to increase with depth. Also there is a decrease of the scattered angular rock debris, which becomes rarer with depth. Finally, fresh highly vesicular glassy pillow lavas become more prominent at depths exceeding 1 300 m.

10.2.4
Volcanic Terrains

The two main types of volcanic terrains recognized on the Macdonald Seamount are volcaniclastics and lava flows. The lava flows consist of highly to moderately vesicular, tubular or bulbous pillow basalts (vesicles up to 1–2 mm in diameter). Moderately fresh pillow buds and trap-door pillow lavas exhibiting radial jointing and thin (<10 mm) glassy margins are abundant. The freshest vesicular basalts were collected from the summit and the northeastern flank of the volcano, between 1 300 and 2 020 m. The volcanics consist of an alkali-basaltic suite comparable to other samples previously reported from shallower (<200 m) depths (Brousse and Richter de Forges 1980). The volcanic ejecta are mainly coarse-grained vesicular lapilli (1 to 100 mm in diameter), which differ from typical basaltic shards by their dull appearance and cryptocrystalline nature (Hekinian et al. 1991). Semiconsolidated lapilli or volcanic sand form spatter cones and coat the walls of small craters.

The eruption of lapilli is apparently accompanied by volcanic bombs with breadcrust and cauliflower surfaces (Hekinian et al. 1991). These highly vesicular bombs (10–20 cm in diameter) were collected near the summit at about 125 m below sea level. Among the lapilli is scattered and unsorted accidental rock debris. These ejecta are rounded or angular irregularly-shaped blocks that vary from 1 to 40 cm in diameter and weigh up to 20–30 kg. They consist of both plutonics and volcanics, including alkali-gabbros and metagabbros, one metadolerite, altered picritic rocks, a troctolite, altered basalts and ankaramites (Table 10.1). Samples of dolerites and altered pillow lavas exhibit stockwork mineralization of sulfides in veinlets and as scattered, isolated crystals. Some angular blocks of fine-grained dolerite are entirely embedded in scoriaceous material like the "core bombs" found in subaerial explosive events (Fisher and Schmincke 1984). These volcanic ejecta are the result of explosive fallout.

Whether the explosions were caused by exsolution of magmatic volatiles or by the boiling of seawater trapped in the inner parts of the volcano is unknown. This is the type of explosive event recorded by the 'T' phases at the Polynesian seismic stations (Talandier and Okal 1984). According to Hekinian et al. (1991), the eruptions probably began by an explosion from inside the volcanic edifice as a result of the high pressure caused by boiling seawater overpressure in pores and fractures. Angular blocks were consecutively ejected then followed by hydromagmatic events. Similar clasts reported by Kokelaar (1986) were also inferred to be formed during explosive activity due to magma-water contact. It is believed that the stage of spatter cone formation on the summit and flanks (down to 180 m depth) of the Macdonald Seamount results from later and quieter volcanic activity than that which produced the lapilli and bombs.

10.3
Petrology

10.3.1
Analytical Techniques

The bulk chemical analyses of the major, rare earth and transition elements were performed by a Jobin Yvon JY 70 spectrometer coupled with a plasma source (induced coupled plasma, ICP) at the geochemical laboratory of the Centre de Recherche Pétrographique et Géochimique (CRPG, Nancy, France). Other bulk and trace element analyses were analyzed by X-ray fluorescence (XRF) techniques at the geology department of the University of Karlsruhe (Germany). The results are given in Tables 10.2 and 10.3. The range of error is derived from repeated analyses on standards using both techniques. More details on the analytical techniques are reported by Govindaraju

Table 10.1. Description and setting of the volcanics and gabbroic ejecta (boldface) sampled from Macdonald Seamount during the 1987 cruise of the F/S Sonne

Sample	Depth (m)	Rocks type	Primary mineralogy	Seconary mineralogy	Description
55	214 – 215	Dolerite	pl[An56], cpx, amph, biot	cc	Dolerite xenolith
55-1E	–	**Gabbro-norite**	pl, cpx, opx	cc	Altered gabbro
55-3A	–	Basanite	pl[An39-53,Or3]		Pillow lava
55-4	–	Picrite	ol[Fo84], pl[An66], cpx, (sp)	serp, cc, tc	Porphyritic pillow lava
55-6A	–	**Ol-ox-gabbro**	pl, cpx, hb, biot, ol, (ap)	tc, sulf, smt, mag, chl	Gabbro cummulate
55-6A[a]	–	**Leuco-diorite**	pl, biot, Qz	clay, sulf, ab	Felsic vein in gabbro
55-6A1	–	**Ol-ox-gabbro**	pl, cpx, hb, biot, ol, ox, (ap)	smt, ab	Layered cummulate
55-6B	–	**Oxide-gabbro**	pl, cpx, hb, biot, ox		Gabbro cummulate
55-6C	–	**Metagabbro**	pl, cpx, (opx), ox, (biot)	ep, gt, act, qz, chl, an	Gabbro cummulate
55-6D	–	**Metagabbro**	pl	ab, ep, chl, sph	Altered massive rock
55-7B	–	**Metagabbro**	pl, altered cpx	chl-smt	Altered massive rock
55-H8	–	**Troctolite**	ol, pl, (cpx), (chr)	serp, chl, trem	Altered massive rock
55-H9	–	**Ol-ox-gabbro**	pl, cpx, hb, biot, ol, ox, (ap)	serp, mag, tc, smt, chl, cc, ab	Gabbro cummulate
55-H10	–	**Ol-ox-gabbro**	pl, cpx, hb, biot, ol, ox	clay, ab, chl	Gabbro cummulate
55-H11	–	**Ol-ox-gabbro**	pl, cpx, hb, biot, ol, ox	serp, mag, tc, chl, ab	Gabbro cummulate
57-2A	1 600 – 2020	Picrite	ol, cpx		Vesicular pillow lava
57-9	–	Basanite	ol, cpx		Vesicular porphyritic lava

Primary mineralogy: gl, glass; ol, olivine; pl, plagioclase; cpx, clinopyroxene; opx, orthopyroxene; hb, hornblende; ox, oxides; sp, spinel; chr, chromite; biot, biotite; ap, apatite; Qz, quartz; (sp) accessory minerals; cc/gl, carbonate alteration of glass.
Secondary mineralogy: cc, carbonate; ab, albite; serp, serpentine; amph, amphibole; trem, tremolite; act, actinolite; ep, epidote; gr-biot, green biotite; tc, talc; Qz, quartz; gt, garnet; chl, chlorite; zeol, zeolite; mag, magnetite.
Fo = forsterite content; An = anorthite content; Ab = albite content; Or = orthoclase content.
[a] Sub-sample included or veining the main host rock sample.

Table 10.1. *Continued*

Sample	Depth (m)	Rocks type	Primary mineralogy	Seconary mineralogy	Description
59-2	125 – 230	Basanite	ol, cpx, pl		Vesicular pillow lava
59-3	–	Mugearite	pl[An29,Or3], cpx		Vesicular porphyritic lava
59-16A	–	**Ox-gabbro**	pl, cpx, biot, opx, hb, ox		Gabbro cummulate
64	850 – 1480	Basanite	ol[Fo80], pl[An73,Or0.5]		Vesicular pillow lava
64-1	–	Alkali-basalt	ol[Fo80], pl[An71-76]		Vesicular pillow lava
64-2	–	Picrite	ol[Fo80], pl[An75]		Vesicular pillow lava
68	1580 – 2020	Basanite	ol[Fo78], pl[An70], cpx		Vesicular pillow lava
68-2	–	Picrite	gl, ol[Fo81-82], cpx, pl		Porphyritic pillow lava
68-3	–	Picrite	Fo80-85, cpx, pl, gl		Vesicular pillow lava
69-7A	500 – 1400	**Oxide-gabbro**	cpx, pl, hb, ox, biot, (ap)	clay	Non-cummulate gabbro
69-7B	–	Altered lava		clay	Altered volcanics
69-H100	–	**Oxide-gabbro**	pl, cpx, ox	ab, ep, zeol	Non-cummulate gabbro
73-2	70 – 200	Altered basanite	pl		Host lava of the xenolith
73-2[a]	–	**Olivine-gabbro**	cpx, pl, ol	ep, trem, cc/gl	Gabbroic xenolith
71	100 – 220	Basanite	gl, ol, cpx, pl		Volcanic bomb
70-2	220 – 450	Basanite	ol, pl		Porphyritic lava

Primary mineralogy: gl, glass; ol, olivine; pl, plagioclase; cpx, clinopyroxene; opx, orthopyroxene; hb, hornblende; ox, oxides; sp, spinel; chr, chromite; biot, biotite; ap, apatite; Qz, quartz; (sp) accessory minerals; cc/gl, carbonate alteration of glass.
Secondary mineralogy: cc, carbonate; ab, albite; serp, serpentine; amph, amphibole; trem, tremolite; act, actinolite; ep, epidote; gr-biot, green biotite; tc, talc, Qz, quartz; gt, garnet; chl, chlorite; zeol, zeolite; mag, magnetite.
Fo = forsterite content; An = anorthite content; Ab = albite content; Or = orthoclase content.
[a] Sub-sample included or veining the main host rock sample.

Table 10.2. Bulk rock analyses of gabbroic rocks from Macdonald Seamount in the South Pacific and error ranges for ICP (CRPG-Nancy) and XRF (University of Karlsruhe)

	Bulk rock analysis							Error range	
	55H8	55-6A bulk	55-6A[a] vein	55-6D	59-16A	55-1E	55	ICP	XRF
SiO_2	40.65	39.52	58.77	47.89	45.14	43.03	43.48	0.190	0.300
TiO_2	1.00	5.46	0.62	3.74	5.01	3.80	4.11	0.009	0.020
Al_2O_3	6.84	15.86	21.14	16.79	17.50	15.12	14.78	0.070	0.100
Fe_2O_3	6.07	6.42	0.91	5.05	2.95	2.52	2.74	0.050	0.060
FeO	6.14	10.66	1.72	5.77	9.00	8.79	9.48	–	–
MnO	0.17	0.22	0.04	0.12	0.14	0.17	0.18	0.004	0.005
MgO	21.55	6.21	0.76	3.58	4.18	4.04	3.68	0.030	0.090
CaO	7.25	9.21	4.14	9.84	11.24	11.89	9.80	0.060	0.040
Na_2O	0.45	2.27	7.00	3.55	3.09	3.18	3.76	0.016	0.070
K_2O	0.10	0.37	2.30	0.65	0.38	0.94	1.24	0.018	0.007
P_2O_5	0.09	0.16	0.17	0.40	0.17	0.42	0.54	0.009	0.007
CO_2	1.76	0.38	0.11	0.83	0.06	4.30	4.24	–	–
H_2O	7.44	0.99	1.15	1.71	0.25	0.69	1.06	–	–
Total	99.51	97.73	98.83	99.92	99.11	98.89	99.08	–	–
Traces (ppm)									
Ba	29	128	1 229	144	121	223	332	13.0	5.0
Sr	66	782	1 226	874	555	581	775	3.6	13.0
Rb	2	36	50	7	7	18	24	1.6	2.0
Nb	16	27	28	28	36	43	60	1.1	1.0
Zr	62	105	332	117	110	211	267	4.3	11.0
Y	11	13	21	17	16	27	–	0.3	1.0
Co	104	72	17	31	41	47	34	2.7	2.0
Cr	1 500	47	8	20	67	79	5	4.0	4.0
Ni	880	96	7	23	53	52	5	4.0	6.0
Cu	94	104	12	83	135	70	106	4.0	2.0
Rb/Sr	0.030	0.046	0.041	0.008	0.013	0.031	0.031	–	–
Zr/Nb	3.9	3.9	–	4.2	3.1	4.9	4.5	–	–

[a] Sample 55-6A is a gabbro with veins of albitized plagioclase and traces of amphiboles.
Sample 55H8 contains cumulates of olivine (>20%) partially serpentinized.
Sample 55 is a dolerite embedded in a vesiculated lava.
The Mg # represents the atomic proportion of $Mg^{2+}/Mg^{2+} + Fe^{2+}$.
The estimation of Fe^{3+} used in the calculations is obtained by the method of Brooks (1976).
The analyses were done by ICP (induced coupled plasma) emission spectrometry at the Centre de Recherche Pétrographique et Géochimique in Nancy. Error ranges are derived from repeated analyses on standards (Govindaraju 1982; Laschek 1985).

Table 10.3. Rare-earth element analyses of bulk rocks from the Macdonald and the Society Hot Spot Volcanoes in the South Pacific

Element	Tholeiite	Gabbros				Dolerite	Picrite			Alkali-basalt		Basanite			Neph.	
	45	55H8	55-6A bulk	55-6A vein	59-16A	55	68-3	55-4	68-2	64-1	3-2	64	68	71	79-2	s.d.
La	2.82	7.47	14.70	41.31	12.60	41.79	29.55	20.55	18.86	21.46	88.80	40.80	34.17	38.15	91.96	3.30
Ce	8.24	21.47	33.00	73.07	34.91	93.21	71.58	52.22	45.97	54.94		93.33	78.07	86.36	179.73	5.72
Nd	6.09	8.67	15.49	23.82	15.04	41.32	29.31	21.89	21.95	25.08	50.90	39.39	36.41	38.13	68.51	3.12
Sm	2.58	2.90	3.59	4.85	3.99	9.42	7.34	5.33	5.50	6.61	10.87	9.13	8.24	8.79	14.07	0.53
Eu	1.36	1.11	1.86	3.16	1.72	3.37	2.84	1.99	2.15	2.16	3.41	3.33	2.61	2.82	5.07	0.18
Gd	2.85	2.42	3.21	4.20	3.21	7.09	5.69	3.98	4.59	5.32	9.87	6.99	5.95	6.72	9.26	0.51
Dy	3.88	1.89	2.46	3.30	2.75	5.62	4.11	2.98	4.15	4.75	6.54	4.99	4.94	5.26	6.29	0.31
Er	2.54	0.90	1.07	1.71	1.19	2.41	1.79	1.29	1.86	2.09	2.62	2.03	2.00	2.22	2.44	0.11
Yb	2.86	0.80	0.90	1.71	1.02	2.13	1.57	1.06	1.70	1.95	1.82	1.70	1.67	2.00	1.86	0.10
Lu	0.33	0.10	0.15	0.20	0.12	0.27	0.21	0.15	0.22	0.24		0.24	0.21	0.29	0.24	0.04

The analyses were done by induced current plasma (ICP) method at the Centre de Recherche Pétrographique et Géochimique (CRPG) in Nancy. The analysis of sample 3-2 is from Devey et al. (1990). Descriptions of samples are found in Table 10.1. Neph, nepheline-tephrite.

(1982) and Laschek (1985). H_2O represents the total of H_2O^+ and $H_2O\%$. FeO was obtained by classical wet method. CO_2 was analyzed by colorimetry. The collection and reduction of data by microprobe techniques were made using the ZAF procedure with a CAMEBAX MBX instrument at IFREMER (CAMEBAX de l'Ouest). The analytical conditions consisted of 15 kV accelerating potential, 15 ηA sample current, and a counting time of 6 s on a focusing beam of about 1 μm. Only representative compositions of the main mineral phases are represented in Tables 10.4 to 10.8.

10.3.2
Rock Descriptions

Most of the volcanics collected have a basaltic composition, containing phenocrysts of plagioclase and variable amounts of olivine (fo_{81-86}) and clinopyroxene. These rock samples have been studied in detail and described by Hekinian and colleagues (see Sect. 14.4.2; Hekinian et al. 1991). Country rock material ejected by the explosions includes a large variety of rock types (Table 10.1). The freshest specimens are found as xenolith inclusions in the flows and consist of basaltic rocks with a composition similar to that of the host rocks. More altered specimens include altered basalts, metamorphosed picrites and troctolites, and breccia. The presence of granulated crystals in the brecciated samples is the result of pressure-induced hydrobrecciation that took place during magma-seawater interaction. The replacement of early formed mineral phases (olivine, clinopyroxene) and the association of secondary silicates (quartz, epidote, actinolite), phyllosilicates (chlorite, smectites,) and carbonates indicate that hydrothermal metamorphism has affected these rocks. The fact that metamorphism could result from the circulation of hydrothermal fluids is supported by the occurrence of veinlets of sulfides (pyrite) and smectite in some metabasalts.

The accidental rock debris consists of large blocks of usually fresh and older altered basaltic flows, fragments of alkali ferro-gabbros, polygenic breccias and occasional dolerites. The gabbroic rocks are among the least altered specimens and consist of fine- to coarse-grained melanocratic rocks, with minor mesocratic gabbros and leucocratic felsic veins or impregnations (Fig. 10.3). A contact zone between a glassy basanite and a gabbro (73-2) suggests that the gabbro was a xenolith carried up by the lava during its ascent to the surface. The size of the primary minerals ranges from 10 mm to about 600 μm. The rocks vary widely in texture and composition and are divided into cumulate and non-cumulate species (Fig. 10.4). They are also classified (Table 10.1) as troctolites (olivine, plagioclase, minor clinopyroxene and chromite), olivine-bearing and olivine-devoid oxide-gabbros (ol-ox-gabbros and ox-gabbros). The oxide-gabbros contain abundant plagioclase, clinopyroxene and ilmenite-magnetite associations, variable amounts of titanium-amphibole (Ti-hornblende) and biotite, and minor orthopyroxene and apatite. The presence of epidote, actinolite, quartz and garnet in veins or vacuoles of some specimens (Fig. 10.5g,h) suggests that metasomatic alteration has affected this gabbroic complex (Table 10.1).

The microprobe analyses of primary plagioclase (Table 10.4), exclusive of altered troctolitic cumuli, show a composition of bytownite (an_{80-70}), with some occurrences of labradorite (around an_{60}). Secondary plagioclase varies from andesine (an_{40}) to pure albite. The clinopyroxene of the troctolite has a restricted range of composition at the boundaries of diopside, Ca-endiopside and Mg-salite (wo_{40-50}, en_{38-47}). Clinopyroxene

Chapter 10 · Intraplate Gabbroic Rock Debris Ejected from the Macdonald Seamount 321

Fig. 10.3. Macrophotographs of gabbroic samples recovered from the Macdonald Seamount; a 55-6A2 is a melanocratic medium-grained olivine-oxide gabbro (ol-ox-gabbro) impregnated by leucocratic material; b 55-6D is an altered medium-grained mesocratic gabbro, whose clinopyroxenes are almost transformed into clay minerals; c 55-H8 is an altered troctolite with a slightly fresher core, where some relicts of olivine and clinopyroxene are observed. Plagioclase is completely altered, but the cumulate protolite texture is preserved; d 55-H9 is a melanocratic coarse-grained ol-ox-gabbro exhibiting some patches of alteration into clays; e 69-7A is a melanocratic, fine- to medium-grained oxide-gabbro (ox-gabbro) exhibiting sub-doleritic texture; f 55-H11 is a melanocratic, fine- to medium-grained ol-ox-gabbro; g 69-5 is a melanocratic, fine- to medium-grained ol-ox-gabbro with intimately intermixed minerals and crisscrossed by a leucocratic felsic veinlet; h 59-A16 is a medium- to coarse-grained mesocratic ox-gabbro

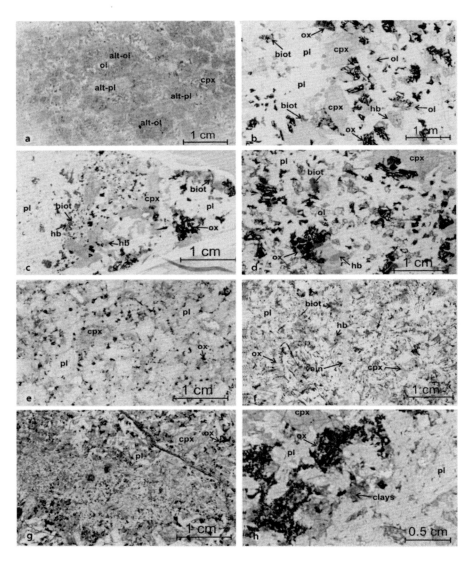

Fig. 10.4. Scanned thin section photographs of gabbroic samples from the Macdonald Seamount; a 55-H8 is a coarse-grained, highly altered troctolitic cumulate showing dark-gray altered olivine pseudomorphs (alt-ol) containing a few relicts of fresh olivine (white, ol), light-gray completely altered plagioclase pseudomorphs (alt-pl) and some fresh crystals of clinopyroxene (cpx, also white but as tiny crystals in some interstices of the pseudomorphs); b 55-H9 is a non-deformed coarse-grained ol-ox-gabbro cumulate showing fresh crystals of white plagioclase (pl), light gray clinopyroxene (cpx), brown hornblende (hb), dark brown biotite (biot), black poikilitic oxides (ox) and gray fractured olivine (ol); c 55-6A1 shows similar mineralogy with grain-size layering from fine- to coarse-grained ol-ox-gabbro cumulate; d 55-6A is a medium-grained, also non-deformed ol-ox-gabbro cumulate with similar composition to b and c; e 55-6D is a non-deformed, medium-grained ox-gabbro cumulate with scarce tiny crystals of oxides; f 69-5 is a fine-grained non-cumulate ox-gabbro, with similar composition as above, but with the later feathery development of quenched oxides; g 69-7A is a fine-grained non-cumulate ox-gabbro sub-doleritic textured non-cumulate ox-gabbro, showing grain-size variations but no layering; h zoomed close-up on a portion of 69-7A, showing local concentration of oxide granules

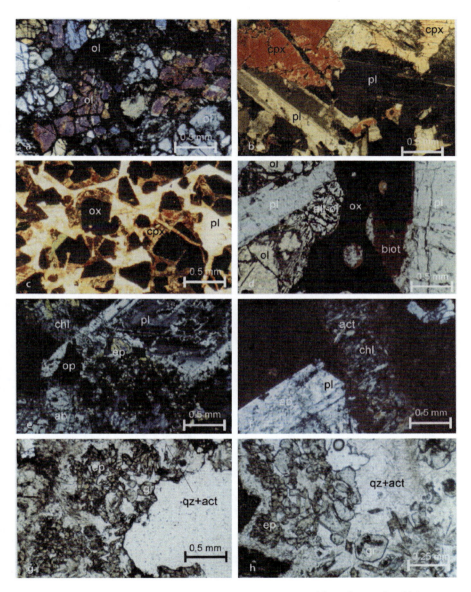

Fig. 10.5. Microphotograph of thin sections from gabbros recovered from the Macdonald Seamount; a fresh relicts of the troctolite 55-H8 showing fractured olivine (ol) without ductile deformation; b detail of ol-ox-gabbro 55-6A1 showing non-deformed plagioclases (pl) and clinopyroxene; c Close-up of a cluster of subhedral oxide crystals in ox-gabbro 69-7A; d ol-ox-gabbro 55-6A showing poikilitic crystal of ilmenite-magnetite (ox) associated with biotite (biot) and surrounded by olivine (ol) altered into smectites and opaques (alt-ol), plagioclase (pl); e a portion of 55-6D metagabbro showing alteration of plagioclase into zeolites (pl), and an opaque grain in a matrix of albite, epidote and chlorite; f vein of actinolite and chlorite, and alteration of plagioclase (pl) into albite (ab) in a metabasalt 55-7A; g Vacuole in 55-6C metagabbro showing an internal rim of quartz and actinolite fibbers (qz + act), containing subhedral crystals of epidote and garnet, and an external rim of chlorite; h close-up on sample 55-6C showing garnet, epidote and quartz + actinolite rimming a vacuole

Table 10.4. Representative microprobe analyses of plagioclase in Macdonald Seamount's gabbroic rocks (FeO* = total Fe as FeO)

	69-7A Ox-gabbro Andesine	69-7A Ox-gabbro Bytownite	69-7A Ox-gabbro Labradorite	69-7A Ox-gabbro Albite	55H8 Troctolite Bytownite	55-6B Ox-baggro Oligoclase	55-6B Ox-gabbro Andesine	55-6B Ox-gabbro Albite	55-6D Ox-gabbro Bytownite
Oxide weight percent									
SiO_2	58.41	48.28	51.92	64.83	50.05	61.66	61.26	57.72	48.10
TiO_2	0.08	0.11	0.16	0.04	0.02	0.06	0.08	0.22	0.13
Al_2O_3	25.74	31.66	29.05	20.98	30.31	23.39	24.85	27.43	31.73
FeO*	–	0.62	0.52	0.54	0.72	0.23	0.02	0.78	0.52
MnO	–	0.02	0.04	0.05	0.03	–	0.03	–	0.01
MgO	0.01	0.08	0.07	0.16	0.23	0.01	0.01	0.07	0.07
CaO	7.87	15.72	12.75	2.66	14.19	5.05	6.58	0.21	15.77
Na_2O	7.11	2.62	4.32	10.23	3.26	7.68	7.73	13.49	2.45
K_2O	0.14	0.09	0.28	0.15	0.12	1.22	0.31	0.09	0.09
Total	99.36	99.21	99.12	99.62	98.93	99.32	100.87	99.99	98.86
Number of ions on the basis of 32(O)									
Si	10.51	8.94	9.55	11.50	9.25	11.05	10.81	10.36	8.93
Ti	0.01	0.02	0.02	0.01	–	0.01	0.01	0.03	0.02
Al	5.46	6.91	6.30	4.39	6.60	4.94	5.17	5.80	6.94
Fe*	–	0.10	0.08	0.08	0.11	0.03	–	0.12	0.08
Mn	–	–	0.01	0.01	–	–	–	–	–
Mg	–	0.02	0.02	0.04	0.06	–	–	0.02	0.02
Ca	1.52	3.12	2.51	0.50	2.81	0.97	1.24	0.04	3.14
Na	2.48	0.94	1.54	3.52	1.17	2.67	2.65	4.69	0.88
K	0.03	0.02	0.07	0.03	0.03	0.28	0.07	0.02	0.02
Total	20.01	20.07	20.09	20.08	20.04	19.95	19.96	21.07	20.03
An	37.66	76.42	60.96	12.45	70.14	24.76	31.43	0.83	77.66
Ab	61.56	23.05	37.43	86.74	29.14	68.12	66.83	98.76	21.80
Or	0.79	0.53	1.61	0.81	0.72	7.12	1.74	0.41	0.54

Table 10.5. Representative microprobe analyses of clinopyroxene in Macdonald Seamount's gabbroic rocks (FeO* = total Fe as FeO)

	69-7A ox-gabbro	69-7A ox-gabbro	55-6B ox-gabbro	55H9 ol-ox-gabbro	55H9 ol-ox-gabbro	55H8 Troctolite	55H8 Troctolite
Oxide weight percent							
SiO_2	47.96	48.38	49.00	46.18	48.04	52.44	54.17
TiO_2	2.36	2.42	2.39	3.65	2.00	1.02	0.17
Al_2O_3	5.06	4.03	4.26	6.22	4.37	2.22	0.93
Cr_2O_3	0.01	–	0.02	0.04	–	0.24	0.09
FeO*	7.56	8.20	7.94	8.53	8.71	6.33	4.00
MnO	0.13	0.22	0.07	0.13	0.19	0.16	0.25
MgO	13.05	13.79	13.46	12.35	13.17	16.78	16.03
CaO	22.65	22.25	22.39	21.57	20.28	21.63	25.84
Na_2O	0.50	0.44	0.42	0.47	0.50	0.28	0.15
K_2O	0.03	–	–	–	0.09	–	0.01
Total	99.30	99.76	99.95	99.16	97.35	101.13	101.65
Number of ions on the basis of 6(O)							
Si	1.81	1.82	1.83	1.75	1.85	1.91	1.96
Ti	0.07	0.07	0.07	0.10	0.06	0.03	0.00
Al	0.22	0.18	0.19	0.28	0.20	0.10	0.04
Cr	–	–	–	–	–	0.01	–
Fe*	0.24	0.26	0.25	0.27	0.28	0.19	0.12
Mn	–	0.01	–	–	0.01	–	0.01
Mg	0.73	0.77	0.75	0.70	0.75	0.91	0.87
Ca	0.92	0.90	0.90	0.88	0.83	0.85	1.00
Na	0.04	0.03	0.03	0.03	0.04	0.02	0.01
K	–	–	–	–	–	–	–
Total	4.03	4.04	4.02	4.02	4.02	4.02	4.02
Wo	48.50	46.52	47.32	47.50	44.67	43.33	50.40
En	38.86	40.09	39.58	37.83	40.35	46.76	43.50
Fs	12.64	13.39	13.10	14.67	14.98	9.90	6.09

Table 10.6. Representative microprobe analyses of Ca-amphiboles in Macdonald Seamount's gabbroic rocks (FeO* = total Fe as FeO; FM = Fe* / (Fe* + Mg))

	55-6B Ox-gabbro	69-7A Ox-gabbro		55-6A1 Ol-ox-gabbro	55H9 Ol-ox-gabbro
	Hornblende	Actinolite	Hornblende	Hornblende	Hornblende
	Oxide weight percents				
SiO_2	40.88	53.96	42.33	42.18	41.17
TiO_2	5.53	0.20	1.78	5.76	5.51
Al_2O_3	11.45	2.51	11.19	10.78	10.36
Cr_2O_3	0.05	–	–	–	–
FeO*	11.76	9.89	11.54	12.34	12.88
MnO	0.24	0.28	0.18	0.23	0.10
MgO	11.84	17.81	14.60	11.13	11.70
CaO	11.71	13.07	11.81	11.47	11.18
Na_2O	2.88	0.67	3.20	3.15	3.08
K_2O	1.26	0.09	0.29	1.24	1.15
Total	97.61	98.47	96.93	98.28	97.13
	Numbers of ions on the basis of 23(O)				
Si	6.13	7.64	6.30	6.28	6.22
Al	2.02	0.42	1.96	1.89	1.84
Cr	0.01	–	–	–	–
Ti	0.62	0.02	0.20	0.64	0.63
Mg	2.64	3.76	3.24	2.47	2.64
Fe*	1.47	1.17	1.44	1.54	1.63
Mn	0.03	0.03	0.02	0.03	0.01
Ca	1.88	1.98	1.88	1.83	1.81
Na	0.84	0.18	0.92	0.91	0.90
K	0.24	0.02	0.05	0.24	0.22
Total	15.89	15.23	16.03	15.82	15.90
Al_{IV}	1.87	0.36	1.70	1.72	1.78
Al_{VI}	0.15	0.06	0.27	0.17	0.07
FM	0.36	0.24	0.31	0.38	0.38

(Table 10.5) of the oxide-bearing gabbros is confined to a small field at the Mg-salite-augite boundary (wo_{45-50}, en_{38-40}). That is not generally the case in ocean-ridge Fe-Ti gabbros, such as for samples recovered from the Hess Deep (Hekinian et al. 1993a). The primary amphiboles (Table 10.6) are kaersutitic hornblendes ($TiO_2 > 5$ wt.%, $Al^{IV} > 1.6$ and Si < 6.5), which are clearly distinguished from the composition of secondary actinolite and actinolitic hornblende ($TiO_2 < 2$ wt.%, $Al^{IV} < 0.6$ and Si > 7.5), as seen in Fig. 10.6. The primary oxides (Table 10.7) have compositions between intermediate

Table 10.7. Representative microprobe analyses of oxides from Macdonald Seamount's gabbroic rocks (FeO* = total Fe as FeO; Ti-magn. = titano-magnetite; Nox() = number of oxygen in the formula unit)

	69-7A Ox-gabbro				55-6B Ox-gabbro	55H9 Ol-ox-gabbro	55-H8 Troctolite
	Ti-magn.		Ilmenite		Ilmenite	Ilmenite	Chromite
SiO$_2$	0.96	1.2	0.86	2.52	0.02	0.83	0.11
TiO$_2$	10.28	14.79	36.04	47.61	51.99	50.8	3.21
Al$_2$O$_3$	3.12	1.49	3.94	0.36	0.12	0.21	17.28
Cr$_2$O$_3$	0.07	0.12	–	0.05	0.1	0.13	35.85
FeO*	80.32	74.95	54.54	46.56	42.94	43.93	35.31
MnO	0.59	0.57	1.05	1.4	0.59	1.35	0.33
MgO	0.36	0.15	0.77	0.94	4.21	2.31	7.24
NiO	–	0.13	0.07	–	0.05	–	0.1
CaO	0.05	0.27	0.05	1.92	0.01	0.06	0.12
Total	95.75	93.67	97.32	101.36	100.03	99.62	99.55
Fe$_2$O$_3$	45.97	37.04	34.57	12.8	9.11	5.05	10.4
FeO	38.95	41.62	23.43	35.04	34.75	39.38	25.95
Nox	(32)	(32)	(24)	(24)	(24)	(24)	(32)
Si	0.28	0.36	0.17	0.48	–	0.16	0.03
Ti	2.28	3.38	5.26	6.83	7.52	7.52	0.63
Al	1.08	0.53	0.9	0.08	0.03	0.05	5.31
Cr	0.02	0.03	–	0.01	0.02	0.02	7.38
Fe^{3+}	10.2	8.48	5.05	1.84	1.32	0.75	2.04
Fe^{2+}	9.6	10.59	3.8	5.59	5.59	6.49	5.65
Mn	0.15	0.15	0.17	0.23	0.1	0.23	0.07
Mg	0.16	0.07	0.22	0.27	1.21	0.68	2.81
Ni	–	0.03	0.01	–	0.01	–	0.02
Ca	0.02	0.09	0.01	0.39	–	0.01	0.03
Total	23.79	23.73	15.6	15.72	15.79	15.91	23.98

ilmenite (ilmenite-magnetite solid solution) and pure ilmenite and Ti-magnetite associations. Biotite (Table 10.8) occurs as red-brown homogenous primary crystals (TiO$_2$ = 3–5 wt.%) adjacent to or included in large poikilitic Fe-Ti-oxides. The rocks are alkali-gabbros (Table 10.2) that allow us to distinguish them from the tholeiitic gabbros commonly encountered on Mid-Ocean Ridges. However, some rare exceptions of alkali-gabbros are found on spreading ridge systems, such as near the 15° N fracture zone on the Mid-Atlantic Ridge (Cannat et al. 1992; Simonov et al. 1999). Although devoid of a major topographic anomaly such as the Azores and Iceland plateaus, this region of the Atlantic spreading center located between 15° N and 3° S is known to present an atypical heterogeneous geochemistry (Sushchevskaya et al. 2002).

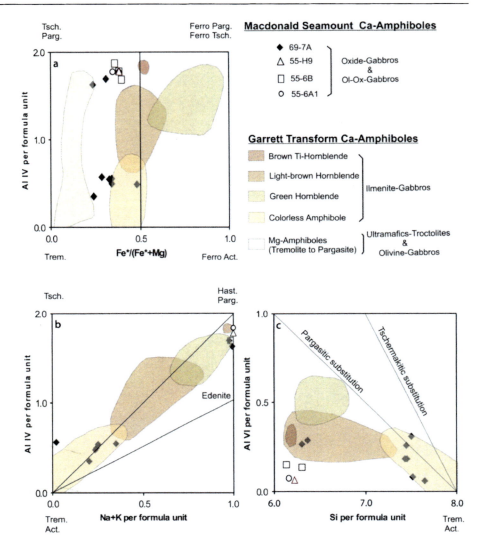

Fig. 10.6. Compositional variation of Ca-amphiboles in gabbros recovered from the top of the Macdonald Seamount; **a** Fe* / (Fe* + Mg) vs. Al^{IV} variation diagram (Fe* = total Fe as Fe^{2+}) showing a large gap of composition between the Al-rich primary Ti-hornblendes and the secondary actinolite to actinolite-hornblende amphiboles. Also shown is the compositional distribution of Ca-amphiboles in gabbros recovered from the Garrett transform fault, East Pacific Rise near latitude 13° S (Bideau et al. 1991). The small most Al^{IV}-rich field represents the composition of primary brown kaersutitic hornblende found as inclusion in clinopyroxenes of ilmenite-gabbros. The other fields include light-brown, mostly secondary hornblendes, blue-green Cl-bearing, Fe-rich hornblendes developing after primary crystals or crystallizing in veins, and late colorless actinolite-hornblendes and actinolites. The field of the Mg-amphiboles (tremolite to pargasite) from the most mafic rocks is also represented for comparison; **b** Na$_2$O + K$_2$O vs. Al^{IV}; and **c** Si vs. Al^{VI} variation diagrams. The field of the Mg-amphiboles from the Garrett transform that overlaps most of the trend of the ilmenite-gabbros and extends close to the pargasite substitution trend is not shown here. *Tsh.* = tschermakite, *Hast.* = hastinsite, *Parg.* = pargasite, *Trem.* = tremolite, *Act.* = actinolite

Chapter 10 · Intraplate Gabbroic Rock Debris Ejected from the Macdonald Seamount 329

Table 10.8. Representative microprobe analyses of biotite from Macdonald Seamount's gabbroic rocks (FeO* = total Fe as FeO)

	69-A Biotite	69-7A Biotite	55-6B Biotite	69-7A Biotite	69-7A Biotite	69-7A Biotite
SiO_2	38.68	38.24	36.24	38.49	34.74	39.13
TiO_2	5.66	3.68	0.21	5.50	4.25	5.71
Al_2O_3	13.03	13.81	13.22	12.10	11.87	12.27
FeO*	10.01	9.49	20.78	9.60	13.25	9.55
MnO	0.08	0.10	0.13	0.09	0.29	0.04
MgO	18.04	20.03	14.48	18.99	15.82	17.62
CaO	0.11	0.07	0.21	0.00	0.49	0.10
Na_2O	0.57	0.51	0.10	0.58	0.79	0.74
K_2O	8.98	9.21	5.36	9.49	6.57	9.73
Total	95.19	95.34	90.74	94.84	88.12	94.90

10.3.2.1
Troctolite Cumulates

This group, represented by only one sample (55-H8), consists of altered ultramafic material. The alteration is nearly complete in the 1–3 cm thick external rim of the specimen, but the core is more preserved (Fig. 10.3c). As attested by the well-preserved texture and some fresh relics of primary minerals in the pseudomorphs, the protolite of the rock was a troctolite cumulate (Figs. 10.3c and 10.4a) composed of large crystals of altered olivine (20–30 vol.%) and plagioclase (up to 0.5 cm in size) and minor (<1 vol.%) chromite (Table 10.7). In addition, fresh, elongated (up to about 0.5 mm) Ca-clinopyroxene (diopside to endiopside, wo_{43-50}, en_{43-47}) and plagioclase (an_{70-74}) laths occur in the interstices of the former minerals, forming a sub-doleritic matrix. This occurrence instead of poikilitic clinopyroxene in an inter-cumulus position suggests quenching of the residual melt during the last stage of magma crystallization. The large cumuli, presumably composed of Ca-plagioclase crystals, are completely saussuritized. Olivine pseudomorphs contain variable associations of talc + magnetite, serpentine + magnetite, mixed-layer smectite-chlorite, and/or smectite (saponite), but some fresh cores of olivine (fo_{84}) are well preserved in the inner part of the specimen (Fig. 10.5a). Finally, the sample is crosscut by a few carbonate veinlets.

10.3.2.2
Olivine-Oxide-Gabbro Cumulates

These samples (55-H11, 55-H9, 55-6A, 55-6A1, 73-2) consist of fine- to coarse-grained gabbros (Figs. 10.3d, 10.3f and 10.5b), which regardless of rare kinkbands, are mostly devoid of ductile deformation (syn- or post-tectonic recrystallization). Sample 73-2 is a xenolith inclusion in basanitic lava. Some specimens have a large variation of grain sizes (1–5 mm), and sample 55-6A1 exhibits a clear-layered texture (Fig. 10.4c).

Sample 55-H9 (Figs. 10.3d and 10.4b) is typical of the coarse-grained gabbros (clinopyroxene crystals up to 0.5 cm). The rocks contain large crystals of titaniferrous augite ($TiO_2 = 1-4$ wt.%; wo_{44-47}, en_{38-40}), plagioclase (an_{61-79}), olivine (fo_{62-65}) and large poikilitic oxide crystals (Fig. 10.4d). The oxide minerals consist of ilmenite-magnetite associations (Table 10.7) containing plagioclase and apatite inclusions and are surrounded by brown kaersutitic hornblende ($TiO_2 = 5-6$ wt.%) and biotite ($TiO_2 = 5-6$ wt.%). Olivine is altered into talc + magnetite and late mixed-layer smectite-chlorite. Some clouded crystals of feldspars are albite. Veinlets of chlorite, white mica and carbonates also occur in the rock. Sample 55-6A contains primary Ti-amphibole, abundant biotite, and scattered olivine (fo_{62}) crystals (Figs. 10.4d and 10.5d) partially altered into talc + magnetite and/or serpentine + magnetite (Fig. 10.5d). Olivine is essentially altered into serpentine, mixed-layers of smectite-chlorite and opaque oxides.

10.3.2.3
Oxide-Gabbro Cumulates

The most evolved types of cumulates are medium-grained oxide-gabbros (55-H10, 55-1E, 59-16A, 55-6A2, 55-6B, 55-6C, 55-6D and 55-7B) without visible ductile deformation or recrystallization (Figs. 10.3a,b and 10.4e). Some of these evolved gabbros (i.e., 55-6A2) contain dikelets (<5 cm thick) or more diffuse impregnations (Fig. 10.3a) of leucodiorite (albitic plagioclase an_{17-11}, minor biotite and quartz). Sample 59-16A is a coarse-grained mesocratic gabbro (Fig. 10.3h). The rocks consist mainly of plagioclase showing variable composition (55-6D, an_{78-11}; 55-6B, an_{24-32}; 59-16A, an_{74-29}), and clinopyroxene. Kaersutitic hornblende ($TiO_2 = 5$ wt.%), biotite ($TiO_2 = 3-4$ wt.%) and opaques (ilmenite-magnetite associations) are also common. Occasional orthopyroxene and poikilitic ilmenite are found to be intimately associated with biotite in samples 59-16A and 55-1E. Albite also occurs throughout the altered portions of sample 59-16A in association with labradorite (an_{74}). Apatite occurs as inclusions in or concentrated near poikilitic oxides. Secondary mineralogy consists of albite, chlorite, titanite and smectite-chlorite mixed-layer clays. Sample 55-6C exhibits several 3 mm large, irregular vacuoles containing an external rim made up of chlorite and followed by subhedral epidote and garnet (hydrogrossular) crystals in a matrix of quartz and acicular actinolite (Fig. 10.5g,h). The interior of the vacuoles is empty in the thin sections but filled by clay minerals in the hand specimen. The metagabbro 55-6D also shows alteration vacuoles containing clay minerals (Fig. 10.3b) and contains fibrous zeolite, as well as abundant albite, chlorite and epidote in the thin sections (Fig. 10.5e). 55-7B is a metagabbro devoid of preserved primary minerals. Plagioclase is completely altered into albite, and pyroxene is altered into chlorite and smectite-chlorite mixed-layer clays.

10.3.2.4
Non-Cumulate Gabbros

These rocks are fine- to medium-grained oxide-bearing gabbros (Fig. 10.3e and 10.3g) exhibiting variable textures (Fig. 10.4f) going from doleritic (69-7A and 69-H100) to feathery or comb-like (69-5). This suggests "quenched" gabbros formed in dykes or close to the top of the magma chamber (isotropic gabbros). Sample 69-5 is crosscut by a leucocratic dikelet of more felsic material (Fig. 10.3g). They are composed of plagioclase (an_{31-81}), clinopyroxene (wo_{46-49}, en_{37-40}), oxides, brown Ti-hornblendes and biotite. Samples 69-7A, 69-7A1 and 69-H100 have doleritic textures but contain clusters of granular to subhedral titano-magnetite, sometimes associated with apatite (Figs. 10.4g,h and 10.5c). These clusters are like the chromitite micropods such as observed occasionally in dunitic material from ophiolite complexes and more rarely in rocks from the ocean floor (Arai and Matsukage 1998). Secondary mineralogy consists of rare sulfides (pyrite and chalcopyrite), chlorite, actinolite, epidote, albite, and veinlets of clay minerals or carbonates.

10.4
Geochemistry

Bulk rock analyses of representative samples from all the petrographic groups are shown in Tables 10.2 and 10.3. The associated lava composition previously discussed by Stoffers et al. (1989) and Hekinian et al. (1991) is given for comparison. The aphiric lavas are quite evolved (MgO = 2–5 wt.%) and show a compositional gap at about 10 wt.% MgO between the most magnesian and the porphyritic lavas. The gabbroic ejecta have compositions that are very close to those of the lavas for most elements. Increasing Al_2O_3- and Sr-contents with decreasing MgO-content in the lavas suggests that plagioclase, although a ubiquitous phenocryst in the rocks, has not strongly fractionated from the melt. According to Hekinian et al. (1991), the positive linear relationship between the two highly incompatible elements Nb and Zr suggests that simple fractional crystallization/accumulation is the most likely controlling factor for the Macdonald geochemistry. Changes in the degree of partial melting or of the source type do not apparently play a major role in differentiating these rocks. The latter observation agrees with the isotopic homogeneity ($^{87}Sr/^{86}Sr = 0.7037$, $^{143}Nd/^{144}Nd = 0.5128$) of the Macdonald lavas so far analyzed (Devey et al. 1990). The most primitive, troctolitic sample (55-H8) has the lowest K_2O (0.10%), Nb (16 ppm) and light rare earth element (LREE) contents and the highest MgO (22%) and Ni (880 ppm) contents. The leucocratic dikelets (55-6A) are depleted in K_2O (0.28 wt.%) and in Ba (121 ppm) and enriched in Zr (332 ppm) when compared to the surrounding bulk rock. The gabbroic compositions of the Macdonald Seamount are plotted on a SNK (SiO_2 vs. $Na_2O + K_2O$) diagram after Le Bas et al. (1986), together with the volcanics collected from the Macdonald Seamount (Fig. 10.7a). The entire series show alkaline affinity and extend from primitive composition (troctolite 55-H8) to the differentiated felsic vein in 55-6A (leucodiorite).

The compositional variation of the Macdonald Seamount's gabbroic clasts is also compared to a compilation of published data concerning gabbroic xenoliths from other intraplate regions and gabbroic rocks sampled from Mid-Ocean Ridges (Figs. 10.7–10.10). Although alkaline and tholeiitic gabbros are often reported from the volcanic islands of the Pacific midplate areas such as Tahiti and Manihiki Islands (Clague 1976; Bardintzeff et al. 1989) and mainly from the Hawaiian Islands of Mauna Kea, Kauai, Mauna Loa, Kilauea and Kahoolawe Island (Clague 1987; Fodor and Moore 1994; Fodor and Vandermeyden 1988; Kennedy et al. 1991; Fodor and Galar 1993, 1997; Fodor et al. 1993; Hoover at al. 1996; Reiners et al. 1999; Johnston et al. 1985; Baten 1997; Gaffney 1999), very little data are available concerning their geochemical composition. Some bulk rock and mineral analyses (Clague et al. 1988) and REE data (Fodor and Vander-

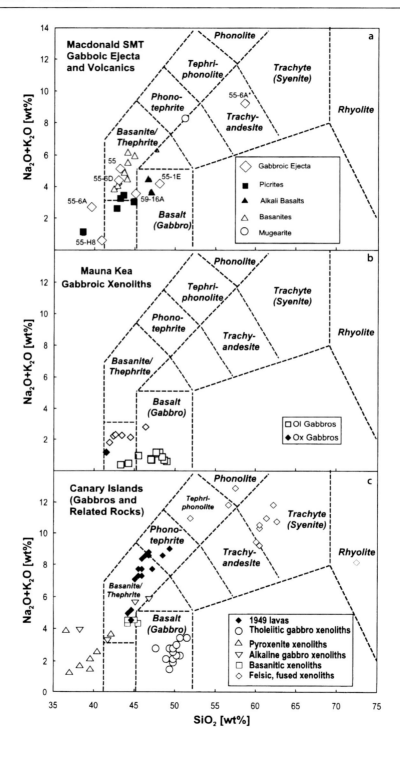

◀ **Fig. 10.7.** Total alkalis (Na$_2$O + K$_2$O) versus silica diagrams (classifications of volcanics by Le Bas et al. 1986); **a** volcanic and gabbroic rock compositions from the Macdonald Seamount; **b** gabbroic xenoliths from Mauna Kea, Hawaii (Fodor and Vandermeyden 1988); **c** lavas and xenoliths of the 1949 eruption in the Canary Islands. Note the wide range in composition of the 1949 lavas (modified after Klügel et al. 1999), the distinct chemical characteristics of the tholeiitic gabbro xenoliths (Schmincke et al. 1998), and the extremely evolved compositions of some felsic xenoliths

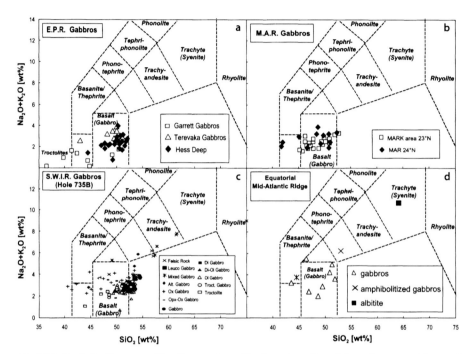

Fig. 10.8. Total alkalis versus silica diagram for gabbroic rocks collected from the ocean ridges and fracture zones; **a** East Pacific Rise data are from the Garrett transform near 13° N (Hébert et al. 1983; Constantin 1999), the Terevaka fracture zone (unpublished data) in the Eastern microplate, and from the Hess Deep (ODP, Leg 147, Hole 894-895; Gillis et al. 1993; Pedersen et al. 1996); **b** Mid-Atlantic Ridge data are from the MARK area near 23° N (ODP, Leg 153; Werner 1997) and from near 24° N (Miyashiro and Shido 1980); **c** Southwest Indian Ridge data are from the Atlantis fracture zone (ODP, Leg 176, Hole 735B; Dick et al. 1999); **d** gabbros and related rocks from the Equatorial Mid-Atlantic transform faults Romanche, Chain and St. Paul (data from Ploshko et al. 1969; Melson and Thompson 1970; Bonatti et al. 1971; Hekinian et al. 2000)

meyden 1988) from Mauna Kea were published (Figs. 10.7b and 10.10a). The rocks present some alkaline affinity, but show a more primitive global composition. The most detailed description of gabbroic xenoliths or ejecta from oceanic intraplate countries found in the literature is probably that concerning the Canary Islands on the North Atlantic eastern margin (Neumann et al. 2000; Klügel et al. 1999; Schmincke et al. 1998). Their REE compositional variations are compared to that of the Macdonald Seamount (Fig. 10.9). This collection of gabbroic xenoliths is interesting, because it allows us to make some distinction concerning the relative original depths of the rocks. Some could represent deep-seated rocks from an ancient oceanic crust, while others have crystallized in a shallow level magmatic reservoir.

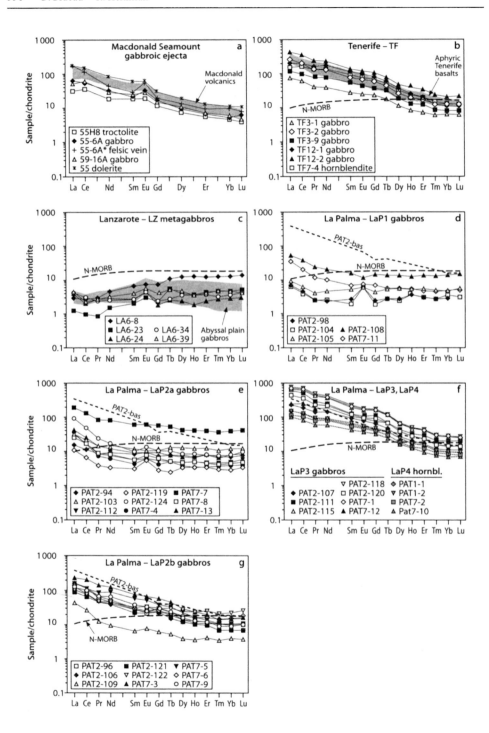

Chapter 10 · Intraplate Gabbroic Rock Debris Ejected from the Macdonald Seamount   335

◄ Fig. 10.9. a Chondrite-normalized REE patterns of the of gabbroic ejecta collected on the Macdonald Seamount (data on chondrite C1 from Sun and McDonough 1989), compared with (b–g) representative gabbro and hornblendite xenoliths from different Canary Islands (Neumann et al. 2000). The xenoliths from La Palma are divided into groups of different origin: group *LaPl* represents old oceanic gabbros mildly affected by reactions and interaction with La Palma magmas, and groups *LaP2a* and *LaP2b* have undergone moderate and high degrees of chemical changes due to infiltration by La Palma magmas, whereas group *LaP3* is believed to represent cumulates/gabbroic rocks formed from La Palma magmas. The field of hydrothermally altered gabbros from the Iberia Abyssal Plain (Seifert et al. 1996, 1997), N-MORB (*long-dashed line*; Sun and McDonough 1989), the host basalt of most xenoliths from La Palma (PAT2-bas; *short-dashed line*), and the field of aphyric basaltic lavas from Tenerife (Neumann et al. 1999) are shown for comparison

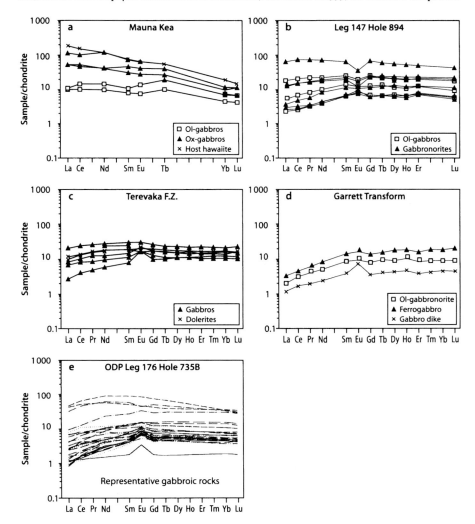

Fig. 10.10. Chondrite-normalized REE patterns of the gabbros and related rocks (data normalized on chondrite C1 from Sun and McDonough (1989) collected from; a Mauna Kea Volcano, Hawaii (Fodor and Vandermeyden 1988); b Hess Deep (ODP, Leg 147, Hole 894, Pedersen et al. 1996); c Teravaka Fracture Zone, Eastern Microplate (unpubl. data); d Garrett Transform, East Pacific Rise near 13° S (Constantin 1999; Hébert et al. 1983); e Atlantis Fracture Zone on the Southwest Indian Ridge (ODP, Leg 176, Hole 735B, Hertogen et al. 2002)

10.5
Discussion

The major aim of this discussion is to decipher geological and petrological features that are likely to indicate the origin of the gabbro ejecta. Do these gabbros represent intraplate hotspot formations or have they originated underneath spreading ridges? Two approaches are proposed, which consist in comparing the gabbroic clasts with rocks from Mid-Ocean Ridge provinces and with rocks from intraplate areas where their origin has been proven. The first approach is relatively easy because of the existence of abundant literature on the subject. The reader is particularly referred to Ocean Drilling Project data obtained from Hole 735B (Atlantis Fracture zone, Southwest Indian Ridge) during Leg 118 (Robinson et al. 1989; Von Herzen et al. 1991) and Leg 176 (Dick et al. 1999; Natland et al. 2002), Leg 153 south of Kane transform on the Mid-Atlantic Ridge (Cannat et al. 1995; Karson et al. 1997), and Leg 147 in the Hess Deep rift valley near the Galapagos-East Pacific Rise triple junction (Gillis et al. 1993; Mével et al. 1996). The second approach is more difficult, because very little has been reported in the literature from hotspot regions.

10.5.1
Comparison with Gabbros Recovered from Mid-Ocean Ridges

The metagabbroic rocks ejected during hydromagmatic explosions at the Macdonald Seamount show various degrees of differentiation from troctolitic to oxide-bearing gabbros. Most are isotropic (no foliation and no layering), some exhibit clear cumulate textures, and others show magmatic grain-size layering. The primary minerals and textures are essentially non-deformed, and secondary mineralogy is related to fracturing and hydration under static conditions. These features are characteristics of ocean-floor metagabbros (Cann 1979; Mével 1987; Mével and Stamoudi 1996) and in particular, for rocks from ultra-fast spreading ridge areas (Bideau et al. 1991; Hekinian et al.1993a; Mével 1996). More precisely, the ocean-floor metabasites are generally devoid of complete strain-induced recrystallization (schistosity, foliation), unlike their continental counterparts. However, high-temperature syn-tectonic (kinkbands, oriented crystals) and post-tectonic recrystallization (granulation) are often observed. Also, static sub-solidus recrystallization (exsolutions in pyroxene, high-temperature coronitic reactions between olivine and plagioclase) is common. This is clearly seen in data recovered from fast- and slow-spreading ridges (e.g., Leg 147 and 153), and particularly true on the ultra-slow spreading Southwest Indian Ridge (Hole 735B). High-temperature deformation is significant in the 1500 meters of gabbros cored in Hole 735B, where abundant deformation features, complex intrusive relationships and very few magmatic textures were described (Dick et al. 2000).

10.5.1.1
Igneous and Metamorphic Characteristics

The textural and mineralogical relationships observed in the thin sections indicate that the lowest temperature of igneous crystallization is represented by the exsolution of ilmenite and Ti-magnetite in opaque oxides. This temperature is obviously sub-solidus but reasonably close to the gabbro solidus. Considering a basalt solidus of about 980 °C (Peck et al. 1966), the required temperature range should be about 900–1 000 °C. At lower temperatures (700–900 °C), recrystallization and possible hydration under high-temperature metamorphism should occur in the brittle-ductile transition zone, as observed in the vicinity of cracking fronts (Mével and Cannat 1991). Granulite (secondary diopside and anorthite) to amphibolite (secondary Al-hornblende and Ca-plagioclase) facies minerals are often observed in veins or in the matrix of ridge gabbros. Metamorphism of a single ocean-crust formation may extend from the granulite-amphibolite facies (about 500–800 °C), through the actinolite facies (about 500–300 °C) and greenschist facies (about 300–200 °C), down to the zeolite facies and weathering at lower temperatures. This can occur in a single sample, where several generations of veins and metamorphic sequences have succeeded each other. Clearly, this is not the case for the Macdonald Seamount gabbroic clasts, where the highest temperature secondary minerals are talc + magnetite, actinolite hornblende, hydrogarnet and secondary Ca-poor plagioclase. While the Ca-amphiboles in ocean-ridge gabbros generally exhibit large, near continuous compositional variations, even in a single specimen, and often in a single crystal as the result of retrograde metamorphism, the amphiboles from the Macdonald Seamount gabbros show a wide compositional gap between the brown magmatic Ti-hornblendes and the field of the actinolites and actinolite hornblendes (Fig. 10.6). Typical examples of such ocean-ridge amphibole trends are found in gabbros from the Garrett transform, which intersects the ultrafast spreading East Pacific Rise near 13° S (Fig. 10.6; Bideau et al. 1991) and in the Leg 147 gabbroic cores from the Hess Deep (Fletcher at al. 1997).

As in the Garrett transform, the Ca-amphiboles of the Macdonald Seamount are found to plot below the pargasitic substitution line in the Si versus Al^{VI} diagram. This line represents the lower limit of high-pressure (>5 kbar) hornblende (Raase 1974). Hexacoordinated Al^{VI} has been suggested to increase with pressure, whereas tetra-coordinated Al^{IV} is temperature dependent (Fleet and Barnett 1978). Quantitative correlations are, however, severely limited by the introduction of significant Fe^{3+} in the lattice. The amount of Fe^{3+} is not known from microprobe analyses, but the amphiboles of the Macdonald gabbros are magnesian, which implies low Fe^{3+}-contents. The primary Ti-amphiboles contain smaller amounts of Al^{VI} than the Garrett kaersutitic hornblendes and have similar Al^{IV}-contents (Fig. 10.6c). This suggests lower pressure and a similar high temperature of crystallization (>800 °C). The low Ti-content (<0.1 cationic proportion) of the secondary actinolitic hornblendes also suggests low-grade metamorphism at the boundary of the greenschist and amphibolite facies (Raase 1974).

10.5.1.2
Temperature of Metamorphism

The observation of low-Ca plagioclase in the presence of actinolite-hornblendes is indicative of the lowest amphibolite facies conditions. This sub-facies, where actinolite is stable with plagioclase instead of albite, is called the actinolite facies (Elthon 1981) and should represent a temperature range of about 300–500 °C.

The other secondary mineral phases encountered in the Macdonald gabbros indicate the existence of an important temperature gap between the igneous and metamorphic phase crystallizations. Talc, which is experimentally synthesized above 500 °C

(Mottl and Holland 1978), is associated with low-titanium magnetite, suggesting a rather low temperature, below 600 °C and probably close to 500 °C. Serpentine minerals have been experimentally synthesized above 300 °C at 2 kbar, but lower pressure and/or lower Al-contents decrease their temperature of stability (Caruso and Chernosky 1979). Serpentinization temperatures of ultramafics deduced from stable isotopes $^{18}O/^{16}O$ were found in the range of 130–235 °C (Wenner and Taylor 1971). This range is believed to be underestimated by about 100 °C (Bottinga and Javoy 1973; Moody 1976), but a much lower $^{18}O/^{16}O$ temperature range (30–180 °C) was also found in equatorial Atlantic ultramafics (Bonatti et al. 1984). The hydration rates of hydrogrossular (Yoder 1950; Carlson 1956; Pistorius and Kennedy 1960) gave a pressure-independent temperature range of 249–360 °C in Californian rocks (Coleman 1967), and 290–450 °C in New Zealand rocks (O'Brien and Rodgers 1973). Based on mineral assemblages observed in rodingites from the equatorial Mid-Atlantic fracture zones, the hydrogarnet crystallization temperature was estimated to be <300 °C (Honnorez and Kirst 1975).

The pistacite-content of epidote is commonly assumed to decrease with increasing metamorphic grade (Brown 1967; Raith 1976; Cavanetta et al. 1980). The scarcity of epidote in ocean-floor metabasites has been attributed to low-pressure conditions (Miyashiro et al. 1971; Liou and Ernst 1979), which is consistent with synthesis experiments (Liou 1973). Observations in various geothermal sites (Steiner 1966; Sigvaldason 1962) recorded the epidote-clinozoïsite at low pressure and low temperature (120–350 °C), under high f_{O2} conditions (Keith et al. 1968; Liou et al. 1974; Liou and Ernst 1979; Holdaway 1972; Liou 1973). The temperature of epidote formation in the dyke complex of the Oman ophiolite, as deduced from studies of microthermometry on fluid inclusions, has an average of 370 °C (Nehlig and Juteau 1988). The other metamorphic assemblages that are typical of the greenschist facies (quartz, albite, chlorite, and actinolite), the zeolite facies and low-temperature alteration (carbonates and clays) indicate temperatures below 300 °C.

In order to be conservative, the lowest temperature estimate of the igneous crystallization process recorded in the Macdonald gabbros is in the range of 800–900 °C, and most probably above 900 °C. Nevertheless, the highest-grade metamorphic recrystallization occurred at much lower temperatures in the range of 600–500 °C, and most probably below 500 °C.

10.5.2
Comparison with Gabbroic Ejecta from Other Intraplate Regions

The gabbros collected on intraplate volcanoes are mostly xenoliths carried up from various levels of the lithosphere by lava extrusion, and can variably originate from a shallow or deep intraplate magma chamber, as well as from the older crust forming the walls of the magma conduits. The cumulate clasts found on the Piton de la Fournaise (Upton et al. 2000) are clearly coming from the magma chamber of the volcano, because they were ejected before the end of their crystallization. This is attested to by the presence of highly vesicular glass in the interstices of the cumulus crystals, and by their friability, which is indicative of drastic quenching (Upton et al. 2000). The troctolite from Macdonald (55-H8) shows sub-doleritic plagioclase-pyroxene phases in the inter-cumulus phase, instead of poikilitic pyroxene. This suggests rather short cooling conditions at the end of crystallization, but this quenching is obviously not as severe as that experienced by the Réunion Island gabbros. At most, it should imply a

Chapter 10 · Intraplate Gabbroic Rock Debris Ejected from the Macdonald Seamount

higher lateral temperature gradient than in ridge environments, as must be the case for an isolated volcano. However, all the studied gabbros were entirely solidified and then fractured and hydrated before their ejection. Unlike the Réunion clasts, their textural characteristics alone are insufficient to definitively distinguish them from older oceanic crust material without ambiguity.

10.5.2.1
The Gabbroic Xenoliths from Hawaii

Hawaiian tholeiitic-stage lavas rarely contain xenoliths. Two picrite flows erupted on the Southwest Rift Zone of Mauna Loa contain xenoliths of dunite, harzburgite, plagioclase-bearing lherzolite and harzburgite, troctolite, gabbro, olivine gabbro, gabbronorite, olivine gabbronorite, and olivine norite (Gaffney 1999). Whole-rock and clinopyroxene REE patterns indicate that the xenolith source is a tholeiitic-stage Mauna Loa magma, rather than underlying mantle or oceanic crust (Gaffney 1999). Poikilitic textures in ultramafic and holocrystalline gabbroic xenoliths are indicative of magma chamber cumulates, whereas open textured xenoliths are interpreted as being crystallization products from a liquid-crystal mush zone at the top of a magma chamber. According to Gaffney (1999) the crystallization of all cumulate xenoliths occurred in a static environment, isolated from dynamic regions of the magma system, which are subject to frequent episodes of magma replenishment and venting. Periods of high magma flux associated with picrite eruptions provide a mechanism to access and carry material from otherwise isolated regions of the magma system where the xenoliths crystallized.

Xenoliths occur in ejecta from the upper slopes and the summit area of Mauna Kea Volcano (Stewart and Brunstad 1999; Hoover and Fodor 1997). The Hawaiian volcanoes have gone through a sequence of four developmental stages characterized by distinct lava types, magma supply rates, and xenolith populations (Clague 1987). These four stages are the alkalic preshield stage, the tholeiitic shield stage, the alkalic postshield stage, and the strongly alkalic rejuvenated stage. Magma supply rates are low in the alkalic preshield, postshield, and rejuvenated stages and high in the tholeiitic shield stage. During the alkalic preshield and rejuvenated stages, primitive alkalic magma carries common mantle xenoliths of lherzolite to the surface, usually with other xenoliths, mainly of dunite or garnet peridotite. During the tholeiitic shield stage, differentiated tholeiitic lava carries rare xenoliths of gabbro and coarse-grained basalt to the surface; these xenoliths are of shallow origin. During the alkalic postshield stage, differentiated alkalic lava carries common crustal xenoliths of dunite, wehrlite, and gabbro to the surface; some of these xenoliths originate within the Cretaceous oceanic crust underlying the volcanoes.

The postshield eruptive stage of Mauna Kea Volcano is divided into an early basaltic stage, the Hamakua Volcanics, containing picrites, ankaramites, alkalic and tholeiitic basalt, and a hawaiite stage, the Laupahoehoe Volcanics, containing only hawaiites and rare mugearites and cumulate gabbroic xenoliths (Kennedy et al. 1991). Cumulate-textured xenoliths of olivine gabbro ($fo_{80} + wo_{44}en_{46} + an_{84}$) and opaque-oxide gabbro ($wo_{44}en_{44}$, an_{60}, magnetite + ilmenite) found in the hawaiite summit cone of Mauna Kea (Fodor and Vandermeyden 1987) enabled modeling of the crystallization of the alkalic basalt suite from a mafic parent to the highly-evolved derivatives (e.g., mugearite). Their depth of origin is inferred from seismic tomography, which showed the magma source beneath Mauna Loa and Kilauea (Okubo et al. 1997). High-velocity bodies (>6.4 km s^{-1}) in the upper 9 km of the crust beneath the summits of the edifice correlating with high magnetic intensity zone are interpreted as solidified gabbro-ultramafic cumulates. The proximity of these high-velocity features to the rift zones is consistent with a ridge-spreading model of the volcanic flank. Southeast of the Hilina fault zone along the south flank of Kilauea, low-velocity material (<6.0 km s^{-1}) is observed extending to depths of 9–11 km, indicating that the Hilina fault may possibly extend as deep as the basal decollement. Along the southeast flank of Mauna Loa, a similar low-velocity zone associated with the Kaoiki fault zone is observed extending to depths of 6–8 km. These two upper crustal low-velocity zones suggest common stages in the evolution of the Hawaiian shield volcanoes, where these fault systems are formed as a result of upper crustal deformation in response to magma injection within the volcanic edifice.

Clague (1987) pointed out that the alkali and tholeiitic stages of activity are consistent with a model in which two magma-storage reservoirs develop at different depths as the magma supply rate increases in the alkalic preshield stage, and subsequently crystallize as the magma supply rate decreases during and after the alkalic postshield stage. The magma reservoirs function as hydraulic filters and remove any dense xenoliths that the ascending magma has carried. At the time of the alkalic preshield and rejuvenated stages, no magma-storage reservoirs exist, so mantle xenoliths of lherzolite are carried up to the surface. In the tholeiitic shield stage, magma reservoirs develop and persist both at the base of the oceanic crust (15–20 km deep) and beneath the caldera complex (3–7 km); xenoliths from greater depths are thought to settle out in these reservoirs.

10.5.2.2
The Gabbroic Xenoliths from the Canary Islands

Since the Hawaiian volcanoes do not have explosive events to extrude accidental debris such as the Macdonald Seamount, the reported occurrences of lower crust and upper mantle-originated material (peridotites, gabbros, dolerites) are xenoliths brought up to the surface by extruded lava. The only well-documented accidental gabbroic rock debris that we found in the literature concerns the Canary Islands. According to Klügel et al. (1999), the eruption of June 24, 1949 on La Palma (Canary Islands) began with phreatomagmatic activity at Duraznero crater on the ridge top, which erupted tephritic lava. On June 8, the Duraznero vents shut down abruptly, and the activity shifted to an off-rift fissure at Llano del Banco. This eruptive center issued initially tephritic aa and later basanitic pahoehoe lava at high rates, producing a lava flow that entered the sea. Two days later, basanite began to erupt at Llano del Banco, in the Hoyo Negro crater north of Duraznero along the rift opened on July 12, and there were ash and bombs of basanitic to phonotephritic composition produced in violent phreatomagmatic explosions (White and Schmincke 1999). The lava contains about 1 vol.% crustal and mantle xenoliths consisting of 40% tholeiitic gabbros from the oceanic crust, 35% alkaline gabbros, and 20% ultramafic cumulates. The almost exclusive occurrence of xenoliths in the last lava flow is consistent with their origin due to wall-rock collapse at depth near the end of eruption. The volcanic evolution of the 1949 eruption is typical of La Palma eruptions generally. Considerable shallow magma migration prior to and during eruption is manifested by strong seismicity, intense faulting, and vent openings that are located three or more km away from each other.

The study of gabbroic xenoliths from the Canary Islands by Neumann et al. (2000) shows that the gabbroic and hornblendite xenoliths fall into three main groups based on their petrography and chemistry: (1) Group 1 consists of highly deformed orthopyroxene-bearing gabbroic rocks that show a strong affinity to N-MORB and oceanic gabbro cumulates in terms of mineral chemistry and REE relations (Fig. 10.9c–e). However, they show mild enrichment in the most incompatible elements (particularly Rb + Ba ± K) relative to intermediate and heavy REE, and their Sr-Nd isotope ratios fall within or close to the N-MORB field; (2) Group 2 includes gabbroic cumulates with zoned clinopyroxenes (Ti-Al-poor cores, Ti-Al-rich rims) and reaction rims of hornblende, biotite and clinopyroxene on other phases. Their trace element and Sr-Nd isotope relations are generally transitional between N-MORB cumulates and the Canary Islands alkali basalts (Fig. 10.9d,e), but they show strong enrichment in Rb, Ba and K with respect to other strongly incompatible elements; (3) Group 3 is composed of non-deformed gabbroic and hornblendite rocks in which hornblende and biotite appear to belong to the primary assemblage. These rocks show strong affinities to the Canary Islands alkali basaltic magmas with respect to the minerals, trace elements, and Sr-Nd isotope chemistry (Fig. 10.9b, 10.9f and 10.9g). The first two groups are interpreted as being fragments of old oceanic crust that have been mildly to strongly metasomatized in reactions with the alkaline magmas of the Canary Islands. The third group represents intrusions/cumulates formed from mafic alkaline Canary Island magmas. Reactions and the formation of cumulates do not represent simple underplating at the mantle/crust boundary but have taken place within the pre-existing oceanic crust and are likely to have significantly thickened the old oceanic crust. The description of the rocks and the comparisons of the various diagrams (Figs. 10.7–10.10) suggest that the Macdonald gabbros have strong affinities with the last group and are clearly different from typical ocean-ridge gabbroic rocks.

10.5.3
Origin of the Macdonald Seamount Gabbroic Clasts

The igneous textural and mineralogical characteristics the Macdonald gabbros are different from those of usual ocean-floor basalts. They indicate crystallization in an undisturbed reservoir, or in an undisturbed part of a magma chamber. Later alteration is the result of fracturing during cooling, perhaps enhanced by magma pressure underneath, and a subsequent hydration. These conditions, however, can occur in localized areas of a Mid-Ocean Ridge magma chamber and particularly beneath a fast-spreading ridge. Thus, the temperature gap separating the metamorphic and the igneous histories of the Macdonald gabbro collection is at least 200 °C without a crystallization event, and could probably exceed 400 °C. It is likely that the gabbros cooled under quiet conditions before fracturing and hydration at a rather low temperature (<500 °C). This fracturing episode has essentially resulted from cooling and hydraulic fracturing under a high thermal gradient. This is also suggested by the presence of doleritic-like textures in inter-cumulus position instead of poikilitic pyroxene in the most mafic rock (troctolite 55-H8). However, the rocks do not exhibit interstitial glass like the gabbroic clasts recovered from the Réunion Island. Thus, the magma reservoir was completely cooled before their ejection. The rocks present geochemical and petrographical characteristics, which are similar to gabbro xenoliths from other intraplate regions, and which are believed to have originated from intraplate magmatism or from older oceanic crust contaminated by intraplate magmas. However, the latter origin seems unlikely because the rocks have a strong chemical affinity with the volcanics extruded on the Macdonald Seamount and should have otherwise recorded intermediate compositions. The gabbroic rocks studied here were collected in January 1987, just after the eruptive activity of 1986. They were most probably ejected at the beginning of several previous volcanic events, because they were not in contact with lava but were further covered by lapilli. Since the part of the magma chamber from which they originated was completely solidified, they must have been formed during a preceding eruptive episode. This episode is most probably the last 1983 manifestation of the Macdonald Seamount activity, because the cumulate formations were not disturbed by "mush-like" crystal-liquid motions, which should have resulted from further replenishments. This implies that about three years of cooling were sufficient for complete igneous crystallization

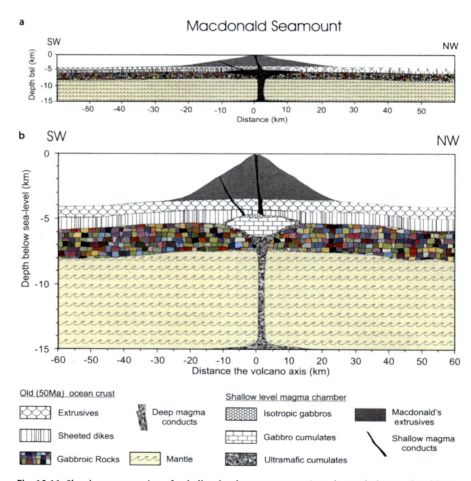

Fig. 10.11. Sketch representation of a shallow level magma reservoir underneath the Macdonald Seamount. This magma chamber or sill is believed to have originated the gabbroic clasts ejected in the caldera during the 1986 explosive volcanic activity; a no vertical exaggeration; b vertical exaggeration × 4

and partial metamorphic recrystallization. Therefore, a high-temperature gradient and vigorous hydrothermal convections were necessary to provide such severe cooling conditions, and this suggests a rather shallow level beneath the volcano.

The fact that most samples were not xenoliths embedded in lavas but accidental rock debris indicates that they were ejected without contact with further lava input. The existence of shallow level magma chambers underneath intraplate volcanoes has been previously pointed out on several occasions. In the Hawaiian volcanic shields, such magma reservoirs can persist both in the decollement zone at the mantle-crust boundary (15–20 km) and beneath the caldera complex (3–7 km). The recent Society and Austral hotspots have not constructed a volcanic shield, and the underlying crust is certainly thinner. Among the accidental debris, there were abundant fragments of altered lavas but few dolerites. This may be a sampling bias, but if it is statically representative, it should indicate that the reservoir was localized within the dyke complex. The presence of such a large magma sill in the upper oceanic crust could contribute to (or be responsible for) the formation of the bulge surrounding the volcano. According to the drilling results on the Galapagos Ridge Hole 504B (e.g., Alt et al. 1989), the thickness of the extrusive is about 700–800 m and the total Layer 2 is >1 500 m. From seismic data on the East Pacific Rise, the roof of the magma chamber is about 1 500–2 000 m (e.g., Perfit and Chadwick 1998). Therefore, the roof of the Macdonald Seamount upper reservoir is assumed to be at about 4.3 to 5 km below the top of the volcano (Fig. 10.11).

10.6
Summary and Conclusions

Several lines of evidence indicate that the accidental gabbroic rock debris recovered from the Macdonald Seamount originates from plume magmatism responsible for the formation of the edifice: The rocks have (1) strong chemical affinities with the lavas and other volcanics forming the volcano; (2) textural and mineralogical features that distinguish them from other common ocean ridge gabbros; and (3) geochemical and petrographical characteristics similar to gabbroic xenoliths or clasts ejected from other intraplate volcanoes, and which have been recognized as material from hotspot magmatism.

The rocks crystallized in an undisturbed part of a magma chamber, because they consist of common cumulate textures that are not plastically deformed and do not show high-temperature metamorphic recrystallization. Some primary textural features and the secondary mineralogy suggest a faster cooling rate under a higher thermal gradient than that existing in ocean ridge environments: Instead of poikilitic diopside in the inter-cumulus position as is usual in ultramafic cumulates, the troctolite shows plagioclase and clinopyroxene forming a sub-doleritic matrix. In addition, the Ca-amphiboles show a wide compositional gap between the primary magmatic hornblendes and the secondary lower-temperature actinolitic amphiboles, whereas typical ocean-ridge lower-crustal rocks show more continuous chemical variations of Ca-amphibole composition. Unlike other alkali-gabbros from Mid-Ocean Ridge systems (Mid-Atlantic Ridge near 15° N) the Macdonald rock's biotite composition is homogeneous, and the recrystallization of brown, primary Ti-biotite to green iron-rich biotite (Cannat et al. 1992) is not observed. This suggests crystallization in a magma chamber or sill at a relatively shallow depth, and then a rapid cooling and thermal hydraulic fracturing at rather low temperatures, mostly below 500 °C.

The gabbroic rocks have a blocky appearance and are not embedded in lavas. It is believed that they were most probably ejected during the 1986 eruptive phase of the Macdonald Seamount. Their residence time in the crust was sufficient for complete solidification and partial alteration before extrusion, indicating that the reservoir formed during an earlier eruptive phase. The cumulate and high-level isotropic gabbros were not in contact with further extruded lavas nor contaminated by late magma intrusions. Thus, this part of the magma reservoir where they cooled was not replenished by more recent magma supplies, or else the reservoir was previously formed during the preceding 1983 eruptive phase. The hydraulic fracturing and a penetration of boiling seawater into the core of the volcano and the underlying oceanic crust caused the early explosions that ejected the gabbroic blocks through the caldera. This phase ended with hydromagmatic explosions and the eruption of lapilli.

Based on a comparison with gabbro xenoliths derived from sequential volcanic stages in Hawaii, similar gabbroic ejecta from the Canary Islands and on seismic studies on Mauna Kea Island, we can infer the origin of the Macdonald Seamount gabbroic clasts. It has been suggested that they formed in a quiet, short-lived and shallow reservoir located in the underlying, roughly fifty-million-year-old oceanic crust at about 5 km beneath the top of the volcano. Then, they underwent relatively fast cooling, as the result of high lateral temperature gradients, in the absence of a further magma supply. This cooling was accompanied by late fracturing and the formation of veins under actinolite-greenschist facies conditions down to lower-temperatures. Finally, the gabbroic clasts were ejected by explosions related to the extrusive activity responsible for the 1986 seismic swarm, several months before the 1987 cruise of the F.S. SONNE.

Acknowledgements

We are grateful to the captain, officers and crew of the FS SONNE during its cruise in January 1987. We are also indebted to Professor Peter Stoffers (University of Kiel, Germany) chief scientist of the Leg SO47 Midplate expedition. The microprobe analyses were performed at IFREMER ("Microsonde de l'Ouest", Brest, France) with the expertise of Marcel Bohn. The sample preparations were done by Ronan Apprioual.

References

Alt JC, Anderson TF, Bonnell L, Muehlenbachs K (1989) Mineralogy, chemistry and stable isotopic composition of hydrothermaly altered sheetes dikes: ODP Hole 504B, Leg 111. In: Becker K, Sakai H, et al. (eds) Proc ODP, Sci Results, vol 111, College Station TX (Ocean Drilling Program), pp 27–40

Arai S, Matsukage K (1998) Petrology of a chromitite micropod from Hess Deep, equatorial Pacific: A comparison between abyssal and alpine-type podiform chromitites. Lithos 43:1–14

Bardintzeff JM, Bonin B, Brousse R, McBirney AR (1989) Plutonic complex of Tahiti-Nui Caldera (Pacific Ocean); Magmatic evolution of gabbroic and theralitic trends. International Geological Congress, Abstr 28(1):86

Baten SK (1997) A petrologic study of the 1924 ejecta from Halemaumau Crater, Kilauea Caldera, Kilauea Volcano, Hawaii. In: Mendelson CV, Mankiewicz C (compiler) Keck Research Symposium in Geology 10:203–206

Bideau D, Hébert R, Hekinian R, Cannat M (1991) Metamorphism of deep-seated rocks from the Garrett ultrafast transform (East Pacific Rise near 13°25' S). J Geophys Res 96:10079–10099

Binard N, Hekinian R, Cheminée J-L, Searle RC, Stoffers P (1991) Morphological and structural studies of the Society and Austral hotspot regions in the south pacific. Tectonophysics 186:293–312

Binard N, Hekinian R, Stoffers P, Cheminée J-L (2004) South Pacific intraplate volcanism: Structure, morphology and style of eruption. Springer-Verlag, this volume

Chapter 10 · Intraplate Gabbroic Rock Debris Ejected from the Macdonald Seamount 345

Bonatti E, Honnorez J, Ferrara G (1971) I. Ultramafic rocks: Peridotites-gabbro-basalt complex from the equatorial Mid-Atlantic Ridge

Bonatti E, Lawrence JR, Morandi N (1984) Serpentinization of ocean-floor peridotites: Temperature dependence on mineralogy and boron content. Earth Planet Sci Lett 70:88–94

Bottinga Y, Javoy M (1973) Comments on oxygen isotope geothermometry. Earth Planet Sci Lett 20: 250–265

Brooks CK (1976) The Fe_2O_3/FeO ratio of basalts analyses: An appeal for standardized procedure. Bull Geol Soc Den 25:117–120

Brousse R, Richter de Forges B (1980) Laves alcalines et differenciées du volcan sousmarin Macdonald. CR Acad Sci Paris Ser D290:lOS5–lOS7

Brown EH (1967) The greenschist facies in part of eastern Otago, New Zealand. Contrib Mineral Petrol 14:259–292

Cann JR (1979) Metamorphism in the ocean crust. In: Talwani MHCG, Hayes DE (eds) Deep drilling results in the Atlantic Ocean: Ocean crust. Am Geophys Union Geodyn Ser, pp 230–238

Cannat M, Bideau D, Bougault H (1992) Serpentinized peridotites and gabbros in the Mid Atlantic Ridge axial valley at 15°37' N and 16°52' N. Earth Planet Sci Lett 109:87–106

Cannat M, Karson JA, Miller DJ, et al. (1995) Proc ODP, Init Repts, vol 153, College Station TX (Ocean Drilling Program)

Carlson ET (1956) Hydrogarnet formation in the system lime-alumina-silica-water. J Res Natl Bur Stand 56:327–335

Caruso LJ, Chernosky JV Jr (1979) The stability of lizardite. Can Mineral 17:757–769

Cavarretta G, Gianelli G, Puxeddu M (1980) Hydrothermal metamorphism in the Larderello geothermal field. Geothermics 9:297–314

Cheminée JL, Hekinian R, Talandier J, Albarède F, Devey CW, Francheteau J, Lancelot Y (1989) Geology of an active hot-spot: Teahitia-Mehetia region of the South Pacific. Marine Geophys Res 11:27–50

Cheminée J-L, Stoffers P, McMurtry G, Richnow H, Puteanus D, Sedwick P (1991) Gas-rich submarine exhalations during the 1989 eruption of Macdonald Seamount. Earth Planet Sci Lett 107:318–327

Clague DA (1976) Petrology of basaltic and gabbroic rocks dredged from the Danger Island Troughs, Manihiki Plateau. In: Initial Repts DSDP 33; Honolulu, Hawaii to Papeete, Tahiti, pp 891–911

Clague DA (1987) Hawaiian xenolith populations, magma supply rates, and development of magma chambers. Bull Volc 49(4):577–587

Clague DA, Dalrymple GB (1988) Age and petrology of alkali postshield and rejuvenated-stage lava from Kauai, Hawaii. Contrib Mineral Petrol 99:202–218

Coleman RG (1967) Low-temperature reaction zones and alpine ultramafic rocks of California, Oregon and Washington. US Geol Surv Bull 1247:49

Constantin M (1999) Gabbroic intrusions and magmatic metasomatism in harzburgites from the Garrett transform: implications for the nature of the mantle-crust transition at fast-spreadind ridges. Contrib Mineral Petro 136:111–130

Davis AS, Clague DA (1990) Gabbroic xenoliths from the northern Gorda Ridge; Implications for magma chamber processes under slow spreading centers. J Geophys Res 95(7):10885–10905

Devey CW, Albarède F, Cheminée J-L, Michard A, Mühe R, Stoffers P (1990) Active submarine volcanism on the Society hotspot swell (West Pacific): A geochemical study. J Geophys Res 95:5049–5067

Dick HJB, Natland JH, Miller DJ, et al. (1999) Proc. ODP, Init Repts, 176 (CD-ROM). Available from: Ocean Drilling Program, Texas A&M University, College Station, TX, USA

Dick HJB, Natland JH, Alt JC, Bach W, Bideau D, Gee JS, Haggas S, Hertogen JGH, Hirth G, Holm PM, Ildefonse B, Iturrino GJ, John BE, Kelley DS, Kikawa E, Kingdon A, LeRoux PJ, Maeda J, Meyer PS, Miller DJ, Naslund HR, Niu Y, Robinson PT, Snow J, Stephen RA, Trimby PW, Wörm H, Yoshinobu A (2000) A long in situ section of the lower ocean crust: Results of ODP Leg 176 drilling at the Southwest Indian Ridge. Earth Planet Sci Lett 179 (1):31–51

Dixon JE, Clague DA, Eissen J-P (1986) Gabbroic xenoliths and host ferrobasalts from the southern Juan de Fuca Ridge. J Geophys Res 91(3):3795–3920

Duncan RA, McDougall I (1976) Linear volcanism in French Polynesia. J Volcanol Geotherm Res 1:197–227

Elthon D (1981) Metamorphism in oceanic spreading centers. In: Emiliani C (ed) The sea, vol. 7: The oceanic lithosphere. Wiley, New York, pp 285–303

Fisher RV, Schmincke H-U (1984) Pyroclastic rocks. Springer-Verlag, Berlin Heidelberg

Fleet ME, Barnett RL (1978) Al^{IV}/Al^{VI} partitioning in calciferous amphiboles from the Frood Mine, Sudbury, Ontario. Can Mineral 16:527–532

Fletcher JM, Stephen CJ, Petersen EU, Skerl L (1997) Greenschist facies hydrothermal alteration of oceanic gabbros: a case study of element mobility and reaction paths. In: Karson JA, Cannat M, Miller DJ, Elthon D (eds) Proc ODP, Sci Res, vol 153, pp 389–398

Fodor RV, Galar P (1993) Hawaiian magma reservoir processes; interpretations from textures and mineral compositions for xenoliths of Mauna Kea Volcano. Abstr Prog, Geol Soc Am 25(6):444–445

Fodor RV, Galar P (1997) A view into the subsurface of Mauna Kea Volcano, Hawaii; Crystallization processes interpreted through the petrology and petrography of gabbroic and ultramafic xenoliths. J Petrol 38(5):581–624

Fodor RV, Moore RB (1994) Petrology of gabbroic xenoliths in 1960 Kilauea basalt; Crystalline remnants of prior (1955) magmatism. Bull Volc 56(1):62–74

Fodor RV, Vandermeyden HJ (1987) Mauna Kea gabbroic xenoliths: cumulates from alkalic-basalt suite fractional crystallization. In Decker RW, Halbig JB, Hazlett RW, Okamura R, Wright TL (eds) Hawaii symposium on How volcanoes work; Abstract volume, p 78

Fodor RV, Vandermeyden HJ (1988) Petrology of gabbroic xenoliths from Mauna Kea volcano, Hawaii. J Geophys Res 93:4435–4452

Gaffney AM (1999) Crystallization and emplacement of Hawaiian tholeiitic-stage xenoliths. Abstr Prog, Geol Soc Am 31(7):180

Govindaraju K (1982) Report (1967–1981) on four ANRT rock reference samples: diorite DR-N, serpentine UB-N, bauxite BX-N and disthene DT-N. Geostandards Newsletters 6(1):91–159

Gillis K, Mével C, Allan J, et al. (1993) Proc ODP, Init Repts, vol 147, College Station TX (Ocean Drilling Program)

Hébert R, Bideau D, Hekinian R (1983) Ultramafic and mafic rocks from the Garret Transform Fault near 13°30' S on the east Pacific Rise: Igenous petrology. Earth Planet Sci Lett 65:107–125

Hekinian R, Hébert R, Maury RC, Berger ET (1985) Orthopyroxene-bearing gabbroic xenoliths in basalts from East Pacific Rise axis near 12°50' N. Bull Mineral 108(5):691–698

Hekinian R, Bideau D, Stoffers P, Cheminée J-L, Mühe R, Puteanus D, Binard N (1991) Submarine intraplate volcanism in the South Pacific: Geological setting and petrology of the Society and the Austral Regions. J Geophvs Res 96:2109–2138

Hekinian R, Bideau D, Francheteau J, Cheminée J-L, Armijo R, Lonsdale P, Blum N (1993a) Petrology of the East Pacific Rise crust and upper mantle exposed in the Hess Deep (eastern equatorial Pacific). J Geophys Res 98:8069–8094

Hekinian R, Hoffert M, Larqué P, Cheminée JL, Stoffers P, Bideau D (1993b) Hydrothermal Fe and Si oxyhydroxides deposits from south pacific intraplate volcanoes and East Pacific Rise axial and off-axial regions. Economic Geology 88:2099–2121

Hekinian R, Juteau T, Gràcia E, Sichler B, Udintsev G, Apprioual R, Ligi M (2000) Submersible observations of Equatorial Atlantic mantle: The St. Paul Fracture Zone region 21:529–560

Herron EM (1972) Seafloor spreading and the Cenozoic history of the east-central Pacific. Geol Soc Am Bull 83:1671–1692

Hertogen J, Emmermann R, Robinson PT, Erzinger J (2002) Lithology, mineralogy, geochemistry of the lower ocean crust, ODP Hole 735B, Southwest Indian Ridge. In: Natland JH, Dick HJB, Miller DJ, Von Herzen RP (eds) Proc ODP, Sci Results, vol 76, 1–82 [CD-ROM]. Available from: Ocean Drilling Program, Texas A&M University, College Station TX 77845-9547, USA

Holdaway MJ (1972) Thermal stability of Al-Fe epidote as a function of fO_2 and Fe content. Contrib Mineral Petrol 37:307–340

Honnorez J, Kirst P (1975) Petrology of rodingites from the equatorial mid-Atlantic fracture zone and their geotectonic significance. Contrib Mineral Petrol 49:233–257

Hoover SR, Fodor RV (1997) Magma-reservoir crystallization processes: small-scale dikes in cumulate gabbros, Mauna Kea Volcano, Hawaii. Bull Volc 59(3):186–197

Hoover S, Ginn F, Fodor RV (1996) Magma-reservoir crystallization processes; small-scale dikes in cumulate gabbro, Mauna Kea Volcano, Hawaii. Abst Prog, Geol Soc Am 28(7):289

Johnson RH (1970) Active submarine volcanism in the Austral Islands. Science 167:977–979

Johnson RH (1980) Seamounts in the Austral Islands region. National Geographic Society Res Reports 12:389–405

Johnson RH, Malahoff A (1971) Relation of Macdonald volcano to migration of volcanism along the Austral chain. J Geophys Res 76:3282–3290

Johnston AD, Stout JH, Murthy VR (1985) Geochemistry and origin of some unusually oxidized alkaline rocks from Kauai, Hawaii. J Volc Geotherm Res 25(3–4):225–248

Jordahl K, Caress D, McNutt M, Bonneville A (2004) Seafloor topography and morphology of the Superswell Region. Springer-Verlag, this volume

Karson JA, Cannat M, Miller DJ, et al. (1997) Proc ODP, Sci Results, vol 153, College Station TX (Ocean Drilling Program)

Keith TEC, Mufflu LJP, Cremer N (1968) Hydrothermal epidote formed in the Salton Sea geothermal system, California. Am Mineral 53:1635–1644

Kennedy AK, Kwon S-T, Frey FA, West HB (1991) The isotopic composition of postshield lavas from Mauna Kea Volcano, Hawaii. Earth Planet Sci Lett 103:339–353

Kokelaar P (1986) Magma-water interactions in subaqueous and emergent basaltic volcanism. Bull Volcanol 48:275–289

Klügel A, Schmincke H-U, White JDL, Hoernle KA (1999) Chronology and volcanology of the 1949 multi-vent rift-zone eruption on La Palma (Canary Islands). J Volc Geotherm Res 94: 1-4, 267-282

Laschek D (1985) Geochemische Untersuchugen an Basalten vom Galapagos Spreading Center und vom East Pacific Rise. Doctoral dissertation, Fakultät für Bio- und Geowissenschaften der Universität (TH) Fridericiana Karlsruhe (Germany), pp 1-133

Le Bas MJ, Le Maître RW, Streckeisen A, Zanettin B (1986) A chemical classification of volcanic rocks based on total alkali-silica diagram. J Petrol 27(3):745-750

Liou JG (1973) Synthesis and stability relations of epidote, $Ca_2 Al_2 Si_3 O_{12}$ (OH). J Petrol 14:381-413

Liou JG, Ernst WG (1979) Oceanic ridge metamorphism of the East Taiwan ophiolite. Contrib Mineral Petrol 68:335-348

Liou JG, Kuniyoshi S, Ito K (1974) Experimental studies of the phase relations between greenschist and amphibolite in a basaltic system. Am J Sci 274:613-632

Melson WG, Thompson G (1970) Layered basic complex in oceanic crust, Romanche fracture, equatorial Atlantic Ocean. Science 168:817-820

Mével C (1987) Evolution of oceanic gabbros from DSDP Leg 82: influence of the fluid phase on metamorphic crystallizations. Earth Planet Sci Lett 83:67-79

Mével C, Cannat M (1991) Lithospheric stretching and hydrothermal processes in oceanic gabbros from slow-spreading ridges. In: Peters T, Nicolas A, Coleman R.G. (eds) Ophiolite genesis and evolution of the oceanic lithosphere. Proceedings of the Ophiolite conference. Kluwer Acad, Dordrecht, Netherlands, pp 293-312

Mével C, Stamoudi C (1996) Hydrothermal alteration of the upper-mantle section at Hess Deep. In: Mével C, Gillis KM, Allan JF, Meyer PS (eds) Proc ODP, Sci Results, vol 147, pp 293-309

Mével C, Gillis KM, Allan JF, Meyer PS (1996) Proc ODP, Sci Results, vol 147, College Station TX (Ocean Drilling Program)

Miyashiro A, Shido F (1980) Differentiation of gabbros in the Mid-Atlantic Ridge near 24° N. Geochem Journ 14:145-154

Miyashiro A, Shido F, Ewing M (1971) Metamorphism in the Mid-Atlantic Ridge near 24° and 30° N. Phil Trans Roy Soc London 268:589-604

Moody JB (1976) Serpentinization: A review. Lithos 9:125-138

Mottl MJ, Holland HD (1978) Chemical exchange during hydrothermal alteration of basalt by seawater: I. Experimental results for major and minor components of seawater. Geochim Cosmochim Acta 42:1103-1115

Natland JH, Dick HJB, Miller DJ, Von Herzen RP (2002) Proc ODP, Sci Res, vol 176 [CD-ROM and Online], available from: Ocean Drilling Program, Texas A&M University, College Station TX 77845-9547, USA

Nehlig P, Juteau T (1988) Flow porosities, permeabilities and preliminary data on fluid inclusions and fossil thermal gradients in the crustal sequence of the Samail ophiolite (Oman). Tectonophysics 151:199-221

Neumann ER, Wulff-Pedersen E, Simonsen SL, Pearson NJ, Martí J, Mitjavila J (1999) Evidence for fractional crystallization of periodically refilled magma chambers in Tenerife, Canary Islands. J Petrol 40:1089-1123

Neumann ER, Sorensen VB, Simonsen SL, Johnsen K (2000) Gabbroic xenoliths from La Palma, Tenerife and Lanzarote, Canary Islands: Evidence for reactions between mafic alkaline Canary Islands melts and old oceanic crust. J Volc Geotherm Res 103:1-4, 313-342

Norris A, Johnson RH (1969) Submarine volcanic eruptions recently located in the Pacific by sofar hydrophones. J Geophys Res 74:650-664

O'Brien JP, Rodgers KA (1973) Xonotlite and rodingites from Wairere, New Zealand. Mineral Mag 39(202):233-240

Okubo PG, Benz HM, Chouet BA (1997) Imaging the crustal magma sources beneath Mauna Loa and Kilauea Volcanoes, Hawaii. US Geoll Survey, Hawaii National Park, HI, United States, Geology (Boulder) 25(10):867-870

Peck DL, Wright TL, Moore JG (1966) Crystallization of tholeiitic basalt in Alae Lava Lake, Hawaii. Bull Volcanol 29:629-655

Pedersen RB, Malpas J, Falloon T (1996) Petrology and geochemistry of gabbroic and related rocks from Site 894, Hess Deep. In: Mével C, Gillis KM, Allan JF, Meyer PS (eds) Proc ODP, Sci Res, vol 147, College Station TX (Ocean Drilling Program), pp 3-19

Perfit MR, Chadwick WW Jr (1998) Magmatism at mid-ocean ridges: Constraints from volcanological and geochemical investigations. In: Buck WR, Delaney PT, Karson JA, Lagabrielle Y (eds) Faulting and magmatism at mid-ocean ridges. , Am Geophys Union, Geophys Monogr 106:59-115

Pistorius CWST, Kennedy GC (1960) Stability relations of grossularite and hydrogrossularite at high temperatures and pressures. Am J Sci 258:247-257

Ploshko VV, Bogdanov YA, Knyazeva DN (1969) Gabbro-amphibolite from the abyssal romance trench, Atlantic region. Doklady Akad Nauk SSSR 192:40–43

Raase P (1974) Al and Ti contents of hornblende; indicators of pressure and temperature of regional metamorphism. Contrib Mineral Petrol 45:231–236

Raith M (1976) The Al-Fe(III) epidote miscibility gap in a metamorphic profile through the penninic series of the Tawn Window, Austria. Contrib Mineral Petrol 57:99–117

Reiners PW, Nelson BK, Izuka SK (1999) Structural and petrologic evolution of the Lihue Basin and eastern Kauai, Hawaii. Geol Soc Am Bull 111(5):674–685

Robinson PT, Von Herzen R, et al. (1989) Proc ODP, Init Repts, vol 118, College Station TX (Ocean Drilling Program)

Sailor RV, Okal EA (1983) Applications of seasat altimeter data in seismotectonic studies of the south-central Pacific. J Geophys Res 88:1572–1580

Schmincke H-U, Klügel A, Hansteen TH, Hoernle K, van den Bogaard P (1998) Samples from the Jurassic ocean crust beneath Gran Canaria, La Palma and Lanzarote (Canary Islands). Earth Planet Sci Lett 163:343–360

Seifert K, Gibson I, Weis D, Brunotte D (1996) Geochemistry of metamorphosed cumulate gabbros from Hole 900A, Iberia Abyssal Plain. In: Whitmarsh RB, Sawyer DS, Klaus A, Masson DG (eds) Poc ODP, Sci Res, vol 149, pp 471–488

Seifert K, Chang C-W, Brunotte D (1997) Evidence from Ocean Drilling Program Leg 149 mafic igneous rocks for oceanic crust in Iberia Abyssal Plain ocean-continent transition zone. J Geophys Res 102:7915–7928

Sigwaldason GE (1962) Epidote and related minerals in two deep geothermal drill holes, Reykjavick and Hveragerdi, Iceland. US Geol Surv Prof Pap 450E:77–84

Simonov VA, Kolobov VYu, Peyve AA (1999) Petrology and geochemistry of geodynamic processes in central Atlantic. In: Dobretson NL (ed) SPC UIGGM, Siberian branch of RAS, Novosibirsk, pp 75–77, in Russian,

Steiner A (1966) On the occurrence of hydrothermal epidote at Wairakei, New Zealand. IZV Akad Nauk USSR, Geol Series 2:167

Stewart LJ, Brunstad KA (1999) Scanning electron microscope study of xenoliths from Mauna Kea Volcano, Hawaii. Abs Prog, Geol Soc Am 31(7):166

Stoffers P, Botz R, Cheminée J-L, Dewey CW, Froger V, Glasby GP, Hartmann M, Hekinian R, Kögler F, Laschek D, Larqué P, Michaelis W, Mühe RK, Puteanus D, Richnow HH (1989) Geology of the Macdonald "hot-spot": Recent submarine eruptions in the South Pacific. Mar Geophys Res 11: 101–112

Sushchevskaya NM, Bonatti E, Peive AA, Kamenetskii VS, Belyatskii BV, Tsekhonya TI, Kononkova NN (2002) Heterogeneity of rift magmatism in the equatorial province of the Mid-Atlantic Ridge (15° N to 3° S). Geochem Intern 40(1):26–50

Sun SS, McDonough WF (1989) Chemical and isotopic systematics of oceanic basalts: implications for mantle composition and processes. In: Saunders AD, Norry MJ (eds) Magmatism in the ocean basins. Geol Soc Spec Pub London 42:313–345

Talandier J (2003) Seismicity of the Society and Austral hot spots in the South Pacific. Seismic detection, monitoring and interpretation of underwater volcanism. This volume

Talandier J, Okal EA (1983) The volcanoseismic swarms of 1981–1983 in the Tahiti-Mehetia area. J Geophys Res 89:11216–11234

Talandier J, Okal EA (1984) New survey of Macdonald eamount, South central Pacific, following volcanoseismic activity, 1977–1983. Geophys Res Lett 1:813–816

Thieâen O, Schmit M, Botz R, Stoffers P (2003) Biogenic methane formation in hotspot. This volume

Upton BGJ, Semet MP, Joron J-L (2000) Cumulate clasts in the Bellecombe ash member, Piton de la Fournaise, Réunion Island, and their bearing on cumulative processes in the petrogenesis of the réunion lavas. J Volc Geotherm Res 104:297–318

Vogt PR, Smoot NC (1984) The Geisha Guyots: Multibeam bathymetry and morphological interpretation. J Geophys Res 89:11085–11107

Von Herzen RP, Robinson PT, et al. (1991) Proc ODP, Sci Rses, vol 118, College Station , TX (Ocean Drilling Program)

Wenner DB, Taylor HP (1971) Temperature of serpentinization of ultramafic rocks based on $^{18}O/^{16}O$ fractionation between co-existing serpentine and magnetite. Contrib Mineral Petrol 32:165–185

Werner C-D (1997) Data report: Geochemistry of rocks and minerals of the gabbro complex from the MARK area. In: Karson JA, Cannat M, Miller DJ, Elthon D (eds), Proc ODP, Sci Res, vol 153, College Station TX (Ocean Drilling Program), pp 491–504

White JDL, Schminkce HU (1999) Phreatomagmatic eruptive and depositional processes during the 1949 eruption on La Palma (Canary Islands). J Volc Geotherm Res 94:203–304

Yoder HS (1950) Stability relations of grossularite. J Geol 58:221–253

Chapter 11

The Foundation Chain: Inferring Hotspot-Plate Interaction from a Weak Seamount Trail

J. M. O'Connor · P. Stoffers · J. R. Wijbrans

11.1
Introduction

The Foundation Chain was first detected using a combination of satellite altimetric and conventional geophysical data (Sandwell 1984; Mammerickx 1992) and described initially as a ~1350 km long chain of seamounts trending approximately in the direction of the absolute motion of the Pacific Plate (Mammerickx 1992) (Fig. 11.1). A significant section of the Foundation Chain lies in a tectonic setting influenced by a change in the direction of sea-floor spreading between 26 and 11 Ma (Herron 1972; Lonsdale 1988; Mayes et al. 1990; Mammerickx 1992). This motion change is reflected in the curvature of the Agassiz Fracture Zone (FZ) and its west-east shift in orientation between the Resolution/Del Cano and Chile FZs (Fig. 11.2). A segment of the Nazca plate was transferred to the Pacific Plate during this period of reorganization to form the short-lived Selkirk Microplate (Fig. 11.2) (Mammerickx 1992; Tebbens and Cande 1997; Tebbens et al. 1997) via a spreading-ridge propagation event between chron 6C (23.4–24 Ma) and Chron 6(0) (20.2 Ma) (Tebbens and Cande 1997).

Predicted sea-floor topography (Smith and Sandwell 1997) reveals that the Foundation Chain changes ~450 km west of the present Pacific-Antarctic spreading axis from a nar-

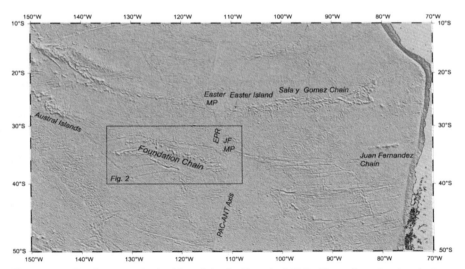

Fig. 11.1. Predicted topography (Smith and Sandwell 1997) of SE Pacific seafloor showing the location of the Foundation Chain. *MP* = microplate; *JF* = Juan Fernandez; *EPR* = East Pacific Rise

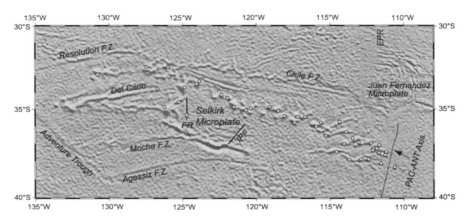

Fig. 11.2 Predicted topography of the Foundation Chain (Smith and Sandwell 1997). F/S SONNE and N/O ATALANTE dredge sites are indicated by *black-rimmed white dots*. ^{40}Ar/^{39}Ar ages are summarized in Table 11.1; details of sample information and analytical data are in O'Connor et al. (1998, 2001, 2002). *IPF* = Inner Pseudo Fault and *FR* = failed rift of Selkirk Microplate (Mammerickx 1992)

row line of individual or clustered seamounts to a much broader region of en échelon Volcanic Elongate Ridges (VERs) (Fig. 11.2). This morphology switch correlates broadly with a change in composition of Foundation lavas from enriched hotspot-like to a more depleted mixed type (Hémond and Devey 1996; Hekinian et al. 1997; Devey et al. 1997; see Sect. 9.3.2.4). Such changing fabric and less hotspot-like composition have been explained by interaction between the inferred Foundation hotspot and the Pacific-Antarctic spreading-axis (Hémond and Devey 1996; Hekinian et al. 1997; Devey et al. 1997).

The Foundation Chain was extensively dredge-sampled (Fig. 11.2) during the F/S SONNE SO100 cruise in 1995 (Devey et al. 1995, 1997), followed by some additional sampling during the 1997 N/O ATALANTE 'Hotline' cruise (Maia et al. 2001). ^{40}Ar/^{39}Ar ages for Foundation rock samples (O'Connor et al. 1998, 2001, 2002) show a linear trend of decreasing Foundation Chain age between ~22 Ma (northwestern end) and ~2 Ma (southeastern end) at a rate of 91 ±2 mm yr^{-1} (Fig. 11.3) (O'Connor et al. 1998). This rate corresponds well with estimates of the migration rate of volcanism along the Hawaiian Chain (O'Connor et al. 1998; Koppers et al. 2001) suggesting – together with the intraplate nature of the Foundation Seamounts (O'Connor et al. 1998) and their hotspot-like composition (Devey et al. 1997; Hekinian et al. 1997, 1999; Hémond and Devey 1996; Hémond et al. 1999) – that Foundation Chain volcanism documents a ~22 Myr history of Pacific Plate motion over the narrow locus of an inferred upwelling Foundation mantle plume (O'Connor et al. 1998, following Morgan 1971). Furthermore, measured Foundation Chain ages suggest that volcanism erupted at a regular interval of ~1 Myr (O'Connor et al. 1998, 2001, 2002).

Hotspots appear to influence a significant part of the global spreading center system. Thus, to understand the global convection system, it is important to determine the processes controlling hotspot-spreading center interactions. Foundation VERs present a valuable chance to directly investigate these processes. ^{40}Ar/^{39}Ar ages show that individual Foundation VERs can occasionally develop synchronously (i.e., coeval volcanism along one continuous ridge) (Fig. 11.4). But surprisingly, the dominant trend in the new Foundation Chain age data is for structurally disconnected sections of different

Chapter 11 · The Foundation Chain: Inferring Hotspot-Plate Interaction from a Weak Seamount Trail   351

VERs to be coeval. These synchronously erupted, yet structurally unconnected, VER sections define a series of NE-SW en échelon elongated zones of hotspot volcanism that crosscut the overall NW-SE trend of the Foundation Chain (O'Connor et al. 2001) (Fig. 11.5). 'Zones' appear to have developed at intervals of ~1 Myr while maintaining a steady state size (~250 km by ~150 km) and orientation (NE-SW), as the Pacific-Antarctic spreading center migrated progressively closer to the Foundation hotspot.

Although such VER development was controlled in part by local factors (e.g., location of nearest spreading center segment, lithospheric stress), long-lived attributes of the Foundation hotspot melting anomaly (e.g., size, orientation, periodicity) appear to have played a significant role. The key to testing this notion is the fact that the Foundation Chain represents a rare, possibly unique case of a hotspot trail crossing a fossil microplate. Prior to encountering the Selkirk Microplate, the Foundation Chain formed as broad zones of scattered, synchronous Foundation volcanism – similar to those identified west of the present Pacific-Antarctic spreading center (Fig. 11.6). However, once the significantly older microplate lithosphere began capping the hotspot about 14 Myr ago (O'Connor et al. 2002), the chain narrowed abruptly into a line of discrete seamounts, only to broaden again about 5 Myr ago when sufficiently young lithosphere once again drifted over the hotspot (Fig. 11.7). Foundation hotspot volcanism can therefore be prevented across elongate hotspot zones, if the capping tectonic plate is too thick for melts to penetrate to the surface (O'Connor et al. 1998, 2002). We infer from this information that Foundation Chain development was controlled primarily by tectonic plate migration over broad hotspot zones of fundamentally constant size and orientation created with an apparent 1 Myr periodicity (O'Connor et al. 2002).

11.2
Sample Preparation and Analytical Procedure

11.2.1
Sample Selection and Preparation

Descriptions and locations of dredge samples are in O'Connor et al. (1998, 2001, 2002). Pieces of selected rock samples were crushed and sieved, following removal with a saw of outer surfaces and as much visible alteration as possible. Altered whole rock samples (500–250 µm) were treated with 7 N HCl prior to treatment for one hour in 1 N HNO$_3$ (in an ultrasonic bath at 50 °C) followed by rinsing in distilled water (Koppers 1998). Plagioclase (250–125 µm) was separated from crushed whole rock using a magnetic separator. Separated plagioclase was treated in 7 N HCl for 30 min, 5–8% HF for 5 min, 1 N HNO$_3$ for one hour and then washed in distilled H$_2$O (Koppers 1998). Complete removal of alteration products sometimes required the repetition of certain steps.

11.2.2
Dating Technique

The theory of ^{40}Ar/^{39}Ar geochronology has previously been described (e.g., Faure 1986, York 1984; McDougall and Harrison 1999). Sample ages are calculated using the standard age equation. The uncertainty in age is calculated by partial differentiation of

Fig. 11.3. a Sample age as a function of distance of dredge site from the present Pacific-Antarctic spreading axis. The *solid line* is the York-2 linear regression to age data representing an average migration rate of volcanism along the Foundation Chain of 91 ±2 mm yr^{-1} (O'Connor et al. 1998). Ages reported in O'Connor et al. (1998) have been recalculated (O'Connor et al. 2001) using the new TCR monitor age (28.34 Ma; Renne et al. 1998). This recalculation leads to a small systematic increase in previously calculated ages, leaving the calculated migration rate reported in O'Connor et al. (1998) unchanged. Analytical error bars (±2σ) are shown. **b** Seafloor age (Lonsdale 1994) as a function of distance from the present-day Pacific-Antarctic spreading center (*dashed line*). *Gray shaded horizontal lines* indicate individual episodes of coeval VER volcanism. **c** *Solid spheres* show correlation between weighted averages of measured Foundation Chain ^{40}Ar/^{39}Ar ages (Table 11.1) and corresponding ages predicted on the basis of an assumed ~1 Myr periodicity. A perfect correlation between measured and predicted ages assuming a 1 Myr periodicity is shown as a regression line with individual episodes indicated by *short crosscutting lines*. Cumulative probability plotting of Foundation ages (e.g., Table 11.1) further supports this inferred ~1 Myr periodicity

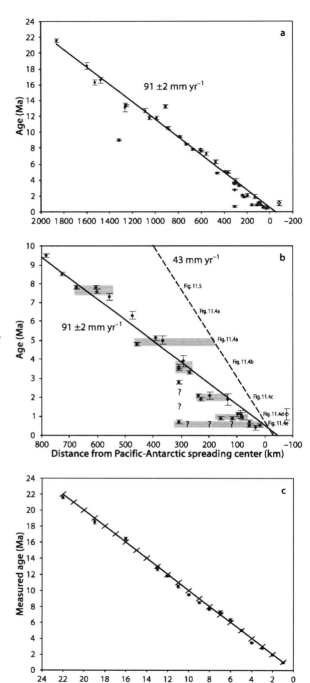

the age equation (Dalrymple and Lanphere 1969; Dalrymple et al. 1981) and includes uncertainties in the determination of the flux monitor J, blank determination, the regression of the intensities of the individual isotope peaks, the correction factors for interfering isotopes, and the mass discrimination correction. The argon laserprobe facility at the Vrije Universiteit Amsterdam has previously been described in considerable detail (Wijbrans et al. 1995). It consists of a 24W argon ion laser, beam optics, a low volume UHV gas inlet system and a Mass Analyser Products Ltd. 215-50 noble gas mass spectrometer. The mass spectrometer is fitted with a modified Nier type electron bombardment source. During data collection, the mass spectrometer is operated with an adapted version of the standard MAP software written in TurboPascal, allowing data collection for all isotopes of argon using a secondary electron multiplier collector operated at a gain of 50 000. Modifications include variable dwell time on each peak during data collection, valve control, laser control and x-y stage control, allowing the data collection during data acquisition for this project in semi-automated mode. Data reduction is described in some detail in Sect. 11.2.4. Mass fractionation was determined by measuring aliquots of air argon at regular intervals and at a later stage of the project by measuring aliquots of ^{38}Ar spiked air (Kuiper 2003).

11.2.3
Irradiation and Analysis

Plagioclase (250–125 µm) and whole rock chips (500–250 µm) were irradiated for either 7, or in the case of younger near-axis seamounts, 1.5 hours with Cd-shielding in the CLICIT facility at the Oregon State University TRIGA reactor. Cd shielding significantly reduces the effects of slow neutrons, leading to a major reduction in the ^{40}Ar produced without affecting the production of ^{39}Ar (e.g., McDougall and Harrison 1999), which in turn reduces the impact of the $(^{40}Ar/^{39}Ar)_K$ correction factor, particularly useful in the case of young and low potassium samples. The K and Ca correction factors for the Cd-shielded position at the OSU TRIGA reactor have previously been determined (Wijbrans et al. 1995). Between 50 and 100 mg of rock chips or plagioclase separate from each sample were wrapped in aluminum foil and stacked in 9 mm ID quartz tubes. Taylor Creek Rhyolite sanidine, TCR 85G003 (28.34 Ma; Renne et al. 1998), was used to measure the flux gradients. TCR was loaded between every four unknowns and also at the top and bottom of each vial. Between four and six replicate analyses of four to six sanidine crystals were made for each monitor position, typically giving uncertainties of 0.1 to 0.3% (sem). Using a best-fit curve between all of the standards allowed the determination of J factors with a similar precision.

High precision ^{40}Ar/^{39}Ar ages have been determined (plagioclase separates and whole rock chips) by incremental heating with the argon laser probe, using a defocused CW laser beam (e.g., York et al. 1981). We encountered difficulties initially in both heating samples with sufficient uniformity and in producing satisfactorily high peak intensities. Following some experimentation, we solved these problems by using large (13 mm) custom-made Cu-sample pans, which allowed us to load significantly more sample in even, single-grain layers. Heating such thin layers of sample (under manual x-y stage control) allowed us to produce predictable analytical results: plagioclase consistently showed excellent plateaus, whereas the whole rock experiments for older

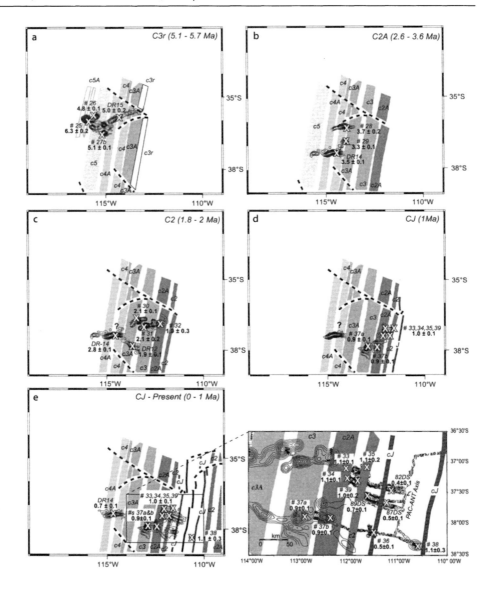

Fig. 11.4. Schematic reconstructions of the temporal and spatial relationship between the Foundation Chain and approaching Pacific-Antarctic spreading center (O'Connor et al. 2001). Bathymetry is interpolated from predicted topography of the Foundation Chain region (Smith and Sandwell 1997). The enlargement of the spreading center region (*e*) incorporates Hydrosweep data collected during the SO100 cruise of the F/S SONNE (Devey et al. 1997). Measured ^{40}Ar/^{39}Ar ages (Table 11.1) are in *bold*. Sea-floor ages and interpolated FZs (*solid dashed lines*) are from Lonsdale (1994), following the timescale of Cande and Kent (1995). C = seafloor isochron; # = seamount number; X = dredge site

samples typically showed elevated ages in the first ~20% of gas released, a good plateau, and anomalous young ages in the final ~10 to 20% when the sample was partially melting. However, due to the significantly younger sample ages involved in this study, combined with frequently low to very low % K_2O (Devey et al. 1997; Hekinian et al. 1997, 1999), we conducted also a significant number of multiple single fusion experiments whenever necessary to reduce analytical uncertainty to acceptable levels. The uncertainty in the system blank was shown to be the single most important factor contributing to the uncertainty in the ages, especially in cases where young, low-K samples were being analyzed. Blanks were run on average between every four blocks of unknown heating steps or fusions. The intensity and uncertainty for the ^{40}Ar, ^{39}Ar, ^{37}Ar and ^{36}Ar blanks for each day of analyses were calculated by regression of blank peak intensities versus time of measurement, allowing prediction of a blank for each experiment by interpolation (Koppers 2002).

11.2.4
Data Reduction

$^{40}Ar/^{39}Ar$ incremental heating data were reduced as both age spectra and isochrons, using the freeware data reduction package ArArCALC developed at the VU (Koppers 2002). It allows the choice between a straight linear regression of peak intensities with respect to inlet time, asymptotic curve fitting minimizing standard deviation and sum of squared residuals as criteria for best fit, and an average of peak intensities. The asymptotic extrapolation was very often the most appropriate, as the highest peak intensities often showed deviations from straight-line behavior with time. A summary of the calculated ages is in Table 11.1. Calculated ages, analytical data as well as the plateau and isochron plots are in O'Connor et al. (1998, 2001, 2002). However for consistency, ages reported in O'Connor et al. (1998) have been recalculated (O'Connor et al. 2001) using the new TCR monitor age (28.34 Ma; Renne et al. 1998). Plateau ages are presented as weighted means over the steps contributing to the plateau. We have used the York-2 least-squared linear fit with correlated errors (York 1969), and both the $^{40}Ar/^{36}Ar$ versus $^{39}Ar/^{36}Ar$ and $^{36}Ar/^{40}Ar$ versus $^{39}Ar/^{40}Ar$ correlation diagrams to calculate isochron ages. Mean squares of weighted deviates (MSWD) (York 1969; Roddick 1978) have been calculated for both the plateau and the isochron ages based on (N-1) and (N-2) degrees of freedom, respectively. If the scatter around the plateaus or isochrons was beyond analytical error at the 95% confidence level (i.e., MSWD > 1), the reported analytical error was multiplied by the MSWD (York 1969; Kullerud 1991). Following Fleck et al. (1977), Lanphere and Dalrymple (1978), Dalrymple et al. (1980) and Pringle (1993), all of the Foundation Chain ages – with very rare exception – pass the following tests and so are accepted as reliable (Table 11.1):

1. A well-defined high-temperature age spectrum plateau is created by three or more concordant (within 2σ), contiguous steps representing at least 50% of the ^{39}Ar released;
2. A well-defined isochron exists for the plateau points, i.e., the mean squares of weighted deviates (MSWD) (the ratio between the scatter about the line and the extent to which the scatter can be explained by analytical uncertainty) are not greater than the cut-off value of 2.5 (following Brooks et al. 1972 and McIntyre et al. 1966);
3. The $^{40}Ar/^{36}Ar$ intercepts found by regression analysis are not significantly different from the atmospheric level of 295.5, i.e., the plateau and isochron ages are concordant.

Table 11.1. Summary of Foundation Chain $^{40}Ar/^{39}Ar$ age data[a]

Sample ID	Smt. number[b]	Smt. name[c]	Latitude °S[d]	Longitude °W[d]	Sample	Type	$\%^{39}Ar$	Plateau (Ma)	2σ	Inverse isochron (Ma)	2σ
SO100 11DS-1	1a	Ampère	32°56.414' / 32°55.453'	130°45.459' / 130°45.982'	wr	IH	64.4	21.6	0.2	21.7	1.1
FH DR1-3		Aristotelis	32°30.33'	127°30.09'	wr	IH	66.7	18.5	0.3	18.4	0.4
SO100 18DS-1	5	Becquerel	32°28.871' / 32°28.870'	126°00.456' / 126°00.981'	wr	IH	67.2	16.6	0.4	16.6	0.4
SO100 17DS-1	5	Becquerel	32°28.474' / 32°28.403'	126°04.069' / 126°04.086'	wr	IH	42.2	16.3	0.3		
Weighted average								16.4	0.2		
FH DR4-1	Del Cano		33°46.69'	126°43.83'	wr	IH	44.5	16.1	0.4	16.3	0.7
FH DR5-2	VER	Boltzmann	34°55.112'	126°13.798'	wr	IH	99.7	16.8	0.3	16.6	1.0
FH DR6-1	VER	Laplace	34°34.000'	125°16.400'	wr	IH	32.3	16.6	0.2	16.1	0.6
FH DR7-16	Failed Rift		35°20.849'	124°45.723'	wr	SFs	100	20.4	0.4	19.5	3.7
FH DR7-16	Failed Rift		35°20.849'	124°45.723'	wr	IH	22.5	20.3	0.6	18.5	3.7
Weighted average								20.4	0.3	19.0	2.6
SO100 28GTV-2	8	Buffon	33°41.735' / 33°41.868'	124°54.612' / 124°54.425'	plag	IH	88.1	9.0	0.1	9.0	0.1
SO100 25DS-1	9	Celsius	33°20.764' / 33°20.891'	123°52.525' / 123°52.752'	wr	IH	72.9	13.3	0.1	13.3	0.1
SO100 26DS-1	9	Celsius	33°31.700' / 33°31.664'	124°06.434' / 124°05.535'	plag	IH	93.3	13.1	0.6	13.2	1.2
Weighted average								13.3	0.1	13.3	0.1

Table 11.1. Continued

Sample ID	Smt. number[b]	Smt. name[c]	Latitude °S[d]	Longitude °W[d]	Sample	Type	%^{39}Ar	Plateau (Ma)	2σ	Inverse isochron (Ma)	2σ
SO100 33DS-1	10	Curie	34°07.384' / 34°08.396'	122°22.038' / 122°21.610'	plag	IH	100	12.7	0.3	12.7	0.6
SO100 38DS-1	11	Da Vinci	34°19.119' / 34°20.326'	121°58.729' / 121°58.455'	plag	IH	99.6	11.8	0.2	11.6	0.9
SO100 41DS-1	12b	Darwin b	34°52.352' / 34°51.896'	121°33.309' / 121°33.353'	plag	IH	99.2	11.8	0.2	11.8	0.3
SO100 45DS-1	13a	Einstein a	35°03.180' / 35°03.224'	120°43.226' / 120°43.198'	plag	IH	89.2	13.6	0.2	13.2	0.8
SO100 46DS-2	16	Fermi	34°57.336' / 34°57.508'	120°24.490' / 120°23.455'	wr	IH	56.5	10.5	0.2	10.3	0.5
SO100 50DS-1	18	Galilei	34°51.492' / 34°51.860'	119°06.724' / 119°06.811'	plag	IH	99.9	9.5	0.2	9.2	1.1
SO100 50DS-1	18	Galilei			wr	IH	68.0	9.5	0.2	8.9	0.9
Weighted average								9.5	0.1	9.0	0.7
SO100 54DS-1	19b	Herschel b	35°06.976' / 35°06.200'	118°33.132' / 118°33.329'	wr	IH	59.1	8.5	0.1	8.4	0.2
SO100 56DS-1	21a	Hubble a	35°22.943' / 35°22.959'	118°05.256' / 118°05.314'	plag	IH	99.6	7.8	0.1	7.8	0.2
SO100 63DS-1	22	Hubboldt	35°48.000' / 35°47.663'	117°26.305' / 117°26.303'	wr	IH	64.5	7.8	0.1	7.7	0.1
SO100 60DS-1	23	Jenner	35°27.003' / 35°26.762'	117°11.833' / 117°12.007'	plag	IH	92.0	7.6	0.1	7.8	3.8

Table 11.1. *Continued*

Sample ID	Smt. number[b]	Smt. name[c]	Latitude °S[d]	Longitude °W[d]	Sample	Type	%³⁹Ar	Plateau (Ma)	2σ	Inverse isochron (Ma)	2σ
SO100 59DS-1	24b	Kepler	35°26.729' 35°26.334'	116°38.813' 116°39.352'	plag	IH	99.9	7.3	0.2	7.3	0.2
SO100 67DS-4	25	Kopernik	36°01.980' 36°01.473'	115°59.276' 115°59.294'	plag	IH	98.4	6.3	0.2	6.3	0.4
SO100 66DS-1	26	Lavoisier	35°47.484' 35°47.454'	115°39.335' 115°38.124'	plag	IH	99.5	4.9	0.2	4.9	0.8
SO100 66DS-1	26	Lavoisier			plag	IH	95.4	4.8	0.2	3.9	1.5
SO100 66DS-1	26	Lavoisier			wr	IH	69.7	4.7	0.2	4.8	0.5
Weighted average								4.8	0.1	4.8	0.4
SO100 69DS-1	27b	Linné B	36°33.804' 36°33.524'	115°16.627' 115°16.230'	wr	IH	100	5.1	0.1	5.1	0.1
SO100 70DS-2	28	Mendel	36°20.907' 36°20.848'	113°55.696' 113°55.790'	wr	IH	76.2	3.7	0.2	3.9	0.3
FH DR13-1	29	Mendeleiev	37°01.600' 37°01.600'	114°02.56' 114°02.56'	wr	IH	70.7	3.3	0.1	3.2	0.2
FH DR13-1	29	Mendeleiev			wr	IH	75.6	3.4	0.2	3.4	0.2
Weighted average								3.3	0.1	3.3	0.1
SO100 71DS-1	30	Mercator	36°40.812' 36°41.770'	113°28.350' 113°26.926'	wr	IH	100	2.1	0.1	2.1	0.1
FH DR11-1	31	Newton	36°55.477' 36°55.477'	113°04.585' 113°04.585'	wr	SFs	100	2.1	0.3	2.0	1.4
FH DR11-1	31	Newton			wr	IH	51.8	2.1	0.3	2.1	0.4
Weighted average								2.1	0.2	2.1	0.4

Table 11.1. *Continued*

Sample ID	Smt. number[b]	Smt. name[c]	Latitude °S[d]	Longitude °W[d]	Sample	Type	%^{39}Ar	Plateau (Ma)	2σ	Inverse iso-chron (Ma)	2σ
SO100 74DS-1	32	Ohm	36°56.987' 36°57.427'	112°12.875' 112°13.765'	wr	SFs	100	2.0	0.4	2.3	1.2
SO100 74DS-1	32	Ohm			wr	IH	100	1.6	0.5	2.0	0.6
Weighted average								1.8	0.3	2.1	0.5
SO100 76DS-1	33	Pascal	37°22.190' 37°22.377'	112°06.098' 112°06.003'	wr	IH	77.4	1.1	0.1	1.1	0.2
SO100 97DS-2	33	Pascal	37°25.016' 37°24.662'	112°03.419' 112°03.882'	wr	SF		1.1	0.4		
SO100 101GTV-2	33	Pascal	37°23.149' 37°23.209'	112°05.940' 112°05.920'	wr	SF		1.1	0.3		
Weighted average								1.1	0.1		
SO100 99DS-5	34	Pasteur	37°15.768' 37°15.328'	112°03.044' 112°03.565'	wr	IH	96.1	1.1	0.1	1.2	0.4
SO100 77DS-1	35	Pauling	37°03.427' 37°04.094'	111°39.605' 111°39.662'	wr	IH	100	1.1	0.2	1.1	0.3
SO100 93DS-1	36	Planck	38°08.063' 38°07.515'	111°32.894' 111°33.403'	wr	IH	100	0.5	0.1	0.5	0.3
SO100 94DS-1	37a	Platon	37°53.367' 37°53.227'	112°53.847' 112°55.644'	wr	IH	80	0.9	0.1	0.9	0.1
SO100 95DS-1	37b	Richter	37°53.312' 37°53.787'	112°25.089' 112°25.912'	wr	IH	100	0.9	0.1	0.9	0.1
SO100 90DS-1	38	Rutherford	38°21.598' 38°20.966'	110°37.905' 110°37.903'	wr	SFs	100	1.1	0.5	0.6	1.1
SO100 90DS-1	38	Rutherford			wr	IH	76.2	1.1	0.4	0.9	0.7
Weighted average								1.1	0.3	0.8	0.6

Table 11.1. *Continued*

Sample ID	Smt. number[b]	Smt. name[c]	Latitude °S[d]	Longitude °W[d]	Sample	Type	%³⁹Ar	Plateau (Ma)	2σ	Inverse iso-chron (Ma)	2σ
SO100 75DS-2	39	Schrödinger	37°16.680' 37°17.107'	111°51.521' 111°51.800'	wr	IH	97.3	1.0	0.2	0.9	0.3
SO100 82DS-1	39	North Ridge	37°27.530' 37°27.134'	111°12.654' 111°12.059'	wr	SFs	100	0.4	0.1	0.5	0.4
SO100 87DS-1	39	South Ridge	37°38.941' 37°39.052'	111°17.397' 111°16.650'	wr	SFs	100	0.5	0.1	0.6	0.6
SO100 87DS-1	39	South Ridge			wr	IH	71.8	0.7	0.2	0.4	0.3
Weighted average								0.5	0.1	0.4	0.3
SO100 89DS-1		South Ridge	37°31.388' 37°31.897'	111°40.707' 111°41.332'	wr	IH	94.5	0.7	0.1	0.7	0.2
FH DR12-8		Wegener	37°55.935'	113°42.372'	wr	IH	80.5	1.9	0.1	1.9	0.2
FH DR14-14		Mohorovicic	37°27.78'	114°34.773'	wr	IH	45.7	3.5	0.1	3.4	0.3
FH DR14-16		Mohorovicic	37°27.78'	114°34.773'	wr	IH	64.2	3.6	0.1	3.6	0.1
FH DR14-17		Mohorovicic	37°27.78'	114°34.773'	wr	IH	63.2	3.5	0.9	1.7	1.0
Weighted average								3.5	0.1	3.6	0.1
FH DR14-5		Mohorovicic	37°27.78'	114°34.773'	wr	IH	59.1	2.8	0.1	2.8	0.4
FH DR14-13		Mohorovicic	37°27.78'	114°34.773'	wr	IH	86.4	0.7	0.03	0.7	0.1
FH DR15-4		Linné Ridge	35°44.983'	114°21.989'	wr	IH	49.3	5.0	0.2	4.6	0.5

[a] Argon isotopic data, age calculation from argon isotopic data, plateau and isochron plots – together with detailed sample information – are in O'Connor et al. 1998, 2001, 2002. Ages in O'Connor et al. 1998 have been recalculated in O'Connor et al. 2001 using new TCR standard age of 28.34 Ma (Renne et al. 1998).
[b] Seamount and ridge numbers assigned during 1995 F/S SONNE cruise (Devey et al. 1997).
[c] Seamount/ridge names (Devey et al. 1997; Maia et al. 2001). Corresponding gravity anomaly in Mammerickx (1992) are in Devey et al. 1997 and O'Connor et al. 1998. $\lambda(^{40}K)_{total} = 5.543 \cdot 10^{-10} \ yr^{-1}$; Correction factors: $^{40}Ar/^{39}Ar$ (K) = 0.00086; $^{36}Ar/^{37}Ar$ (Ca) = 0.00026; $^{39}Ar/^{37}Ar$ (Ca) = 0.00067.
wr = whole rock; *plag* = plagioclase; *IH* = incremental heating; *SFs* = multiple single fusions; *SF* = single fusion.
[d] In case where two sets of coordinates are given the top and bottom sets refers to dredge location on-bottom and off-bottom, respectively.

Chapter 11 · The Foundation Chain: Inferring Hotspot-Plate Interaction from a Weak Seamount Trail 361

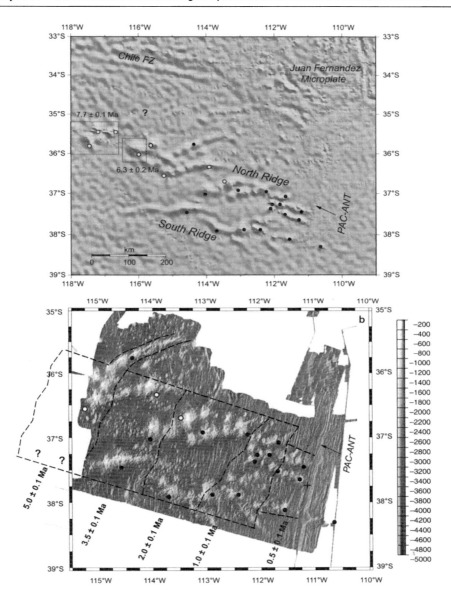

Fig. 11.5. a Predicted topography of the Foundation Chain region (Smith and Sandwell 1997). Solid circles show locations of 'SO100' (Devey et al. 1997) or 'Hotline' (Maia et al. 2001) cruise dredge sites for which $^{40}Ar/^{39}Ar$ ages have been determined. 'North' and 'South' lines indicate the apparent bifurcation of the Foundation Chain. *Box* labeled 7.7 Ma shows a cluster of coeval seamounts (the most western seamount in this cluster is not shown) that might be linked to the development of the oldest Pacific-Antarctic Foundation VER (indicated by a *question mark*). *PAC-ANT* = Pacific-Antarctic spreading center. **b** Shaded multibeam bathymetry of the Foundation Chain VERs (Maia et al. 2000, 2001). *Dashed lines* outline inferred en échelon NE-SW elongate 'zones' of coevally erupted VER volcanism. The weighted average of all ages measured for samples recovered from within each 'zone' (Table 11.2) is shown. *Plus symbol* denotes point on Pacific-Antarctic spreading center (37°45' S, 111°7.5' W) from which sample site distances along the Foundation Chain (Figs. 11.3 and 11.7) have been calculated

Table 11.2. Coeval Foundation Chain volcanism

Seamount/sample	Plateau (Ma)	±2σ	Inverse isochron (Ma)	±2σ
Del Cano	16.1	0.4	16.3	0.7
VER	16.8	0.3	16.6	1.0
VER	16.6	0.2	16.1	0.6
5	16.6	0.4	16.6	0.4
Weighted average	16.6	0.1	16.4	0.3
21a	7.8	0.1	7.8	0.2
22	7.8	0.1	7.7	0.1
23	7.6	0.1	7.8	3.8
Weighted average	7.7	0.1	7.7	0.1
24	7.3	0.2	7.3	0.2
25	6.3	0.2	6.3	0.4
26	4.8	0.1	4.8	0.4
27b	5.1	0.1	5.1	0.1
FH DR15-4	5.0	0.2	4.6	0.5
Weighted average	5.0	0.1	5.1	0.1
28	3.7	0.2	3.9	0.3
29	3.3	0.1	3.3	0.1
FH DR14	3.5	0.1	3.6	0.1
Weighted average	3.5	0.1	3.5	0.1
FH DR14	2.8	0.1	2.8	0.4
30	2.1	0.1	2.1	0.1
31	2.1	0.2	2.1	0.4
32	1.9	0.3	2.2	0.5
FH DR12-8	1.9	0.1	1.9	0.2
Weighted average	2.0	0.1	2.1	0.1
33	1.1	0.1	1.1	0.2
34	1.1	0.1	0.9	0.5
35	1.1	0.2	1.1	0.3
39	1.0	0.2	0.9	0.3
Weighted average	1.1	0.1	1.0	0.1
37a	0.9	0.1	0.9	0.1
37b	0.9	0.1	0.9	0.1
Weighted average	0.9	0.1	0.9	0.1
33	1.1	0.1	1.1	0.2
34	1.1	0.1	0.9	0.5
35	1.1	0.2	1.1	0.3
39	1.0	0.2	0.9	0.3
37a	0.9	0.1	0.9	0.1
37b	0.9	0.1	0.9	0.1
Weighted average	1.0	0.1	0.9	0.1
FH DR14-13	0.7	0.03	0.7	0.1
SO100 89DS-1	0.7	0.1	0.7	0.2
Weighted average	0.7	0.03	0.7	0.1
SO100 82DS-1	0.4	0.1	0.5	0.4
SO100 87DS-1	0.5	0.1	0.4	0.3
36	0.5	0.1	0.5	0.3
Weighted average	0.5	0.1	0.5	0.2

11.3
Results

11.3.1
Migration of Volcanism Along the Foundation Chain

The 1 900 km distribution of dredge sample $^{40}Ar/^{39}Ar$ ages shows that volcanism has migrated along the Foundation Seamount chain at a constant rate of 91 ±2 mm yr^{-1} for at least the past 22 Myr (Fig. 11.3).

The distribution of $^{40}Ar/^{39}Ar$ dated seamounts predicts that the present Foundation hotspot melting anomaly (or at least its easternmost extent) is presently located under, or very close to, the Pacific-Antarctic spreading-axis (Fig. 11.3).

Comparison between seamount and sea-floor ages reveals that the Foundation Chain erupted primarily in the interior of tectonic plates (Fig. 11.7). Thus, linear migration of intraplate volcanism along the Foundation Chain is compatible with Pacific Plate drifting over a stationary hotspot (Morgan 1971).

Foundation $^{40}Ar/^{39}Ar$ ages (Table 11.1) indicate that Foundation Seamounts and VERs developed in a series of discrete magmatic episodes at intervals of ~1 Myr (Fig. 11.3). This apparent periodicity might, however, be an artifact of dredge sampling of late stage volcanism coating seamount/ridge flanks. Nevertheless, we believe it is more likely that the isolated Foundation Chain seamounts or seamount clusters erupted relatively rapidly (~1 Myr) considering that (1) episodicity is evident in both seamounts and VERs and (2) the significantly lower volumes of magma in individual Foundation Chain seamounts and ridges compared to chains of larger seamounts such as Hawaii that are known to span at least 5 Myr of magmatism (e.g., Clague and Dalrymple 1989).

11.3.2
Hotspot-Spreading Center / Microplate Interaction

Between ~22 and ~14 Ma, the Foundation Chain was forming on the Pacific Plate west of the 'Failed Rift' of the Selkirk Microplate (Fig. 11.7). The age of Pacific sea floor migrating over the hotspot decreased systematically during this interval (~7.5 to ~5.5 Ma) (O'Connor et al. 1998, 2002).

This tectonic configuration changed when the 'Failed Rift' representing the western boundary of the microplate migrated over the hotspot at ~14 Ma (Fig. 11.7). The arrival of the 'Failed Rift' marked the onset of microplate migration over the hotspot melting anomaly, which continued until about 11 Ma.

A $^{40}Ar/^{39}Ar$ age for a rock sample dredged from the 'failed' spreading center of the Selkirk Microplate (Mammerickx 1992; Tebbens and Cande 1997) showed that it jumped eastward (failed) at 20.4 ±0.3 Ma (O'Connor et al. 2002) (Table 11.1). However, the spreading center north of the 'failed rift' probably continued spreading normally after microplate development so that much 'younger' seafloor was migrating over the northern flank of the Foundation hotspot (O'Connor et al. 2002) (Figs. 11.6 and 11.7).

East of the Selkirk Microplate, seafloor migrating over the Foundation melting anomaly decreased systematically in age from 11 to 0 Ma at the present-day Pacific-Antarctic spreading axis (Fig. 11.7).

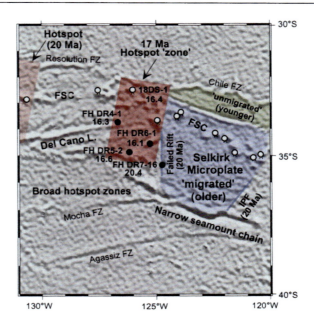

Fig. 11.6. *Dark red zone* labeled '*17 Ma hotspot zone*' indicates the broad swath of scattered, structurally disconnected coeval Foundation hotspot magmatism produced during the ~17 Ma event (pulse?) of the Foundation hotspot melting anomaly (mantle plume?). Seafloor east of this ~17 Ma inferred hotspot zone/event/pulse labeled '*migrated*' became significantly older at 20 Ma (Table 11.1) when the Selkirk Microplate was created by the transfer of a segment of the Nazca plate by an eastward spreading center jump (failure) and initiation of a 'new rift' (Mammerickx 1992; Tebbens and Cande 1997; O'Connor et al. 2002). The location of the hotspot during this 20 Ma event is indicated by *light red zone*. Seafloor labeled younger '*unmigrated*' to the north of the Selkirk Microplate is significantly younger than the Selkirk Microplate because it was produced by continuous uninterrupted spreading at an unmigrated segment of the spreading axis north of the 'Failed Rift' (Fig. 11.7). Variability in age of lithosphere drifting over, or close to, the Foundation hotspot resulting from microplate creation is shown in Fig. 11.7. Age data for dredge sites indicated by *solid dots* and *open circles* are in O'Connor et al. (2002 and 1998, respectively). Measured ages are shown below dredge sample numbers. *FSC* = Foundation Seamount Chain; *IPF* = Inner Pseudo Fault; *FZ* = Fracture Zone; *bold blue lines* = Failed Rift and IPF

Systematic changes in the age (i.e., thickness, strength, temperature) of Pacific lithosphere migrating over the Foundation hotspot have been occurring since the start of the Foundation Chain's creation (Fig. 11.7) – as local spreading boundaries (responsible for the formation of the seafloor on which the Foundation Chain was later erupted) have migrated systematically towards the Foundation hotspot (i.e., ~23 mm yr^{-1} and ~48 mm yr^{-1} west and east of the failed rift of the Selkirk Microplate, respectively) (Mammerickx 1992; O'Connor et al. 1998).

11.3.3
Volcanic Elongated Ridges (VERs)

Individual VERs, as defined on the basis of structural morphology (i.e., volcanism along one continuous ridge), may occasionally develop synchronously (Fig. 11.4).

However, the dominant trend is for structurally disconnected sections of VERs to be coeval (Figs. 11.4 and 11.5). These synchronously erupted, yet structurally uncon-

Chapter 11 · The Foundation Chain: Inferring Hotspot-Plate Interaction from a Weak Seamount Trail   365

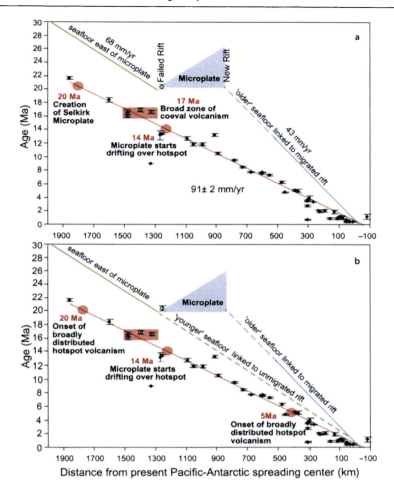

Fig. 11.7. a $^{40}Ar/^{39}Ar$ and 'migrated' sea-floor age as a function of distance from the present Pacific-Antarctic spreading axis. The *solid red line* is the York-2 linear regression fit representing an average rate for the migration of volcanism along the chain of 91 ±2 mm yr^{-1} (Fig. 11.3). As the Selkirk Microplate reached the active hotspot region at about 14 Ma, much older seafloor (≥6 Myr) began to cap the Foundation melting anomaly. This 'older'/'migrated' seafloor placed a sufficiently thick/strong lid over most of the region impacted by by the melting anomaly (pulsed plume?) to prevent the migration of hotspot melts to the seafloor during the interval ~14 Ma to ~11 Ma, resulting in the transition from broad zones of hotspot volcanism to a narrow chain of seamounts. *Blue solid lines* denote 'older' seafloor, *unbroken blue lines* indicate known sea-floor ages (Mam-merickx 1992; Lonsdale 1994), and *dashed blue line* indicates extrapolated sea-floor ages. The *red swath* indicates the broad zone of scattered coeval hotspot volcanism (*open spheres*) created by the proposed 17 Ma pulse of the Foundation plume crosscutting the narrow Foundation Chain – as revealed by predicted topography maps (Fig. 11.2). Analytical error bars are ±2σ. **b** 'Migrated' sea-floor age as a function of distance from the present Pacific-Antarctic spreading center – other details are the same as in *Part a*. *Long green dashed line* indicates younger 'unmigrated' sea floor created by the spreading center north of the 'Failed Rift' (extrapolation of known sea-floor ages west of the Selkirk Microplate). The spreading center north of the 'Failed Rift' probably continued spreading normally so that 'younger' (≤6 Ma) seafloor could move over the northern flank of the Foundation hotspot. This younger, 'unmigrated' seafloor drifting by the northern flank of the Foundation melting anomaly was thin enough to allow the passage of melts along a narrow strip leading to the creation of a correspondingly narrow chain of Foundation Seamounts

nected VER sections define a series of en échelon elongated zones of coeval volcanism at different stages of the Foundation Chain's development (Figs. 11.5 and 11.6). Elongated zones maintained a steady state size (~250 km by ~150 km) and NE-SW orientation, developing at intervals of approximately 1 Myr (Fig. 11.5).

Overall, VER volcanism generated during each successive episode or in each zone of coeval volcanism was emplaced progressively onto younger seafloor and closer to the Pacific-Antarctic spreading center as it migrated steadily nearer to the Foundation hotspot (Fig. 11.4).

The main direction of the somewhat sinuous VERs changed progressively with time from NE, to E-W, and most recently to SE. This change in direction correlates well with the location of the nearest spreading center segment to the hotspot (Fig. 11.4).

VER structures tend to develop predominantly along the northern and southern ends of the elongate en échelon 'zones'. This resulted in the apparent bifurcation of the Foundation Chain into distinct 'North' and 'South' lines of volcanism, a distinction that disappeared by ~1 Ma (Fig. 11.5). The NE-SW orientation of these zones resulted in the northern region of each elongated zone of coeval volcanism being emplaced closer to the Pacific-Antarctic spreading center than the corresponding southern region, i.e., the 'South' line developed systematically in a more unambiguously intraplate setting when compared to the 'North' line (Fig. 11.4).

The oldest Foundation Chain VER is possibly a minor NE trending 7.7 ±0.1 Ma old volcanic ridge (Fig. 11.5). Volcanism erupted next at 6.3 ±0.2 Ma in the form of a single large intraplate seamount without any associated VER (Figs. 11.4 and 11.5).

In contrast, a prominent, somewhat sinuous coeval ~200 km long VER formed at 5.0 ±0.1 Ma, trending NE from a cluster of coeval intraplate seamounts as far as the Pacific-Antarctic spreading axis (Fig. 11.4). The point of intersection was both the closest spreading boundary segment and the active tip of a propagating spreading axis segment (Lonsdale 1994).

By 3.5 ±0.1 Ma, scattered coeval Foundation Chain volcanism was being emplaced in a basically NE-SW elongated zone that was equidistant from all local spreading boundaries (Fig. 11.4). Although the individual VER segments defining this zone were orientated to the NE toward the spreading center, no volcanic connection was established. Subsequent VER development involved eruption of structurally disconnected coeval VER segments defining elongated zones of hotspot volcanism located successively closer to the Pacific-Antarctic spreading center (Fig. 11.4). Although the NE-SW orientation of the elongated zones remained unchanged between ~3.5 Ma and the present, the overall orientation of individual VER segments changed from NE to SE (Fig. 11.5).

The most recent Foundation Chain volcanism developed in the form of three SE-orientated parallel VERs extending almost as far as the present Pacific-Antarctic spreading axis (Fig. 11.4). The approximate midpoints of these three VERs are coeval (i.e., 0.5 ±0.1 Ma) (Fig. 11.4). The 0.7 ±0.1 Ma sample dredged at the western end of the central ridge is similar in age (within analytical uncertainty), indicating that these three ridges are probably also rapidly formed coeval VERs.

Ages for five samples dredged from the most western dredge site located along the 'South' line indicate at least three different phases of volcanism, i.e., 3.5 ±0.1 Ma, 2.8 ±0.1 Ma, and 0.7 ±0.1 Ma (Fig. 11.4) (O'Connor et al. 2001).

A broad swath of scattered, structurally disconnected coeval Foundation hotspot magmatism was emplaced west of the Selkirk Microplate ~17 Ma (Fig. 11.6).

11.4
Discussion

11.4.1
VERs and the Pacific-Antarctic Spreading Axis

The main process responsible for VER development is the continuation of the 91 ± 2 mm yr^{-1} lithospheric plate migration over the Foundation hotspot that began at least 22 Ma. We base this conclusion on the persistence during VER development of the linear trend of decreasing Foundation Chain age (O'Connor et al. 1998), their development in an intraplate setting (O'Connor et al. 1998), and hotspot-influenced geochemistry (Devey et al. 1997; Hekinian et al. 1997, 1999; Hémond and Devey 1997; Hémond et al. 1999). However, other second-order processes are likely to have controlled the onset and development of the Foundation Chain's VERs. The most obvious notion is an interaction between the Foundation hotspot and the Pacific-Antarctic spreading axis (Devey et al. 1997; Hekinian et al. 1997, 1999; O'Connor et al. 1998; Maia et al. 2000; Hémond and Devey 1996; Hémond et al. 1999).

Due to the long-standing Pacific-Antarctic spreading axis migration toward the Foundation hotspot, lithosphere migrating toward/over the hotspot eventually became young/thin/weak/hot enough to trigger the onset of VER development (e.g., O'Connor et al. 1998). The most unequivocal evidence is the synchronously developed ~200 km long NE-orientated 5 Ma VER connecting Foundation intraplate volcanism with the Pacific-Antarctic spreading axis (Fig. 11.4). This provides strong evidence that individual VERs can occasionally form syn-chronously (i.e., as lines of continuous volcanism), connecting intraplate volcanism with the Pacific-Antarctic spreading axis. The correlation between changes in VER orientation and the minimum distance to the Pacific-Antarctic spreading axis further supports hotspot-spreading axis interaction (Fig. 11.4). Models envisioning hotspot-spreading axis interaction in terms of sublithospheric channeling of melt anomaly material towards spreading axes (Schilling 1985; Schilling et al. 1985, 1991) can draw support from these results.

However, other more dominant trends in VER chronology cannot be explained by current models of hotspot-spreading axis interaction. The most prominent of such trends is for coeval hotspot volcanism to erupt in a series of NE-SW elongated 'zones' crosscutting the main NW-SE trend of the 22 Ma Foundation Chain (O'Connor et al. 2001) (Fig. 11.5). The transition from isolated seamount chain to NE-SW elongated 'zones' of VER hotspot volcanism is marked by the broadening of the Foundation Chain, en échelon VER distribution, and the significantly greater scatter of VER ages about the overall linear trend of decreasing Foundation Chain age. These zones are considerably larger in scale than individual VERs and maintained a steady state orientation and size, despite the fact that the Pacific-Antarctic spreading axis was migrating progressively closer to the hotspot.

VER volcanism developed preferentially along the north and south sides of these elongated zones, leading to the bifurcation of the Foundation Chain into seemingly continuous 'North' and 'South' lines of volcanism. Other such chains near the East Pacific Rise have been attributed to thinner lithosphere and a more coherent stress field near spreading axes (Hieronymus and Bercovici 2000). Straight and long lines of seamounts (e.g., 'North' and 'South' lines) might therefore be similarly explained by a strong and coherent nearby spreading center stress field with control of their alignments attributed to the direction of the most tensile principal tectonic stress (Hieronymus and Bercovici 2000). Furthermore, control of the volume and location of intraplate volcanism due to stress in the lithosphere has been shown in the case of the Ngatemato Seamounts (McNutt et al. 1997). Compared to the 'South' line, the 'North' line has ~2–3 times greater seamount volume (Maia et al. 2001), a greater number of individual VERs showing more pronounced changes in orientation (i.e., NE to SE), and possibly a less pronounced hotspot-like geochemistry (e.g., Hémond et al. 1999). These and other such compositional differences noted in Maia et al. (2001) can be explained, therefore, by the greater proximity of the 'North' line to the Pacific-Antarctic spreading axis, resulting in greater partial melting, more dilution of the hotspot geochemical signature, and greater perturbations to local stress control due to transform faults, overlapping spreading centers and propagators (Hieronymus and Bercovici 2000). Thus, the pronounced shift 1 Myr ago from distinct 'North' and 'South' lines to an elongated zone of significantly more continuous volcanism can be explained by the thinning/ weakening of the lithosphere together with enhanced interaction between the Foundation hotspot and the encroaching Pacific-Antarctic spreading axis. We infer further support for this notion from the three parallel, SE-orientated VERs that apparently developed synchronously at 0.5 ±0.1 Ma between the 1 Ma elongated 'zone' and the present Pacific-Antarctic spreading axis (Fig. 11.4).

The second important trend in our age data not explicable by current VER and hotspot-spreading interaction models is the interval of approximately 1 Myr between the developments of successive elongated zones of hotspot volcanism (O'Connor et al. 2001). This trend has also been detected in measured ages for the Foundation Seamounts, irrespective of tectonic setting (O'Connor et al. 1998). Persistence of this trend during transition in morphology from isolated seamounts to VERs suggests that it is an intrinsic characteristic of the Foundation hotspot and/or its hypothesized causal mantle plume. The ~3 Myr of volcanic activity at one dredge site can be explained by, for example, rejuvenation of the Foundation Chain triggered by local lithospheric weaknesses associated with the earlier loading of the Foundation Chain (McNutt et al. 1997). However, it could also point to a significantly greater (i.e., >300 km) distribution of Foundation melt anomaly material west of the Pacific-Antarctic spreading axis than indicated by the inferred ~150 km wide elongated zones of hotspot volcanism. We speculate that this could reflect the second order flow of material under the lithosphere away from the elongated hotspot zones (e.g., Maia et al. 2000).

11.4.2
Foundation VERs and the Selkirk Microplate

A broad swath of ~17 Myr old scattered, structurally disconnected coeval Foundation Seamounts and VERs was emplaced west of the Selkirk Microplate (Fig. 11.6) (O'Connor et al. 2002). This 'zone' is comparable in scale to the series of similarly sized elongated (~250 km by ~150 km) swaths of disconnected coeval hotspot volcanism created since at least 5 Ma at the young end of the chain (Fig. 11.5) (O'Connor et al. 2001). However, once the Selkirk Microplate began migrating over the Foundation hotspot at ~14 Ma, the Foundation Chain abruptly started forming as a narrow line of seamounts (O'Connor

Chapter 11 · The Foundation Chain: Inferring Hotspot-Plate Interaction from a Weak Seamount Trail   369

et al. 1998). We attribute this switch in morphology to differences in the age and thus the physical properties of the lithosphere migrating across the inferred broad Foundation hotspot 'zone'. A $^{40}Ar/^{39}Ar$ age for the 'failed rift' bounding the Selkirk Microplate to the west (Mammerickx 1992; Tebbens and Cande 1997) showed that it jumped eastward at 20.4 ±0.3 Ma (O'Connor et al. 2002) (Fig. 11.6). This jump created a large offset in the age of the seafloor that subsequently migrated over the Foundation hotspot, resulting in old/thick microplate (transferred Nazca plate) migrating over most of the hotspot and younger/thinner ('unmigrated') seafloor ($\leq$6 Myr) across the northern flank (Fig. 11.6). We also consider it possible that younger/thinner seafloor flanking the hotspot to the north was thermally reset to younger ages due to preferential channeling of hotspot material (e.g., Morgan 1978; Schilling et al. 1985: Schilling 1985, 1991) to the much nearer 'unmigrated' younger spreading center. Hotspot magmatism between ~14 and ~11 Ma was therefore most likely restricted to a narrow region along the northern boundary of the Selkirk Microplate (Figs. 11.6 and 11.7).

Once the old/thick microplate lithosphere had drifted past the hotspot, it was followed by seafloor formed at the spreading center segment that 'jumped' eastward at ~20 Ma to create the Selkirk Microplate (Fig. 11.7). This lithosphere was again too old/thick to allow the passage of significant amounts of hotspot melt (Fig. 11.7). Nonetheless, narrow chain development continued, because younger ('unmigrated') seafloor was migrating over the northern regions of the hotspot. This situation persisted until ~5 Ma, when the Pacific-Antarctic spreading axis had migrated sufficiently close so that younger/thinner lithosphere ($\leq$6 Myr) reached the hotspot, so facilitating the onset of VER development/broadening of the Foundation Chain (Fig. 11.7). Thus, we explain broadening and narrowing of the Foundation Chain since at least 22 Ma in terms of the age, and consequently thickness and strength, of the seafloor migrating over the Foundation hotspotmelting anomaly, which acts as a threshold parameter controlling the mode of hotspot volcanism.

11.4.3
Pacific Plate Motion

A linear velocity of 91 ±2 mm yr^{-1} for plate motion over the Foundation hotspot melting anomaly agrees with that predicted by selected Euler poles for Pacific Plate motion over fixed hotspots (O'Connor et al. 1998; Koppers et al. 2001) (Figs. 11.3 and 11.6). The fact that single reconstruction poles can predict the plate velocities (and azimuths) derived from measured ages distributed along the Hawaiian and Foundation Chains adds confidence to the assumption of fixed plume-hotspots (at least for the past 22 Myr of Pacific Plate motion in the case of the Hawaiian and Foundation Chains).

Furthermore, if the previously proposed relationship between the Foundation hotspot and older Ngatemato Chain Seamounts (McNutt et al. 1997; O'Connor et al. 1998) is correct (Fig. 11.1), then the Foundation melting anomaly was active for at least the past ~34 Myr. Assuming that a stationary Foundation plume-hotspot is in fact responsible for the formation of the Ngatemato-Foundation Chains and that the plume has indeed been active for the last ~34 Myr, then the question arises as to why a more continuous chain of ~3 000 km was not created. As in the case of the Foundation Chain, the lack of significant hotspot volcanism between the Foundation Chain and the Ngatemato and Taukina Seamounts (McNutt et al. 1997; Maia et al. 2001) can be explained by migration of sea floor that was too old and consequently too thick and strong (≥6 Ma in the case of the Foundation hotspot) to allow any Foundation hotspot magma to reach the lithospheric surface.

11.4.4
Implications for Plume-Hotspot Theory

Key observations that need to be reconciled with standard plume-hotspot theory are (1) the broadening and narrowing of the Foundation Chain due to (2) the appearance and disappearance, respectively, of broad elongated 'zones' of scattered coeval volcanism, and (3) the persistence of ~1 Myr episodicity/periodicity of Foundation Chain volcanism – irrespective of tectonic setting or chain mode of volcanism (VER or seamount). We propose that these observations are related primarily to the long-term dynamics of a Foundation plume for at least 22 Myr and possibly 34 Myr. In contrast, localized distribution of volcanism across each 'zone' linked to an 'event' or 'pulse' was controlled by local factors such as lithospheric age and stress (e.g., McNutt et al. 1997; Hieronymus and Bercovici 2000; Maia et al. 2001; O'Connor et al. 2001, 2002). Likewise, the involvement of pre-existing lithospheric structures acting as 'weak zones' facilitating decompression melting of hot or geochemically anomalous plume material is strongly indicated by the fact that volcanic lineaments south of the Foundation Chain – e.g., Del Cano Lineament – are parallel to the local fracture zone direction (Fig. 11.6).

A physical process which could episodically/periodically bring up masses of plume material extremely fast from depth and which is capable of focusing the hot material into broadly elongated zones (~250 km by ~150 km) – at least on initial impact against the base of the drifting Pacific Plate – is therefore required. Depth-dependent properties have been shown to theoretically play an important role in controlling plume dynamics (Hansen et al. 1993). For example, 'ultrafast' focused mantle plumes are theoretically possible in the upper mantle during thermal convection with a non-Newtonian temperature- and depth-dependent rheology operating at a reasonably effective Rayleigh number on the order of 10^6 (Larsen and Yuen 1997; Larsen et al. 1999). Such strong depth-dependence in viscosity also results in masses of plume material pulsing at intervals ranging from a few Myr (Larsen et al. 1999) to about 10 Myr (Larsen and Yuen 1997) with calm periods in between (Larsen et al. 1999). These current models indicate that plumes can rise through the upper mantle in much less than one Myr (e.g., Larsen et al. 1999), thus providing a mechanism by which hot plume material can be brought from the transition zone to the lithosphere extremely quickly ($m\,yr^{-1}$) in an otherwise slowly convecting mantle ($cm\,yr^{-1}$). Following impact against the base of the lithosphere, this material can flow laterally at rates as high as ~0.5 m yr^{-1}, thereby creating the potential of coeval magmatism scattered over large areas (Larsen and Yuen 1997; Sleep 1997; Larsen and Saunders 1998; Larsen et al. 1999).

Thus, our observations about the temporal-spatial development of the Foundation Chain are compatible with non-Newtonian plume theory and also indicate that SE Pacific mantle plumes are more tightly focused and faster-pulsing than that which has been presently incorporated into numerical models. This points in turn to even greater viscosity stratification (or changes in the creep law) across the upper mantle compared to, for example, the region of the North Atlantic influenced by the Iceland plume (Larsen et al. 1999; O'Connor et al. 2000).

11.5
Conclusions

Linear migration of intraplate – often geochemically enriched – volcanism at a rate of 91 ± 2 mm yr^{-1} along the Foundation Chain for at least the past 22 Myr is compatible with drifting of the Pacific Plate over a narrow long-lived hotspot melting anomaly. Due to the isolated nature of the relatively small, likely rapidly created Foundation Chain seamounts, we have been able to distinguish second-order volcanism (using high precision ages) from that created via the first-order influence of the Foundation hotspot – the inferred product of a stationary mantle plume upwelling from depth under the drifting Pacific plate. Therefore, despite changes in morphology and geochemistry, the migration of volcanism along the Foundation Chain can be interpreted as a record of the absolute motion path of the Pacific Plate (except for three seamount ages close to the outer edges of the Selkirk Microplate – discussed in O'Connor et al. 1998). Similarity between rates of propagation of volcanism along the Hawaiian and Foundation Chains supports a stationary Foundation versus Hawaiian plume-hotspot, at least for the past 22 Myr. On a more localized scale, the Foundation Chain developed as a line of relatively small, rapidly erupted (~1 Myr) individual seamounts or clusters at a rate of approximately once every Myr.

The transition from a narrow line of seamounts to a broad region of volcanic elongate ridges (VERs) about 5 Ma was assumed initially to be the result of interaction between the Foundation hotspot and the encroaching Pacific-Antarctic spreading-center. Some of our data support this notion by showing that volcanism along morphologically distinct VERs can develop occasionally as rapidly formed continuous lines of coeval volcanism extending from a region of intraplate volcanism to the Pacific-Antarctic spreading center. However, a significantly more dominant trend is for coeval yet structurally disconnected segments of Foundation Chain VERs to develop in a series of en échelon, NE-SW elongate 'zones' of coeval hotspot volcanism. These elongated zones developed at intervals of approximately 1 Myr while maintaining a basically steady state orientation and size as the Pacific-Antarctic spreading center migrated continually closer to the Foundation hotspot.

Our age data indicate that Foundation Chain development between ~22 Ma and ~14 Ma was also in the form of broad zones of scattered, synchronous Foundation volcanism, which are very similar to those identified west of the present Pacific-Antarctic spreading center. But once the significantly older Selkirk Microplate lithosphere (Tebbens and Cande 1997) began capping the hotspot melting anomaly about 14 Ma, the Foundation Chain narrowed into a line of discrete seamounts, only to broaden again about 5 Ma when sufficiently young lithosphere drifted once again over the hotspot. Thus, Foundation hotspot volcanism could have been be prevented if the capping tectonic plate is too thick (≥ 6 Myr in the case of Foundation) for hotspot melts to penetrate to the surface (O'Connor et al. 1998, 2001, 2002). The lack of a seamount chain connecting the Foundation and the Ngatemato Chains (McNutt et al. 1997) can be similarly explained, thus supporting the notion that the Pacific Plate has drifted a distance of at least 3 400 km over a Foundation hotspot melting anomaly during the last ~34 Myr.

Creation of broad zones of synchronous Foundation magmatism at regular ~1 Myr intervals leads us – in combination with recent numerical plume modeling (e.g., Larsen and Yuen 1997; Larsen et al. 1999) – to propose that the Foundation Chain might be the product of a stationary plume pulsing hot masses against the base of the Pacific Plate from depth with an apparent ~1 Myr periodicity. Assuming the validity of the hypothesis of deep mantle plumes (Morgan 1971), our model for Foundation Chain development has implications for future investigations of Pacific midplate volcanism. We propose that hotspot melting anomalies such as Foundation, spreading on impact with the lithosphere, might influence very wide areas so that apparently unconnected hotspot volcanism can be produced simultaneously across wide swaths, often cross-cutting seamount chains. Thus, variations in the age, structure and stress patterns of tectonic plates drifting over hotspot melting anomalies (pulsing mantle plumes?), might influence if, where and how hotspot volcanism develops on the Pacific Plate. This modified plume-hotspot theory might also explain widespread scattered midplate volcanism (e.g., VERs) revealed by satellite altimetry mapping as well as randomly distributed reheating events warming and raising Pacific lithosphere (Smith and Sandwell 1997), given that many other mantle plumes are similarly pulsing large masses of plume material (not necessarily with the same periodicity or mass) into broad regions impacting the base of the Pacific lithosphere.

Acknowledgements

We thank Captains H. Andresen and J.-C. Gourmelon and crews of the F/S SONNE and N/O ATALANTE and the 'SO100' and 'Hotline' scientific parties for unstinting efforts in making our dredge-sampling programs so successful. A. Koppers developed the freeware ArArCALC for blank handling and isochron and plateau calculations (Koppers 2002). This work was supported by BMBF (Bundesministerium für Bildung, Wissenschaft, Forschung und Technologie) projects 03G0100A, 03G0100A0 and 03G0157A. The 'Hotline' cruise was supported by INSU-CNRS. NSG contribution 20040201.

References

Brooks C, Hart SR, Wendt I (1972) Realistic use of two-error regression treatments as applied to rubidium-strontium data. Rev Geophys Space Phys 10:551–557
Cande SC, Kent DV (1995) Revised calibration of the geomagnetic polarity time scale for the Late Cretaceous and Cenozoic. J Geophys Res 100:6093–6095
Clague DA, Dalrymple GB (1989) Tectonics, geochronology and origin of the Hawaiian-Emperor volcanic chain. In: Winterer EL, Hussong DM, Decker RW (eds) The geology of North America. Vol. N, The Eastern Pacific Ocean and Hawaii. Geol Soc Am Boulder, pp 188–217
Dalrymple GB, Lanphere MA (1969) Potassium argon dating. W. H. Freeman Co, San Francisco
Dalrymple GB, Lanphere MA, Clague DA (1980) Conventional and $^{40}Ar/^{39}Ar$ ages of volcanic rocks from Ojin (Site 430), Nintoku (Site 432), and Suiko (Site 433) Seamounts and the chronology of volcanic propagation along the Hawaiian-Emperor Chain. Init Rep Deep Sea Drill Proj 55:659–676
Dalrymple GB, Alexander Jr. EC, Lanphere MA, Kraker GP (1981) Irradiation of samples for $^{40}Ar/^{39}Ar$ dating using the Geological Survey TRIGA reactor. U.S. Geol Surv Prof Pap 1176 San Francisco
Devey CW, SO100 Scientific Party (1995) The Foundation Seamount chain. Cruise Report 75, Geol Pal Inst University of Kiel
Devey CW, Hekinian R, Ackermand D, Binard N, Francke B, Hémond C, Kapsimalis V, Lorenc S, Maia M, Möller H, Perot K, Pracht J, Rogers T, Stattegger K, Steinke S, Victor P (1997) The Foundation Seamount chain: A first survey and sampling. Mar Geol 137:191–200
Faure G (1986) Principles of isotope geology, 2nd ed. Wiley, New York
Fleck RJ, Sutter JF, Elliot DH (1977) Interpretation of discordant $^{40}Ar/^{39}Ar$ age spectra of Mesozoic tholeiites from Antarctica. Geochim Cosmochim Acta 41:15–32
Hansen U, Yuen DA, Kroening SE, Larsen TB (1993) Dynamical consequences of depth-dependent thermal expansivity and viscosity on mantle circulation and thermal structure. Phys Earth Planet Inter 77:205–223

Hekinian R, Stoffers P, Devey CW, Ackermand D, Hémond C, O'Connor JM, Binard N, Maia M (1997) Intraplate versus oceanic ridge volcanism on the Pacific Antarctic ridge near 37° S–111° W. J Geophys Res 102:12265–12286

Hekinian R, Stoffers P, Ackermand D, Revillon S, Maia M, Bohn M (1999) Ridge-hotspot interaction: The Pacific-Antarctic Ridge and the Foundation Seamounts. Mar Geol 160:199–223

Hémond C, Devey CW (1996) The Foundation Seamount chain, Southeastern Pacific: First isotopic evidence of a newly discovered hotspot track. VM Goldschmidt Conf J Conf Abstr

Hémond C, Maia M, Gente P (1999) The Foundation Seamounts: Past and present ridge hotspot interactions? EOS Trans Am Geophys Union 80:1056

Herron EM (1972) Sea-floor spreading and the Cenozoic history of the east-central Pacific. Geol Soc Am Bull 83:1671–1692

Hieronymus CF, Bercovici D (2000) Non-hotspot formation of volcanic chains: Control of tectonic and flexural stresses on magma transport. Earth Planet Sci Lett 181:539–554

Koppers AAP (1998) $^{40}Ar/^{39}Ar$ geochronology and isotope geochemistry of the West Pacific seamount province: Implications for absolute Pacific Plate motions and the motion of hotspots. Ph.D thesis, Vrije Universiteit, Amsterdam

Koppers AAP (2002) ArArCALC – software for $^{40}Ar/^{39}Ar$ age calculations. Computers Geosciences 5:605–619

Koppers AAP, Phipps Morgan J, Morgan JW, Staudigel H (2001) Testing the fixed hotspot hypothesis using $^{40}Ar/^{39}Ar$ age progressions along seamount trails. Earth Planet Sci Lett 185:237–252

Kuiper K (2003) Direct intercalibration of radio-isotopic and astronomical time in the Mediterranean Neogene. Ph.D thesis, Vrije Universiteit, Amsterdam

Kullerud L (1991) On the calculation of isochrons. Chem Geol 87:115–124

Lanphere MA, Dalrymple GB (1978) The use of $^{40}Ar/^{39}Ar$ data in evaluation of disturbed K-Ar systems. Short Papers 4th Int. Conf Geochronol Cosmochronol Isot Geol 78-701:241–243

Larsen HC, Saunders AD (1998) Tectonism and volcanism at the southeast Greenland rifted margin: A record of plume impact and later continental rupture. Proc ODP Sci Res 152:503–533

Larsen TB, Yuen DA (1997) Ultrafast upwelling bursting through the upper mantle. Earth Planet Sci Lett 146:393–399

Larsen TB, Yuen DA, Storey M (1999) Ultrafast mantle plumes and implication for flood basalt volcanism in the Northern Atlantic Region. Tectonophysics 311:31–43

Lonsdale P (1988) Geography and history of the Louisville hotspot chain in the southwest Pacific. J Geophys Res 93:3078–3104

Lonsdale P (1994) Geomorphology and structural segmentation of the crest of the southern (Pacific-Antarctic) East Pacific Rise. J Geophys Res 99:4683–4702

Maia M, et al. (2000) The Pacific-Antarctic Ridge-Foundation hotspot interaction: A case study of a ridge approaching a hotspot. Mar Geol 167:61–84

Maia M, Hémond C, Gente P (2001) Contrasted interactions between plume, upper mantle and lithosphere: Foundation chain case. Geochem Geophys Geosyst 1:2000GC000117

Mammerickx J (1992) The Foundation seamounts: Tectonic setting of a newly discovered seamount chain in the South Pacific. Earth Planet Sci Lett 113:293–306

Mayes CL, Lawyer LA, Sandwell DT (1990) Tectonic history and new isochron chart of the South Pacific. J Geophys Res 95:8543–8567

McDougall I, Harrison TM (1999) Geochronology and thermochronology by the $^{40}Ar/^{39}Ar$ method. University Press, Oxford

McIntyre GA, Brooks C, Compston W, Turek A (1966) The statistical assessment of Rb–Sr isochrons. J Geophys Res 71:5459–5468

McNutt MK, Caress DW, Reynolds J, Jordahl KA, Duncan RA (1997) Failure of plume theory to explain the southern Austral islands. Nature 389:479–482

Morgan WJ (1971) Convection plumes in the lower mantle. Nature 230:42–43

Morgan WJ (1978) Rodriguez, Darwin, Amsterdam: A second type of hotspot island. J Geophys Res 83:5355–5360

O'Connor JM, Stoffers P, Wijbrans JR (1998) Migration rate of volcanism along the Foundation Chain, SE Pacific. Earth Planet Sci Lett 164:41–59

O'Connor JM, Stoffers P, Wijbrans JR (2000) Evidence from episodic seamount volcanism for pulsing of the Iceland plume in the past 70 Myr. Nature 408:954–958

O'Connor JM, Stoffers P, Wijbrans JR (2001) En Echelon volcanic elongate ridges connecting intraplate Foundation Chain volcanism to the Pacific-Antarctic spreading center. Earth Planet Sci Lett 192: 633–648

O'Connor JM, Stoffers P, Wijbrans JR (2002) Pulsing of a focused mantle plume Geophys Res Lett 29:10.1029/2002GL014681

Pringle MS (1993) Age progresive volcanism in the Musician Seamounts: A test of the hot spot hypothesis for the Late Cretaceous. In: Pringle MS, Sager WW, Sliter WV, Stein S (eds) The Mesozoic

Pacific: Geology, tectonics, and volcanism. Geophys Monogr Series 77:187-215

Renne PR, Swisher CC, Karner DB, Owens TL, de Paulo DJ (1998) Intercalibration of standards, absolute ages and uncertainties in $^{40}Ar/^{39}Ar$ dating. Chem Geol 145:117-152

Roddick JC (1978) The application of isochron diagrams in ^{40}Ar-^{39}Ar dating: A discussion. Earth Planet Sci Lett 41:233-244

Sandwell DT (1984) A detailed view of the South Pacific geoid from satellite altimetry. J Geophys Res 89:1089-1104

Schilling J-G (1985) Upper mantle heterogeneities and dynamics. Nature 314:62-67

Schilling J-G (1991) Fluxes and excess temperatures of mantle plumes inferred from their interaction with migrating mid-ocean ridges. Nature 352:397-403

Schilling J-G, Thompson G, Kingsley R, Humphris S (1985) Hotspot-migrating ridge interaction in the South Atlantic. Nature 313:187-191

Sleep NH (1997) Lateral flow and ponding of starting plume material. J Geophys Res 102:10001-10012

Smith WHF, Sandwell DT (1994) Bathymetric prediction from dense satellite altimetry and sparse shipboard bathymetry. J Geophys Res 99:21803-21824

Smith WHF, Sandwell DT (1997) Global sea floor topography from satellite altimetry and ship depth soundings. Science 277:1956-1962

Tebbens SF, Cande SC (1997) Southeast Pacific tectonic evolution from Early Oligocene to Present. J Geophys Res 102:12061-12084

Tebbens SF, Cande SC, Kovacs L, Parra JC, LaBrecque JL, Vergara H (1997) The Chile Ridge: A tectonic framework. J Geophys Res 102:12035-12059

Wijbrans JR, Pringle MS, Koppers AAP, Scheveers R (1995) Argon geochronology of small samples using the Vulkaan argon laserprobe. Proc Kon Ned Akad Wet 98:185-218

York D (1969) Least-squares fitting of a straight line with correlated errors. Earth Planet Sci Lett 5:320-324

York D (1984) Cooling histories from $^{40}Ar/^{39}Ar$ age spectra: Implications for Precambrian plate tectonics. Annu Rev Earth Planet Sci 12:383-409

York D, Hall CM, Yanase Y, Hanes JA, Kenyon WJ (1981) Laser-probe $^{40}Ar/^{39}Ar$ dating of terrestrial minerals with a ontinuous laser. Geophys Res Lett 8:1136-1136

Chapter 12

Hydrothermal Iron and Manganese Crusts from the Pitcairn Hotspot Region

J. C. Scholten · S. D. Scott · D. Garbe-Schönberg · J. Fietzke · T. Blanz · C. B. Kennedy

12.1
Introduction

Submarine iron and manganese deposits have a widespread occurrence in the oceanic environment. Genetically they can be subdivided into three discrete types (Boström 1983; Usui and Terashima 1997): (1) hydrogenetic, (2) diagenetic, and (3) hydrothermal. Hydrogenetic deposits occur as crusts on seamounts and other volcanic outcrops and as nodules on abyssal sediments via the direct precipitation of iron-manganese oxides and hydroxides from seawater (Koschinsky and Halbach 1995). Since these oxides and hydroxides have high adsorption capabilities, hydrogenetic crusts are characterized by relatively high trace element contents (e.g., Pb, Co, Ni) and slow growth rates (on the order of mm Ma^{-1}; Segl et al. 1989). Mineralogically, they are composed of vernadite (Fe-rich δ-MnO$_2$) and X-ray amorphous iron oxyhydroxides (δ-FeOOH) (Hein et al. 1999). The growth and the composition of diagenetic iron-manganese nodules are controlled by diagenetic element supply from the sediments. These nodules are characterized by high growth rates (on the order of 10–200 mm ka^{-1}) and high Mn/Fe ratios as well as low trace element contents (Reyss 1982). The third type of deposit, the hydrothermal iron-manganese crust, is ubiquitous along Mid-Oceanic Ridges and back arc spreading centers. They are characterized by high growth rates (cm ka^{-1}) and low trace element content. Their origin is closely related to the emanation of metal-rich hydrothermal fluids. These fluids are the result of hydrothermal convection cells that are fueled by the heat of a subsurface magma. When cold seawater penetrates the crust, its chemical composition gradually changes due to reactions with the host rocks: the temperature increases while the oxygen content and pH drop. The solution is now able to mobilize elements such as copper, zinc, iron, and manganese from the rocks (Alt 1995). As temperatures increase, the fluids become buoyant and rise rapidly. If the fluids ascend without significant subsurface mixing with seawater (focused fluid flow), most of the metal load precipitates as Cu, Zn and Fe sulfides at the sea floor as a result of mixing with oxygen-rich seawater. At a greater distance from the hydrothermal vent site, the more mobile elements, like Mn and to a lesser extent Fe, form Fe-Mn oxide deposits. Subsurface mixing of hydrothermal fluids with seawater results in stockwork-type mineralization that depletes the fluid in sulfur and results in Fe-Mn oxide deposits precipitating at the sites of fluid venting on the sea floor.

Whereas much attention has been paid to hydrothermal Fe and Mn crusts associated with Mid-Oceanic Ridge systems (cf. Boström et al. 1969; Cronan et al. 1982; Hekinian et al. 1993; Binns et al. 1993), hydrothermal mineralization associated with hotspot volcanism has been less well investigated. These hotspots are the result of plumes rising from the core-mantle boundary. The ascending magma and the related volcanic activity on the sea floor acts as a heat source driving the hydrothermal circulation. Since these volcanoes are generally dormant for about 99% of their lifetime (Latter 1987), the potential for long lasting hydrothermal activity at hotspots is assumed to be low. Nevertheless, hydrothermal deposits occur along the Hawaiian volcanic chain at Loihi (De Carlo et al. 1983; Malahoff et al. 1982; Karl et al. 1988), at the Tahiti hotspot, which is part of the Society Islands, at Macdonald Seamount and at the Pitcairn hotspots (Puteanus et al. 1991; Hekinian et al. 1993; Stoffers et al. 1993; Hodkinson et al. 1994; Glasby et al. 1997). Manganese and iron oxyhydroxides as well as Si oxyhydroxides are the principal forms of hydrothermal mineralization observed at these sites. At the Pitcairn Seamounts, hydrothermal deposits differ from others in that cogenetic Fe and Mn crusts are found within the same volcanic edifices. Moreover, thanks to extensive investigations during the SONNE 65 cruise (1989) and the Polynaut diving cruise (1999) with the submersible *Nautile*, the Pitcairn Seamounts are among the best-investigated hotspot volcanoes. This chapter reviews the controls of hydrothermal crust formation at Pitcairn Seamounts using results from previous investigations (Stoffers et al. 1993; Hodkinson et al. 1994; Glasby et al. 1997; Jeschke 1991) from the SONNE cruise 65 complemented by previously unpublished data from the Polynaut cruise.

12.2
Geological Setting

Duncan et al. (1974) proposed that the Pitcairn Island chain, consisting of the Gambiers, Mururoa, and the Duke of Gloucester Island, were all part of a volcanic chain generated by a hotspot, which has an assumed present position southeast of Pitcairn Island (see Sect. 5.4). During the SONNE 65 (1989) cruise, several seamounts with recent volcanism were indeed found about 70–100 km east-southeast of Pitcairn Island (Stoffers and Scientific Party 1990; see Sect. 5.4). Observations and sampling of the seamounts Adams (*Volcano #1*), Bounty (*Volcano #2*) (Stoffers and Scientific Party 1990) and *Volcano #5* (see Sect. 5.4.1) (Fig. 12.1) also called *Pitcairn #5* in Chap. 5 (see Sect. 5.4.1.7) revealed that these structures are covered with fresh lava with almost no sediment. Therefore, it was suggested that these seamounts are the result of recent volcanic hotspot activity. While fresh lava and active venting of hydrothermal fluids occur at the summit of Bounty, Adams, with its altered rocks and cover of coral sand, seems to be older. The composition of the volcanic edifices varies from picritic basalt, alkali basalt, trachybasalt, trachyandesite to trachyte (Hekinian et al. 2003). Field observations indicate that several cycles of magmatic eruptions formed both the Adams and Bounty Seamounts, which have the most extensive eruptions. These are also the sites where the compositions of lava flows are more variable in comparison to the smaller edifices (e.g., *Volcano #5*), which are mainly composed of evolved silicic lava (Hekinian et al. 2003; see Sect. 5.4.1.7 and Fig. 5.12b).

Images recovered by a deep-towed TV-camera during the SONNE cruise 65 as well as in situ observations by scientists diving with *Nautile* revealed that Bounty is covered with Fe crusts from about 500 m water depth to the summit at 420 m (Fig. 12.2). Some crusts form little chimneys. Shimmering water indicates active venting of low-temperature fluids (~14–19 °C). Mn crusts were only observed on the lower slope in water depths of ~2 000–3 000 m. At Adams, isolated fragments of Fe crusts occur on the upper slope (~600–400 m) with no indications of active venting. Most of the Mn crusts at Adams were found on the lower slope (1 000–2 500 m).

Chapter 12 · **Hydrothermal Iron and Manganese Crusts from the Pitcairn Hotspot Region**

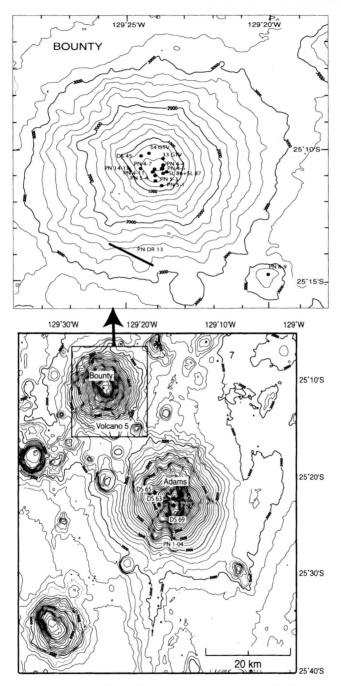

Fig. 12.1. Bathymetric map of the Pitcairn Seamounts Bounty, Adams and *Volcano #5* with sampling locations (modified after Hekinian et al. 2003). Samples from the Bounty's summit consist of Fe crusts; Mn crusts were collected at locations PN DR 13, *Volcano #5* and at Adams (PN 1-04, DS 63, DS 65, DS 69). The only Fe crusts sampled at Adams were collected by piston coring (71 SL, 72 SL)

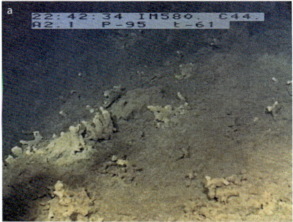

Fig. 12.2. Bottom photograph (courtesy of IFREMER) taken with the submersible *Nautile* (dive PN6) on the summit of Bounty Volcano at 580–582 m depth. **a** Small chimneys on a blanket of iron oxides. The field of vision is approximately 1.2 m. **b** A Niskin water sampler (*left*) and a field of iron oxides with small chimneys. The field of view is approximately 1 m

12.3
Sample Description

During the research campaigns SONNE 65 (1989) and Polynaut (1999), hydrothermal crusts were sampled by means of dredging (abbreviations PN DR and DS), TV guided grabs (abbreviation GTV) and when diving with *Nautile* (abbreviation, Polynaut *Nautile* = PN). The majority of the Fe crusts were retrieved from the summit region of Bounty (Fig. 12.1); the only Fe crusts sampled from Adams were obtained by piston coring (abbreviation SL; Fig. 12.1). Mixed Fe-Mn crusts were sampled on the southern slope of Bounty (location PN DR 13) and from *Volcano #5* (PN 8-9), whereas Mn crusts were found at Bounty (PN DR 13) and Adams (PN 1-04, 63 DS, 65 DS, 69 DS).

The Fe crusts range in thickness from 5 to 80 mm and have an orange brown to reddish brown color (Fig. 12.3a). Generally, they all have a similar appearance with shapes ranging from flat slabs to irregular. Some of the crusts display internal structures such as fine laminae with occasional darker color. Occasionally, a mixture of Fe and Mn crusts is seen where the Mn crusts form massive black layers

Chapter 12 · **Hydrothermal Iron and Manganese Crusts from the Pitcairn Hotspot Region**

Fig. 12.3. Photographs of Fe crusts from the summit of Bounty; **a** sample PN 5-3 shows *white lines with numbers* indicating subsampling of the crusts; the fine lamination comprises layers one to three; **b** a mixed Fe and Mn crust (PN DR 13c) showing *white lines* indicating the subsampling of the crust. Layers one and two consist of Fe, whereas three to eight are manganese

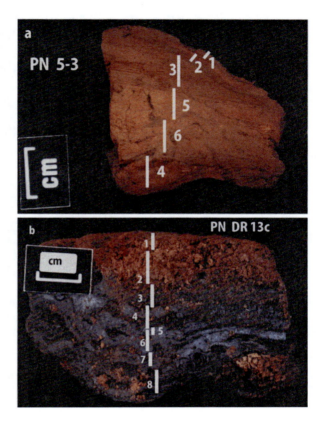

(Fig. 12.3b). The Mn crusts have a gray to black color and occur as flat slabs up to 5 cm thick. Most of the Mn crusts consist of various layers, and, according to Hodkinson et al. (1994), three sub-types can be distinguished (Fig. 12.4a,b). Type I consists of friable layers with sandy and porous textures. It is comprised of volcaniclastic material cemented by Mn oxides. Type II is a massive Mn crust; some of it contains columnar growth structures and submillimeter laminations that vary in color from black to light gray. Type III is uncommon and comprises closely packed botryoids with shiny surfaces (Fig. 12.4b).

12.3.1
Mineralogy

Most of the X-ray powder diffraction patterns of the iron oxyhydroxides from Pitcairn show only broad humps in the vicinity of 2.6 and 1.5 Å (Fig. 12.5), characteristic of the partially ordered phase two-XRD-line ferrihydrite (Boyd and Scott 1999), hereafter referred to simply as ferrihydrite. Other mineral phases are minor amounts of goethite as well as traces of calcite and volcaniclastic minerals (feldspars, pyroxene). Mossbauer studies confirm that ferrihydrite is a poorly crystalline phase and the bulk of the iron is ferric (Stoffers et al. 1993). Rapid oxidation of Fe^{2+} in hydrothermal so-

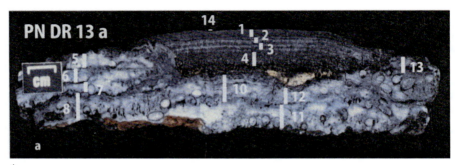

Fig. 12.4a. Photograph of hydrothermal Mn-crust section PN DR 13a collected on Bounty Seamount. *White lines with numbers* indicate subsampling of the crust. Layers one to four correspond to a type I Mn crust consisting of sandy volcanoclastic material cemented by Mn oxides. The other layers comprise type II Mn-crust

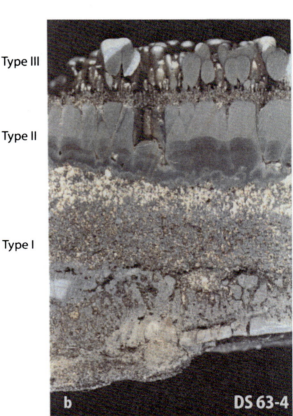

Fig. 12.4b. Photograph a section of Mn-crust DS 63-4 recovered from Adams (Jeschke 1991); the crust consists of the three types of Mn layers (type I, II and III) typical for Mn crusts from the Pitcairn Seamount

lutions is believed to be the primary formation pathway of ferrihydrite (Chukrov 1974), suggesting a rapid deposit of the Fe crusts. Goethite forms through aging of ferrihydrite and/or reactions of Fe^{2+} solutions with ferrihydrite. In mixed Fe-Mn crusts, a paragenesis of goethite with 10 Å manganite was observed. The dominant minerals in the Mn crusts are 7 Å manganite, 10 Å manganite and $\delta\text{-}MnO_2$, but the identification of the latter mineral by XRD is equivocal due to overlapping reflections with 7 Å manganite and 10 Å manganite. Type I crust contains volcaniclastic minerals, including feldspars, pyroxene and traces of calcite and quartz.

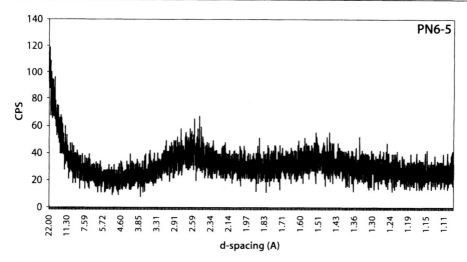

Fig. 12.5. X-ray diffraction pattern of iron oxide crust PN 6-5; d-spacing values of 2.6 Å and 1.5 Å correspond to 2-line ferrihydrite

12.3.2
Age Dating

The uranium-thorium disequilibrium method can provide reliable ages for the past 300 ka if certain precautions are taken. This method is based on the incorporation of seawater-derived uranium into hydrothermal precipitates during the contact of hydrothermal fluids with seawater. The growth of ^{230}Th from the decay of ^{234}U is used for age determinations (Lalou et al. 1993). One prerequisite for reliable age dating is that no ^{230}Th enters the hydrothermal precipitate during or after its formation. Evidence for this is found in the very low ^{232}Th concentrations in the analyzed samples, ^{232}Th being an isotope of terrigenous origin. Furthermore, the hydrothermal precipitate should be a "closed-system" for uranium, i.e., no uranium enters or leaves the precipitate. For slowly growing hydrogenetic manganese crusts, diffusion of uranium may be an important factor influencing age determination (Henderson and Burton 1999). Nevertheless, if ^{230}Th/^{234}U dating of individual cogenetic samples results in comparable apparent ages, these can be assumed to represent "true" ages.

Table 12.1 presents isotope measurements for the Fe and Mn crusts from the Pitcairn Seamounts and the calculated apparent ages. For the Mn and Fe crusts, the ages range from 2.9 to 130 ka and from 0.29 to 3.1 ka, respectively. All the Mn crusts having apparent ages >~100 ka are characterized by relatively low uranium content (^{234}U ≤ 2.4 dpm g^{-1}), and a loss of uranium seems likely for these samples. Such a loss increases the ages with the consequence that the calculated apparent age estimates for these crusts should be regarded as maxima. For the rest of the Mn crusts, their apparent ages range between 2.9 and 20.8 ka. Since the apparent ages of subsamples, e.g., PN DR 13c and PN DR 13a crusts, are relatively similar and even show a chronology, these apparent ages can be regarded as "true" ages. The Fe crusts are, in comparison to the Mn crusts, younger (<3.1 ka) and the small differences in the ages of the Fe crust subsamples (e.g., PN 5-3) indicate their very rapid formation.

Table 12.1. Isotope measurements and derived apparent ages of Mn and Fe crusts from Pitcairn Seamounts (with 1 σ errors)

Sample	^{232}Th	+/-	^{230}Th	+/-	^{234}U	+/-	^{234}U/^{238}U	+/-	Age[a]	+/-
	(dpm g^{-1})		(dpm g^{-1})		(dpm g^{-1})				(kyr)	
Mn crusts										
DS 63/2 31-36[b]	b.d.	–	0.170	0.009	0.24	0.019	1.16	0.07	131	12
DS 69/1 12-18[b]	b.d.	–	0.030	0.006	0.52	0.04	1.16	0.06	6.7	0.6
DS 69/1 19-24[b]	b.d	–	0.030	0.002	1.06	0.06	1.20	0.06	2.9	0.2
PN 1-04/1[c]	1.016	0.022	1.54	0.36	1.64	0.071	1.13	0.069	–[e]	–
PN 1-04/2[d]	0.557	0.003	3.043	0.005	9.773	0.023	1.157	0.004	–	–
PN DR 13a/5[c]	0.059	0.001	1.510	0.015	2.400	0.074	1.215	0.053	103	9.7
PN DR 13a/7[c]	0.158	0.004	0.760	0.018	5.194	0.186	1.226	0.062	17.1	1.1
PN DR 13a/8[d]	0.007	0.001	0.797	0.004	4.537	0.022	1.254	0.007	20.8	0.2
PN DR 13a/11[c]	0.092	0.002	0.465	0.011	6.070	0.174	1.279	0.052	8.7	0.5
PN DR 13a/13[c]	0.671	0.015	1.095	0.025	1.214	0.048	1.107	0.062	–[e]	–
PNDR 13b/3[d]	0.003	0.001	0.328	0.001	4.182	0.020	1.286	0.008	8.9	0.07
PNDR 13b/4[d]	0.009	0.001	0.476	0.003	5.103	0.023	1.300	0.007	10.6	0.1
PN DR13c/3[d]	0.002	0.001	0.189	0.001	3.37	0.017	1.256	0.008	6.2	0.05
PN DR13c/7[c]	0.029	0.001	0.700	0.007	9.20	0.372	1.220	0.070	8.6	0.7
PN DR13c/8[d]	0.017	0.001	0.896	0.005	7.62	0.027	1.261	0.006	13.5	0.1
Fe crusts										
PN 4-4-1[c]	0.039	0.001	b.d.	–	1.48	0.072	1.31	0.090	–	–
PN 5-3/1[d]	0.001	0.0001	0.022	0.0003	3.302	0.016	1.192	0.007	0.71	0.010
PN 5-3/3[d]	0.002	0.0001	0.007	0.0002	2.516	0.012	1.178	0.007	0.29	0.007
PN 5-3/4[d]	0.002	0.0001	0.010	0.0002	3.168	0.010	1.163	0.004	0.34	0.006
PN DR 13c/2[c]	0.114	0.001	1.01	0.010	0.42	0.033	1,23	0,135	–[e]	–
33 DS/1[c]	0.331	0.025	0.247	0.011	7.74	0.30	1.10	0.034	–[e]	–
33 DS/2[c]	b.d.	–	b.d.	–	6.44	0.20	1.13	0.028	–	–
33 DS/3[c]	b.d.	–	0.051	0.005	4.64	0.22	1.14	0.043	1.2	0.15
33 DS/4[c]	b.d.	–	0.102	0.008	3.61	0.21	1.16	0.053	3.1	0.15

[a] Apparant ages calculated following Kaufman and Broecker (1965).
[b] Hodkinson et al. (1994), data obtained by means of alpha spectrometry.
[c] Data obtained by means of alpha spectrometry.
[d] Measurements by means of double focussing sector field ICP-MS.
[e] No age determination due to high exogenous thorium and/or low uranium concentrations; *dpm* = decays per minute; *b.d.* = below detection.

12.3.3
Biomineralization

Ferrihydrite deposits from low-temperature hydrothermal vents such as on the Loihi (Emerson et al. 2002), Franklin (Binns et al. 1993; Boyd and Scott 1999, 2001) and Axial

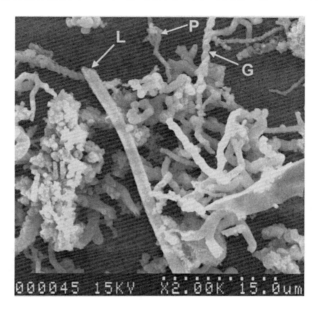

Fig. 12.6. Scanning electron microscope image of typical *Gallionella ferruginea* (G), *Leptothrix ochracea* (L), and PV-1 (P) from a sea-floor low-temperature hydrothermal vent site in the caldera of Axial Volcano, Juan de Fuca Ridge

(Kennedy et al. 2003) submarine volcanoes and the Southern Explorer mid-ocean spreading ridge (Fortin et al. 1998) commonly teem with mineralized bacterial forms. Based on morphological features, the rod-like sheath of *Leptothrix ochracea*, the distinctive helical stalk of *Gallionella ferruginea* and the non-helical filamentous structures of the newly described PV-1 bacterium (Emerson and Moyer 2002) (Fig. 12.6) are the most common iron oxidizing bacteria in ferrihydrite (Kennedy et al. 2003). With respect to ferrihydrite formation at neutral pH, iron oxidizing bacteria play at least two roles. First, they have been shown to increase the rate of iron oxidation over the rate of strictly inorganic oxidation (Søgaard et al. 2001; Kasama and Murakami 2001). Second, they inherently lower the degree of supersaturation required for ferrihydrite precipitation by behaving as geochemically reactive solids for heterogeneous surface nucleation (Kennedy et al. 2003). The outer surface of the bacteria is commonly encrusted with ferrihydrite and silica in these environments. The silica can act to stabilize ferrihydrite on the sea floor, inhibiting its conversion to more stable mineral forms such as goethite or hematite (Carlson and Schwertmann 1981; Boyd and Scott 1999). In the cited examples, bacterially produced ferrihydrite is the most common mode of occurrence of iron oxyhydroxide.

The situation on Adams and Bounty Volcanoes is different. Here, there is a dominance of spherical aggregates of ferrihydrite and a relative lack of mineralized bacterial structures (Fig. 12.7a). *Gallionella ferruginea* is totally absent in all the samples examined. There are only a few isolated sheath structures that appear to resemble *Leptothrix ochracea* (Fig. 12.7b). It is apparent that inorganic and not bacterial processes dominate the formation of the iron oxides found on the Pitcairn volcanoes. The absence of a biogenic influence on iron oxide precipitation is further supported by carbon isotopic data (Table 12.2). The $\delta^{13}C$ values are consistent with those normally found for vent fluids (−4 to −11‰) but are well outside the range for vent organisms (−10 to −35‰) (Shanks et al. 1995).

Fig. 12.7. Scanning electron microscope images of; **a** typical iron oxyhydroxide globular aggregates; **b** rare sheath of what appears to be *Leptothrix ochracea* from Pitcairn

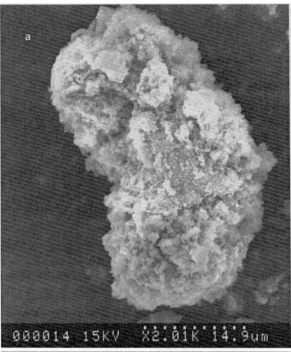

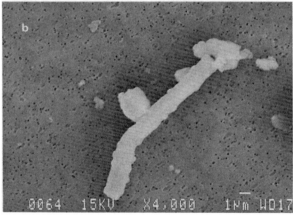

The absence of abundant iron oxidizing bacteria at Pitcairn together with the total absence of iron sulfides suggests that the vent fluids were already oxidized and carried only ferric iron when they reached the sea floor. Mixing of this fluid with more oxygenated seawater caused rapid hydrolysis and precipitation of the iron as two-XRD-line ferrihydrite without the mediation of iron oxidizing bacteria. In the absence of a bacterial template on which to form, the ferrihydrite formed disorganized globular structures. Pitcairn hotspot offers for future study an interesting contrast to those more prevalent sites where bacterial intervention is responsible for the precipitation of iron oxides.

Chapter 12 · **Hydrothermal Iron and Manganese Crusts from the Pitcairn Hotspot Region**

Table 12.2. Carbon isotopic data for Mn and Fe crust samples from Bounty Seamount

Sample		$\delta^{13}C_{PDB}$ (‰)
Mn crust	PNDR13-8	−7.4
Fe crust	PN5-03	−8.6
Fe crust	PN6-05	−8.2

12.4 Chemical Composition

The chemical composition of the iron and manganese deposits is indicative of their genesis. Low concentrations of Cu, Ni and Co distinguish hydrothermal Fe and Mn deposits from hydrogenous and diagenetic deposits, which have higher trace metal contents. In a ternary diagram of Fe-Mn-(Cu + Ni + Co) × 10, all samples from Pitcairn Seamounts (Adams, Bounty, *Volcano #5*) plot in the field of hydrothermal deposits (Fig. 12.8) (see Appendix for chemical compositions of crusts sampled during Polynaut). The Fe content of the Fe crusts ranges from 8.8 to 45.8 wt.% with Fe/Mn ratios between 7 an 1100 (Mn/Fe = 0.0009 to 0.1) (Table 12.3). In the Mn crusts, the Mn content ranges from 9.4 to 52.6 wt.% and Fe/Mn ratios from 0.007 to 0.85 (Mn/Fe ratio = 1.2–140) (Table 12.3). Apart from their Fe/Mn ratios, the crusts can also be distinguished by their trace element contents (Fig. 12.9). The Fe and Mn crusts do not

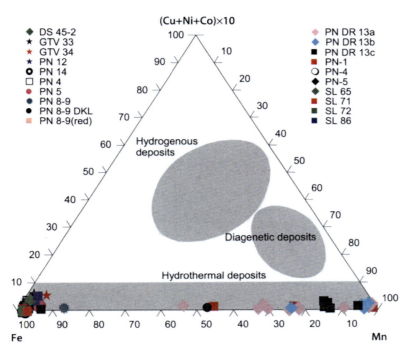

Fig. 12.8. Ternary diagram Fe-Mn-(Cu + Ni + Co) × 10 for Fe crusts (data from Stoffers et al. 1993 and from the Polynaut cruise) and Mn crusts (data from Hodkinson et al. 1994 and the Polynaut cruise); all samples plot in the field of hydrothermal deposits (cf. Boström 1983)

Table 12.3. Mean geochemical data of Pitcairn hotspot crusts and comparative data of other crust deposits

Location	Fe (%)	Mn (%)	Si (%)	Ca (%)	Ti (%)	Co (ppm)	Ni (ppm)	Cu (ppm)	Zn (ppm)	As (ppm)	Ba (ppm)	Mo (ppm)	Pb (ppm)	Th (ppm)	U (ppm)	ΣREE (ppm)
Fe-crusts Pitcairn																
Min.	8.82	0.02	5.19	–	0.02	0.84	2.12	1.17	5.17	12.0	16.6	40.6	0.06	0.00	0.33	0.89
Max.	45.8	4.14	33.3	–	0.32	231	225	288	544	473	360	240	52.0	8.46	13.6	253
Mean	34.7	0.36	10.9	–	0.1	30.7	23.7	49.4	68.0	97.0	89.2	1.55	5.69	0.63	3.81	34.0
Number of samples, n	48	48	19	–	23	48	48	48	48	48	48	30	48	48	48	48
Teahitia[a]	35.4	0.22	15.7	–	–	55.3	22.2	119	69.4	409	676	137	8.20	1.60	13.7	102
Moua Pihaa[b]	13.6	0.19	26.9	–	–	57.6	139	109	147	329	5 373	23.6	12.6	3.40	4.30	194
Loihi[c]	19.8	0.89	18.8	3.40	0.37	64.4	233	344	765	42.1	–	–	454	–	–	–
Macdonald[d]	16.3	0.15	19.0	2.84	2.37	55.1	111	143	119	969	24 420	20.3	–	2.50	2.75	126
Off axis seamount, EPR[e]	31.1	5.00	8.79	2.22	0.13	112	186	12 075	435	–	–	–	–	–	–	–
Franklin Seamount[f]	21.2	2.4	17.5	0.97	0.07	75.4	277	43.9	49	1 662	450	204	10.6	0.31	7.93	20.6
Mn-crusts Pitcairn																
Max.	18.2	52.6	11.9	15.4	0.42	125	399	194	856	42.8	1 918	1 109	40	3.71	11.1	204
Min.	0.03	9.44	0.05	1.64	0.02	0.46	6.09	6.86	12	0.10	329	46.6	0.09	<0.01	0.75	2.04
Mean	2.25	29.6	2.66	6.40	0.32	31.0	138	30.4	112	13.5	767	379	1.58	1.07	3.96	64.1
Number of samples, n	102	102	77	77	102	40	102	102	102	25	25	25	25	25	25	25
Mean type I[g]	1.50	26.7	–	7.00	0.19	52	193	35	80	5.00	720	127	30	–	–	–
Mean type II[h]	0.07	44.0	–	2.02	0.02	30	126	28	21	1.19	933	146	15	–	–	–
Mean type III[i]	0.06	46.4	–	1.07	0.01	66	64	23	15	0.2	737	151	18	–	–	–
Tonga-Kermadec[j]	0.48	40.2	–	1.62	0.04	10.8	115	47.2	52.7	106	390	958	<10	–	–	–
Valu Fa Ridge[k]	0.68	45.0	–	1.63	0.03	16.2	45.3	124	69.3	34.3	567	1 548	<10	–	–	–
Hawaiian Archipelago[l]	1.82	41.8	–	1.17	0.55	880	3 694	1 342	2 337	–	9 800	–	186	–	–	–
Galapagos[m]	0.66	47.0	–	0.98	0.01	13.0	125	80.0	90.0	–	–	–	–	–	–	–
Hydrogenous Pitcairn[n]	20.5	12.8	–	3.31	0.98	3 183	1 767	703	543	282	2 400	540	–	–	–	1 318
NW-Pacific, hydrogenetic[o]	15.1	22.1	3.7	4.10	0.77	6 372	5 403	1 075	680	165	1 695	455	1 777	–	–	–

[a] Stoffers et al. (1993); $n = 20$; [b] Stoffers et al. (1993); $n = 16$; [c] DeCarlo et al. (1983); $n = 7$; [d] Puteanus et al. (1991); Stoffers et al. (1993); $n = 8$ (5); [e] cf. Puteanus et al. (1991), $n = 5$; [f] Binns et al. (1993), $n = 7$; [g] Hodkinson et al. (1994), $n = 10$; [h] Hodkinson et al. (1994), $n = 10$; [i] Hodkinson et al. (1994), $n = 2$; [j] Rogers et al. (2001); $n = 20$; [k] Rogers et al. (2001); $n = 6$; [l] Carlo et al. (1987); $n = 1$; [m] Moorby and Cronan (1983); [n] Hodkinson et al. (1994), $n = 5$; [o] Hein et al. (1999), $n = 1 478$.

Chapter 12 · **Hydrothermal Iron and Manganese Crusts from the Pitcairn Hotspot Region**

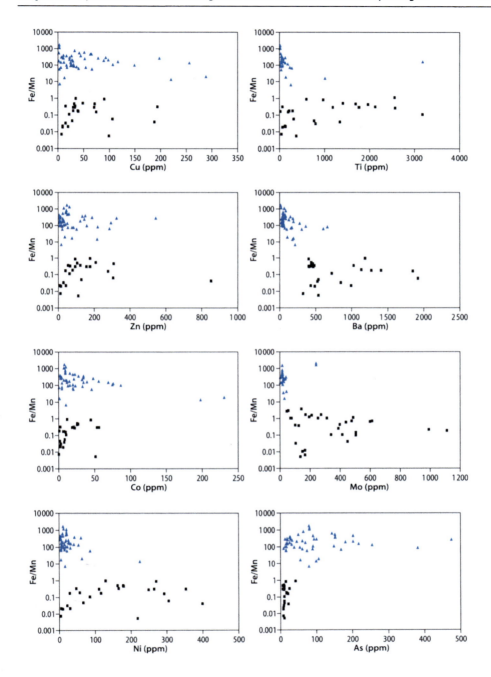

Fig. 12.9. Fe/Mn ratios versus trace element concentrations for crusts derived from the SONNE 65 and Polynaut cruises. Samples with Fe/Mn ratios >1 are Fe crusts (*triangles*), samples with Fe/Mn < 1 are Mn crusts (*squares*). The Co, Zn and Cu contents in the crusts do not significantly differ between Fe and Mn crusts. Concentrations of As are higher in the Fe crusts, whereas the Mn crusts are relatively enriched in Ni, Ti, Ba and Mo

differ significantly with respect to their Co, Zn and Cu concentrations, but the amount of As is much higher in the Fe crusts. In contrast, Ni, Ti, Ba and Mo are higher in the Mn crusts compared to the Fe crusts. As will be discussed in more detail below, different scavenging abilities of Fe and Mn as well as different amounts of volcaniclastic material are responsible for most of the observed differences.

12.4.1
Fe Crusts

A comparison of the average composition of Fe crusts from the Pitcairn Seamounts (Adams, Bounty, *Volcano #5*) with those from other hotspot deposits is given in Table 12.3. Much of the variations in the Fe concentrations between the various locations can be attributed to variable Si content. For instance, the Fe concentrations in Moua Pihaa crusts are about half of those observed in the Pitcairn Fe crusts, but the Si concentrations in Moua Pihaa are about twice as high as in the Pitcairn Fe crusts. High Si contents can arise from several sources: contamination by volcaniclastic material (Boyd et al. 1993) or siliceous marine particles such as diatoms and sponge spicules (Kennedy et al. 2003, Fig. 12.2b), co-precipitation of nontronite or amorphous silica with ferrihydrite (Fortin et al. 1998; Kennedy et al. 2003) and "essential" Si in the ferrihydrite structure (Boyd and Scott 1999).

In general, trace element concentrations of Pitcairn Fe crusts are at the lower end of the typical concentrations observed at other localities (Table 12.3). In comparison to Teahitia and Macdonald crusts, average Co, Ni, and Zn concentrations in the Pitcairn Fe crusts are in the same range, whereas the Ba, Mo and As concentrations are lower. In comparison to Loihi, As and Co are in the same range but the concentrations of Ni, Cu, Zn, and Pb are lower in Pitcairn crusts. Stoffers et al. (1993) suggested that the water depth of the volcanoes (Bounty: ~420 m; Teahitia ~1500 m; Macdonald ~150 m; Loihi ~1000 m) may affect the temperature of the discharging hydrothermal fluids and thus the ability of the solutions to transport trace metals. The temperatures of hydrothermal fluids at Bounty (~14–19 °C) are, however, in the same range as those observed at the other intraplate volcanoes of Teahitia (15.8–30 °C) and Loihi (30–70 °C).

Principal Component Analysis (PCA) was conducted in order to obtain a more detailed understanding of the composition of the Fe crusts recovered from the Pitcairn Seamounts of Bounty, Adams and *Volcano #5*. The PCA calculates a number of principal components, i.e., linear combinations of analyses that together explain the variance in the data set. The first principal component or factor explains most of the variance, while the second is the next highest. Both factors span a two-dimensional plane, and the relation between the samples and the two factors can be visualized by a so-called bi-plot (Gabriel 1971) (Fig. 12.10). All the variances are explained by the two factors when a sample reaches the drawn unit circle. Two factors explain 56% of the variance for the Fe crusts from Pitcairn (Fig. 12.10). Factor I, which explains 35% of the variance, has high loadings for Fe, Th, Cu, Zn, and Ba. Apart from Th, all the other elements are characteristic of hydrothermal deposits (Herzig and Hannington 1995). For instance, Ba-rich Fe crusts have been reported from Macdonald, Teahitia and Moua Pihaa (Stoffers et al. 1993) as well as from the Izu-Bonin Arc (Urabe and Kuskabe 1990) and the western Woodlark Basin (Binns et al. 1993). The mineral barite has, however, not been identified in the Pitcairn crusts. The grouping of Pb, Cu, Zn, and Ba with Th, which is almost absent in

Chapter 12 · Hydrothermal Iron and Manganese Crusts from the Pitcairn Hotspot Region

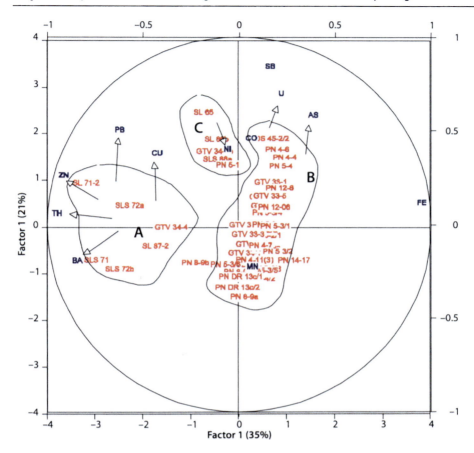

Fig. 12.10. Principal component analysis of Fe crusts. The two factors that explain the variance in the samples span a two-dimensional plane on which the position of each sample and associated elements are plotted. Group A is characterized by relatively high amounts of volcaniclastic material. Group B clusters all samples where the hydrothermal influence is highest. Group C has an intermediate composition between A and C

hydrothermal fluids (Chen et al. 1986), suggests an association of these elements with volcaniclastic material. Most samples with high loadings for factor I (group A in Fig. 12.10) originated from the subsurface and were recovered with a gravity corer. A mix of volcaniclastic material and hydrothermal Fe crusts characterizes the lithology of these cores, as is also evident from the relatively high Si content in these samples (>14% Si). Factor II is characterized by high loadings for Co, As, U, Sb, and Ni, and most of the Fe crusts investigated cluster near these elements (Fig. 12.10, group B). An association of these elements with hydrothermal Fe crusts and iron rich sediments has been previously described from Loihi Seamount (De Carlo et al. 1983), Hellenic Volcanic Arc (Varnavas and Cronan 1991), Franklin Seamount in the western Woodlark Basin (Boyd et al. 1993), and in metalliferous sediments from the East Pacific Rise (Kunzendorf et al. 1984). Iron oxides have high sorption capacities and, therefore, scavenging of these elements from hydrothermal plumes is likely (German et al. 1991; Rudnicki and Elderfield

1993). For example, Boyd et al. (1993) showed that the trace element content of Fe crusts reflected the geochemistry of the local environment: high values of Au and its surrogate element, As, in the vicinity of Au-rich barite spires at Franklin Seamount and high values of Zn in the vicinity of black smokers at 11–13° N EPR.

To what extent the trace elements in Fe crusts have a hydrothermal and/or seawater origin still remains unclear. The concentrations of As in the hydrothermal solution sampled at Bounty (0.74 ppb; Table 12.A.1 in Appendix) are slightly lower than that of seawater (3.7 ppb), whereas the Co content is much higher in the fluids (0.01 ppm) compared to seawater (0.003 ppb). In hydrothermal solutions, the uranium concentrations are generally much lower than in seawater (Chen et al. 1986). The average $^{234}U/^{238}U$ ratio in the Fe crusts of 1.21 ±0.08 is similar to that of seawater ($^{234}U/^{238}U = 1.14$ ±0.03; (Ku et al. 1977) so that a seawater origin of uranium is very likely. The third group (C) of Fe crusts plots near Ni (Fig. 12.10). Glasby et al. (1997) found an association of Ni with the hydrogenous Mn oxide component, but a volcaniclastic origin is also likely. The clustering of group C between groups A and B may suggest an intermediate composition between Fe crusts with relatively high amounts of volcaniclastic material and the purest Fe crusts.

12.4.2
Mn Crusts

A comparison of Mn crusts from the Pitcairn Seamounts with those from other midplate volcanoes is difficult due to the fact that trace amounts of a hydrogenous Mn component may significantly alter the trace metal concentrations. Such a hydrogenous phase was observed, for example, in hydrothermal Mn crusts from the Hawaiian hotspot Archipelago (De Carlo et al. 1983), and consequently these crusts have higher trace metal contents than those from Pitcairn (Table 12.3). Moreover, the element concentrations of the Pitcairn Mn crusts depend on the type (types I, II or III) of crust considered. Jeschke (1991) reported high Mn and low Fe, Cu, Ni, and Zn concentrations in type II and III crusts whereas in type I crusts, i.e., those crusts with high amounts of volcaniclastic material, the opposite relation was observed (Fig. 12.11). When considering only the purest hydrothermal crusts (types II and III), Mn and the trace element concentrations (Cu, Ni, Co, Zn) are in the same range as reported from Tonga-Kermadec, Valu Fa Ridge and Galapagos spreading center (Table 12.3).

The compositional relations between hydrothermal and volcaniclastic components are well resolved in a PCA for the Pitcairn Mn crusts in which two factors explain 66% of the variance. Factor I has high loadings for Mn, Mo, Co, Ti, Sc, Pb, Th, Rb, and Y; factor II has high loadings for Fe, As, Sr, and Ba. In a bi-plot, the samples cluster in three different groups (Fig. 12.12). One group of samples (group D) with high loading for factor I clusters near the elements Co, Ti, Cu, Th, and Pb. These samples are all characterized by volcaniclastic layers impregnated by Mn (type I crust) and factor I; therefore, this describes the mixed hydrothermal and volcanic composition of the Mn crusts. A group of samples (group E, Fig. 12.12) has high loadings for the second factor (As, Ba, U, and Fe). The association of Fe with U and As was also observed for some Fe crusts (group B, Fig. 12.10). Most of the Mn crust samples in group E (e.g., samples PN 8 9 DKL; PN DR 13c/6; PN 13 c/8) are comprised of a mixture of Fe and Mn crusts (see Fig. 12.3b) and hence their grouping with the elements Fe, As and U can be explained by scavenging mechanisms similar to those already discussed for the Fe crusts.

Chapter 12 · Hydrothermal Iron and Manganese Crusts from the Pitcairn Hotspot Region

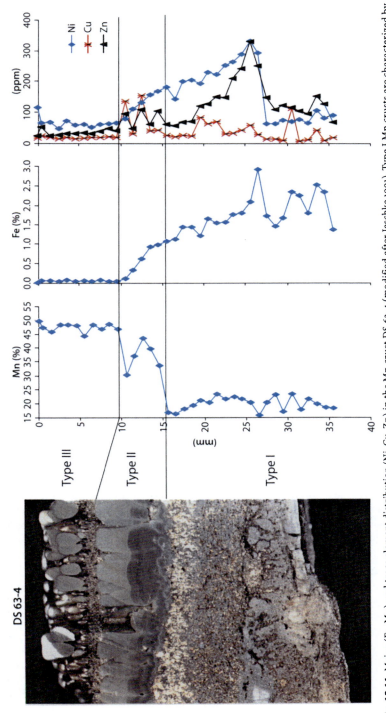

Fig. 12.11. Major (Fe, Mn) and trace element distribution (Ni, Cu, Zn) in the Mn crust DS 63-4 (modified after Jeschke 1991). Type I Mn crusts are characterized by having relatively high trace element and Fe concentrations due to the presence of volcanoclastic material, whereas the amount of these elements is low for type II and type III crusts

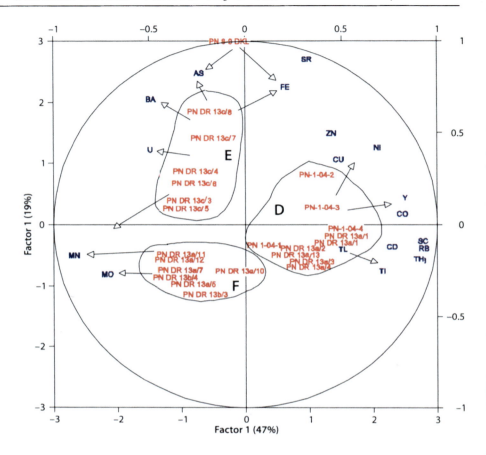

Fig. 12.12. Principal component analysis of Mn crusts. Samples that cluster in group D comprise type I Mn crusts. Group E represents samples that are characterized by mixed Fe and Mn crusts; Group F samples are those in which the hydrothermal influence is highest

Sequential leaching experiments of Mn crusts indicated that most of the barium is associated with the Mn oxides except for the type I crusts, where Ba was believed to occur as barite (Hodkinson et al. 1994). Hence, a hydrothermal origin of Ba is possible, although the Ba concentrations in the hydrothermal fluids sampled at Bounty are similar to that of normal seawater (~0.02 ppm). Whatever the origin of Ba may be, scavenging is most likely the primary process responsible for its enrichment because Mn oxides have been shown to be very effective adsorbents for this element (DeLange et al. 1990). One group of samples (group F) plots near Mn and Mo (Fig. 12.12). Hydrothermal Mn crusts rich in Mo were previously described from the Tonga-Kermadec Ridge and the Valu Fa Ridge. According to Hein et al. (1996), this enrichment may be due to hydrothermal leaching of basement rocks. Although the average amount of Mo in the Pitcairn Mn crusts is at the lower end in comparison to other hydrothermal intraplate deposits, sequential leaching indicated a hydrothermal origin for the Mo in Pitcairn Mn crusts (Hodkinson et al. 1994), and the hydrothermal solution sampled at Bounty shows an enrichment of Mo relative to seawater (Table 12.A.1 in Appendix).

12.4.3
Rare Earth Elements (REE)

The rare earths form a very coherent group of elements. Their most stable oxidation state is REE^{3+}, but other important states occur for Eu (Eu^{2+}) and Ce (Ce^{4+}). The gradual filling of the inner 4f shell with increasing atomic number leads to a gradual change in the chemical properties of the REE (De Baar et al. 1985), and this is why REE distribution patterns have been very useful for establishing the genesis of magmatic and sedimentary deposits (Fleet 1983). The REE pattern of seawater, for instance, is characterized by a negative Ce anomaly (i.e., $Ce/Ce^* < 1$; $Ce / Ce^* = Ce / ((La + Pr) / 2)$), which is the result of oxidation of Ce^{3+} to Ce^{4+} followed by a rapid removal from the water column, a behavior in contrast to the other REE. The preferential removal of Ce relative to the other REE leads to positive Ce anomalies in hydrogenetic Mn-deposits. In hydrothermal fluids, the identification of a positive Eu anomaly ($Eu / Eu^* > 1$, $Eu / Eu^* = Eu((Sm + Gd) / 2)$) is a very important criterion for the understanding of a hydrothermal system (Michard and Albarède 1986; Campbell et al. 1988). In hydrothermal fluids with temperatures <250 °C, Eu occurs in its oxidized state Eu^{3+}. At higher temperatures, Eu is present in the divalent state, and the resulting larger ionic radius differentiates Eu from other REE, causing a positive Eu anomaly in vent fluids. Most of the hydrothermal fluids from Mid-Oceanic Ridges show a positive Eu anomaly (Michard 1989), and this was also observed in hydrothermally enriched plume waters at Macdonald Seamount (Stüben et al. 1992). The REE patterns of Teahitia and Bounty vent waters, in contrast, display a small negative Eu anomaly (Michard et al. 1993; this study, unpublished data).

In order to compare REE data between different settings, it is the convention to normalize REE to the REE concentrations of North American Shale (NASC; cf. De Carlo and McMurtry 1992; Stoffers et al. 1993; Glasby et al. 1997) or chondrite (Hodkinson et al. 1994). However, as pointed out by Glasby et al. (1997), there is no agreement on how to normalize REE data. Because of the relatively low Eu concentration in comparison to Sm and Gd in NAS, normalizing to NAS may create a positive Eu anomaly that may lead to erroneous conclusions. In the following discussion, the REE data are C1 chondrite-normalized after Sun and McDonough (1989).

The REE concentrations in the Fe crusts (Table 12.A.2 in Appendix) are lower than those of the volcanic rocks sampled at Pitcairn Seamounts (Hekinian et al. 2003) but higher than those in the hydrothermal fluid at Bounty (Fig. 12.13). The C1 chondrite-normalized REE patterns of the Fe crusts generally show an enrichment of the light REE (LREE, La–Nd) relative to the others. Most of the Fe crusts exhibit negative Ce and Eu anomalies, but for samples having high REE concentrations (e.g., sample PN 8-9b), these features are not observed. Positive Eu-anomalies (NAS normalized) were previously reported for Pitcairn Fe crusts (Stoffers et al. 1993), but as already pointed out, these may be the result of the normalizing procedure.

According to the mineralogical composition, three potential sources may determine the REE patterns in the crusts: volcaniclastic (detrital) material, hydrothermal solutions and seawater. Because concentrations of REE in hydrothermal fluids and seawater are low, any admixture of volcaniclastic material will significantly increase the REE concentrations in fast-growing Fe crusts and this will change the patterns in the direction of those of the volcanic rocks found at Pitcairn Seamounts (Fig. 12.13). Apart from the up to two orders of magnitude higher REE concentrations in the hydrothermal fluids compared

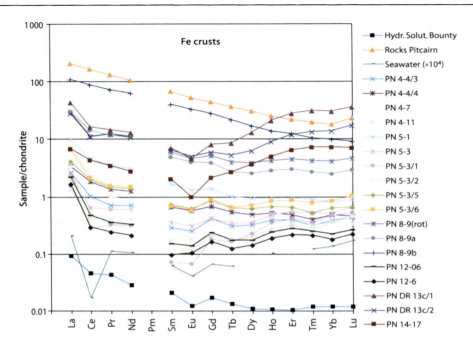

Fig. 12.13. Selected C1 chondrite normalized REE patterns of Fe crusts (Polynaut data) and comparison to hydrothermal fluids from Bounty's summit (this study, unpublished data), average volcanic rocks (Hekinian et al. 2003) and seawater (De Baar et al. 1985). The negative Ce anomaly for crusts with relatively low REE concentrations suggests a major seawater influence

to seawater, the REE pattern of the hydrothermal solution shows an enrichment of LREE and a weak negative Ce anomaly compared to seawater. Both have a slightly negative Eu anomaly, but the Yb/Sm ratio of 2.4 in seawater indicates a depletion of medium REE (MREE, Sm–Tb) relative to heavy REE (HREE, Ho–Lu). On the other hand, the MREE are relatively enriched in the hydrothermal solution (Yb/Sm = 0.56), which may be the result of the chemical reactions of the solutions with the volcanic rocks (Yb/Sm = 0.27). Most of the Fe crusts have Yb/Sm ratios >1, and this together with the slight negative Ce anomalies indicates that seawater is a dominant factor for the distribution of REE in those crusts having low contents of volcanic material. Iron oxides are known to be very good adsorbers of rare earth elements from seawater (Koeppenkastrop and DeCarl 1993) as was shown by the increasing REE/Fe ratios with increasing distance from the hydrothermal vent sites at the Mid-Atlantic Ridge (German et al. 1990). In comparison to other Fe crusts associated with intraplate volcanoes, the total REE concentrations in Pitcairn Fe crusts are low (Table 12.3), suggesting a rapid deposition upon interaction of the hydrothermal fluids with seawater.

The C1 chondrite-normalized REE concentrations of the Mn crusts from Pitcairn volcanoes vary over three orders of magnitude (Fig. 12.14). In all the Mn crusts, the LREE are enriched relative to the HREE. Due to the high amounts of volcaniclastic components, type I Mn crusts exhibit the highest concentrations of REE, and their patterns (e.g., PN-1-04, Fig. 12.14) are similar to those of the volcanic rocks sampled at Pitcairn Seamounts (Hekinian et al. 2003). The REE patterns of those Mn crusts with relatively

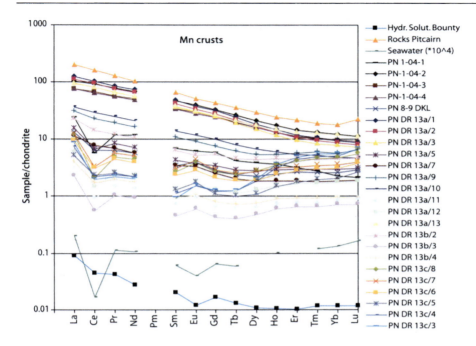

Fig. 12.14. Selected C1 chondrite normalized REE pattern of Mn crusts (Polynaut data) and their comparison with hydrothermal fluids from Bounty's summit (this study, unpublished data), average volcanic rocks (Hekinian et al. 2003) and seawater (De Baar et al. 1985). Some of the Mn crusts show a positive Eu anomaly indicative of high-temperature hydrothermal activity

low REE concentrations show slight negative Ce and positive Eu anomalies. High-resolution studies of REE distribution by laser ablation (Glasby et al. 1997) indicated that the Eu and Ce anomalies vary significantly with depth within a Mn crust. Lower REE contents coincide with positive Eu and negative Ce anomalies, and it was suggested that such sections of the crust correspond to periods of extensive high-temperature hydrothermal activity. Some sections of the type I Mn crusts investigated by Glasby et al. (1997) and Hodkinson et al. (1994) also displayed a positive Ce anomaly, which is indicative of a hydrogenous origin. The authors attributed the presence of a hydrogenous component to a greater age and a closer proximity of type I Mn crusts to seawater.

The appearance of a positive Eu anomaly (on a C1 chondrite-normalized basis) in Pitcairn Mn crusts but not in the Fe crusts suggests a general difference in their process of formation.

12.5
Formation of Fe and Mn Crusts

In order to understand the factors controlling the formation of Fe and Mn crusts at Pitcairn Seamounts, it is necessary to consider the nature of hotspot volcanoes and the associated modes of hydrothermal activity. Drawing an analogy to observations at White Island Volcano (Giggenbach 1987), Puteanus et al. (1991) suggested that during periods of volcanic quiescence, the circulation of seawater at the summit of the volcano is the dominant process causing a seawater-dominated hydrothermal discharge. During volcanic eruptions, the magmatic component becomes dominant with the release of CO_2 and other volatiles. This phase is accompanied by the deposition of lapilli. A further model for the evolution of hydrothermal fluids was presented by Butterfield et al. (1990). They suggested (based on observations at the Juan de Fuca Ridge) that during the ascent of a fluid in the host rocks, the pressure decreases, causing the fluid to boil. As a consequence, the fluid separates into a vapor phase of low salinity that is also low in iron due to precipitation of iron sulfides. The residual highly saline brine phase is enriched in metals. Such a phase separation is depth dependent and was observed in a water depth at about 1500 m at Axial Volcano (Juan de Fuca Ridge) but also in 400 m water depth at Grimsey Hydrothermal field, where fluid temperatures were about 250 °C (Hannington et al. 2001).

Low-temperature venting was observed at the Loihi intraplate volcano (temperature ~30–70 °C, Hilton 1998). The chemistry of the emanating fluids was, apart from their high CO_2 and Fe content, comparable to that observed at the Galapagos rift system (Edmond et al. 1979) and the Axial Volcano (Chase et al. 1985). It was suggested that the Loihi fluids were a mixture of high-temperature fluids (>200 °C) with seawater, with a later overprint by low-temperature reactions (Sedwick et al. 1992). At the Pitcairn Seamounts, there were no indications of high-temperature venting during the diving campaign Polynaut 1999. Shimmering water with temperatures of about 14–19 °C exiting iron oxide chimneys on the summit of Bounty (~420 m) was the only indication of recent active venting. Although there is no evidence of high-temperature fluids and related mineralization on the Pitcairn Seamounts, Stoffers et al. (1993) assumed that boiling fluids and subsequent phase separation might have produced sulfide mineralization at greater depths (>~1500 m) within the volcanic pile. Accordingly, the low-temperature venting at Pitcairn may be the result of a hydrothermal circulation cell that develops only in shallow water depths. An alternative but related explanation is that massive basalt flows on the flanks of the volcanoes (Hekinian et al. 2003) inhibited a deep incursion of seawater. Instead, seawater penetrated only the upper portion of the volcano so it was only weakly heated. It could extract iron from the basalts but not the other metals. Glasby et al. (1997) surmised that the intrusion of seawater into the upper part of the volcano leads to a development of an oxycline above which iron oxides are formed, whereas the formation of Mn oxides is believed to be mediated by microbial activity and thus this formation occurs more distally from the actual vent sites.

The understanding of the modes of formation of the Mn crusts at Pitcairn Seamounts is further complicated due to the fact that the growth direction of the crusts investigated in relation to the seabed is not known. Based on the similar morphological appearance of Pitcairn Mn crusts to those sampled by submersible at the Bonin arc (Usui and Nishimura 1992) and Tonga Ridge (Hein et al. 1990), Hodkinson et al. (1994) proposed a downward growth of the crusts. Low-temperature hydrothermal fluids are thought to ascend through volcanic sands and precipitate Mn oxides, causing a consolidation of the sands. These layers then act as a cap for the ascending fluids so that below the sands, dense Mn-layers develop. As pointed out by Glasby et al. (1997), a similar morphological appearance is not a sufficient argument to rigorously test this hypothesis of downward growth. An upward growth was suggested for Mn crusts from the Tyrrhenian Sea, for example, although these crusts have a similar appearance to those from Pitcairn Seamounts.

Chapter 12 · Hydrothermal Iron and Manganese Crusts from the Pitcairn Hotspot Region

Irrespective of the actual growth direction, there are several lines of evidence suggesting that Mn and Fe crusts precipitate from fluids of different compositions. A crust (PN DR 13c; Fig. 12.3b) consisting of iron and manganese layers was collected at Bounty on the Polynaut cruise. The concentrations of Mn and Fe as well as some trace elements (As, Cu, Ni, Zn) in this crust show a distinctive distribution, which was also observed in the separate Fe and Mn crusts. The highest concentrations of Cu, Ni, and Zn are associated with the Mn crusts, whereas As is relatively enriched in the Fe crust (Fig. 12.15). Interestingly, within the Mn-rich layers, there is a pronounced decrease in Cu, Zn and Ni (from subsample 7 to subsamples 5, 6), although the Mn concentrations in these layers remain relatively constant. Such a change could not be attributed to changes in the amount of volcaniclastic material, because this would also have affected the concentrations of Mn and other elements. As the sequence of ages for sublayers in crust PN DR 13c (age of sublayer 8 > 7 > 3; Fig. 12.15) suggests, it seems possible that the observed element distribution in crust PN DR 13c records a temporal change of the hydrothermal activity at Bounty. Sublayers 8 and 7 of crust PN DR 13c were formed when hydrothermal activity was relatively vigorous as indicated by the positive Eu anomaly (Fig. 12.14) and the high concentrations in Cu, Zn and Ni (Fig. 12.15), elements typically enriched in high-temperature hydrothermal fluids (Von Damm et al. 1985). Decrease in the activity resulted in low trace element concentrations in the crusts. The formation of the Fe crusts (sublayers 1 and 2; Fig. 12.15) would thus resemble a phase with (recent) low-temperature activity.

This model implies that the Mn crusts at Pitcairn Seamounts are older than the Fe crusts and that the composition of the hydrothermal fluids has changed over the passage of time. Indeed, all the available age information (Table 12.1) suggests that with the exception of the youngest Mn crust (DS 69 /1 19-24, 2.9 kyr) and the oldest Fe crust (33 DS/4, 3.1 kyr), which were sampled at different seamounts (Adams and Bounty, respectively), the Mn crusts are relatively older than the Fe crusts. The REE pattern of the hydrothermal fluids sampled at Bounty's summit do not display a positive Eu anomaly in contrast to the purest Mn crusts (i.e., those crusts with high Mn and low ΣREE concentrations). Furthermore, the composition of the vent fluids at Bounty's summit indicates a molar Fe/Mn ratio of about 57, which is similar to the average Fe/Mn ratio of the host rocks (Fe/Mn ~ 54) of the Pitcairn Seamounts, but this ratio is up to two orders of magnitude lower than in those hydrothermal systems with recent Mn oxide precipitation such as at the Galapagos Spreading Center (cf. Von Damm et al. 1995). This indicates that the Mn crusts were formed by a hydrothermal fluid that is different from the one recently sampled at Bounty's summit and that the chemical composition of the hydrothermal fluids, rather than the distance from the vent site, controls the formation of Fe and Mn crusts at Pitcairn.

12.6
Conclusions

In conclusion, dating in combination with the geochemical data (e.g., REE pattern) indicates that the formation of Fe and Mn crusts found at the Pitcairn Seamounts is not a result of a chemical fractionation of a single hydrothermal fluid. It rather seems that the Mn crusts precipitated during past periods of relatively intense hydrothermal activity, whereas the Fe crusts formed during (recent) periods of low activity. It

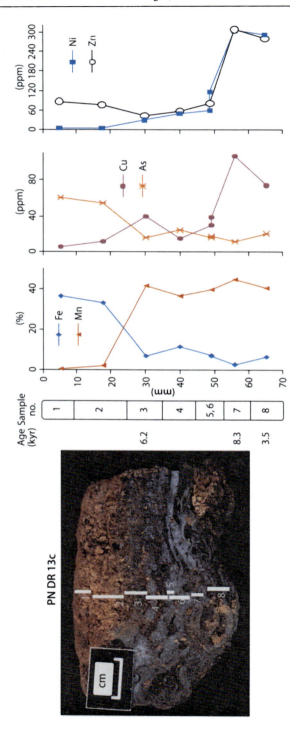

Fig. 12.15. Major and trace element distribution in mixed Fe and Mn crust PN DR 13c. This crust probably records the temporal change of hydrothermal activity that was highest at the beginning (samples 7 and 8) and then decreased, with Fe precipitation being a period of lowest hydrothermal activity

Chapter 12 · Hydrothermal Iron and Manganese Crusts from the Pitcairn Hotspot Region

would be nice to determine to what extent the high level of hydrothermal activity was caused by volcanic eruptions, as the presence of volcanic material in the Mn crust would suggest, but this will only be possible if the ages of both the volcanic rocks and hydrothermal deposits are acquired. Such data are not yet available and remain an ambitious objective for future studies to elucidate the relation between magmatic and hydrothermal processes.

Acknowledgements

We are grateful to the captains and the crews of R.V. SONNE, R.V. ATALANTE and submersible *Nautile*. Special thanks to J.-L. Cheminée † and R. Hekinian for their support. This study would not have been possible without the expertise and continual support of P. Stoffers. This research project was funded by the Natural Sciences and Engineering Research Council (NSERC) of Canada and the Bundesministerium für Forschung und Technologie (BMBF) (grants no. 3R397A 5 and 03G0543A).

References

Alt JC (1995) Subseafloor processes in mid-oceanic ridge hydrothermal systems. Amer Geophys Monograph 91:85–114

Binns RA, Scott SD, Bogdanov YA, Lisitzin AP, Gordeev VV, Gurvich EG, Finlayson EJ, Boyd T, Dotter LE, Wheller GE, Muravyev KG (1993) Hydrothermal oxide and gold-rich sulfate deposits of Franklin Seamount, Western Woodlark Basin, Papua New Guinea. Econ Geol 88:2122–2153

Boström K (1983) Genesis of ferromanganese deposits – diagnostic criteria for recent and old deposits. In: Rona P, Boström K, Laubier L, Smith KL (eds) Hydrothermal processes at seafloor spreading centers. NATO Conference Series 12:473–490

Boström K, Peterson MNA, Joensuu O, Fisher DE (1969) Aluminium-poor ferromanganoan sediments on active oceanic ridges. J Geophys Res 74:3261–3270

Boyd T, Scott SD (1999) Two-XRD-line ferrihydrite and Fe-Si-Mn oxyhydroxide mineralization from Franklin Seamount, Western Woodlark Basin, Papua New Guinea. Can Mineralog 37:973–990

Boyd T, Scott SD (2001) Microbial and hydrothermal aspects of ferric oxyhydroxides and ferrosic hydroxides: the example of Franklin Seamount, Western Woodlark Basin, Papua New Guinea. Geochem Trans (electronic journal) 7, http://www.rsc.org.is/journals/current/geochem/geocon.htm

Boyd T, Scott SD, Hekinian R (1993) Trace element patterns in Fe-Si-Mn oxyhydroxides at three hydrothermally active seafloor regions. Resource Geol Spec issue 17:83–95

Butterfield DA, Massoth GJ, McDuff RE, Lupton J, Lilley M (1990) Geochemistry of hydrothermal fluids from axial seamount hydrothermal emissions study vent field, Juan de Fuca Ridge: Subseafloor boiling and subsequent fluid-rock interaction. J Geophys Res 95:12895–12921

Campbell AC, Palmer MR, Klinkhammer GP, Bowers TS, Edmond JM, Lawrence JR, Casey JF, Thompson G, Humpris S, Rona P, Karson JA (1988) Chemistry of hot springs on the Mid-Atlantic Ridge. Nature 335:514–519

Chase RL, Delaney JR, Karsten JL, Johnson HP, Juniper SK, Lupton JE, Scott D, Tunicliffe V, Hammond SR, McDuff RE (1985) Hydrothermal vents on an axis seamount of the Juan de Fuca Ridge. Nature 313:212–214

Chen JH, Wasserburg GJ, Von Damm KL, Edmond J (1986) The U-Th systematics in hot springs on the East Pacific Rise at 21° N and Guaymas Basin. Geochim Cosmochim Acta 50:2467–2479

Chukrov FV (1973) On the genesis of thermal sedimentary iron ore deposits. Min Deposita 8:138–147

Cronan DS, Glasby GP, Moorby SA, Thomson J, Knedler KE, McDougall JC (1982) A submarine hydrothermal manganese deposit from the south-west Pacific island arc. Nature 298:456–458

De Baar HJW, Bacon MP, Brewer PG, Bruland K (1985) Rare earth elements in the Pacific and Atlantic Oceans. Geochim Cosmoschim Acta 49:1943–1959

DeCarlo EH, McMurtry GM (1992) Rare earth element geochemistry of ferromanganese crusts from the Hawaiian Archipelago, central Pacific Chem Geol 95:235–250

DeCarlo EH, McMurtry GM, Yeh H-W (1983) Geochemistry of hydrothermal deposits from Loihi submarine volcano, Hawaii. Earth Planet Sci Lett 66:438–449

DeCarlo EH, McMurtry GM, Kim KH (1987) Geochemistry of ferromanganese crusts from the Hawaiian Archipelago – I. Northern survey areas. Deep Sea Res 34:441–467

DeLange GJ, Catalano G, Klinkhammer GP, Luther III GW (1990) The interface between oxic seawater and the anoxic Bannock brine; Its sharpness and the consequences for the redox-related cycling of Mn and Ba. Mar Chem 21:205–217

Duncan RC, McDougall I, Carter RM, Coombs DS (1974) Pitcairn Island – another Pacific hotspot? Nature 251:679–682

Edmond JM, Measures C, McDuff RE, Chan LH, Collier R, Grant B, Gordon LI, Corliss JB (1979) Ridge crest hydrothermal activity and the balances of the major and minor elements in the ocean: The Galapagos data. Earth Planet Sci Lett 46:1–18

Emerson D, Moyer CL (2002) Neutrophilic Fe-oxidizing bacteria are abundant at the Loihi Seamount hydrothermal vents and play a major role in Fe oxide deposition. Appl Environ Microbiol 68: 3085–3093

Fleet AJ (1983) Hydrothermal and hydrogenous ferro-mangnese deposits: Do they from a continuum? The rare earth element evidence. In: Rona P, Boström K, Laubier L, Smith KL (eds) Hydrothermal processes at seafloor spreading centers. NATO Conference Series 12:535–555

Fortin D, Ferris FG, Scott SD (1998) Formation of Fe-silicates and Fe-oxides on bacterial surfaces in samples collected near hydrothermal vents on the Southern Explorer Ridge in the northeast Pacific Ocean. Am Mineralogist 83:1399–1408

Gabriel KR (1971) The biplot graphic display of matrices with application to principal component analysis. Biometrika 58(3):453–467

German CR, Klinkhammer GP, Edmond JM, Mitra A, Elderfield H (1990) Hydrothermal scavenging of rare-earth elements in the ocean. Nature 345:516–518

German CR, Campbell AC, Edmond JM (1991) Hydrothermal scavenging at the Mid-Atlantic Ridge: Modification of trace element dissolved fluxes. Earth Planet Sci Lett 107:101–114

Giggenbach WF (1987) Redox processes governing the chemistry of fumarolic gas discharges from White Island. New Zealand Appl Geochem 2:143–161

Glasby GP, Stüben D, Jeschke G, Stoffers P, Garbe-Schönberg C-D (1997) A model for the formation of the hydrothermal manganese crusts from the Pitcairn Island hotspot. Geochim Cosmochim Acta 61:4583–4597

Hannington M, Herzig P, Stoffers P, Scholten J, Botz R, Garbe-Schönberg D, Jonasson IR, Roest W, Scientific Party (2001) First observation of high-temperature submarine hydrothermal vents and massive anhydrite deposits off the north coast of Iceland. Marine Geology 177:199–220

Hein JR, Gibbs AE, Clague DA, Torresan M (1996) Hydrothermal mineralization along submarine rift zones. Hawaii Mar Georesource Geotechnol 14:177–203

Hein JR, Koschinsky A, Bau M, Manheim FT (1999) Cobalt-rich ferromanganese crusts in the Pacific. In: Cronan D (ed) Handbook of mineral deposits. CRC Marine Science Series:239–280

Hekinian R, Hoffert M, Larque P, Cheminee JL, Stoffers P, Bideau D (1993) Hydrothermal Fe and Si oxyhydroxide deposits from the South Pacific intraplate volcanoes and East Pacific Rise axial and off-axial regions. Econ Geol 88:2099–2121

Hekinian R, Cheminée J-L, Dubois J, Stoffers P, Scott S, Guivel C, Garbe-Schönberg D, Devey C, Bourdon B, Lackschewitz K (2003) The Pitcairn hotspot in the South Pacific: Distribution and composition of submarine volcanic sequences. J Volc Geoth Res 121:219–245

Henderson G, Burton K (1999) Using $^{234}U/^{238}U$ to asses diffusion rates of isotope tracers in ferromanganese crusts. Earth Planet Sci Lett 170:169–179

Herzig PM, Hannington MD (1995) Polymetallic massive sulphides at the modern seafloor. A review. Ore Geology Reviews 10:95–115

Hodkinson RA, Stoffers P, Scholten J, Cronan DS, Jeschke G, Roger TDS (1994) Geochemistry of hydrothermal manganese deposits from the Pitcairn Island hotspot, southeastern Pacific. Geochim Cosmochim Acta 58:5011–5029

Jeschke G (1991) Charakterisierung und Vergleich hydrogenetischer und hydrothermaler Manganerzkrusten vom Pitcairn Hot Spot (zentraler Südpazifik) und den Gesellschaftsinseln anhand hochauflösender geochemischer Spurenanalytik. Thesis University Kiel (in German)

Karl DM, McMurtry GM, Malahoff A, Garcia MO (1988) Loihi Seamount, Hawaii: A mid-plate volcano with a distinctive hydrothermal system. Nature 335:532–535

Kaufmann A, Broecker W (1965) Comparison of ^{230}Th and ^{14}C Ages for carbonate materials from Lakes Lahontan and Bonneville. J Geophys Res 70:4039–4055

Kennedy CB, Scott SD, Ferris FG (2003) Ultrastructure and potential sub-seafloor evidence of bacteriogenic iron oxides from Axial Volcano, Juan de Fuca Ridge, north-east Pacific Ocean. FEMS Microbiol Ecol 43:247–254

Koeppenkastrop D, DeCarl E (1993) Uptake of rare earth elements from solution by metal oxides. Environ Sci Technol 27:1796–1802

Koschinsky A, Halbach P (1995) Sequential leaching of marine ferromanganese precipitates: Genetic implications. Geochim Cosmochim Acta 59:5113–5132

Chapter 12 · Hydrothermal Iron and Manganese Crusts from the Pitcairn Hotspot Region

Ku T-L, Knauss KG, Mathieu GG (1977) Uranium in open ocean: Concentration and isotopic composition. Deep-Sea Res 24:1005–1017

Kunzendorf H, Walter P, Stoffers P, Gwozdz R (1984) Metal variation in divergent plate boundary sediments from the Pacific. Chem Geol 47:113–133

Lalou C, Reyss JL, Brichet E (1993) Actinide-series disequilibrium as a tool to etablish the chronology of deep-sea hydrothermal activity. Geochim Cosmochim Acta 57:1221–1231

Latter JH (1987) Volcanoes and volcanic risks in the circum-Pacific. Pac Rim Congr 87:745–752

Malahoff A, McMurtry GM, Wiltshire JC, Yeh H-W (1982) Geology and geochemistry of hydrothermal deposits from active submarine volcano Loihi, Hawaii. Nature 298:234–239

Michard A (1989) Rare earth element systematics in hydrothermal fluids. Geochim Cosmochim Acta 53:745–750

Michard A, Albarede F (1986) The REE content of some hydrothermal fluids. Chemical Geology 55:51–60

Michard A, Michard G, Stüben D, Stoffers P, Cheminee J-L, Binard N (1993) Submarine thermal springs associated with young volcanoes: The Teahitia vents, Society Island, Pacific Ocean. Geochim Cosmochim Acta 57:4977–4986

Moorby BS, Cronan DS (1983) The geochemistry of hydrothermal and pelagic sediments from the Galapagos hydrothermal mound D. S. D. P. Leg 70. Mineral Mag 47:291–300

Puteanus D, Glasby GP, Stoffers P, Kunzendorf H (1991) Hydrothermal iron-rich deposits from the Teahitia-Mehitia and Macdonald hot spot areas, Southwest Pacific. Marine Geol 98:389–409

Reyss JL (1982) Rapid growth of a deep-sea manganese nodule. Nature 295:401–403

Rogers TDS, Hodkinson RA, Cronan DS (2001) Hydrothermal manganese deposits from the Tonga-Kermadec ridge and Lau Basin region, southwest Pacific. Mar Georesource Geotechnol 19:245–268

Rudnicki MD, Elderfield H (1993) A chemical model of the buoyant and neutrally buoyant plume above the TAG vent field, 26 degrees N, Mid-Atlantic Ridge. Geochim Cosmochim Acta 57: 2939–2957

Sedwick PN, McMurtry GM, MacDougall JD (1992) Chemistry of hydrothermal solutions from Pele's Vents, Loihi Seamount, Hawaii. Geochim Cosmochim Acta 56:3643–3667

Segl M, Mangini A, Beer J, Bonani G, Suter M, Wölfli W (1989) Growth rate variations of manganese nodules and crusts induced by paleoceanographic events. Paleoceanography 4:511–530

Shanks WC, Bohlke JK, Seal RR (1995) Stable isotopes in mid-ocean ridge hydrothermal systems; Interactions between fluids, minerals and organisms. In: Seafloor Hydrothermal Systems: Physical, Chemical, Biological, and Geological Interactions. Geophysical Monograph 91: American Geophysical Union, pp 194–221

Stoffers P, Scientific Party (1990) Active Pitcairn hotspot found. Marine Geol 95:51–55

Stoffers P, Glasby GP, Stüben D, Renner RM, Pierre TG, Webb J, Cardile CM (1993) Comparative mineralogy and geochemistry of hydrothermal iron-rich crusts from the Pitcairn, Teahitia-Mehetia, and Macdonald hot spot areas of the S. W. Pacific. Mar Geores Geotechn 11:45–86

Stüben D, Stoffers P, Cheminée J-L, Hartmann M, McMurtry GM, Richnow H-H, Jenisch A, Michaelis W (1992) Manganese, methane, iron, zinc, and nickel anomalies in hydrothermal plumes from Teahitia and Macdonald Volcanoes. Geochim Cosmochim Acta 56:3693–3704

Urabe T, Kuskabe M (1990) Barite silica chimneys from the Sumisu Rift, Izu-Bonin Arc: Possible analog to hematitic chert associated with Kuroko deposits. Earth Planet Sci Lett 100:283–290

Usui A, Nishimura A (1992) Submersible observations of hydrothermal manganese deposits on the Kaikata Seamount, Izu-Ogasawara (Bonin) Arc. Marine Geol 106:230–216

Usui A, Terashima S (1997) Deposition of hydrogenetic and hydrothermal manganese minerals in the Ogasawara (Bonin) Arc Area, Northwest Pacific. Mar Georesource Geotechnol 15:127–154

Varnavas SP, Cronan DS (1991) Hydrothermal metallogenic processes off the islands of Nisiros and Kos in the Hellenic Volcanic Arc. Marine Geol 99:109–133

Von Damm KL, Edmond JM, Grant B, Measures CI (1985) Chemistry of submarine hydrothermal solutions at 21° N, East Pacific Rise. Geochim Cosmochim Acta 49:2197–2210

Appendix

Table 12.A1: Major and trace element concentrations of hydrothermal fluid from Bounty (PN 5-5), and of Fe and Mn crusts sampled during Polynaut cruises at Pitcairn Seamounts (on anhydrous basis); average composition of seawater from Open University (1994). Table 12.A2: REE concentrations of hydrothermal fluids from Bounty (PN 5-5), seawater (from DeBaar et al. 1985), and of Fe and Mn crusts from Polynaut cruise (hydrothermal fluids and seawater in ppt, crusts in ppm).

Table 12.A1. Major and trace element concentrations of hydrothermal fluid from Bounty (PN5-5), and of Fe and Mn crusts sampled during POLYNAUT cruises at Pitcarin seamounts (on anhydrous basis); average composition of seawater from Open University, 1994

	Fe (ppm)	Mn (ppm)	Co (ppm)	Ni (ppm)	Cu (ppm)	Zn (ppm)	As (ppm)	Rb (ppm)	Sr (ppm)	Mo (ppm)	Cs (ppm)	Ba (ppm)	Tl (ppm)	Pb (ppm)	^{238}U (ppm)
Seawater	5.50E-05	1.00E-04	3.00E-06	4.80E-04	1.00E-04	5.00E-04	3.70E-03	0.12	8	0.010	4.00E-04	0.02	1.00E-08	2.00E-06	3.20E-03
Hydrothermal fluid, Bounty	19.58	0.34	0.01	0.08	b.d	b.d.	7.47E-04	0.134	7.95	0.011	4.53E-04	0.022	6.79E-04	b.d.	2.24E-03

Sample	Fe (%)	Mn (%)	Sc (ppm)	Ti (ppm)	Co (ppm)	Ni (ppm)	Cu (ppm)	Zn (ppm)	As (ppm)	Rb (ppm)	Sr (ppm)	Mo (ppm)	Ag (ppm)	Cd (ppm)	Cs (ppm)	Ba (ppm)	Tl (ppm)	Pb (ppm)	^{232}Th (ppm)	^{238}U (ppm)
Fe crusts																				
PN 4-4	39.52	0.45	0.32	20	33.41	31.64	30.36	21.87	380.6	1.33	703.3	22.72	<0.02	0.77	<0.02	62.25	0.98	0.79	0.00	6.70
PN 4-4/1	37.74	0.12	0.47	127	3.91	5.29	5.88	14.48	18.90	1.48	572.2	12.10	<0.02	0.08	<0.02	49.90	0.04	0.13	0.02	1.89
PN 4-4/2	38.16	0.13	0.38	68	2.68	3.05	8.98	6.50	13.15	1.31	604.9	12.31	<0.02	0.08	<0.02	51.87	0.03	0.21	<0.02	1.64
PN 4-4/3	39.29	0.11	0.34	30	2.90	2.95	16.09	6.53	25.33	1.06	594.5	14.00	<0.02	0.23	<0.02	39.81	0.07	0.07	<0.02	2.38
PN 4-4/4	38.29	0.26	0.42	126	12.21	7.94	23.94	6.17	19.15	1.22	587.2	19.88	<0.02	0.12	<0.02	51.77	0.05	0.13	0.02	1.74
PN 4-4/5	37.78	0.36	0.59	65	16.04	10.44	15.83	7.89	16.22	1.06	646.0	17.81	<0.02	0.18	<0.02	56.12	0.09	0.07	<0.02	2.25
PN 4-4/6	40.28	0.43	0.32	56	13.09	12.11	26.91	9.14	37.39	0.94	612.5	19.68	<0.02	0.37	<0.02	51.25	0.13	0.08	<0.02	4.24
PN 4-6	41.36	0.32	0.24	182	76.10	54.84	25.09	52.09	254.2	1.68	719.8	28.87	<0.02	0.47	<0.02	55.14	0.48	0.46	0.03	7.44
PN 4-7	35.41	0.20	0.11	89	55.86	26.77	1.47	24.34	68.72	2.17	423.7	37.47	<0.02	0.05	<0.02	36.42	0.05	0.06	0.02	1.95
PN 4-11	34.41	0.06	0.17	17	16.34	19.50	1.17	25.91	49.93	2.72	540.1	12.46	<0.02	0.02	<0.02	55.81	0.01	0.12	<0.02	0.66
PN 4-11(3)	33.25	0.03	0.35	13	8.00	21.04	2.01	31.52	79.88	2.64	526.6	15.28	<0.02	0.03	<0.02	62.61	0.01	0.15	<0.02	0.79
PN 5-1	28.85	1.45	0.23	135	231.2	64.91	288.4	36.54	105.7	2.94	796.0	42.78	<0.02	0.24	<0.02	106.0	0.17	0.29	0.03	0.33
PN 5-3	33.18	0.31	0.44	26	20.12	18.94	29.34	15.72	148.3	1.75	636.7	19.28	<0.02	0.49	<0.02	44.01	0.40	0.14	<0.02	4.19
PN 5-3/1	40.49	0.20	0.19	33	7.79	7.15	27.22	9.47	202.0	0.91	608.6	17.55	<0.02	0.35	<0.02	47.87	0.26	0.13	<0.02	3.92
PN 5-3/2	42.26	0.15	0.24	41	4.79	4.43	21.95	7.57	100.2	1.11	643.1	16.51	<0.02	0.30	<0.02	45.00	0.13	0.07	<0.02	3.42
PN 5-3/3	40.48	0.10	0.30	40	1.95	2.12	9.69	5.17	26.66	1.13	620.1	14.61	<0.02	0.24	<0.02	40.50	0.07	0.09	<0.02	2.55
PN 5-3/4	40.60	0.21	0.53	34	11.83	11.32	88.44	30.99	180.0	0.81	659.0	16.86	<0.02	0.42	<0.02	52.02	0.16	0.21	<0.02	4.70
PN 5-3/5	39.17	0.15	0.52	82	3.64	3.25	9.78	9.31	20.03	1.21	629.5	14.19	<0.02	0.21	<0.02	43.20	0.05	0.40	<0.02	2.47
PN 5-3/6	35.11	0.17	0.93	82	10.77	8.35	97.19	29.56	44.95	1.42	612.7	10.54	<0.02	0.09	<0.02	52.36	0.05	0.30	<0.02	2.15
PN 5-4	39.39	0.26	0.01	35	68.98	42.40	4.56	15.66	217.1	1.40	730.4	33.33	<0.02	0.78	<0.02	51.29	0.36	0.13	<0.02	7.43
PN 6-5 Steve	47.42	0.24	–	–	–	–	–	–	–	–	–	–	–	–	–	–	–	–	–	–
PN 8-9(rot)	34.82	0.48	0.42	277	0.99	6.20	17.30	77.34	88.78	2.89	1016	1.55	<0.02	0.07	<0.02	158.4	0.02	0.39	0.07	0.79
PN 8-9a	28.80	4.14	0.23	251	10.23	17.28	3.30	18.05	78.50	2.20	977.4	27.93	<0.02	0.30	<0.02	214.4	1.04	0.26	0.04	0.97
PN 8-9b	17.90	0.10	12.97	3 187	15.50	29.69	29.13	123.0	24.76	27.13	638.4	1.63	0.11	0.15	0.54	194.1	0.05	2.28	2.25	0.71
PN 12-06	41.80	0.02	<0.02	12	8.19	11.43	1.86	49.66	79.02	1.46	140.3	236.9	<0.02	0.20	<0.02	16.57	0.01	0.26	<0.02	10.41

Table 12.A1. Continued

Sample	Fe (%)	Mn (%)	Sc (ppm)	Ti (ppm)	Co (ppm)	Ni (ppm)	Cu (ppm)	Zn (ppm)	As (ppm)	Rb (ppm)	Sr (ppm)	Mo (ppm)	Ag (ppm)	Cd (ppm)	Cs (ppm)	Ba (ppm)	Tl (ppm)	Pb (ppm)	^{232}Th (ppm)	^{238}U (ppm)
Fe crusts																				
PN 12-6	45.79	0.04	<0.02	10	10.79	12.86	1.55	62.91	80.06	1.55	127.1	240.3	<0.02	0.20	<0.02	18.00	0.01	0.19	<0.02	13.61
PN 14-17	33.65	0.59	0.07	30	20.80	22.06	1.88	21.67	72.82	3.44	626.6	17.54	<0.02	0.61	0.03	100.2	0.78	0.15	0.05	1.62
PN DR 13a/9	20.95	17.78	2.28	2556	11.78	55.72	25.75	81.59	44.02	5.25	644.7	143.2	0.07	0.38	0.05	638.2	1.17	2.98	0.69	2.26
PN DR 13c/1	36.33	0.48	0.31	226	0.84	5.57	4.98	85.67	60.20	2.72	704.4	7.24	<0.02	0.04	0.05	170.9	0.15	1.02	0.06	0.57
PN DR 13c/2	33.31	1.95	0.86	1014	1.45	5.87	11.82	76.58	54.27	4.35	649.5	21.55	<0.02	0.02	0.06	188.7	0.09	0.83	0.27	0.62
Mn crusts																				
PN 1-04-1	0.25	47.65	0.45	374	51.99	219.6	98.86	112.8	10.59	6.22	542.1	102.9	<0.02	3.19	<0.02	539.4	17.89	1.67	0.06	1.76
PN-1-04-2	7.96	9.44	15.71	975	45.67	270.7	89.90	97.40	15.32	17.04	658.3	46.60	0.16	1.75	0.19	403.9	0.47	2.26	2.41	11.05
PN-1-04-3	4.79	16.21	11.78	1186	53.63	265.2	193.9	69.17	7.46	10.62	705.7	69.25	0.12	2.64	0.10	404.7	0.37	2.33	1.73	1.22
PN-1-04-4	4.64	15.86	11.88	1769	56.88	354.0	70.45	158.5	6.95	10.57	682.7	75.55	<0.02	8.83	0.10	438.6	8.60	2.52	1.71	0.75
PN 8-9 DKL	18.24	20.05	0.73	605	12.60	129.8	34.30	180.8	42.81	3.16	1171	56.53	<0.02	0.33	<0.02	1183	0.57	0.43	0.11	3.33
PN DR 13a/1	6.24	12.51	10.22	1419	26.39	163.4	48.25	209.6	9.86	18.65	577.7	168.1	0.18	4.26	0.18	442.9	9.18	4.98	3.47	1.45
PN DR 13a/2	5.55	17.03	8.94	2125	21.82	165.5	34.19	124.3	11.11	17.30	554.5	194.6	0.16	4.65	0.16	485.5	2.36	3.72	3.20	1.64
PN DR 13a/3	5.50	19.82	8.46	2570	23.24	249.8	32.38	97.07	9.25	14.96	522.6	312.5	0.15	2.37	0.14	470.0	0.94	3.33	2.66	1.42
PN DR 13a/4	6.35	13.57	10.73	1703	28.07	178.8	32.08	107.9	6.86	18.29	564.5	272.1	0.17	0.62	0.18	437.2	0.67	3.59	3.21	1.11
PN DR 13a/7	2.11	46.08	0.88	765	2.11	68.76	28.25	131.4	11.63	3.13	353.9	989.1	<0.02	1.98	<0.02	542.3	4.45	0.54	0.18	2.46
PN DR 13a/10	1.53	49.45	0.75	806	3.45	23.38	14.73	30.44	9.88	1.84	406.8	506.0	0.05	0.30	0.07	852.1	0.04	0.33	0.21	5.61
PN DR 13a/11	3.88	36.19	3.42	3186	10.96	86.84	21.41	65.70	10.74	6.46	447.0	397.9	0.05	0.52	0.05	726.6	0.52	2.23	0.95	3.12
PN DR 13a/12	0.71	38.58	0.17	72	7.24	11.49	8.76	16.31	7.94	1.01	367.2	505.1	<0.02	0.29	<0.02	488.8	0.05	0.14	0.02	5.90
PN DR 13a/13	1.05	52.63	0.41	134	2.54	32.11	20.29	46.27	8.90	1.77	438.2	339.2	<0.02	0.29	<0.02	996.1	0.06	0.37	0.02	4.84
PN DR 13a/14	6.19	21.62	8.55	1781	19.70	112.1	29.75	179.5	8.61	15.85	508.2	253.9	0.13	2.19	0.26	414.9	3.20	2.43	2.63	1.52
PN DR 13a/15	5.99	13.63	10.57	1967	26.42	180.5	70.90	311.2	11.28	19.86	603.7	212.2	0.19	4.28	0.20	463.6	15.79	4.09	3.71	1.43
PN DR 13b/2	1.69	44.22	0.66	1343	8.51	399.3	188.0	855.8	22.86	4.14	453.2	1109	0.04	13.73	0.01	524.0	58.45	1.33	0.26	2.06
PN DR 13b/3	0.31	45.20	0.21	49	0.46	6.09	6.86	14.42	9.05	1.36	402.4	448.0	<0.02	0.50	<0.02	329.0	0.26	0.09	0.00	4.33
PN DR 13b/4	1.02	48.50	0.27	123	2.33	6.47	9.03	12.81	8.96	1.10	358.3	415.2	<0.02	0.37	<0.02	431.3	0.42	0.18	0.03	5.25
PN DR 13c/3	6.80	41.59	<0.02	23	8.31	30.59	39.64	41.89	16.45	1.49	523.1	599.0	0.03	0.40	<0.02	1267	0.12	0.12	<0.02	3.63
PN DR 13c/4	11.69	36.32	0.02	69	3.12	48.67	14.96	58.02	24.68	1.37	560.5	488.6	<0.02	0.51	<0.02	1028	0.18	0.65	0.01	4.32
PN DR 13c/5	7.06	39.49	0.14	215	1.16	58.34	29.80	81.72	16.49	1.87	534.6	612.2	0.08	0.49	<0.02	1135	0.17	0.51	0.06	5.60
PN DR 13c/6	6.87	39.84	0.20	299	7.22	117.6	38.80	82.56	17.73	1.86	583.6	478.5	<0.02	0.92	<0.02	1408	0.16	0.82	0.06	7.39
PN DR 13c/7	2.59	44.86	0.24	317	9.44	305.4	106.1	306.4	11.50	2.96	717.5	388.8	<0.02	0.85	<0.02	1918	1.00	0.43	0.07	7.93
PN DR 13c/8	6.19	40.61	0.26	192	10.56	293.6	74.07	280.4	20.91	3.43	724.4	439.6	<0.02	0.85	<0.02	1844	0.71	0.32	0.04	9.79

Table 12.A2. REE concentrations of hydrothermal fluids from Bounty (PN5-5), seawater (from DeBaar et al. 1985), and of Fe and Mn-crusts from POLYNAUT cruise; hydrothermal fluids and seawater in ppt, crusts in ppm

Sample	La	Ce	Pr	Nd	Sm	Eu	Gd	Tb	Dy	Ho	Er	Tm	Yb	Lu
Seawater	4.8	1.0	1.1	4.9	0.9	0.2	1.3	0.2	–	0.5	–	0.3	2.2	0.4
Hydrothermal fluids, Bounty	214	277	41	130	32	7.0	35	5.0	28	6.0	17	3.0	20	3.0
Fe crusts														
PN 4-4	0.59	0.15	0.06	0.31	0.06	0.02	0.11	0.02	0.16	0.05	0.18	0.03	0.19	0.04
PN 4-4/1	1.07	1.01	0.11	0.52	0.08	0.02	0.11	0.01	0.11	0.03	0.08	0.01	0.08	0.02
PN 4-4/2	0.87	0.71	0.08	0.35	0.04	0.01	0.08	0.01	0.08	0.02	0.07	0.01	0.07	0.01
PN 4-4/3	0.62	0.62	0.07	0.33	0.04	0.01	0.09	0.01	0.08	0.02	0.06	0.01	0.06	0.01
PN 4-4/4	0.80	1.11	0.13	0.58	0.10	0.03	0.14	0.02	0.12	0.03	0.08	0.01	0.08	0.01
PN 4-4/5	0.95	0.99	0.12	0.58	0.09	0.03	0.16	0.02	0.15	0.04	0.10	0.01	0.11	0.02
PN 4-4/6	0.57	0.52	0.06	0.30	0.03	0.01	0.07	0.01	0.07	0.02	0.06	0.01	0.06	0.01
PN 4-6	0.65	0.68	0.11	0.48	0.10	0.03	0.12	0.02	0.12	0.03	0.11	0.02	0.10	0.02
PN 4-7	0.78	0.81	0.10	0.42	0.08	0.03	0.11	0.02	0.11	0.03	0.09	0.01	0.08	0.02
PN 4-11	0.31	0.26	0.03	0.15	0.03	0.01	0.04	0.01	0.05	0.01	0.05	0.01	0.06	0.01
PN 4-11(3)	0.58	0.60	0.08	0.33	0.06	0.02	0.10	0.02	0.11	0.03	0.10	0.02	0.11	0.02
PN 5-1	2.31	2.83	0.34	1.44	0.25	0.08	0.29	0.04	0.24	0.05	0.16	0.02	0.14	0.02
PN 5-3	0.59	0.39	0.06	0.28	0.05	0.02	0.09	0.01	0.09	0.03	0.09	0.01	0.09	0.02
PN 5-3/1	0.37	0.17	0.03	0.15	0.01	0.00	0.04	0.01	0.06	0.02	0.07	0.01	0.08	0.01
PN 5-3/2	0.46	0.32	0.04	0.19	0.01	0.01	0.04	0.01	0.05	0.02	0.05	0.01	0.06	0.01
PN 5-3/3	0.59	0.63	0.07	0.33	0.04	0.01	0.08	0.01	0.08	0.02	0.07	0.01	0.06	0.01
PN 5-3/4	0.88	0.56	0.07	0.35	0.04	0.02	0.10	0.01	0.11	0.03	0.09	0.01	0.10	0.02
PN 5-3/5	0.97	1.22	0.14	0.65	0.11	0.04	0.17	0.02	0.15	0.04	0.11	0.01	0.11	0.02
PN 5-3/6	1.59	1.32	0.15	0.70	0.10	0.03	0.19	0.02	0.18	0.05	0.14	0.02	0.14	0.03
PN 5-4	0.76	0.36	0.05	0.24	0.04	0.01	0.08	0.01	0.09	0.03	0.10	0.01	0.09	0.02
PN 8-9(rot)	6.99	6.99	1.17	5.29	0.91	0.27	1.07	0.15	0.98	0.24	0.75	0.11	0.71	0.12
PN 8-9a	7.21	8.72	1.09	5.02	0.75	0.23	0.80	0.10	0.65	0.16	0.50	0.07	0.43	0.08
PN 8-9b	26.26	54.25	6.86	29.25	6.15	1.88	5.62	0.80	4.28	0.79	2.06	0.27	1.65	0.23
PN 12-06	0.53	0.29	0.03	0.16	0.02	0.01	0.05	0.01	0.04	0.01	0.05	0.01	0.04	0.01
PN 12-6	0.40	0.18	0.02	0.10	0.01	0.01	0.03	0.00	0.04	0.01	0.04	0.01	0.03	0.01
PN DR 13c/1	10.18	9.85	1.39	6.16	1.07	0.27	1.67	0.32	3.26	1.19	4.59	0.79	5.11	0.90

Chapter 12 · Hydrothermal Iron and Manganese Crusts from the Pitcairn Hotspot Region

Table 12.A2. *Continued*

Sample	La	Ce	Pr	Nd	Sm	Eu	Gd	Tb	Dy	Ho	Er	Tm	Yb	Lu
Fe crusts														
PN DR 13c/2	6.56	6.71	1.17	5.05	1.00	0.29	1.22	0.20	1.60	0.52	1.96	0.34	2.31	0.44
PN 14-17	1.63	2.66	0.33	1.31	0.31	0.06	0.45	0.10	0.93	0.28	1.08	0.18	1.22	0.18
Mn-crusts														
PN 1-04-1	5.74	3.45	1.13	5.45	1.06	0.36	1.21	0.16	0.97	0.21	0.53	0.07	0.39	0.05
PN-1-04-2	24.59	57.37	7.46	32.18	7.27	2.32	6.87	0.99	5.40	0.97	2.37	0.34	2.07	0.28
PN-1-04-3	18.36	41.31	5.53	23.90	5.39	1.71	5.15	0.75	4.08	0.75	1.81	0.26	1.60	0.22
PN-1-04-4	17.96	39.66	5.30	23.10	5.20	1.63	5.00	0.73	4.05	0.74	1.83	0.26	1.66	0.23
PN 8-9 DKL	3.16	3.85	0.61	2.68	0.53	0.23	0.58	0.09	0.57	0.14	0.44	0.07	0.44	0.07
PN DR 13a/1	29.70	64.29	8.18	34.30	7.39	2.22	6.56	0.91	4.73	0.83	1.94	0.27	1.62	0.22
PN DR 13a/2	26.73	58.41	7.36	30.75	6.53	1.96	5.81	0.81	4.25	0.74	1.75	0.24	1.49	0.20
PN DR 13a/3	22.66	49.96	6.34	26.52	5.69	1.75	5.12	0.71	3.72	0.65	1.54	0.21	1.31	0.18
PN DR 13a/4	27.46	61.49	7.74	32.40	7.01	2.15	6.35	0.87	4.51	0.78	1.82	0.25	1.52	0.20
PN DR 13a/5	3.72	4.10	0.80	3.36	0.67	0.22	0.71	0.11	0.72	0.17	0.49	0.08	0.50	0.08
PN DR 13a/7	2.70	4.72	0.65	2.66	0.53	0.20	0.52	0.08	0.47	0.10	0.30	0.05	0.32	0.05
PN DR 13a/9	7.37	14.03	1.85	7.75	1.63	0.53	1.53	0.23	1.38	0.31	0.95	0.15	0.96	0.17
PN DR 13a/10	8.83	18.16	2.37	9.92	2.12	0.68	2.03	0.30	1.69	0.34	0.88	0.13	0.83	0.12
PN DR 13a/11	0.75	0.56	0.14	0.63	0.11	0.05	0.14	0.02	0.16	0.04	0.13	0.02	0.15	0.02
PN DR 13a/12	1.39	0.89	0.25	1.09	0.20	0.10	0.27	0.04	0.33	0.09	0.28	0.04	0.30	0.05
PN DR 13a/13	21.93	49.29	6.19	26.26	5.86	1.80	5.41	0.79	4.42	0.83	2.14	0.32	2.02	0.27
PN DR 13a/14	31.31	68.09	8.60	35.99	7.60	2.34	6.97	0.96	5.00	0.88	2.09	0.29	1.78	0.24
PN DR 13b/2	5.50	8.71	1.15	4.95	0.98	0.30	1.11	0.17	1.12	0.26	0.74	0.11	0.78	0.12
PN DR 13b/3	0.53	0.34	0.10	0.43	0.07	0.04	0.09	0.02	0.12	0.03	0.11	0.02	0.12	0.02
PN DR 13b/4	0.94	1.04	0.19	0.81	0.14	0.06	0.17	0.03	0.19	0.05	0.15	0.02	0.16	0.02
PN DR 13c/3	2.01	1.19	0.21	0.94	0.14	0.09	0.26	0.05	0.51	0.20	0.81	0.14	0.93	0.17
PN DR 13b/4	1.66	1.36	0.23	1.01	0.17	0.09	0.25	0.05	0.47	0.18	0.73	0.13	0.87	0.17
PN DR 13c/5	1.21	1.43	0.24	1.03	0.20	0.10	0.22	0.04	0.28	0.08	0.29	0.05	0.35	0.07
PN DR 13c/6	2.16	1.89	0.42	1.82	0.36	0.16	0.43	0.07	0.49	0.13	0.46	0.08	0.52	0.10
PN DR 13c/7	2.90	1.99	0.53	2.40	0.47	0.25	0.60	0.10	0.67	0.18	0.54	0.09	0.61	0.10
PN DR 13c/8	2.78	1.27	0.45	2.06	0.41	0.22	0.55	0.09	0.71	0.20	0.68	0.12	0.80	0.14

Chapter 13

Methane Venting into the Water Column Above the Pitcairn and the Society – Austral Seamounts, South Pacific

O. Thießen · M. Schmidt · R. Botz · M. Schmitt · P. Stoffers

13.1
Introduction

In the past, marine hydrothermal systems were studied by numerous work groups that focused on different research aspects. For instance, mechanisms of fluid-associated particle transport and diverse hydrothermal mineralizations have been studied by von Damm and Bischoff (1987), Stüben et al. (1992), Halbach et al. (1993), Rona and Scott (1993), Butterfield et al. (1994), Hannington et al. (1995), Glasby et al. (1997), Scholten et al. (see Sect. 12.2) and others. Moreover, biologists found that hydrothermal vent systems can host chemosynthetic organisms (Tunnicliffe 1991; Jannasch 1995; Nelson and Fischer 1995; Dando et al. 1995). Hydrocarbons observed in hydrothermal systems were ascribed to abiogenic and biogenic formation processes. In particular, numerous hydrothermal vents on deep-seated sea floor show CH_4 of abiogenic origin (Welhan 1988; Charlou et al. 1996, 2002). Hydrocarbon formation could occur via diverse processes like water-/rock reactions (including serpentinization processes) investigated by Seyfried and Dibble (1980), Seyfried and Janecky (1985), Alt (1995), and Charlou et al. (2002). Moreover, hydrothermal trace gases can also be introduced by mantle emanations (Craig and Lupton 1981; Welhan 1988) or formed in the Earth's crust by thermocatalytic (Simoneit 1983; Michaelis et al. 1990) or abiogenic reactions (Apps 1985; Sherwood Lollar et al. 1993). On the other hand, biogenic methane formation (and/or hydrocarbon oxidation) may be caused by microbial activities within the vents and at or near the sediment surface at temperatures below 113 °C (Huber et al. 1990; Burggraf et al. 1990). These biogenic gases of relatively shallow origin may be superimposed on hydrothermal trace gas components formed by abiogenic reactions in the Earth's crust.

The active hydrothermal systems in the Pacific Ocean are related to various tectonic settings such as the East Pacific Rise (Hekinian et al. 1984; Charlou et al. 1991, 1996; Urabe et al. 1995; Ishibashi et al. 1997), the Galapagos spreading center (Corliss et al. 1979), marginal basins (Lonsdale et al. 1980; Both et al. 1986; Fouquet et al. 1991; Ishibashi et al. 1994, 1995) and submarine intraplate hotspot related volcanoes such as Loihi (Malahoff et al. 1982; Gamo et al. 1987; Sakai et al. 1987; Karl et al. 1988; Sedwick et al. 1992), Teahitia (Hoffert et al. 1987; Stüben et al. 1992; Michard et al. 1993), Macdonald (Cheminée et al. 1991; Stüben et al. 1992), and Bounty Seamounts (Stoffers and Hekinian 1990; see Sect. 12.2).

Two research cruises Midplate II (RV SONNE in 1989) and Polynaut (NO ATALANTE and the submersible *Nautile* in 1999) focused on investigations of recent volcanic activity and associated hydrothermalism at submarine volcanic centers in the Society, the Austral Chain and near Pitcairn Island, respectively (10 to 40° S and 110 to 160° W) (see Sect. 5.1 and Fig. 5.1). It is well known that submarine hydrothermal activity is responsible for a significant input of elements into the marine environment. In particular, hydrothermal tracer substances such as Mn, CH_4 and ^3He can be detected in ocean water even far away from active hydrothermal vent sources (Ishibashi et al. 1994, 1997; Stüben et al. 1992; Charlou et al. 1991). The distribution of methane in the water column that derived from hydrothermal vent systems of the Teahitia-, Bounty- and Macdonald Seamounts were studied during the above-mentioned cruises (Table 13.1). Also, we sampled the exsolved gases for later land-based stable isotope investigations of the gas compounds that were sampled. This technique has proved to be of great value when studying the mode of formation of gaseous hydrocarbons on Earth (Schoell 1980; Welhan 1988; Whiticar 1990). Subsequent hydrocarbon oxidation can be recognized due to significant kinetic enrichments of heavy isotopes in the residual hydrocarbon fraction, whereas the CO_2 becomes enriched in ^{12}C (Whiticar and Faber 1986).

Table 13.1. Sampling stations of various hotspot seamounts

Area	Seamount latitude/longitude	CTD station	Dive station
Austral	Macdonald Smt. 28°59' S/140°15' W Summit depth ~40 m	So65 94MS So65 96MS So65 98MS So65 99MS So65 100MS So65 101MS So65 103MS So65 104MS So65 105MS	So65 107OFOS – – – – – – – –
Pitcairn	Adams Smt. 25°23' S/129°16' W Summit depth ~60 m	So65 36MS So65 38MS –	– – –
	Bounty Smt. 25°11' S/129°24' W Summit depth ~430 m	So65 32MS So65 41MS So 65 42MS So65 53MS So65 54MS So65 77MS Pol MS02 Pol MS03 Pol MS04 Pol MS05 Pol MS09	– – – – – – Nautile PN04 Nautile PN05 Nautile PN06 Nautile PN11 Nautile PN12
	Christian Smt. 25°35' S/129°31' W Summit depth ~2000 m	Pol MS06 – –	– – –
Society	Teahitia Smt. 17°34' S/148°49' W Summit depth ~1400 m	So65 118MS Pol MS13 Pol MS14 Pol MS15	– Nautile PN18 Nautile PN19 –

Nautile = manned submersible, *PN* = Polynaut, Nautile dive and station #, *So* = RV SONNE cruise # and sample #, *OFOS* = deep towed camera station and water sampling, *Pol* = Polynaut cruise 1999.

13.2
Geological Setting

Details on the structure, morphology and volcanic activities of the Society, Austral and Pitcairn hotspots are found in Crough (1983) and Hekinian et al. (1991) (see also Sects. 1.1, 1.3–1.4 and 5.2–5.4). Among the numerous (>25) volcanoes situated on the Society hotspot, the three major edifices are the Teahitia Seamount (rising up to 1400 m below sea level), Mehetia Island (450 m above sea level), and the Moua Pihaa Seamount (rising up to 160 m below sea level) (see Sects. 5.2 and 5.2.3). Hydrothermal activity was previously observed near the summit of Teahitia Seamount (Cheminée et al. 1989; Stüben et al. 1992; Michard et al. 1993).

The Pitcairn hotspot area is characterized by many (>90) small (<500 m high) volcanoes as well as two major submarine edifices (e.g., Adams and Bounty Seamounts) (see Sect. 5.4). Adams Seamount rises from 3500 m up to 59 m water depth. Its summit is covered with coral and coral sand. Underwater observations and methane concentration measurements in the water column did not reveal any indications of recent hydrothermal activity (Polynaut Cruise Report 2000). The Bounty Seamount, however, which rises to approximately 430 m water depth, appears to be younger than Adams Seamount and its summit is covered with hydrothermal sediments (Stoffers and Hekinian 1990; Polynaut Cruise Report 2000).

The Austral hotspot is characterized essentially by the presently active Macdonald and the more ancient Rà Seamounts. The approximately 3000 m high Rà Seamount, which rises up to 1040 m water depth, is located about forty nautical miles NNW of Macdonald but did not show any indication of recent volcanic and hydrothermal activities (Stoffers et al. 1987). However, seismic recordings revealed that Macdonald Seamount, which rises from 4250 m to 40 m water depth, has been active through the 1980s (Talandier 1989; see Sect. 2.4). A phreatomagmatic eruption occurred in January 1989 and hydrothermal activity was observed (Stoffers et al. 1989; Stoffers and Hekinian 1990; Cheminée et al. 1991; Stüben et al. 1992).

13.3
Methods

Water sampling was performed at individual water depths using a rosette water sampler CTD device (SeaSun Technology, Germany and Meerestechnik ME, Germany during the Polynaut cruise and during the research cruise with RV SONNE, respectively) equipped with twelve Niskin bottles of 5 and 10 l volume and for comparison purposes, a gas-tight water sampler was also used (Schmitt 1989). Hydrographic parameters (temperature, pressure, conductivity, sound velocity, oxygen content, and light transmission) were continuously recorded while lowering the CTD system (Stoffers and Hekinian 1990; Polynaut Cruise Report 2000). We decided to report the temperature and pH values together with the dissolved CH_4 concentrations, as these parameters reliably indicate hydrothermal activity in the area; however, salinity/density/conductivity anomalies were not observed during the Polynaut cruise.

During the SONNE 65 cruise, emanating hydrothermal fluids (shimmering water) were directly sampled using a deep-towed camera (OFOS = Ocean Floor Observation System), which was equipped with a Seabird CTD, in order to locate and directly sample sea-floor sites where active venting occurred. Water sampling was done by controlled release of 5 l

Niskin bottles, which were connected to the OFOS. At >500 m water depths, positioning of the OFOS was controlled with an AMF-sealink transponder system relative to the ship's position. At <500 m water depths, the ship's positioning system was used directly (since in this case the deviations of the OFOS system were smaller than the accuracy of the transponder network). During the R.V. L'ATALANTE 1999 cruise, hydrothermal vent fluids were also directly sampled by the submersible *Nautile* using a 1.8 l Niskin bottle operated by the pilot. Temperature was measured "in situ", placing the *Nautile* temperature probe directly in the venting fluids near/in the sediment surface.

On board the R.V. SONNE and L'ATALANTE, respectively, the water samples were immediately transferred into 1 l glass bottles (stored at 4 °C for a maximum time of four hours) followed by subsequent degassing in an ultrasonic-vacuum system (Schmitt et al. 1991). Using this technique, the detection limit of hydrocarbons in seawater is <5 nl l^{-1} with a precision of ±13% in the CH_4 concentration range 5–50 nl l^{-1}. Analyses of hydrocarbons were performed using a DANI-Educational gas chromatograph equipped with a 30 m long, 0.32 mm diameter Al_2O_3/KCl column, a flame ionization detector (FID) and a SHIMADZU C-R6A integrator; the carrier gas was nitrogen. Based on daily replicate standard analyses, analytical precision was within ±5%. Gas samples with relatively high hydrocarbon concentrations were stored in evacuated glass vessels for later stable isotope analyses. Preparation of the gas samples for isotope measurements included gas chromatography (DANI-Educational GC, specifications see above), online transfer (He-carrier gas) of individual hydrocarbons into a combustion line (copper oxide, $T = 900$ °C), and stable isotope analyses of CO_2 using a Finnigan MAT 251 isotope ratio mass spectrometer (Faber et al. 1998). Three CH_4 samples with sufficient amounts (8–31µl l^{-1}) were analyzed for their hydrogen isotope composition according to the techniques described by Dumke et al. (1989). The precision of analyses, including conversion of CH_4, was ±0.5‰ ($\delta^{13}C$) and 4‰ (δD). Isotope values for international reference standards obtained in the GCA laboratory were –29.81‰ PDB (NBS-22) and –28.15‰ PDB (NBS-21). The isotope results are expressed in the common δ-notation relative to the PDB ($\delta^{13}C$) and SMOW (δD) standards.

13.4
Results and Discussion

13.4.1
Water Column Characteristics and Methane Distribution

The methane content of ocean water is determined by diverse CH_4 sources such as biological activity in sediments and the water column, thermal hydrocarbon production in organic-rich sediments, abiogenic reactions in the deep subsurface (e.g., lower crust or mantle), and subsequent exchange with the atmosphere. The major sink in the ocean water column is bacterial oxidation (Whiticar and Faber 1986), but depending on the location, the atmosphere may act as a source or sink for dissolved methane (e.g., Faber et al. 1994; Bange et al. 1996; Rehder et al. 1998). Methane concentration in sea surface water that is in equilibrium with atmospheric methane is controlled by temperature and salinity (Wiesenburg and Guinasso 1979). The sea surface water in the area under investigation has temperatures between 20.7 and 26.3 °C at 36‰ salinity. Hence, the CH_4 concentrations in South Pacific Ocean surface waters at equilibrium conditions with

Chapter 13 · Methane Venting into the Water Column Above Pitcairn and Society Seamounts

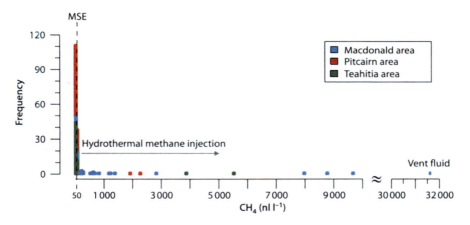

Fig. 13.1. Methane concentrations detected on the Bounty, Macdonald, and Teahitia Seamounts are summarized by their frequency (50 nl l^{-1} interval)

the atmosphere range between 43 and 48 nl l^{-1}. We refer to these equilibrium CH$_4$ concentrations in surface seawater as MSE = "Methane Sea Surface Equilibrium Concentration". Methane concentrations measured on the Bounty, Macdonald, and Teahitia Seamounts from the South Pacific are summarized in Table 13.2 and shown in the histogram of Fig. 13.1. The methane concentrations measured in the water columns and venting fluids exceed the MSE of about 50 nl l^{-1} by several orders of magnitude at all three seamounts (Fig. 13.1). The various methane concentration profiles are discussed in detail for Pitcairn, Macdonald and the Teahitia area.

13.4.1.1
CH$_4$ Concentration Profiles

In the Pitcairn area, most of the water profiles (Figs. 13.2–13.4, Table 13.2) show methane concentrations in the range of about ±20–30 nl l^{-1} off the background concentration of 48 nl l^{-1}. However, weak methane anomalies are recognized near the summit of the Bounty Seamount at a water depth of 300 to 500 m (MS02, MS05, MS09 in Figs. 13.2 and 13.4; 32MS, 42MS in Fig. 13.4) and near the summit of the SSW-located Christian Seamount (MS06 in Fig. 13.2). Methane anomalies near Bounty's summit are associated with weak hydrothermal activity, where yellow to reddish iron oxide sediments were discovered during seafloor observations with the *Nautile* (Polynaut Cruise Report 2000). Shimmering water also occurred at small holes, very small (<10 cm high) and soft chimneys, or veins. Temperatures of 14.4 to 19.1 °C measured directly in the venting fluids were high compared to ambient seawater with a temperature of about 10 °C. However, corresponding temperature or pH anomalies were not detected in the hydrocast stations (Figs. 13.2–13.4) probably caused by efficient mixing of venting fluids with ambient seawater.

About six months after the eruptive phase of Macdonald Seamount in 1989 (Stoffers et al. 1989; Cheminée et al. 1991; see Sect. 2.4), a number of CTD stations showed large methane anomalies near Macdonald's summit (99MS; 100MS; 101MS; 103MS; 104MS in Fig. 13.5, Table 13.5). The strongest methane enrichments (up to 9 691 nl l^{-1}) were measured at 191 m depth on CTD station 104MS and correspond to a pH 6.3 anomaly (Fig. 13.5, Table 13.2).

412 O. Thießen · M. Schmidt · R. Botz · M. Schmitt · P. Stoffers

Table 13.2. Measured methane concentrations, pH and temperature of seawater and hydrothermal fluid at the Macdonald, Teahitia, Adams, Bounty, and Christian Seamounts

Station	Latitude/longitude	Depth (m)	CH$_4$ conc. (nl l^{-1})	pH	Temperature (°C)
Macdonald					
So65 94MS	28°58.90′ S/140°14.96′ W	7	31	8.24	–
		24	39	8.24	–
		39	27	8.24	–
		55	27	8.25	–
		73	29	8.25	–
		94	36	8.21	–
		114	47	8.21	–
		124	65	8.18	–
		134	57	8.19	–
		145	80	8.18	–
		149	71	8.16	–
		154	83	8.16	–
So65 96MS	28°58.97′ S/140°14.87′ W	19	30	8.26	–
		39	42	8.27	–
		60	36	8.25	–
		70	36	8.29	–
		76	28	8.25	–
		79	31	8.24	–
		83	41	8.27	–
So65 98MS	28°58.66′ S/140°14.68′ W	95	45	8.25	–
		105	39	8.24	–
		123	37	8.22	–
		135	52	8.20	–
		137	76	8.20	–
		143	63	8.20	–
		148	54	8.18	–
		152	76	8.16	–
		155	66	8.16	–
		159	88	8.16	–
So65 99MS	28°58.71′ S/140°14.71′ W	97	36	–	20.41
		136	109	–	19.00
		145	67	–	18.44
		155	72	–	17.82
		171	105	–	17.59
		219	85	–	16.63
		257	66	–	16.23
		287	37	–	14.19
		299	36	–	14.14
		304	47	–	14.12
		309	44	–	14.08
So65 100MS	28°58.76′ S/140°15.73′ W	29	31	8.28	–
		59	31	8.28	–
		106	30	8.25	–
		115	37	8.25	–
		122	41	8.22	–
		135	88	8.21	–
		146	204	8.06	–
		151	389	7.99	–
		187	1 240	7.40	–
		192	603	7.37	–
		198	1 182	7.43	–
So65 101MS	28°58.76′ S/140°15.39′ W	47	30	8.29	20.98

Chapter 13 · Methane Venting into the Water Column Above Pitcairn and Society Seamounts

Table 13.2. *Continued*

Station	Latitude/longitude	Depth (m)	CH_4 conc. (nl l^{-1})	pH	Temperature (°C)
Macdonald					
So65 101MS	*Continued*	59	21	8.28	20.98
		79	30	8.27	20.95
		100	30	8.26	20.07
		109	31	8.25	19.72
		121	43	8.24	19.44
		130	54	8.22	19.44
		140	249	8.20	19.02
		158	270	8.05	18.63
		162	188	8.05	18.56
		167	215	8.04	18.52
So65 103MS	28°58.92' S/140°15.39' W	188	1358	7.30	–
So65 104MS	28°58.85' S/140°15.09' W	48	35	8.25	21.00
		82	24	8.34	20.97
		104	40	8.22	19.80
		118	69	8.20	19.41
		129	44	8.20	19.16
		164	584	7.68	18.42
		179	541	7.47	18.11
		180	2831	6.66	18.07
		186	9691	6.35	17.94
		190	8774	6.30	17.79
So65 105MS	28°58.83' S/140°15.12' W	187	7957	6.80	–
So65 107OFOS	28°58.82' S/140°15.24' W	226	31615	5.80	17.70
Teahitia					
So65 118MS	17°34.50' S/148°49.30' W	279	4	–	–
		714	8	–	–
		1296	9	–	–
		1346	4	–	–
		1380	7	–	–
		1420	5	–	–
		1448	5	–	–
		1482	3	–	–
		1520	4	–	–
		1564	3	–	–
		1615	4	–	–
		1624	6	–	–
Polynaut MS13	17°33.79' S/148°49.40' W	53	98	–	26.22
		494	85	–	8.73
		791	52	–	5.29
		1100	36	7.81	3.78
		1304	28	7.80	3.21
		1448	47	7.79	2.87
		1501	46	7.77	2.78
		1512	28	7.78	2.76
		1534	31	7.80	2.72
		1557	51	7.81	2.67
		1572	45	7.83	2.63
Polynaut MS14	17°34.36' S/148°48.95' W	393	64	–	12.20
		790	28	–	5.17
		1200	33	7.80	3.34
		1303	38	–	3.08

Table 13.2. *Continued*

Station	Latitude/longitude	Depth (m)	CH$_4$ conc. (nl l^{-1})	pH	Temperature (°C)
Teahitia					
Polynaut MS14	*Continued*	1355	33	7.79	3.02
		1385	43	–	2.99
		1406	20	7.80	2.96
		1426	34	–	2.92
		1445	36	7.73	2.79
		1457	21	7.72	2.77
		1459	34	–	2.77
Polynaut MS15	17°34.17' S/148°48.97' W	50	65	8.27	26.28
		394	55	7.95	13.04
		792	33	7.76	5.23
		1304	33	7.74	3.06
		1487	30	7.73	2.76
		1528	39	7.73	2.74
		1549	28	7.73	2.68
		1559	40	7.71	2.63
		1569	53	7.70	2.64
		1580	56	7.70	2.65
		1597	72	7.65	2.65
Nautile PN18	17°33.92' S/148°49.17' W	1523	3886	5.33	15.80
Nautile PN19	17°34.40' S/148°48.95' W	1455	5528	5.14	30.00
Adams					
So 65 36MS	25°22.90' S/129°15.80' W	40	44	–	–
		59	32	–	–
		77	19	–	–
		95	19	–	–
		104	16	–	–
		132	18	–	–
		140	15	–	–
		146	19	–	–
So 65 38MS	25°23.64' S/129°15.57' W	31	30	–	22.66
		102	28	–	21.65
		124	28	–	20.58
		149	20	–	19.78
		199	28	–	18.00
		300	26	–	15.54
		350	27	–	13.06
		401	22	–	10.87
		428	18	–	10.01
		452	20	–	8.87
		477	12	–	8.35
		501	9	–	7.87
Bounty					
So 65 32MS	25°10.51' S/129°23.87' W	51	16	–	22.85
		99	19	–	22.32
		200	15	–	18.60
		238	18	–	17.42
		270	17	–	16.14
		300	5	–	15.22
		318	22	–	14.76
		387	48	–	12.04
		396	45	–	11.78

Chapter 13 · Methane Venting into the Water Column Above Pitcairn and Society Seamounts   415

Table 13.2. *Continued*

Station	Latitude/longitude	Depth (m)	CH$_4$ conc. (nl l^{-1})	pH	Temperature (°C)
Bounty					
So 65 32MS	*Continued*	408	21	–	10.87
		412	14	–	10.58
		419	12	–	10.17
So 65 41MS	25°11.32' S/129°23.57' W	199	33	–	18.55
		300	27	–	15.52
		350	21	–	13.72
		391	26	–	11.55
		429	28	–	9.50
		456	17	–	8.94
		512	34	–	7.64
		552	8	–	7.11
		582	12	–	6.47
		595	10	–	6.38
So 65 42MS	25°10.80' S/129°23.66' W	53	25	–	22.62
		157	24	–	19.76
		249	52	–	16.98
		349	18	–	12.91
		393	22	–	11.16
		462	14	–	9.00
		525	14	–	7.58
		563	9	–	7.08
		591	10	–	6.47
		632	8	–	6.29
		642	11	–	6.15
So 65 53MS	25°10.80' S/129°24.03' W	175	27	–	19.24
		259	29	–	16.33
		298	23	–	15.49
		335	26	–	13.64
		370	23	–	12.56
		389	18	–	11.81
		409	22	–	11.20
		427	23	–	10.93
		447	18	–	9.94
		465	14	–	9.20
		491	11	–	8.28
So 65 54MS	25°10.52' S/129°23.68' W	314	26	–	15.01
		334	29	–	13.92
		353	20	–	13.08
		375	22	–	12.03
		394	40	–	11.50
		416	35	–	11.17
So 65 77MS	25°10.70' S/129°24.38' W	200	30	8.19	18.32
		251	20	8.17	17.12
		330	25	8.10	14.52
		351	21	8.08	12.84
		358	20	8.07	12.46
		370	18	8.05	11.17
		374	25	8.04	10.05
		381	24	8.04	11.02
		387	29	8.03	–
		391	23	8.02	–
		397	21	8.02	–
		402	22	8.01	–

Table 13.2. *Continued*

Station	Latitude/longitude	Depth (m)	CH$_4$ conc. (nl l^{-1})	pH	Temperature (°C)
Bounty					
Polynaut MS02	25°11.07' S/129°23.49' W	350	89	–	12.92
		400	–	–	10.98
		450	47	7.93	9.26
		500	36	–	7.61
		550	29	–	7.01
		600	41	7.89	6.34
		650	60	7.89	5.91
		700	47	7.87	5.81
		771	45	7.87	5.42
Polynaut MS03	25°11.06' S/129°23.45' W	50	99	8.24	21.31
		100	82	–	21.10
		200	801	8.23	18.41
		250	79	8.13	17.43
		300	63	8.08	15.18
		330	59	8.04	14.13
		350	64	8.01	13.21
		370	72	7.99	12.51
		400	72	–	11.11
		430	81	7.90	9.54
		450	62	–	9.22
Polynaut MS04	25°11.04' S/129°24.00' W	427	41	–	9.86
		490	51	7.87	8.21
		579	30	7.88	6.72
Polynaut MS05	25°10.97' S/129°23.50' W	530	77	7.91	7.22
		555	75	7.92	6.75
		594	72	7.90	6.55
		602	117	7.91	6.40
		617	80	7.90	6.31
		644	81	7.90	6.21
Polynaut MS09	25°11.94' S/129°24.31' W	200	69	–	17.96
		296	103	–	15.05
		394	79	–	11.25
		445	58	–	8.92
		523	55	–	7.15
		542	52	–	6.93
		562	31	–	6.65
		575	80	–	6.53
		582	67	–	6.43
		602	50	–	6.35
Nautile PN04	25°10.63' S/129°24.30' W	430	101	7.89	–
Nautile PN05	25°11.08' S/129°24.10' W	444	79	7.41	19.10
Nautile PN06	25°10.88' S/129°23.62' W	579	2266	7.03	14.50
Nautile PN11	25°10.86' S/129°23.60' W	585	1934	6.94	14.50
Nautile PN12	25°11.08' S/129°24.10' W	445	110	7.74	19.10
Christian					
Polynaut MS06	25°34.93' S/129°31.03' W	490	47	–	8.66
		1005	22	–	4.27
		1508	24	–	2.75
		1695	30	–	2.42
		1801	36	–	2.31

Table 13.2. *Continued*

Station	Latitude/longitude	Depth (m)	CH_4 conc. (nl l^{-1})	pH	Temperature (°C)
Christian					
Polynaut MS06	*Continued*	1 856	34	–	2.21
		1 900	51	–	2.16
		1 913	58	–	2.17
		1 945	62	–	2.16
		1 966	50	–	2.17
		1 978	36	–	2.17
		1 996	33	–	2.14
Pitcairn area					
Polynaut MS08	25° 14.30' S/129°24.58' W	2 300	43	–	1.93
		2 350	90	–	1.92
		2 400	61	–	1.88
		2 450	24	–	1.86
		2 470	30	–	1.86
		2 500	63	–	1.85
		2 510	57	–	1.84
		2 520	43	–	1.84

In some areas of the sea floor (stations OFOS 107 and MS105, Table 13.2, Fig. 13.5) near the summit of Macdonald Seamount, bubble fields and small (1 to 2 °C) temperature anomalies in bottom seawater were observed. This indicates gas oversaturation and/or boiling conditions at the shallow depth of about 200 m under the assumed high temperature of 200 °C (Cheminée et al. 1991) within fractured systems of the submarine edifice. CO_2 is the major gas component (Cheminée et al. 1991), and its dissolution in seawater occurs when hydrothermal fluids are cooling down on their way to the sea surface due to mixing with ambient seawater ($T_{\text{bottom water}} \sim 16.5$ °C). Locally low pH values (up to 5.8) of bottom water (104MS, OFOS 107 in Table 13.2) reflect these high CO_2 contents (Cheminée et al. 1991). The drop in pH is always associated with injections of hydrothermal methane (up to 31 615 nl l^{-1}; Fig. 13.5) clearly exceeding the local CH_4 background values at stations MS100, 101, 103, 104, 105, and OFOS 107 (Fig. 13.5).

On the Teahitia Seamount, the most active area is located near the southernmost summit (Fig. 13.6). Almost the whole area is covered with Fe-oxide-rich sediments. Intensive venting and shimmering water occurred at numerous small ridges with cm- to m-large chimneys or at fissures in surface sediments. However, we did not detect as comparably large CH_4 anomalies near the sea floor above Teahitia Seamount (Fig. 13.6) as we found on the Macdonald. Moreover, at Teahitia gas oversaturation conditions did not prevail and visible gas bubble formation was not observed. This is probably due to a lower total gas content at Teahitia (but note that the pH values (5.33 and 5.14) of the hydrothermal fluids for Teahitia are lower than the pH values of the Macdonald, indicating significantly higher CO_2 contents at Teahitia when compared with the Macdonald hydrothermal fluids) and/or the higher hydrostatic pressure (1 455–1 523 m water depths) at the Teahitia Seamount. Direct sampling of vent fluids (shimmering water) by the submersible showed that the maximum CH_4 concentration measured was 5 528 nl l^{-1}, which is significantly (about 100 times) more enriched relative to the background value. Fluid temperatures up to 30 °C were measured directly in the vent, which is about 26 °C higher than ambient seawater at the corresponding water depth.

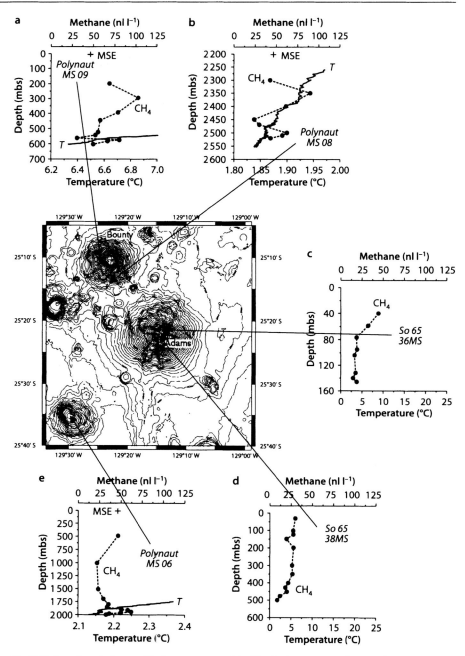

Fig. 13.2. Sampling stations and bathymetry map of the Pitcairn hotspot area showing the Bounty (summit depth ~430 m), Adams (summit depth ~59 m), and Christian (summit depth ~2000 m) volcanoes. Methane concentration and temperature profiles are plotted vs. depth at; a Polynaut MS 09 station; b Polynaut MS 08 station; c SONNE 65 36MS station; d SONNE 65 38MS station; e Polynaut MS 06 station

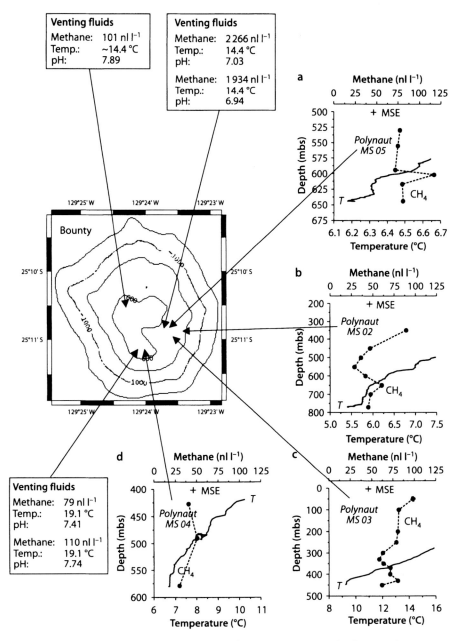

Fig. 13.3. Sampling stations and bathymetry map of the Bounty (summit depth ~430 m) Seamount, Pitcairn hotspot area. Methane concentration and temperature profiles are plotted vs. depth at; **a** Polynaut MS 05 station; **b** Polynaut MS 02 station; **c** Polynaut MS 03 station; **d** Polynaut MS 04 station; methane concentrations, temperatures and pH of fluids from *Nautile* dives during the 1999 Polynaut cruise

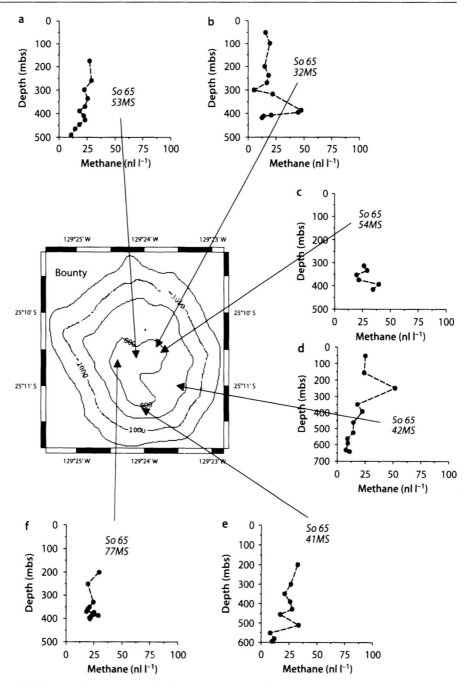

Fig. 13.4. Sampling stations and bathymetry map of the Bounty (summit depth ~430 m) Seamount, Pitcairn hotspot area. Methane concentration profiles are plotted vs. depth at; a SONNE 65 53MS station; b SONNE 65 32MS station; c SONNE 65 54MS station; d SONNE 65 42MS station; e SONNE 65 41MS station; f SONNE 65 77MS station

Chapter 13 · Methane Venting into the Water Column Above Pitcairn and Society Seamounts   421

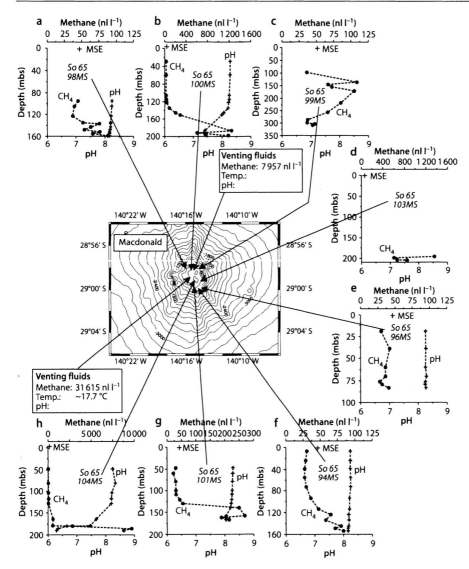

Fig. 13.5. Bathymetry map of the summit area of the Macdonald (summit depth ~40 m) Seamount, Austral hotspot area. Methane concentration profiles are plotted vs. depth at; a SONNE 65 98MS station; b SONNE 65 100MS station; c SONNE 65 99MS station; d SONNE 65 103MS station; e SONNE 65 96MS station; f SONNE 65 94MS station; g SONNE 65 101MS station; h SONNE 65 104MS station; methane concentrations and temperatures from deep-towed camera stations (OFOS) during the SONNE 65 cruise

Strong currents occur at the summit of Teahitia, and intensive mixing of ambient seawater with venting fluids is responsible for the fact that methane enrichments in the water column can only be found directly near the summit (Fig. 13.6).

A comparable mixing situation was found for the Pitcairn area, where we observed emanating hydrothermal fluids on the Bounty Seamount with CH_4 concentrations of up to 2266 nl l^{-1} near the vent (at a water depth of 579 m), and with methane anoma-

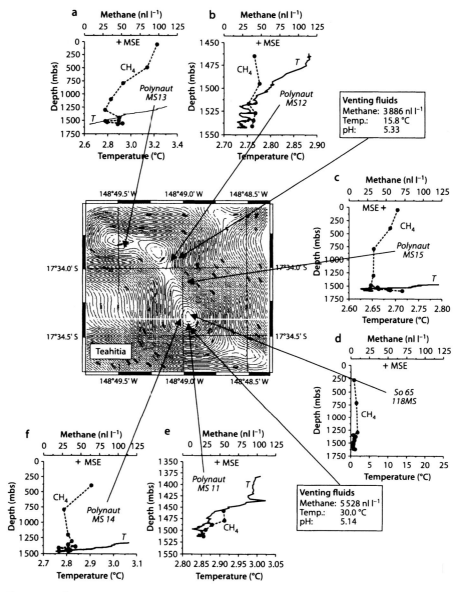

Fig. 13.6. Bathymetry map of the summit area of the Teahitia (summit depth ~1400 m) Seamount Society hotspot area. Methane concentration and temperature profiles are plotted vs. depth at; **a** Polynaut MS 13 station; **b** Polynaut MS 12 station; **c** Polynaut MS 15 station; **d** SONNE 65 118MS station; **e** Polynaut MS 11 station; **f** Polynaut MS 14 station; methane concentrations, temperatures and pH of fluids from *Nautile* dives during the Polynaut cruise

lies in the water column only directly at the summit (Fig. 13.3). It is noteworthy that the pH values of the vent fluids at the sea floor of the Bounty Seamount are only slightly lowered (pH = 6.94–7.74) when compared to ambient seawater (pH = 7.9 of 580 m water depth), indicating only small dissolved CO_2 contents (~2000 nl l^{-1}; calculated for pH 7).

13.4.2
Origin of Hydrothermal Methane

The Macdonald, Teahitia and Bounty Seamounts inject hydrothermal gas into the water column. The fluids released from Macdonald were found to be enriched in CO_2 and CH_4 compared to venting fluids at Galapagos and EPR vent sites (Cheminée et al. 1991), and the origin of gases was related to "shallow summit magma degassing". Higher hydrocarbons such as ethane and propane have not been detected in the gases from the seamounts, indicating sediment-free subsurface conditions where thermal influences on organic matter not only produce methane but also the higher-ordered hydrocarbons (Simoneit at al. 1983). Theoretically, abiogenic reactions in the Earth's crust such as hydrolysis of mafic rocks (Apps 1985; Sherwood Lollar et al. 1993), Fischer Tropsch synthesis (Lancet and Anders 1970) or mantle emanations (Craig et al. 1981), and/or biogenic gas formation could be responsible for the occurrence of hydrothermal CH_4 in the venting fluids of the three seamounts under consideration. In a $\delta^{13}C$-δD diagram (Fig. 13.7), the stable isotope data of hydrothermal methane from Macdonald Seamount (since this is the only location where enough CH_4 for δD analyses occurred) plot in between the two empirical fields "biogenic methane" and that which is commonly referred to as "abiogenic/mantle" methane (Whiticar 1999). Based on the Richet et al. (1977) equation, an isotopic equilibrium temperature near 200 °C is calculated from the carbon isotope values of CO_2 ($\delta^{13}C = -5.2‰$) and CH_4 ($\delta^{13}C = -38.6‰$). A similar temperature is calculated based on the $\delta D_{H_2O\text{-}CH_4}$ equilibrium (Lyon and Hulston 1984). This temperature shows that bacterial activity alone cannot account for the CH_4 formation, since biological activity above 113 °C is unlikely to occur (Blöchl et al. 1997). If a (bacterial) CH_4 formation temperature near 80 °C is assumed within the vents (Huber et al. 1990) near isotope equilibrium (Botz et al. 1996) with the CO_2 ($\delta^{13}C_{CO_2} = -5.2‰$), then the primary (unoxidized) CH_4 should have an isotope value of $-55‰$ (Richet at al. 1977).

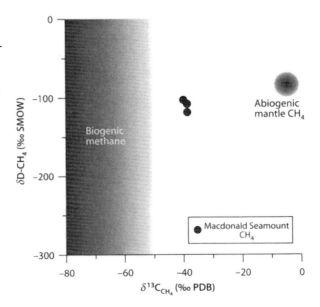

Fig. 13.7. δD and $\delta^{13}C$ values of methane (Table 13.3) of the Macdonald Seamount gas samples are plotted in a modified Welhan (1988) diagram. δD and $\delta^{13}C$ values of methane from biogenic and abiogenic origin are shaded in *gray*

Venting CH_4 with this isotope value was not found at the Macdonald Seamount. However, one CH_4 sample was taken directly by the submersible *Nautile* from an active vent field at the Bounty Seamount. This CH_4 sample had an isotope value of −56‰, which is typical for bacterial CH_4 formation by CO_2 reduction at around 88 °C.

Bacterial CH_4 oxidation reduces the CH_4 content of the water and leads to an increase in the heavy isotopes of the residual CH_4, which follows a Rayleigh-type equation (Fritz and Fontes 1980). The theoretical change of both $\delta^{13}C$ and the concentration of methane in the course of oxidation (fractionation factor $\alpha = 1.005$; Tsunogai et al. 2000) is plotted in Fig. 13.8 and compared to the measured data at the Macdonald Seamount (Fig. 13.8, Table 13.3). It is shown by the calculated and measured data that CH_4 quantities and isotope values are related to each other and furthermore, that early oxidation steps strongly affect the CH_4 concentration without significantly changing the isotope composition of the residual methane (Fig. 13.8, Table 13.3). Moreover, mixing processes of hydrothermal fluids with seawater with a low methane concentration also lower the concentration values as indicated by the strong decrease in measured methane concentrations. This effect is not taken into account by the theoretical Rayleigh fractionation (Fig. 13.8). However, calculated and measured methane data show that during late stage oxidation when most (>90%) of the CH_4 has already been oxidized to CO_2 (e.g., 101MS with a low CH_4 concentration of 270 nl l^{-1}; Table 13.2), significant ^{13}C isotope enrichment ($\delta^{13}C = -28.2‰$) of the residual methane occurs. It is concluded, therefore, that the isotope value of venting CH_4 ($\delta^{13}C = -38.6‰$) at Macdonald is probably not much affected by oxidation. Thus, we have to assume a ^{13}C-rich CH_4 component admixed to the biogenic CH_4. This isotopically heavy CH_4 is known to be present in the tectonically active region at 17–19° S, East Pacific Rise (near −22 to −24‰; Charlou et al. 1996). Based on quantitative estimations, Macdonald's gas probably contains mixtures of two end-members, e.g., 50% biogenic CH_4 ($\delta^{13}C = -55‰$ from bacterial CH_4 production at 80 °C) and 50% of abiogenic CH_4 with an assumed mean $\delta^{13}C = -23‰$ (Charlou et al. 1996).

Warm vent fluids from the Teahitia Seamount contain CH_4 with significantly less negative $\delta^{13}C$-values of −29 and −22‰ (Table 13.3). This range of values is similar to

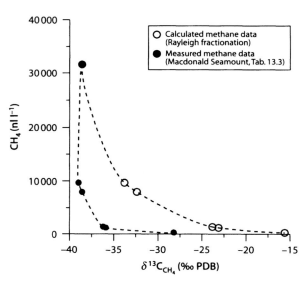

Fig. 13.8. Concentrations and $\delta^{13}C$ values of methane measured (*filled dots*) on the Macdonald Seamount. The calculated data (*dots*) shows the theoretical concentration decrease with increasing $\delta^{13}C_{CH_4}$ values assuming a Rayleigh type fractionation for the methane oxidation process ($\alpha = 1.005$; Tsunogai et al. 2000)

Chapter 13 · Methane Venting into the Water Column Above Pitcairn and Society Seamounts

Table 13.3. δD and $\delta^{13}C$ values of methane and $\delta^{13}C$ value of carbon dioxide detected in gas samples from the Macdonald, Bounty, and Teahitia Seamounts

Station	CH_4 (nl l^{-1})	$\delta^{13}C$-CH_4	δD-CH_4	$\delta^{13}C$-CO_2
Macdonald				
So65 100MS	1 182	-35.9	-	-
So65 101MS	270	-28.2	-	-
So65 103MS	1 358	-36.2	-	-
So65 104MS	9 691	-39.0	-103	-
So65 105MS	7 957	-38.6	-106	-
So65 107OFOS	31 615	-38.6	-117	-5.2
Bounty				
Nautile PN06	2 266	-56.0	-	-
Nautile PN11	1 934	-	-	-
Teahitia				
Nautile PN18	3 886	-22.0	-	-
Nautile PN19	5 528	-29.0	-	-

the values published for abiogenic CH_4 (see above). Thus, based on isotope evidence alone, it cannot be decided whether the Teahitia Seamount injects hydrothermal CH_4 of abiogenic origin (EPR-type) into the water column or alternatively also represents an environment where biogenic CH_4 formation and oxidation processes are important. The CH_4 concentrations of the Teahitia Seamount are remarkably high for possible medium to late stage (which is required for the strong isotope shift to -22‰, see above) oxidation processes. However, the geochemical similarity (Garbe-Schönberg, personal communication) of hydrothermal fluids from all three investigated seamounts (Macdonald, Bounty, Teahitia) does not indicate an extraordinary situation for the Teahitia Seamount.

13.5
Conclusions

Hydrothermal system venting on submarine seamounts in hotspot areas in the South Pacific are recognized by temperature-, pH- and CH_4 anomalies in the bottom seawater. Bubbling vent fields were detected near the summit of the active Macdonald Seamount, where seawater boiling occurs at shallow water depth (<200 m) and high temperature (>200 °C). In contrast, the greater water depth prevents boiling at Teahitia and Bounty Seamounts, where shimmering water emerges through the sea floor. The low pH values (to 5.8) associated with hydrothermal fluid injections are largely caused by magmatic ($\delta^{13}C = -5.2‰$) CO_2.

CH_4 is enriched in the hydrothermal fluids of the Macdonald Seamount up to 31 615 nl l^{-1}. Shimmering water emerging on the sea floor of the two other seamounts (Bounty and Teahitia) contains lower CH_4 concentrations (1 934 to 5 528 nl l^{-1}). Carbon isotope values of CH_4 fall in the range between -22 and -56‰ PDB. The low carbon isotope value of -56‰ for hydrothermal CH_4 from the Bounty Seamount most likely reflects bacterial CH_4 formation by CO_2 reduction at a temperature near 90 °C. The carbon isotope values of Macdonald hydrothermal CH_4 are less negative (between

−28 and −39‰), and together with the δD values of −103 to −117‰ SMOW, they reflect a mixture of abiogenic CH_4 originated in the deep subsurface mixed with biogenic CH_4 formed in the vent system at temperatures below 100 °C. However, CH_4 oxidation causes secondary enrichment of the heavy isotopes in the residual CH_4 fraction. Hence, based on the CH_4 isotope values alone, it unclear as to whether hydrothermal CH_4 from the Teahitia Seamount is derived from abiogenic subsurface reactions and/or reflects an oxidized CH_4 of unknown origin. Nevertheless, significant amounts of hydrothermal CH_4 from active seamounts in the South Pacific are derived from biogenic processes. Our findings are substantiated by microbiologists who found hyperthermophilic archaebacteria within the crater of the erupting Macdonald Seamount.

Acknowledgements

We are grateful to R. Hekinian for valuable discussions regarding hotspot volcanism in the South Pacific. This manuscript was improved by comments and suggestions made by J. L. Charlou. We acknowledge the crews of RV SONNE, NO L'ATALANTE and submersible *Nautile* (IFREMER, Brest) for their support during the sampling campaigns. This research project was funded by the Bundesministerium für Forschung und Technologie (BMFT) through grant no. 03G0543A to Kiel University.

References

Alt JC (1995) Subseafloor processes in Mid-Ocean Ridge hydrothermal Systems. Amer Geophys Union/ Geophys Monograph 91:85–114

Apps JA (1985) Methane formation during hydrolysis by mafic rocks. Lawrence Berkeley Lab Annu Rep 84:13–17

Bange HW, Rapsomanikis S, Andreae MO (1996) The Aegean Sea as a source of atmospheric nitrous oxide and methane. Marine Chemistry 53:41–49

Binard N, Stoffers P, Hekinian R, Cheminée J-L (2004) South Pacific intraplate volcanism: Structure, morphology and style of eruption. *this volume*

Blöchl E, Rachel R, Burggraf S, Hafenbradl D, Jannasch HW, Stetter KO (1997) *Pyrolobus fumarii*, gen and sp nov, represents a novel group of archaea, extending the upper temperature limit for life to 113 °C. Extremophiles 1:14–21

Both R, Crook K, Taylor B, Brogan S, Chappell B, Frankel E, Liu L, Sinton J, Tiffin D (1986) Hydrothermal chimneys and associated fauna in the Manus Back-Arc Basin, Papua New Guinea. Eos, Trans Amer Geophys Union 67:489–490

Botz R, Pokojski HD, Schmitt M, Thomm M (1996) Carbon isotope fractionation during bacterial methanogenesis by CO_2 reduction. Org Chem 25:255–262

Burggraf S, Fricke H, Neuner A, Kristjansson J, Rouvier P, Mandelco L, Woese CR, Stetter KO (1990) *Methanococcus igneus* sp nov, a novel hyperthermophilic methanogen from a shallow submarine hydrothermal system. System Appl Microbiol 13:263–269

Butterfield DA, Massoth GJ (1994) Geochemistry of north Cleft segment vent fluids: Temporal changes in chlorinity and their possible relation to recent volcanism. J Geophys Res 99:4951–4969

Charlou JL, Bougault H, Appriou P, Jean-Baptiste P, Etoubleau J, Biroleau A (1991) Water column anomalies associated with hydrothermal activity between 11°40' and 13° N on the east Pacific Rise: Discrepancies between tracers. Deep Sea Res 38:569–596

Charlou JL, Fouquet Y, Donval JP, Auzende JM, Jean-Baptiste P, Stievenard M, Michel S (1996) Mineral and gas chemistry of hydrothermal fluids on an ultrafast spreading ridge: East Pacific Rise, 17° to 19° S (Naudur cruise, 1993) phase seperation processes controlled by volcanic and tectonic activity. J Geophys Res 101:15899–15919

Charlou JL, Donval JP, Fouquet Y, Jean-Baptiste P, Holm N (2002) Geochemistry of high H_2 and CH_4 vent fluids issuing from ultramafic rocks at the Rainbow hydrothermal field (36°14' N, MAR). Chem Geol 191:345–359

Cheminée J-L, Stoffers P, McMurtry GM, Richnow H, Puteanus D, Sedwick P (1991) Gas-rich submarine exhalations during the 1989 eruption of Macdonald Seamount. Earth Plant Sci Lett 107:318–327

Chapter 13 · Methane Venting into the Water Column Above Pitcairn and Society Seamounts   427

Corliss JB, Dymond J, Gordon LI, Edmond JM, Von Herzen RP, Ballard RD, Green K, Williams D, Brainbridge A, Crane K, Van Adel TH (1979) Submarine thermal springs on the Galapagos Rift. Science 203:1073–1083

Craig H, Lupton JE (1981) Helium-3 and mantle volatiles in the ocean and oceanic crust. In: Emiliani C (ed): The sea. Wiley, New York, pp 391–428

Crough ST (1983) Hotspots swells. Annu Rev Earth Planet Sci 11:165–193

Dando PR, Hughes JA, Thiermann F (1995) Preliminary observations on biological communities at shallow hydrothermal vent in the Aegean Sea. In: Parson LM, Walker CL, Dixon DR (eds) Hydrothermal vents and processes. Geological Society, London, Special Publication 87:301–317

Dumke I, Faber E, Poggenburg J (1989) Determination of stable carbon and hydrogen isotopes of light hydrocarbons. Analytical Chemistry 61:2149–2154

Faber E, Gerling P, Berner U, Sohns E (1994) Methane in ocean waters; Concentration and carbon isotope variability at East Pacific Rise and in the Arabian Sea. Environmental Monitoring and Assessment 31:139–144

Fouquet Y, Von Stackelberg U, Charlou JL, Donval JP, Erzinger J, Foucher JP, Herzig P, Mühe R, Soakai S, Wiedicke M, Whitechurch H (1991) Hydrothermal activity and metallogenesis in the Lau back-arc basin. Nature 349:778–781

Fritz P, Fontes JC (1980) Handbook of environmental isotope geochemistry, vol. 1. Elsevier Scientific Publishing Company, Amsterdam

Gamo T, Ishibashi JJ, Sakai H (1987) Methane anomalies in seawater above the Loihi submarine summit area, Hawaii. Geochim Cosmochim Acta 51:2857–2864

Glasby GP, Stüben D, Jeschke G, Stoffers P, Garbe-Schönberg D (1997) A model for the formation of the hydrothermal manganese crusts from the Pitcairn Island hot spot. Geochim Cosmochim Acta 61:4583–4597

Halbach P, Pracejus B, Märten A (1993) Geology and mineralogy of massive sulfide ores from the Central Okinawa Trough, Japan. Econ Geol 88:2206–2221

Hannington MD, Jonasson IR, Herzig P, Petersen S (1995) Physical and chemical processes of seafloor mineralization at Mid-Ocean Ridges. Amer Geophys Union/Geophys Monograph 91:115–157

Hekinian R, Renard V, Cheminée J-L (1984) Hydrothermal deposits on the East Pacific Rise near 13° N: Geological setting and distribution of active sulfide chimneys. In: Rona PA, Bostrom K, Laubier L, Smith KL (eds) Hydrothermal processes at seafloor spreading centers. Plenum Publishing Corporation 571–602

Hekinian R, Bideau D, Stoffers P, Cheminée J-L, Mühe R, Puteanus D, Binard N (1991) Submarine intraplate volcanism in the South Pacific: Geological setting and petrology of the Society and Austral regions. J Geophys Res 96:2109–2138

Hoffert M, Cheminée J-L, Person A, Larque P (1987) Dépot hydrothermal associé au volcanisme sous-marine intraplaque: prélèvements effectués avec la Cyana sur le volcan actif de Teahitia (Polynésie Francaise). C R Acad Sci, Paris, 304:829–832

Huber R, Stoffers P, Cheminée J-L, Richnow HH, Stetter KO (1990) Hyperthermophilic archaebacteria within the crater and open-sea plume of erupting Macdonald Seamount. Nature 345:179–182

Ishibashi J, Wakita H, Nojiri Y, Grimaud D, Jean-Baptiste P, Gamo T, Auzende JM, Urabe T (1994) Helium and carbon geochemistry of hydrothermal fluids from the North Fiji Basin spreading ridge (southwest Pacific). Earth Plan Sci Letters 128:183–197

Ishibashi J, Sano Y, Wakita H, Gamo T, Tsutsumi M, Sakai H (1995) Helium and carbon geochemistry of hydrothermal fluids from the Mid-Okinawa Trough back arc basin, southwest of Japan. Chem Geol 123:1–15

Ishibashi J, Wakita H, Okamura K, Nakayamy E, Feely RA, Lebon GT, Baker ET, Marumo K (1997) Hydrothermal methane and manganese variation in the plume over the superfast spreading southern East Pacific Rise. Geochim Cosmochim Acta 61:485–500

Jannasch H (1995) Microbial interactions with hydrothermal fluids. In: Humphris S, Zierenberg R, Mullineaux L, Thomson R (eds) Physical, chemical, biological and geological interactions within seafloor. Hydrothermal Systems 91:273–296

Karl DM, McMurtry GM, Malahoff A, Garcia MO (1988) Loihi Seamount, Hawaii: A mid-plate volcano with a distinctive hydrothermal system. Nature 335:532–535

Lancet HS, Anders E (1970) Carbon isotope fractionation in the Fischer-Tropsch synthesis of methane. Science 170:980–982

Lonsdale PF, Bischoff JL, Burns VM, Kastner M, Sweeny RE (1980) A high temperature hydrothermal deposit on the seabed at the Gulf of California spreading center. Earth Planet Sci Lett 49:8–20

Lyon GL, Hulston JR (1984) Carbon and hydrogen isotopic compositions of New Zealand geothermal gases. Geochim Cosmochim Acta 48:1161–1171

Malahoff A, McMurtry G, Wishire JC, Yehy HW (1982) Geology and chemistry of hydrothermal deposits from active submarine volcano Loihi, Hawaii. Nature 298:234–239

Michaelis W, Jenisch A, Richnow HH (1990) Hydrothermal petroleum generation in Red Sea sediments from the Kebrit and Shaban Deeps. Appl Geochem 5:103–114

Michard A, Michard G, Stüben D, Stoffers P, Cheminée J-L, Binard N (1993) Submarine thermal springs associated with young volcanoes: The Teahitia vents, Society Island, Pacific Ocean. Geochim Cosmochim Acta 57:4977–4986

Nelson DC, Fisher CR (1995) Chemoautotrophic and methanotrophic endosymbiotic bacteria at vents and seeps. In: Karl DM (ed) Microbiology of deep sea hydrothermal vent habitats. CRC Press, Boca Raton

Polynaut Cruise Report (2000) La campagne Polynaut de 1999. Rapport de Mission, Departement des Observatoire Volcanologiques, Institut de Physique du Globe, Paris, 4 Place Jussieu, Paris

Rehder G, Keir RS, Suess E, Pohlmann T (1998) The multiple sources and patterns of methane in North Sea waters. Aquatic Geochemistry 4:403–427

Richet P, Bottinga Y, Javoy M (1977) A review of hydrogen, carbon, nitrogen, oxygen, sulphur and chlorine stable isotope fractionation among gaseous molecules. Annu Rev Earth Planet Sci 5:65–110

Rona PA, Scott SD (1993) A special issue on sea-floor hydrothermal mineralization; New perspectives; preface. Economic Geology and the Bulletin of the Society of Economic Geologists 88:1933–1974

Sakai H, Tsubota H, Nakai T, Ishibashi J, Akagi T, Gamo T, Tilbrock B, Igarashi G, Kodera M, Shitashima K, Nalamura S, Fujioka K, Watanabe M, McMurtry GM, Malahoff A, Oruma M (1987) Hydrothermal acticity on the summit of Loihi Seamount, Hawaii. Geochem J 21:11–21

Schmitt M (1989) "Gaswasserschöpfer" Internal Report. Geochemische Analysen, Sehnde-Ilten

Schmitt M, Faber E, Botz R, Stoffers P (1991) Extraction of methane from seawater using ultrasonic vacuum degassing. Anal Chem 63:529–532

Schoell M (1980) The hydrogen and carbon isotopic composition of methane from natural gases of various origins. Geochim Cosmochim Acta 44:649–661

Sedwick PN, McMurtry GM, Macdougall JD (1992) Chemistry of hydrothermal solutions from Pele's Vents, Loihi Seamount, Hawaii. Geochim Cosmochim Acta 57:5087–5097

Seyfried WE, Dibble WE (1980) Seawater-peridotite interaction at 300 °C and 500 bars; Implications for the origin ofoceanic serpentinites. Geochimica et Cosmochimica Acta 44:309–322

Seyfried WE, Janecky DR (1985) Sr and Ca exchange during hydrothermal alteration of basalt/diabase. Eos, Transaction Amer Geophy Union 66:921

Sherwood-Lollar B, Frape SK, Weise SM, Fritz P, Macko SA, Welhan JA (1993) Abiogenic methanogenesis in crystalline rocks. Geochim Cosmochim Acta 57:5087–5097

Simoneit BRT (1983) Effects of hydrothermal activity on sedimentary organic matter: Guaymas Basin, Gulf of California – Petroleum genesis and protokerogen degradation. In: Rona PA, Boström K, Laubier L (eds) Hydrothermal processes at seafloor spreading centers. Plenum Press, New York, pp 451–471

Smith WHF, Sandwell DT (1997) Global seafloor topography from satellite altimetry and ship depth soundings. Science 277:1957–1962

Stoffers P, Hekinian R (1990) Cruise report Sonne 65 – Midplate II, Hot Spot Vulkanismus im zentralen Südpazifik. Berichte-Reports, Geol Paläont Inst Univ Kiel, 40

Stoffers P, et al. (1987) Cruise Report SO-47: Midplate volcanism, Central South Pacific, French Polynesia. Berichte-Reports, Geol Paläont Inst Univ Kiel, 19

Stoffers P, Botz R, Cheminée J-L, Devey CW, Froger V, Glasby GP, Hartmann M, Hekinian R, Kögler F, Laschek D, Larque P, Michaelis W, Mühe R, Puteanus D, Richnow HH (1989) Geology of MacDonald Seamount region, Austral Islands: Recent hotspot volcanism in the South Pacific. Marine Geophysical Researches 11:101–112

Stüben D, Stoffers P, Cheminée J-L, Hartmann M, McMurtry G, Richnow HH, Jenisch A, Michaelis W (1992) Manganese, methane, iron, zinc, and nickel anomalies in hydrothermal plumes from Teahitia and Macdonald Volcanoes. Geochim Cosmochim Acta 56:3693–3704

Talandier J (1989) Submarine volcanic activity; Detection, monitoring, and interpretation. Eos, Transaction, Amer Geophys Union 70:568–569

Tsunogai U, Yoshida N, Ishibashi J, Gamo T (2000) Carbon isotopic distribution of methane in deep-sea hydrothermal plume, Myojin Knoll Caldera, Izu-Bonin arc: Implications for microbial methane oxidation in the oceans and applications to heat flux estimation. Geochimica et Cosmochimica Acta 64:2439–2452

Tunnicliffe V (1991) The biology of hydrothermal vents: ecology and ecolution. Oceanography and Mar Biol Annual Rev 29:319–407

Urabe T, Baker ET, Ishibashi J, Feely RA, Marumo K, Massoth GJ, Maruyama A, Shitashima K, Okamura K, Lupton JE, Sonoda A, Yamazaki T, Aoki M, Gendron J, Greene R, Kaiho Y, Kisimoto K, Lebon G, Matsumoto T, Nakamura K, Nishizawa A, Okano O, Paradis G, Roe K, Shibata T, Dennant D, Vance T, Walker SL, Yabuki T, Ytow N (1995) The effect of magmatic activity on hydrothermal venting along the superfast-spreading East Pacific Rise. Science 269:1092–1095

Von Damm KL, Bischoff JL (1987) Chemistry of hydrothermal solutions from the southern Juan de Fuca Ridge. J Geophys Res 92:11334–11346

Welhan JA (1988) Origins of methane in hydrothermal systems. Chem Geol 71:183–198

Whiticar MJ (1990) A geochemical perspective of natural gas and atmospheric methane. Org Chem 16:759–768

Whiticar MJ (1999) Carbon and hydrogen isotope systematics of bacterial formation and oxidation of methane. Chem Geol 161:291–314

Whiticar MJ, Faber E (1986) Methane oxidation in sediment and water column environments isotope evidence. Org Geochem 10:759–768

Wiesenburg DA, Guinasso J (1979) Equilibrium solubilities of methane, carbon monoxide, and hydrogen in water and seawater. J Chem Eng Data 24:356–360

Chapter 14

Petrology of Young Submarine Hotspot Lava: Composition and Classification

Roger Hekinian

14.1
Introduction

Underwater volcanism is the predominant phenomenon taking place on Earth. About 71% of the Earth's sea-floor surface is the site of volcanism, which mainly extrudes basaltic rocks. The mineralogical and chemical compositions of oceanic rocks are important factors that affect the formation of the oceanic crust, because the rock's composition influences its rate of extrusion, the morphology of lava flows, the cooling rate, the mode of emplacement, etc. The magmatic history of the various geological provinces is also controlled by the processes of partial melting of a heterogeneous mantle and of crystal-liquid fractionation within the magmatic reservoir. Thus, the magmatic history (the rocks' petrology) can be better understood if we first clearly define and understand both the mineral and chemical composition and the morphology of rocks found in different geological settings and provinces of the ocean floor.

Isotopic and trace element analyses of ocean floor basalts (Tatsamuto et al. 1965; Schilling 1971, 1973, 1985) have provided evidence that the Earth's mantle is heterogeneous. The lava erupted on the ridge axis along the various segments of the EPR consists essentially of basalts called Mid-Ocean Ridge Basalts (MORBs). MORBs are subdivided into three categories: normal (N-MORB), transitional (T-MORB) and enriched (E-MORB), which are probably derived from the partial melting of a "Depleted MORB Mantle" (DMM). Because of their lower alkali contents, MORBs can be differentiated from alkali-enriched Ocean Island Basalts (OIBs). The alkali-rich volcanics from intraplate (hotspot) regions fall in the category of OIBs and differ from those of spreading ridges. Nevertheless, some mildly alkali-enriched and silica-enriched ($SiO_2 > 53\%$) extrusives have also been encountered along spreading ridges and were attributed to the influence of a hotspot-generated mantle plume rising in the vicinity of spreading ridge segments (Hekinian et al. 1999; Stoffers et al. 2002; Sect. 8.2.6, 8.2.7, 9.3.2.3 and 9.3.2.4).

Partial melting of the heterogeneous mantle and crystal-liquid fractionation taking place in the lower lithosphere and asthenosphere are important processes leading to the formation of oceanic crust. Hoffman and White (1982) have suggested that enriched melts are due to the mixing of sediment and mantle materials as a result of recycling subducted oceanic lithosphere into the mantle. The heterogeneous mantle is believed to be composed of at least three theoretically enriched mantle compositions called EM1 (Enriched Mantle type I), EM2 (Enriched Mantle type II) and high μ ($^{238}U/^{204}Pb$, HIMU) (Zindler and Hart 1986; Sect. 8.3.6). These enriched mantle (EM) sources are probably derived from the melting of ancient continental crust and/or detrital (terrigenous) sediments transported and subducted in the mantle (Hoffman and White 1982), while

432   R. Hekinian

the HIMU-enriched mantle source is more likely to have its origin in oceanic crust that was subsequently recycled. These various enriched mantles could explain why hotspot volcanics are generally more enriched in incompatible elements in comparison to MORBs. In fact, this classification in terms of mantle enrichment is another way of explaining the differences observed between the volcanics recovered from various oceanic provinces.

The present work deals with the petrological description and classification of volcanic rocks from intraplate provinces compared to spreading ridge samples. A particular emphasis is given to silica-enriched ($SiO_2 > 51\%$) lava erupted in intraplate hotspot provinces and regions of ridge-hotspot interactions (i.e., the Foundation Seamount chain and Pacific-Antarctic Ridge, PAR), because silica-enriched rocks are important for understanding the evolution of magmatic processes during fractionation. The petrological relationship between various hotspot provinces and their individual volcanic edifices is shown. The various rock types are classified according to their morphological appearance in their structural context. This approach was used for recognizing the volcanic stratigraphy in areas of the ocean floor where detailed observations and sampling have been able to be performed (i.e., Pitcairn and Society hotspots). For more information on the morpho-structural setting and rock geochemistry of intraplate hotspot provinces, the reader is referred to this volume's Chapt. 5 by Binard et al. (Sect. 5.2, 5.3 and 5.4), Chap. 8 by Devey and Haase (Sect. 8.2) and Chap. 9 by Niu and Hekinian (Sect. 9.3.2.3 and 9.3.2.4).

14.2
Composition and Description of Oceanic Rocks

The diversity of erupted flow types encountered on the ocean floor is as significant as what is observed in subaerial volcanism. The mineralogy, as well as the bulk rock and glassy chilled margin compositional variability, is used to identify and classify the various rock types encountered in intraplate hotspots and to compare these rocks to samples from spreading-ridge provinces. The information presented below in Fig. 14.1, 14.2, and 14.3 as well as in Tables 14.1, 14.2 and 14.3 is compiled from data from Pacific hotspot volcanoes (Clague 1988; Hekinian et al. 1991; Binard 1991, 1992; Woodhead and Devey 1993; Fretzdorff et al. 1996; Haase and Devey 1996; Pan and Batiza 1998; Ackermand et al. 1998; Haase 2002; Hekinian et al. 2003) and from ridge-hotspot interaction zones (Clague et al. 1981; Devey et al. 1997; Stoffers et al. 2002; and unpublished data).

14.2.1
Common Mineral Constituents

The variation in mineral composition is due to the magma's evolution during solidification. The crystal growth in a melt will change the composition of the remaining liquid. The duration of crystal growth in a liquid might take considerable time, even hundreds of years (Turner et al. 2003), and this will give rise to crystal-liquid segregation during magmatic upwelling. The identification of altered rocks using only bulk rock chemistry is not significant because of the elements' mobility. However, mineral crystals are often preserved, and their analyses help to classify the various rock types encountered on the sea floor. This is mainly true for plagioclase, a major constituent,

Chapter 14 · Petrology of Young Submarine Hotspot Lava: Composition and Classification   433

which is sensitive to the degree of alkalinity of the solidifying melt. In fact, a single plagioclase crystal can vary in its composition from its center to its margin, and this variation takes place during crystallization when the melt surrounding the crystal has changed its composition during solidification. In order to minimize the effect of crystal-liquid interaction and to homogenize the data set, the data on phenocrysts and/or microphenocrysts were, when possible, chosen from analyses performed on the center of the crystals.

The chronological order of the mineral crystallization in basaltic lava erupted under normal conditions on the ocean floor is Cr-spinel, olivine, plagioclase, clinopyroxene, Ti-magnetite and apatite (Sect. 14.2.2.1). Because the submarine lavas are rapidly quenched when at contact with seawater, most collected samples are poorly crystallized; therefore, only the earliest formed minerals are seen. Usually olivine and spinel are the earliest minerals to crystallize in the alkali-enriched melts (non-evolved alkali-basalts and basanites). Plagioclase and Na- and K-feldspathoids (i.e., nepheline) are the next phase to crystallize. Clinopyroxene is also a common phase that is found in most types of hotspot rocks.

14.2.1.1
Feldspar

Feldspar is a general term including plagioclase, orthoclase, alkali-feldspar and sanidine. The feldspar found in the volcanics from the hotspot and EPR areas is composed of both plagioclase and alkali-feldspar; however, the most common feldspar found in ocean floor volcanics remains plagioclase.

The mineralogy of the two types of feldspar reported here is as follows: (1) *Alkalifeldspars* consist essentially of orthoclase and sanidine and are commonly found on hotspot volcanoes associated with basanite, tephrite, trachybasalt and trachyte rock types (Sect. 14.2.2.3). The Society and the Austral hotspots contain trachytes and phonolites with a common occurrence of orthoclase-rich feldspar when compared to other hotspot volcanics. (2) The *Plagioclase* composition of the various oceanic rocks shows two main trends of variability that are *(i)* the trend enriched in normative orthoclase (*or*) and sodic plagioclase (albite = *ab*) and *(ii)* the trend depleted in *or* with an increasing *ab* content (Fig. 14.1a). The change in trend between the spreading-ridge volcanics and that of the hotspot volcanics is marked by the changes in the labradorite-andesine (an_{75-70}, ab_{22-25}, $or_{0.45-0.7}$) composition of the plagioclase. The trachytes and trachy-andesites contain a more sodic plagioclase (ab_{59-70}, an_{1-8}, or_{25-38}) and alkali-feldspar such as orthoclase than do the trachybasalts and alkali basalts from the hotspot regions (ab_{22-45}, $or_{0.45-2.2}$) (Fig. 14.1a). Most of the trachytes from hotspot volcanoes have a K_2O-enriched (>1%) plagioclase content and show an increase of their *or* compared to their *ab* content (Fig. 14.1a).

The plagioclase of N-MORBs is depleted in its normative orthoclase (an_{73-85}, ab_{14-26}, $or_{0-0.3}$) content and consists mainly of labradorite and bytownite (Hekinian et al. 1996). The plagioclase of the silicic lavas consists of andesine with a higher *or* content (ab_{45-65}, $or_{0.3-1.0}$) than the MORBs; however, this silicic-lava plagioclase shows a low trend of variability for the *ab-or* contents when compared to other enriched lava from the hotspots (Hekinian et al. 1999) (Fig. 14.1a). The plagioclase (ab_{35-40}, $or_{0.3-0.7}$) from the Foundation seamounts' alkali basalts have a relatively depleted normative orthoclase

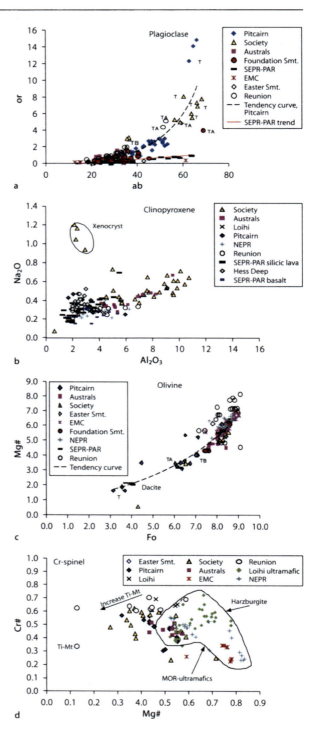

Fig. 14.1. Compositional variability of plagioclase (a), clinopyroxene (b), olivine (c), and chrome-spinel (d) in oceanic rocks collected from hotspots and Mid-Ocean Ridge (MOR) spreading centers. The data are from Hawkins et al. (1983), Clague (1988) (Loihi), Allan et al. (1998) (EPR), Binard (1991) (Society, Australs, Pitcairn), Fretzdorff and Haase (2002) (Easter Seamount chain), Fretzdorff et al. (1996) (Reunion), Hekinian et al. (1995, 1999, 2003) (EMC, Foundation Seamount, SEPR-PAR) Coogan et al. (2003) (Hess deep, Galapagos region), and unpublished data. The trends for the various volcanic provinces are shown. *SEPR-PAR* = South East Pacific Rise Pacific-Antarctic Ridge, *NEPR* = North East Pacific Rise, *EMC* = Easter Microplate Crough Seamount. Abbreviations: *T* = trachyte, *TA* = trachyandesite, *TB* = trachy basalt, *Ti-Mt* = titano-magnetite. The data from Reunion are from the sub-marine flanks of the island (Fretzdorff et al. 1996). The Cr-spinel from the Loihi Seamount (Hawaii) is found in dunite and harzburgite (Clague 1988)

Chapter 14 · Petrology of Young Submarine Hotspot Lava: Composition and Classification   435

$(ab_{20-40}, or_{0.4-0.8})$ content in their plagioclase when compared to the other hotspot alkali basalts (Hekinian et al. 1999). The microprobe data obtained on fresh plagioclase laths from the Foundation Seamounts in the vicinity (<300 km) of the SEPR-PAR axis indicate that the laths are depleted in their orthoclase content (an_{64}, $ab_{35.7}$, $or_{0.3}$) and are similar to those of N- and T-MORBs recovered from the EPR axis (Hekinian et al. 1999). The dacites and andesites from the SEPR-PAR axis itself depart from the hotspot alkali basalts, trachy-andesites and trachytes due to their lower *or* (ab_{53-60}, $or_{0.50-0.85}$) (Hekinian et al. 1999) (Fig. 14.1a). Similarly, data from the Easter-Microplate/Crough Seamount (CEM) area and the few data we have from the Easter Seamount chain also show that the range of plagioclase composition is the same as that of the Foundation Seamounts near the SEPR-PAR axis (Fig. 14.1a).

14.2.1.2
Clinopyroxene

The clinopyroxene (cpx) minerals in hotspot and ridge crest volcanics differ in their major and trace element constituents. The cpx from hotspot regions is more enriched in moderately compatible elements such as Na, K, and Ti (Binard 1991; Allan 1989). For example, the Na_2O versus Al_2O_3 variation trend shows lower values of these oxides for the EPR with respect to those of the Pacific hotspots. Interestingly, the Reunion (island flanks) (Fretzdorff et al. 1996) volcanics fall within the cpx values for the spreading-ridge basalts with Na_2O = 0.1–0.3% and Al_2O_3 of 1–5% (Fig. 14.1b). The hotspot volcanics show the most extensive range of variability in the chemical composition of their clinopyroxene with Na_2O (0.1–0.8) and Al_2O_3 (1–12%) (Fig. 14.1b). It is also observed that the Society hotspot volcanics are the most enriched in these oxides. This agrees with the bulk-rock enrichment for their incompatible elements and corroborates their more alkaline nature (Sect. 14.2.2).

There is also a noticeable range in the content of the hotspot volcanics' mineral assemblages such as diopside-augite-salite ($wo_{41.1-50.1}$, $en_{34.5-48.3}$, fs_{7-22}). Silicic rocks (dacites and andesites) from the SEPR-PAR axis contain microphenocrysts and microlites of augite to ferroaugite clinopyroxene with wo_{29-39}, en_{30-45}, fs_{20-42} and wo_{25-41}, en_{33-45}, fs_{21-31} respectively. They have a higher Na_2O (>0.25%) content for their clinopyroxene than does the cpx of the MORBs found along the SEPR-PAR axis (Fig. 14.1b). Also, xenocrysts of clinopyroxene (wo_{30-46}, en_{33-60}, fs_{10-24}) found in basanites and alkali basalts have low Al_2O_3 (1–3%) and Na_2O (0.7–1.6%) contents (Fig. 14.1b).

14.2.1.3
Olivine

The olivine minerals show a trend of variability in their Mg# versus Forsterite (Fo) content when a comparison is made between the intraplate-hotspot and spreading-ridge volcanics. The olivine composition varies between a forsterite-poor fayalite (Fo_{20}) to forsterite-enriched (Fo_{90}) (Fig. 14.1c). The olivine composition of the spreading-ridge samples is more restricted with Fo_{70}-Fo_{89} than that of the hotspot volcanics, which varies from Fo_{30} to Fo_{90} (Fig. 14.1c). The highest forsterite (Fo_{90}) content in olivine occurs in the xenoliths hosted in the alkali basalts from the Society (Binard 1991). However, the evolved rock types such as the trachybasalts, trachy-andesites, trachytes, and andesites contain an olivine, which is more depleted in forsterite (Fo_{30-70}) than the MORBs and alkali basalts (Fig. 14.1c). The olivine from the Society and Austral hotspots is generally less evolved than that found in samples from the Pitcairn hotspot (Fig. 14.1c). Nevertheless, the olivine found in samples from all three hotspots has a moderate to high Fo composition, which varies from the rims (Fo_{65-71}) to the centers (Fo_{76-85}) of the phenocrysts.

14.2.1.4
Cr-Spinel

Chrome-spinel has an approximate formula $(Mg, Fe^{2+})(Al, Cr, Fe^{3+})_2O_4$ with minor amounts of Ti, Mn and Ca. As noted by Allan et al. (1989), the abundance of spinel is independent of the host rock's Cr content. This means that the Cr-spinel composition is a good indication of mantle source material; it is also most commonly found in the least evolved melts (MORBs, alkali basalts, picritic basalts). The presence of Cr-spinel may suggest that such rocks were derived from the partial melting of a mantle source that was enriched in chrome content. The content in Cr-spinel for the hotspot volcanoes shows a large range in variability. This is seen on the Cr# ($Cr^{2+}/Cr^{2+} + Al^{2+}$) and the Mg# ($Mg^{2+}/Mg^{2+} + Fe^{2+}$) variation diagram (Fig. 14.1d).

Cr-spinel, seen in the groundmass as inclusions in the olivine as well as in the plagioclase, is generally found in equilibrium with the least evolved MORBs, picritic basalts, alkali basalts and trachybasalts. The solubility of Cr in a basaltic liquid is usually between 200 and 300 ppm. It was found (Graham et al. 1998) that the lower the Cr content of the liquid, the lower the liquidus temperature of the Cr-rich spinel would be at partial pressure and at a fixed oxygen level. Also, the Cr content in lava will influence the spinel's composition (Graham et al. 1998). An increase in Fe in the melt will affect the concentration of Cr in the spinel and give rise to Ti-magnetite (Ti-Mt) enriched solutions (Fig. 14.1d). The composition of Cr-spinel (as seen in its Cr# and Mg#) covers the entire field of ultramafic rocks (harzburgites, dunites) from the Pacific and Atlantic provinces (Fig. 14.1d). The Loihi Seamount (Clague 1988) has alkali basalts containing harzburgite and dunites with a variable Cr# (0.3–06)-Mg# (0.4–0.8); these Loihi samples are in the same field as other intraplate volcanics (Fig. 14.1d). The EMC samples have the lowest Cr# content (0.1–0.3), and their Mg# (0.6–0.8) is as high as for the samples from the EPR off-axis seamounts (10° N, Allan et al. 1989) (Fig. 14.1d).

14.2.2
Rock Types

Based on mineralogy and chemistry, a simplified terminology has been adopted here for the various types of rocks encountered in similar regional environments. In the following paragraphs, I will enumerate some of the most common rock types erupted during hotspot and spreading ridge volcanism. Their composition is expressed by their compatible components [Mg#, SiO_2, ($Na_2O + K_2O$)] and their less compatible elements and ratios [Zr, Nd, $(Ce)_N$, K/Ti, Zr/Y, Nb/Zr, $(Ce/Yb)_N$, $(La/Sm)_N$], all of which are shown in Tables 14.1, 14.2 and 14.3 and in Fig. 14.2a–2c, 14.3a–c, 14.4a–4c, 14.5b, and 14.6a (see also Sect. 8.2.2 and Fig. 8.8).

Chapter 14 · Petrology of Young Submarine Hotspot Lava: Composition and Classification

14.2.2.1
Basaltic Rocks

14.2.2.1.1
Spreading-Ridge Axis Basalts (MORBs)

The most common types of volcanics erupted on the ocean floor consist of basalts (MORBs), which are referred to as normal (N-MORB), transitional (T-MORB) and enriched (E-MORB) and they are found on all ridge segments (SEPR, NEPR and SEPR-PAR). These previously defined basalt types (Hekinian et al. 1989; Thompson et al. 1989; Batiza and Niu 1992; Bach et al. 1994; Hekinian et al. 1999) are characterized by low amounts of $Na_2O + K_2O$ (1–4%) and low ratios for K/Ti < 0.15, Zr/Y < 3, Nb/Zr <0.07, $(La/Sm)_N$ < 0.6 and $(Ce/Yb)_N$ < 0.7 for the N-MORB and by higher ratios for K/Ti = 0.25–0.5, Zr/Y = 4–5, $(La/Sm)_N$ = 1–2 and $(Ce/Yb)_N$ = 1–2 for the E-MORBs (Tables 14.1, 14.2 and 14.3, Fig. 14.2a,b, 14.3a–c). The T-MORBs have intermediate values for these incompatible element ratios (Hekinian et al. 1989). The MORBs consist of Ca-rich plagioclase (an_{73-85}, $or_{0.01-0.06}$), Mg-enriched olivine (Fo_{76-88}), Na- and Al-depleted clinopyroxene (Na_2O < 3%, Al_2O_3 < %) and variable Cr-spinel composition (Fig. 14.1a–c). They are hypersthene normative as opposed to the undersaturated alkalic lava suites (Table 14.1). The niodynium and strontium isotopic ratios (Graham et al. 1988; Bach et al. 1994) of lava erupted on the NEPR and SEPR axis are between 0.5129–0.5132 (Nd) and 0.7020–0.7030 (Sr) (Fig. 14.5c).

14.2.2.1.2
Intraplate Hotspot Basaltic Rock

The intraplate extrusive flows were classified into separate categories according to their mineral associations, their composition and morphology. Based on their major element, trace element and REE distributions, several different types of hotspot alkali-enriched basaltic lava have been recognized, including *picritic alkali basalts, tephrites, basanites and alkali basalts*. The geochemistry of these basaltic lavas has been previously well documented (Hekinian et al. 1991; Binard 1991; Devey et al. 1990; Stoffers et al. 1989; Woodhead and Devey 1993; Garcia et al. 1993; Ackermand et al. 1998; Hekinian et al. 2003).

Picritic alkali basalt. These rock types are essentially seen as pillow lavas and contain 7–15% of early formed olivine and spinel. They have high values of Ni (300–500 ppm) and Mg# (Mg# = 100 × Mg^{2+} / Mg^{2+} + Fe^{2+}) (65–80, bulk rock) and low SiO_2 (>47%), $Na_2O + K_2O$ (2–5%), Ba (160–300 ppm), Zr (140–300 ppm), K/Ti (>0.6), Zr/Y (8–10) and $(La/Sm)_N$ (1–3) values that are similar to the alkali basalts (Hekinian et al. 1991, 1999) (Tables 14.1–14.3 and Fig. 14.2b, 14.5a). These rock types are composed of Ca-enriched plagioclase (an_{70-85}), Mg-enriched olivine (Fo_{80-89}) and Cr-spinel (Fig. 14.1c,d). They differ from the picritic basalt of MORB types because of the higher orthoclase content of their plagioclase (ab_{15-30}, or_{1-9}) and the presence of bulk rock normative nepheline (1–5%) (Hekinian et al. 1991; Binard 1991) (Tables 14.1 and 14.3).

Tephrite. Tephrites are vesicular pillow lavas essentially containing plagioclase, clino-pyroxene, and kaersutite. They are less common than the other types of hotspot basaltic rocks and have low SiO_2 (<44%), high Zr (300–500 ppm), high Ba (400–800 ppm), low Ni

Table 14.1. Major oxides bulk analyses of selected representative rock types from hotspots and oceanic spreading ridges

Region	Oceanic spreading ridge				Society hotspot					Pitcairn hotspot		
Province	SEPR-PAR	SEPR-PAR	SEPR-PAR	GSC	Mouha-Pihaa	Mouha-Pihaa	Teahitia	Teahitia	Rocard	Bounty	Bounty	Adams
Rock type	MORB	Dacite	Andesite	Rhyolite	Tephrite	Basanite	Alk. basalt	Picr. B.	Phonolite	TB	TA	T
Sample	SO100-85-1	SO100-91-1	SO100-92-2	997	SO47-29-2	SO47-28-1	CHDR9-I	CHDR9-3	CHDR4-2	PNDR3-01	PNDR12-1	PNDR1-1
SiO_2 (wt.%)	49.74	60.20	55.94	70.71	42.52	43.55	47.15	43.89	58.05	51.86	55.91	59.58
TiO_2	1.56	1.19	1.41	0.61	4.48	3.41	2.98	2.84	0.83	1.80	1.18	0.57
Al_2O_3	14.89	13.25	13.82	12.30	14.76	13.72	12.80	9.87	17.33	16.97	16.38	17.33
Fe_2O_3	1.12	10.10	1.66	–	4.98	4.06	2.61	2.32	1.30	3.10	–	–
FeO	9.04	9.10	8.49	5.30	8.21	8.26	8.31	9.97	3.96	8.82	8.86	5.99
MnO	0.17	0.16	0.16	–	0.23	0.19	0.15	0.17	0.21	0.19	0.22	0.17
MgO	7.48	1.94	3.08	0.43	5.63	6.33	9.91	15.98	1.30	2.52	1.36	0.50
CaO	12.01	5.10	6.31	2.92	9.51	12.18	8.80	9.59	2.04	5.20	3.85	2.29
Na_2O	2.82	4.76	4.90	4.14	4.61	3.01	2.80	2.25	6.30	5.33	4.99	5.86
K2O	0.17	1.05	0.82	1.30	1.90	1.07	2.21	1.16	5.75	3.18	3.73	5.04
P_2O_5	0.27	0.31	0.40	0.05	0.81	0.52	0.59	0.44	0.54	1.14	0.54	0.18
LOI	−0.44	1.52	1.73	–	–	–	–	–	–	0.59	1.41	0.68
Total	98.83	99.58	98.72	97.76	99.36	96.30	98.31	99.53	97.07	99.81	99.41	98.85
K/Ti	0.16	1.31	0.86	3.16	0.63	0.47	1.10	0.61	10.29	2.45	4.69	12.25
$Na_2O + K_2O$	2.99	5.81	5.72	5.44	6.51	4.08	5.01	3.41	12.05	8.51	8.72	10.90
Mg#	59.59	29.69	41.81	13.84	57.60	60.28	70.26	76.05	39.40	36.15	23.30	14.19

Chapter 14 · Petrology of Young Submarine Hotspot Lava: Composition and Classification

Table 14.1. *Continued*

Region	Oceanic spreading ridge				Society hotspot					Pitcairn hotspot		
Province	SEPR-PAR	SEPR-PAR	SEPR-PAR	GSC	Mouha-Pihaa	Mouha-Pihaa	Teahitia	Teahitia	Rocard	Bounty	Bounty	Adams
Rock type	MORB	Dacite	Andesite	Rhyolite	Tephrite	Basanite	Alk. basalt	Picr. B.	Phonolite	TB	TA	T
Sample	SO100-85-1	SO100-91-1	SO100-92-2	997	SO47-29-2	SO47-28-1	CHDR9-I	CHDR9-3	CHDR4-2	PNDR3-01	PNDR12-1	PNDR1-1
CIPW norm												
q (wt.%)	0.00	12.34	5.09	32.24	0.00	0.00	0.00	0.00	0.00	0.00	0.00	0.00
or	1.00	6.21	4.85	7.68	11.23	6.32	13.06	6.86	33.98	18.79	22.05	29.79
ab	23.86	40.28	41.46	35.03	13.37	14.61	20.39	9.84	42.15	38.45	42.22	49.34
an	27.47	11.69	13.29	11.14	13.97	20.76	15.83	13.40	2.02	12.99	11.28	6.09
lc	0.00	0.00	0.00	0.00	0.00	0.00	0.00	0.00	0.00	0.00	0.00	0.00
ne	0.00	0.00	0.00	0.00	13.89	5.88	1.79	4.98	6.05	3.60	0.00	0.13
di	24.84	9.89	12.94	2.64	22.29	29.21	19.22	25.02	3.84	4.52	3.70	3.57
hy	8.76	12.31	13.35	6.96	0.00	0.00	0.00	0.00	0.00	0.00	10.41	0.00
ol	8.12	0.00	0.00	0.00	5.29	5.94	17.21	28.60	4.86	11.20	2.55	6.19
mt	1.62	1.47	2.40	0.85	7.22	5.89	3.78	3.36	1.88	4.49	1.43	0.97
il	2.96	2.26	2.68	1.16	8.51	6.48	5.66	5.39	1.58	3.42	2.24	1.08
ap	0.59	0.68	0.87	0.11	1.77	1.14	1.29	0.96	1.18	2.49	1.18	0.39
Total	99.23	97.12	96.94	97.81	97.54	96.23	98.23	98.42	97.54	99.95	97.05	97.56

The isotope analyses are from Devey et al. (1990), Hémond (1994), and Devey and Haase (2003). Galapagos Spreading Center (GSC) data is from Clague et al. (1981). South East Pacific Rise-Pacific Antarctic (SEPR-PAR) Ridge isotope data is from Maia et al. (2001). The major oxides, trace and lanthanides are from Devey et al. (1990), Hekinian et al. (1991), Hémond et al. (1994) and Hekinian et al. (2003).

$FeO^* = Fe_2O_3 / 1.11$. Total volatiles, LOI = Lost on ignition, SO = FS SONNE cruise (1987, 1995). CH = RV J. CHARCOT (1986). PN = Polynaut cruise, RV L'Atalante (1999).

Mg# = atomic proportion of $Mg^{2+}/Mg^{2+} + Fe^{2+}$. T = trachybasalt, picr. B. = picritic basalt. (–) = no data.

CIPW = initials of the names Cross, Iddings, Pirson and Washington who invinted the system of classification based on their normative minerals.

Table 14.2. Bulk rock trace element analyses of selected representative rock types from hotspots and oceanic spreading ridges

Region	Oceanic spreading ridge				Society hotspot					Pitcairn hotspot		
Province	SEPR-PAR	SEPR-PAR	SEPR-PAR	GSC	Mouha-Pihaa	Mouha-Pihaa	Teahitia	Teahitia	Rocard	Bounty	Bounty	Adams
Rock type	MORB	Dacite	Andesite	Rhyolite	Tephrite	Basanite	Alk. basalt	Picr. B.	Phonolite	TB	TA	T
Sample	SO100-85-1	SO100-91-1	SO100-92-2	997	SO47-29-2	SO47-28-1	CHDR9-I	CHDR9-3	CHDR4-2	PNDR3-01	PNDR12-1	PNDR1-1
Rb (ppm)	2.5	16.5	12.4	32.0	53.0	34.0	41.9	26.9	138.0	47.1	38.1	76.5
Sr	143.0	111.0	109.0	38.0	1031.0	661.0	709.0	546.0	103.0	738.8	459.5	265.3
Ba	28.8	129.0	94.7	–	525.0	291.0	576.0	345.0	260.0	738.8	847.3	1400.2
Y	28.8	115.0	100.0	165.0	43.0	34.0	28.6	25.8	67.0	44.0	52.9	53.6
Zr	100.0	735.0	580.0	365.0	370.0	271.0	328.0	230.0	1170.0	519.7	605.9	819.4
Nb	6.4	32.0	24.7	–	62.0	41.0	40.4	808.0	129.0	66.1	73.5	94.3
Sc	42.0	21.9	19.9	130.0	–	–	–	78.0	–	–	–	–
V	308.0	83.0	128.0	115.0	302.0	248.0	230.0	230.0	–	176.6	73.5	94.3
Cr	190.0	14.0	54.2	28.0	–	53.0	507.0	808.0	–	611.3	0.2	1.9
Co	40.0	18.0	22.0	–	39.0	51.0	56.0	78.0	–	44.0	2.5	1.1
Ni	67.0	8.0	21.9	10.0	8.0	84.0	275.0	540.0	–	519.7	0.3	2.0
Zn	89.8	100.6	102.9	100.0	143.0	116.0	125.0		201.0	738.8	203.9	189.9
Cu	77.2	33.1	31.9	22.0	9.0	130.0	49.0	58.0		66.1	2.7	3.4
Pb	–	–	–	–	8.0	5.0	5.3	3.4	17.0	–	–	–
Zr/Y	3.47	6.39	5.80	2.21	8.60	7.97	11.47	8.91	17.46	11.82	11.45	15.28
Nb/Zr	0.06	0.04	0.04	–	0.17	0.15	0.12	3.51	0.11	0.13	0.12	0.12

The isotope analyses are from Devey et al. (1990), Hémond (1994), and Devey and Haase (2003). Galapagos Spreading Center (GSC) data is from Clague et al. (1981). South East Pacific Rise-Pacific Antarctic (SEPR-PAR) Ridge isotope data is from Maia et al. (2001). The major oxides, trace and lanthanides are from Devey et al. (1990), Hekinian et al. (1991), Hémond et al. (1994) and Hekinian et al. (2003). FeO* = Fe_2O_3/1.11. Total volatiles, LOI = Lost on ignition, SO = FS SONNE cruise (1987, 1995). CH = RV J. CHARCOT (1986). PN = Polynaut cruise, RV L'Atalante (1999). Mg# = atomic proportion of Mg^{2+}/Mg^{2+} + Fe^{2+}. T = trachyte. TA = trachyandesite, TB = trachybasalt, picr B. = picritic basalt, (–) = no data. CIPW = initials of the names Cross, Iddings, Pirson and Washington who invinted the system of classification based on their normative minerals.

Chapter 14 · Petrology of Young Submarine Hotspot Lava: Composition and Classification 441

Table 14.2. Continued

Region	Oceanic spreading ridge				Society hotspot					Pitcairn hotspot		
Province	SEPR-PAR	SEPR-PAR	SEPR-PAR	GSC	Mouha-Pihaa	Mouha-Pihaa	Teahitia	Teahitia	Rocard	Bounty	Bounty	Adams
Rock type	MORB	Dacite	Andesite	Rhyolite	Tephrite	Basanite	Alk. basalt	Picr. B.	Phonolite	TB	TA	T
Sample	SO100-85-1	SO100-91-1	SO100-92-2	997	SO47-29-2	SO47-28-1	CHDR9-1	CHDR9-3	CHDR4-2	PNDR3-01	PNDR12-1	PNDR1-1
La (ppm)	5.94	38.00	30.91	26.50	49.50	25.70	31.00	32.70	–	82.99	89.08	102.35
Ce	15.42	91.00	75.00	68.00	110.50	60.70	98.50	78.20	243.00	165.54	178.43	195.57
Pr	–	–	–	–	–	–	–	–	–	20.46	22.09	23.14
Nd	11.44	57.00	47.89	41.00	60.20	35.50	51.00	42.50	110.40	79.69	85.30	84.03
Sm	3.53	–	12.74	12.40	12.50	8.00	10.64	9.30	19.97	15.60	16.86	16.15
Eu	1.33	2.90	2.73	2.60	3.92	2.61	3.34	2.91	4.78	4.47	5.09	4.44
Gd	3.49	–	12.57	14.00	11.20	7.38	9.19	8.36	14.97	12.89	14.21	13.06
Tb	–	–	–	2.70	–	–	–	–	–	1.82	2.08	1.99
Dy	4.36	18.80	15.13	–	8.24	5.76	6.25	5.72	11.58	9.78	11.45	11.34
Ho	–	–	–	4.80	–	–	–	–	–	1.77	2.11	2.14
Er	2.59	12.50	8.95	–	3.61	2.67	2.52	2.28	5.57	4.35	5.36	5.65
Tm	–	–	–	–	–	–	–	–	–	0.58	0.72	0.79
Yb	2.53	12.10	10.72	13.70	2.70	2.80	1.76	1.54	–	3.47	4.46	5.02
Lu	–	–	–	2.05	–	–	–	–	–	0.48	0.63	0.72
$(Ce/Yb)_N$	1.69	2.09	1.94	1.37	11.37	6.02	15.55	14.11	–	13.26	11.11	10.82
$(La/Sm)_N$	1.09	–	1.57	1.38	2.56	2.07	1.88	2.27	–	3.44	3.41	4.09
$^{87}Sr/^{86}Sr$	0.70272	0.70285	0.70279	–	0.70361	0.70380	0.70551	0.70340	0.70602	0.70507	0.70520	0.70520
$^{143}Nd/^{144}Nd$	0.51303	0.51302	0.51306	–	0.51299	0.51293	0.51271	0.51280	0.51260	0.51253	0.51249	0.51247
$^{206}Pb/^{204}Pb$	19.067	19.065	18.954	–	19.3	–	19.12	19.14	19.21	–	–	–

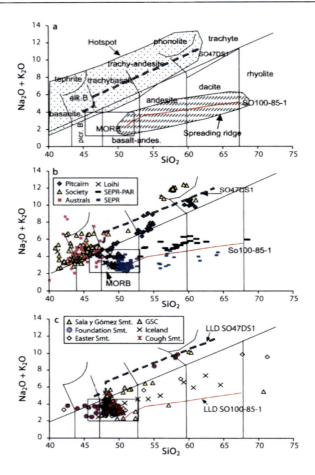

Fig. 14.2. Composition of the volcanics encountered on submarine edifices from hotspots and spreading-ridge axis. The chemical analyses from the hotspot area are compiled from Hekinian et al. (1991), Binard (1991), Binard et al. (1992), Woodhead and Devey (1993), Ackermand et al. (1998), Hekinian et al. (2003) and from unpublished data (see also Sect. 8.2.2 and Fig. 8.6). The East Pacific Rise (EPR) sample distribution are from Hekinian and Walker (1987), Hekinian et al. (1989), Batiza and Niu (1992), Mahoney et al. (1994), Bach et al. (1994) and Batiza et al. (1996). The South East Pacific Rise-Pacific-Antarctic Ridge (SEPR-PAR) data are from Devey et al. (1997) and Hekinian et al. (1999). The *liquid line* of descent defining the fractionation trend was calculated (Weaver and Langmuir 1990; Nielsen and Delong 1992) from a basanite (*solid line*, 47DS1, Hekinian et al. 1991), a picritic basalt (*dashed-line*, PN3-05) and a T-MORB (SO 100-85-1), (Table 14.1). **a** SiO_2 and $Na_2O + K_2O$ variation diagram shows the general and global fields of various types of volcanics from hotspots and the ridge axis. These fields are redrawn after Middlemost (1980). The rock classification adopted here extends the field of the trachybasalt into the conventional trachybasalt-andesite field (Middlemost 1980; Le Bas and Streckeisen 1986). The *small box* encloses the field of East Pacific N-, T- and E-MORBs (data compiled by Hekinian). **b** SiO_2 and $Na_2O + K_2O$ variation diagram shows the detailed plot of samples from the Pacific Ocean hotspots. Data from the Loihi Seamount (Hawaii) is from Garcia et al. (1993). Other references are the same as in Fig. 14.1b,c. **c** SiO_2 and $Na_2O + K_2O$ diagram of South Pacific seamount chains such as the Easter-Microplate-Crough-Seamount (EMC) (Hekinian et al. 1995b), Sala y Gómez (Pan and Batiza 1998), Easter Island Seamounts (Easter Smt.) (Fretzdorff et al. 1996; Haase and Devey 1996; Haase 2002), Galapagos Spreading Center (GSC) (Byerly 1980; Clague 1981), Foundation (Foundation Smt.) (Hekinian et al. 1999) and Iceland (Carmichael 1964). The trends of crystal-liquid fractionation were obtained using Weaver and Langmuir (1990) and Nielsen and Delong (1992) algorithms applied to a MORB (SO100-85-01) and basanite (SO47DS1) parents. The legend is the same as in Fig. 14.1a–c

Chapter 14 · **Petrology of Young Submarine Hotspot Lava: Composition and Classification**   443

Fig. 14.3. a Zr/Y-Zr, b (CE/Yb)$_N$-(Ce)$_N$ and c Nb/Zr-Nb variation diagrams show the trend of incompatible element variation for hotspots. The fields of spreading-ridge samples from the North and South East Pacific Rise (EPR) are *contoured*. The distribution of EPR (Hekinian and Walker 1987; Hekinian et al. 1989; Mahoney et al. 1994; Bach et al. 1994) samples is shown in the *empty fields*. The *single melting curve* represents the combination of several calculated melting trends obtained from the modeling of different mantle sources (garnet lherzolite 66AL-1, spinel-lherzolite R255, and spinel-clinopyroxenite SC73-2P (Irving 1980; Frey 1980; Frey et al. 1985; Hekinian and Bideau 1995). The melting trends comprising most MORBs and alkali basalts are obtained from a mixed source composed of 15% clinopyroxene + 85% spinel-lherzolite (Hekinian and Bideau 1995; Hekinian et al. 1999). The calculated (Weaver and Langmuir 1990; Nielsen and Delong, 1992) trends of crystal-liquid fractionation and of partial melting are shown. The parent sources chosen for the hotspot lava are an alkali basalt (SO100-28-01), a T-MORB (SO100-85-01) and an N-MORB (SO100-74-3) (Hekinian et al. 1999). T = trachyte, TA = trachy-andesite, bas = basanite, TB = trachy basalt. The normalizing factors are related to C1 (Chondrite) values (La = 0.237, Ce = 0.612, Sm = 0.153, Yb = 0.170, Sun and McDonough 1989)

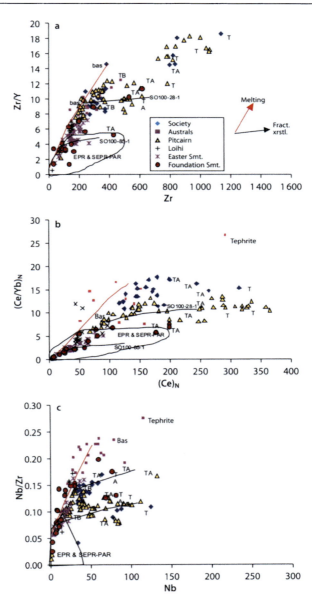

(<50 ppm), high Na$_2$O + K$_2$O (6–9%), high Zr/Y (8–12), (Ce/Yb)$_N$ (10–30), Nb/Zr (0.15–0.28), (La/Sm)$_N$ (2–6) and moderate Mg# (39–62) contents (Binard 1991; Hekinian et al. 1999) (Tables 14.1–14.3, Fig. 14.2a,b, 14.3a–c, 14.5a). The tephrites are comparable to the nephelinite from Tubai reported by Juteau and Maury (1997). Tephrites have a normative nepheline content higher (10–25%) than the other alkalic lava types (Table 14.1) (Hekinian et al. 1991; Binard et al. 1991). In addition, they contain normative leucite (Table 14.1) as well as modal nepheline and orthoclase as phenocrysts and microphenocrysts.

Basanite. The samples of basanites were recovered from vesicular pillow lava and lava tubes with radial jointing. They are composed essentially of clinopyroxene, plagioclase and olivine phenocrysts (9–15%) (Hekinian et al. 1991). When compared to the alkali basalts (see below), basanites have a lower SiO_2 (<46%), high total alkali content (Na_2O + K_2O = 3.5–6%), relatively moderate Ni (70–100 ppm), moderate to high Mg# (40–65), high Ba (200–600 ppm), Zr (200–400 ppm), high $(La/Sm)_N$ (2–5), which is comparable to alkali basalt, $(Ce/Yb)_N$ (6–15), Nb/Zr (0.15–0.26) and Zr/Y (8–10) (Tables 14.1–14.3, Fig. 14.2b, 14.3a–c, 14.5a). Basanites also differ from the tephrites because of their lower incompatible element content and their mineralogy. Their modal plagioclase is relatively depleted in orthoclase content (ab_{20-35}, or $_{0.3-0.8}$) and contains more normative nepheline (5–10%) compared to the alkali basalts and basanites (Table 14.1 and Hekinian et al. 1991).

Alkali basalt. Alkali basalts mainly form pillow lavas and giant tubes with well-developed radial joints. They are enriched in SiO_2 (45–48%) Na_2O + K_2O (3.5–5%), have a variable Mg# (50–69), Ba (250–400 ppm), K/Ti (>0.4), Zr/Y (2–13), Nb/Zr (0.06–0.15) and $(Ce/Yb)_N$ (5–15). They have moderate Ni (30–100), Zr (170–300 ppm), and $(La/Sm)_N$ (1–3) values (Hekinian et al. 1991, 1999) (Tables 14.1–14.3, Fig. 14.2a–c, 14.3a–c). They consist of Ca-rich plagioclase (ab_{19-45}, $or_{0.4-2.5}$), Mg-enriched olivine (Fo_{75-83}) and Cr-spinel (Fig. 14.1a,c,d). The clinopyroxene is often more enriched in TiO_2 and Na_2O (0.15–0.6%) than normal MORBs. The alkali basalts have a relatively low (1–4%) normative nepheline content when compared to the alkali enriched rocks (Table 14.1) (Hekinian et al. 1991; Binard 1991).

14.2.2.2
Silicic Rocks

Oceanic silicic (SiO_2 > 53%) lava differs from subaerial silica-rich rock because of its thick glassy crusts and its shape. Silicic rocks are less abundant than basaltic lava, and it is often difficult to distinguish them on the ocean floor without analytical data. Because of their high viscosity, they are more elongated with a smoother and/or a more scoriaceous surface than normal MORB flows. They rarely show corrugated ridges of cooling surface and are mainly characterized by thick (5–20 cm thick) glassy aphyric margins with large cavities (1–2 cm long). Some surfaces appear scoriaceous due to the agglutination (welded glassy material) of their glassy crusts during quenching (Sect. 5.2.1 and 5.2.3.2 and Fig. 5.8b,h).

Several types of silicic rocks occur in different oceanic environments: (1) trachytes, trachy-andesites and trachybasalt are mainly found in the intraplate provinces of hotspot origin and (2) andesites, dacites, and rhyolites are most commonly found in accreting ridge systems and back arc basins. Although island arcs and oceanic islands have erupted a fair amount of silicic lava, very little is known about silica-enriched flows along the spreading-ridge systems of the world's ocean. The most detailed visual observations of submarine silica-enriched lava are reported from the Pitcairn and Society hotspots and from the SEPR-PAR spreading axis (Hekinian et al. 2003; Hekinian et al. 1999; Stoffers et al. 2002). In the following paragraph, some information is given concerning silicic samples recovered from the South and North EPR, which are believed to be further away from hotspot-centered ridge segments.

Chapter 14 · Petrology of Young Submarine Hotspot Lava: Composition and Classification

14.2.2.2.1
Spreading-Ridge Silicic Lava

In the Eastern Pacific, only a few occurrences of silicic lava are documented. They are located on a 25–30 Myr old crust on the flank of the North-East Pacific Rise (NEPR) at 9°30' N and 10°30' N (Langmuir et al. 1986; Thompson et al. 1989; Batiza et al. 1996). Other samples have been recovered from the 95° W propagator of the Galapagos Spreading Center (Clague et al. 1981), along the South-East Pacific Rise (SEPR) near 14°–21° S (Sinton et al. 1991; Bach et al. 1994), and some silicic lavas have also been found on the Mid-Atlantic Ridge (Aumento 1969). Silicic lavas have also been recently recovered from the South-East Pacific Rise Pacific-Antarctic Ridge (SEPR-PAR) axial domes distributed along a distance of 290 km between 37°11' S and 39°48' S (Hekinian et al. 1999; Stoffers et al. 2002). Spreading-ridge silicic rock types are classified as being *rhyolites, dacites, andesites*, and *basaltic andesites*.

Rhyolite. Two rhyolites (and/or rhyodacites) reported from the Galapagos spreading ridge (Clague et al. 1981; Christie and Sinton 1981; Byerly 1980) contain high SiO_2 (70–73%), a high ratio of $(La/Sm)_N$ (1–2), $(Ce/Yb)_N$ (1–2), and Zr/Y (2–3), high total alkali content ($Na_2O + K_2O = 5$–8%), and a low Mg# (<30) (Clague et al. 1981) (Tables 14.1–14.3, Fig. 14.2a,c, 14.4a). Also, diorites and syenites reported from dredge hauls on the MAR (Aumento 1969) have major and minor element compositions comparable to the other rhyolites.

Dacite. Dacites consist of silica-enriched ($SiO_2 = 59$–66%) and $Na_2O + K_2O$ (3–5%) enriched lavas, with a low Mg# (Mg# = $100 \times Mg^{2+}/Mg^{2+} + Fe^{2+}$) (16–35), high Zr = 300–700 ppm, and high ratios of Zr/Y (>2), Nb/Zr (0.03–0.07) $(La/Sm)_N$ (0.7–2) and $(Ce/Yb)_N$ (1.8–3) (Tables 14.1–14.3, Fig. 14.2b,c, 14.3a–c, 14.4a) (Hekinian et al. 1999). They differ from hotspot region trachytes because of their mineralogy (amphibole) and their lower total alkali content ($Na_2O + K_2O < 5$%). The few visual observations reported (Hekinian et al. 1999, Stoffers et al. 2002) showed that dacites are dark-gray glassy aphyric lava flows with a "bread crumb" surface texture. This type of lava consists of plagioclase (andesine, an_{35-50}, or $or_{0.8-1.2}$) Mg-poor olivine (<Fo_{40}) when present, and titanomagnetite. Dacites have thick (about 22 cm) glassy crusts and form massive flows with a lobate appearance. Large empty cavities (up to 1 cm in length) and small vesicles (10–18%) oriented in the main flow directions are abundant.

Andesite. Andesites are moderately enriched in silica ($SiO_2 = 53$–59%) and $Na_2O + K_2O$ (3–5.5%), have a low Mg# (19–45), high Zr = 150–600 ppm, and high ratios of Zr/Y (1–5), Nb/Zr (0.04–0.08), $(La/Sm)_N$ (0.75–2) $(Ce/Yb)_N$ (1.5–3) (Tables 14.1–14.3, Fig. 14.2b,c, 14.3a–c, 14.4a,b). The dredged samples of freshly erupted andesite from the PAR axis and deep-tow observations made in 1996 (Hekinian et al. 1999) and later in 2001 (Stoffers et al. 2002) show the presence of large fragmented lobate flows on top of domed shaped structures. The samples from these flows show iridescent (oily) glassy crust (10–15 cm thick). The glassy surface was layered in different colors going from deep blue to light brown and greenish grey when first brought on deck of the ship. After a few weeks, the various colors faded away and gave rise to a homogeneous dark-grey normal glassy appearance. The glassy rocks consist of spherules with centers of crystallization (mainly plagioclase needles) and interstitial small vesicles throughout the spherules. Plagioclase phenocrysts consisting of oligoclase-andesine (an_{20-50}, $or_{0.5-1}$) and Fe-rich olivine (Fo_{40-70}) crystals were also encountered (Fig. 14.1a,c).

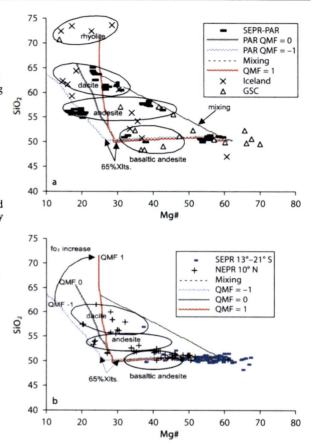

Fig. 14.4. SiO$_2$ versus Mg# plot showing: a Pacific-Antarctic Ridge (*PAR*) volcanics and tholeiites and andesites from the Thingmuli Volcano in Iceland (Carmichael 1964) and from the Galapagos Spreading Center (*GSC*) (Clague et al. 1981) compared to: b other glassy samples from the East Pacific Rise in the South (*SEPR*) (Sinton et al. 1991; Bach et al. 1994) and in the North (*NEPR*) (Batiza et al. 1996; Batiza pers. comm.). The trends of crystal fractionation calculated (Nielsen and Delong 1992) at oxygen fugacity (*fO$_2$*) equivalent to a quartz-magnetite-fayalite (*QMF*) buffer (*QMF* = 0 and 1) and lower than a *QMF* buffer (*QMF* = –1, more reduced) are shown. The calculated mixing curve shows a close fit when 75% MORB is mixed with 25% dacite melt. The legends are as in Fig. 14.2a–c, 14.3a–c. *Xrst.* = crystalline phase solidified

Basaltic andesite. Basaltic andesites consist of relatively high SiO$_2$ (51–53%), high Na$_2$O + K$_2$O contents (1.5–4%), high amounts of zirconium (Zr = 120–350 ppm), and higher ratios for Zr/Y (3–4), Nb/Zr (0.02–0.06), (La/Sm)$_N$ (0.6–1.3) (Ce/Yb)$_N$ (0.7–3), but there is a lower Mg# (33–50) than found in the MORBs (Mg# 55–72) (Table 14.3, Fig. 14.2b,c, 14.3a,c, 14.4a,b). Basaltic andesites have more sodic plagioclase (ab$_{30–40}$, or$_{0.40–0.1}$) and are more depleted in the forsterite (Fo$_{65–80}$) of their olivine with respect to MORBs (Fig. 14.1a,c).

14.2.2.2.2
Relationship Between Spreading-Ridge Silicic Lava and Basalts

Understanding the relationship between silica-enriched lava erupted together with MORBs might give some indication about their origin. To explain any relationship, I have modeled the co-variation of their Mg# with their SiO$_2$ content during the fractional crystallization of a PAR type MORB parent (sample SO100-85-1) (Fig. 14.4a,b). The spreading-ridge volcanics are plotted in order to show the variation in their composition, going from a basalt (MORB) to basaltic-andesite to andesite to dacite and finally to rhyolite (Fig. 14.4a,b).

Chapter 14 · Petrology of Young Submarine Hotspot Lava: Composition and Classification   447

The calculated trends of the plotted spreading-ridge volcanics' liquid lines of descent (LLD) are flat, and the melt increases at a nearly constant Mg# until ~65% crystallization, but thereafter the Mg# becomes more sensitive to the melts' oxidation-state. At high oxygen fugacity, titano-magnetite (Ti-Mt) is crystallized and the SiO_2 content (>51%) of the melt increases drastically. At a variable oxygen fugacity (fO_2) (Sacks et al. 1980; Juster and Grove 1989), Ti-Mt is removed and the SiO_2 content continues to increase with a decline in Mg#. Other silicic lavas (andesite) with intermediate Mg# (45–35) and high SiO_2 contents (58–64%) are best explained by magma mixing between highly fractionated magmas formed at high oxygen fugacity and unevolved (silica poor) basaltic melts. The oxidation of FeO to Fe_2O_3 and the crystallization of Ti-Mt are both controlled by the oxygen fugacity (fO_2) (Juster and Grove 1989). The degree of Fe oxide enrichment also varies with the oxygen fugacity. Thus, increasing the oxidation state in a melt will precipitate more titanium-iron phases and drive the liquid towards an enrichment in silica. This crystal fractionation could take place in a solidification zone surrounding the magma chamber (e.g., Nielsen and DeLong 1992). Relatively buoyant residual silicic magma could migrate upwards along the margin of this solidification zone, potentially interacting with large volumes of altered wallrock and thereby increasing its oxidation state.

Another alternative hypothesis explaining the eruption of andesitic lavas with intermediate composition is the partial melting of oceanic plagiogranite as suggested by Sigurdsson and Sparks (1981) for Icelandic silicic lava. The mixing between a MORB and a residue of silica-enriched (evolved) melt will give rise to eruptions of andesites with intermediate Mg# (38–42) and lower SiO_2 (58–60%) contents than the other more fractionated silicic melt of dacites and andesites (Fig. 14.4a,b). However, as shown in Sect. 14.3, silicic lava has the same $^{87}Sr/^{86}Sr$ and $^{143}Nd/^{144}Nd$ ratios as the associated MORB (Table 14.2, Fig. 14.5c). Hence it is likely that the silicic andesite and dacite have the same origin as the basalt from which they were derived through a process of crystal fractionation.

14.2.2.2.3
Hotspot (Intraplate) Silicic Lava

Based on their surface morphology and their composition, the silicic lava from intra-plate hotspot edifices are more alkali-enriched lava with respect to the spreading-center silicic lava and are divided into *trachybasalts*, *trachy-andesites*, *trachytes* and *phonolites*.

Trachybasalt. Trachybasalts are pillow lava having typical radial jointing. They consist of SiO_2 (49–53%), $Na_2O + K_2O$ (4–7%), Mg# (45–60) Ba (300–800 ppm), Ni (<100 ppm), Zr = 300–500 ppm, Zr/Y (10–13), Nb/Zr (0.06–0.15) $(La/Sm)_N$ (2–3), $(Ce/Yb)_N$ (7–17) and Ni (4–100 ppm) (Hekinian et al. 1991, 1999) (Tables 14.1–14.3, Fig. 14.2a–c, 14.3a–c). They differ from other alkali lavas (picritic basalt, alkali basalt and basanite) in their more sodic-plagioclase (ab_{30-45}, or_{2-4}) and in the forsterite content of their olivine (Fo_{70-80}) (Fig. 14.1a,c).

Trachy-andesite. Trachy-andesites occur mostly as blocky/tabular and flattened giant tubular flows. They are characterized by relatively high values of SiO_2 (53–59%), $Na_2O + K_2O$ (4–8%), Ba (500–1 000 ppm), Zr/Y (9–15), Nb/Zr (0.1–0.2), $(La/Sm)_N$ (2.3–4.5), $(Ce/Yb)_N$ (9–17) and Zr (400–800 ppm) and intermediate Mg# (20–45) (Tables 14.1–14.3,

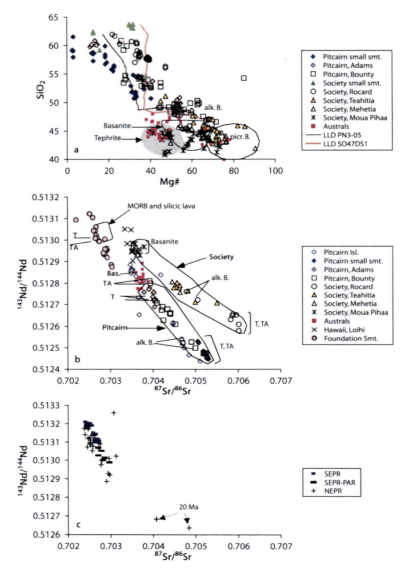

Fig. 14.5. Composition of volcanic rocks from submarine edifices from Pacific hotspots is compared to spreading-ridge volcanics. **a** SiO$_2$ versus Mg# plot using previously published data analyses (Binard et al. 1992a, 1992b; Devey et al. 1990; Hémond et al. 1994; Woodhead and Devey 1993; Hekinian et al. 2003; Devey et al. 2003). The *fields* show the range of composition for the large volcanoes from Pitcairn and Society hotspots. Rocks with high Mg# (>75) contain olivine xenocrysts. **b** and **c** ^{87}Sr/^{86}Sr versus ^{143}Nd/^{144}Nd isotope diagram for Pacific hotspot volcanoes (Sect. 8.2.2 and Fig. 8.8) and the East Pacific Rise (*EPR*). The two *NEPR* samples having the highest Sr isotopic values are from an off-axis seamount built on a 20-Myr crust (Graham et al. 1988). The data are compiled after Graham et al. (1988), Devey et al. (1990), and Mahoney et al. (1994) for the spreading-ridge segments (*NEPR* at 9°–10° N, *SEPR-PAR* at 37°–39° S and *SEPR* at 14°–21° S) and from Garcia et al. (1993) Hémond et al. (1994), Woodhead and Devey (1993), and Devey et al. (2003) for the intraplate hotspot rocks. The legends of the figures are the same as in Fig. 14.2b,c, 14.3a–c. *alk. B.* = alkali basalt, *picr. B.* = picritic alkali basalt

Chapter 14 · Petrology of Young Submarine Hotspot Lava: Composition and Classification   449

Fig. 14.2b, 14.3a–c). They have a plagioclase composition of ab_{50-70}, or_{4-16}, and the bulk rock contains about 20–30% normative orthoclase. The trachy-andesites' clinopyroxene is relatively enriched in Al_2O_3 (6–10%) and Na_2O (5–7%) with respect to the alkali-basaltic lava described later on in this chapter (Fig. 14.1a,b). The forsterite content of the occasional olivine is about Fo_{60-75} (Fig. 14.2b).

Trachyte. Trachytes are light-grey tabular and blocky flows with elongated vesicles (Sect. 5.2.3.1 and Fig. 5.8h). They are aphyric and have the highest SiO_2 (59–64%), $Na_2O + K_2O$ (7–12%), Ba (>1 000 ppm), Zr/Y (>15), Nb/Zr (0.12–0.25), $(La/Sm)_N$ (3–6), $(Ce/Yb)_N$ (10–20) and Zr (>700 ppm) values (Hekinian et al. 1999) (Tables 14.1–14.3, Fig. 14.2a–c, 14.3a–c). They have a low Mg# (19–35) like other silicic lava such as dacites from the spreading ridges (Sect. 14.2.2.2). The trachyte's glassy matrix contains microlites of plagioclase laths and has a fluidal texture. The K-feldspar is commonly an orthoclase as was seen in the trachyte (ab_{75-60}, or_{20-40}) reported by Binard (1991). Also, the sodic-plagioclase (ab_{55-72}, or_{6-17}) falls in the range of that seen for the trachy-andesites (Fig. 14.1a). The olivine, when present, is Mg-depleted (Fo_{40-30}) (Fig. 14.3b).

Phonolite. Phonolites are also grey, tabular and blocky flows comparable to the trachytes from which they differ because of their higher total alkali content ($Na_2O + K_2O = 11$–14%), SiO_2 (57–63%) and higher normative nepheline (about 6%) (Table 14.1, Fig. 14.2b). As was seen in the trachytes and trachy-andesites, plagioclase is enriched in both normative and modal orthoclase (or_{20-40}) content and the rocks also contain K-feldspar (orthoclase) and nepheline (Table 14.1 and Binard 1991).

14.3
Relationship Between Intraplate-Hotspot and Spreading-Ridge Magmatism

In addition to their mineralogical differences (mostly plagioclase composition), intraplate hotspot volcanoes show a more extensive range of major elements (SiO_2, $Na_2O + K_2O$, Mg#), incompatible element ratios [Zr/Y, $(Ce/Yb)_N$ and Nb/Zr] and isotopic variation (see Sect. 8.2.2) than do the volcanics from spreading ridges. The incompatible trace element variations such as Zr/Y, $(Ce/Yb)_N$ and Nb/Zr ratios show two trends of variability: (1) One variability trend is related to melting processes due either to the source and/or the degree of melting and (2) the second trend is due to the extent of the rocks' crystal-liquid fractionation (Fig. 14.3a, 14.4b,c). In order to enhance our appreciation of the differences in the compositional variation for the Zr/Y-Zr, $(Ce/Yb)_N$-$Ce)_N$ and Nb/Zr-Nb relationships between the EPR MORBs when they are compared to the intraplate alkali basaltic and silicic lava, a single continuous "melting curve" has been drawn to help the reader distinguish between the field of non-evolved (potential parent) rock and the field of the more evolved rocks (Fig. 14.3a–c). This continuous melting curve represents three melting trends calculated from several ultramafic rocks (garnet- and spinel-lherzolite and spinel clinopyroxenites) (Hekinian and Bideau 1995; Hekinian et al. 1999) (Fig. 14.3a–c).

In fact, the crystal-liquid fractionation derived from different parental sources along the melting curve could give rise to all the various types of evolved lava. The extreme Zr enrichment of the Society and Pitcairn hotspot lavas corresponds to the presence of highly evolved trachyte and trachy-andesite lava (Fig. 14.3a–c). The Foundation and the Easter Seamounts show less enrichment in Zr/Y (<8), $(Ce/Yb)_N$ (<7) and Nb/Zr (<0.17) ratios than the Society and Pitcairn hotspots lavas. Also, the samples having lower values of these ratios fall in the field of spreading-ridge volcanics. Similarly, the NEPR, SEPR and SEPR-PAR MORBs and silica-enriched lava are relatively depleted in these incompatible element ratios [Zr/Y < 4 $(Ce/Yb)_N$ < 5 and Nb/Zr < 0.1] as well as in their Sr isotopic ratios ($^{87}Sr/^{86}Sr$ < 0.7031) (Sect. 8.2.6, Fig. 14.5b,c). It is also observed that the Foundation Seamounts show similar strontium isotope ratios to those of the NEPR, SEPR and SEPR-PAR. Otherwise, the NEPR, SEPR and SEPR-PAR spreading ridges show more restricted $^{87}Sr/^{86}Sr$ (0.7030–0.7040) and $^{143}Nd/^{144}Nd$ (0.5129) ratios than most of the hotspots with the exception of a few off-axis eruptions, which took place on a 20 Myr crust (Graham et al. 1998) and which might have been caused by a hotspot type of volcanism (Fig. 14.5c).

14.4
Compositional Differences Among Hotspots

There are also compositional differences among the various hotspots themselves. Thus, the Loihi volcanics have a lower $^{87}Sr/^{86}Sr$ ratio (0.7030–0.7040) than those of the rocks from the Society and Pitcairn regions (0.7030–0.7060) and the Loihi rocks have a higher $^{143}Nd/^{144}Nd$ ratio (0.5129) (Garcia et al. 1993; Sect. 8.2.6). The Loihi alkali basalts and tholeiites (Garcia et al. 1993) fall in the same field of Nd and Sr isotopes as the basanites found on Moua Pihaa (Society) (Fig. 14.5b). In addition, the Society hotspot differs from the Pitcairn because of its higher trend of Sr isotopic variation (Sect. 8.3.2, Fig. 8.13, Devey et al. 2003, Fig. 14.5b). Also, the Society and the Austral hotspots show a higher extent of $(Ce/Yb)_N$ (10–20) and Nb/Zr (0.15–0.25) (Fig. 14.3b,c). The Austral hotspot (Macdonald and Rà Seamounts) has erupted basanite, alkali basalts, tephrite and picritic-alkali basalts depleted in silica (SiO_2 < 49%) and having a variable total alkali ($Na_2O + K_2O$ > 2 up to 9%) content (Fig. 14.2b). The Society hotspot edifices have erupted a more extensive variety of lavas from silica-depleted (basanite, tephrites) to alkali- and silica-enriched lavas going from alkali-basalts to trachy-andesites, trachytes and phonolites (Fig. 14.2b).

Two trends of variability are also shown by the SiO_2 (40–65%) and Mg# (20–90) variation as well as by their isotopic $^{87}Sr/^{86}Sr$ (0.7035–0.7040) and $^{143}Nd/^{144}Nd$ (0.51240–0.1295) compositions (Sect. 8.2.6, 8.2.7 and Fig. 8.13, Fig. 14.5a,b). This was attributed to the fact that the volcanics originated from two separate mantle sources (Devey et al. 2003). The SiO_2-Mg# variation trend separating the two hotspot provinces (Society and Pitcairn) is less pronounced than that shown by the isotopes. Nevertheless, the two provinces show a marked difference between their silicic lavas namely for the samples with low Mg# (<40) and high SiO_2 (>50%) contents (Fig. 14.5a). Another major difference between the two hotspot regions resides in the fact that in the Society hotspot, the large edifices such as Moua Pihaa, Mehetia and Teahitia have erupted more undersaturated lava (tephrite, basanite) (Fig. 14.2b, 14.5a). This is also true for the large edifices (Rà and Macdonald Seamounts) from the Austral hotspot.

The Pitcairn and the Society hotspots were defined as "typical examples" of enriched EM1 and EM2 mantle, respectively (Devey et al. 2003). Also, some of the Foundation,

Chapter 14 · Petrology of Young Submarine Hotspot Lava: Composition and Classification 451

the Austral and the Easter Seamount (Sect. 8.2.7) chain volcanics are mildly alkaline and show depleted Sr isotopic ratios (<0.7040) with a high HIMU (recycled oceanic crust) type of mantle source signature. Devey and Haase (Sect. 8.2.1–8.2.3, 8.2.6 and 8.2.7) present more detailed results and discussion on the study of mantle sources for these hotspots. In fact, several different parental melts (issued from a heterogeneous mantle source but having undergone some extent of crystal-liquid fractionation) could give rise to the variety of the observed volcanics including even the most evolved silica-enriched samples. Thus, when taking a basanite (sample SO47DS1 from the Society), a picritic alkali-basalt (PN3-05 from Pitcairn), and a MORB (SO100-85-1 from the SEPR-PAR) as the parental source melt, several trends of crystal-liquid fractionation are observed between the hotspot and spreading-ridge lavas (Fig. 14.2b, 14.3c, 14.6).

14.4.1
Relationship Between Large and Small Hotspot Edifices

The structural relationship between the various sized edifices and the surrounding sea floor within the same hotspot is presented by Binard et al. (Sect. 5.6.3, Fig. 5.20). Also, the compositional variability of the individual edifices within the same hotspot area is shown to be as extensive as the differences observed between the various hotspots (see Sect. 8.2.2, 8.2.6 and 8.2.7, Fig. 8.9).

The small edifices (<500 m high, Volcanoes #3, #4, #5, #6) and the intermediate-sized Rocard Seamount (500–1 000 m high) from the Society hotspot and samples from the Pitcairn region (i.e., dive site PN13 and dredge site PNDR6 on Adams, see Sect. 5.2.3.5, 5.4.1.3, Fig. 5.14b; Hekinian et al. 2003) are essentially made up of trachy-andesites and trachytes. These silica-enriched lavas plot on a field of high $^{87}Sr/^{86}Sr$ and low $^{143}Nd/^{144}Nd$ ratios on the Sr-Nd variation diagram (Woodhead and Devey 1993; Woodhead and McCulloch 1989; Devey et al. 2003; Sect. 8.2.2, Fig. 8.9 and Sect. 8.2.6, Fig. 8.13; Fig. 14.5b). On the other hand, the larger edifices (1 000–3 000 m high) from the Society and Pitcairn hotspots consist of a variety of rock types (alkali basalt, trachybasalt, trachy-andesite, trachyte and phonolite) and show a large range of their major oxides ($SiO_2 = 40$–65%, $Na_2O + K_2O = 4$–12%) and their Mg# variation (10–70) (Fig. 14.2b, 14.6a). They also show larger isotope ratios ($^{87}Sr/^{86}Sr = 0.7030$–0.7050, $^{143}Nd/^{144}Nd = 0.5130$–0.5124), which overlap the field of the small edifices (Fig. 14.5b). Some of these overlapping fields include the alkali-basalt rocks found on the large edifices (from Pitcairn Island and the Adams Seamount), which have the same isotopic values ($^{87}Sr/^{86}Sr = 0.7046$–0.0755, $^{143}Nd/^{144}Nd = 0.525$–0.5124) as the most evolved silicic lava from the small edifices of the Pitcairn hotspot (Fig. 14.5b).

As shown by Woodhead et al. (1993), Devey et al. (2003) and Devey and Haase (Sect. 8.2.6, 8.2.7), the source rocks for the Society hotspot volcanoes are different for the various edifices because the samples show variable lead, strontium and oxygen isotope ratios. It is believed that these compositional changes could reflect the heterogeneity of the melting sources and could also be related to the crystal-liquid fractionation processes at each site. This agrees with a concept involving several magmatic pulses generating separate pathways and/or replenishing the same magmatic reservoir (see Sect. 5.6.3, Fig. 5.20). However, it is not excluded that focussed magmatism underneath the individual small edifices has taken place independently from the magmatism that built the larger volcanoes.

452 R. Hekinian

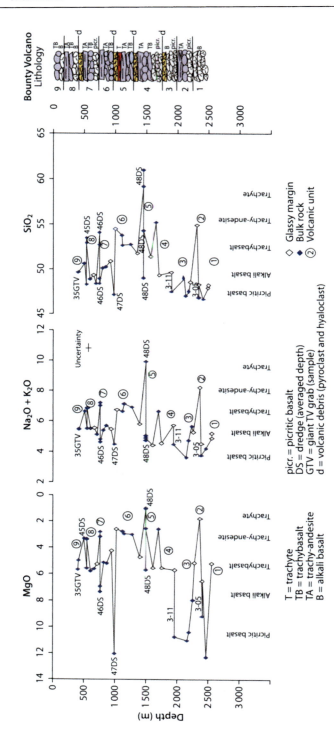

Fig. 14.6. Compositional variation of lava showing the volcanic sequences observed on the Bounty Seamount. The individual volcanic events recognized are shown with visually estimated tie-lines (*solid lines*) between analyzed samples. The *dots* are samples collected by submersible, except for a few samples collected by dredging (see Sect. 5.4.1.2 and Fig. 5.14a; Hekinian et al. 2003). The surface ship stations (dredge = *DS* and camera television grab = *GTV*) are indicated. The differences in composition between the bulk rock and microprobe analyses of the glassy rim in samples 3–05 and 3–11 are due to their high content in olivine phenocrysts (up to 15%). A schematic stratigraphic column with the various volcanic lithologies is shown on the right hand side of the diagram (Hekinian et al. 2003). For further explanation, see text Sect. 14.4.1 and Sect. 5.4.1

Table 14.3. Range of compositional variation of East Pacific Rise and hotspot submarine lava

Regions	East Pacific Rise				GSC	Hotspots (Society, Australs, Pitcairn, Easter Seamounts, Loihi)							
Rock types	MORB	Basaltic andesite	Andesite	Dacite	Rhyolite	Tephrite	Basanite	Alk. basalt	Picr. basalt	Trachy-basalt	Trachy-andesite	Trachyte	Phonolite
SiO_2 (%)	48–50	51–53	53–59	59–66	70–74	40–44	42–46	45–48	41–47	49–53	53–59	59–64	57–63
$Na_2O + K_2O$ (%)	0.8–3.5	1.5–4	3–5.5	3–5	5–8	6–9	3.5–5.8	3.5–5	2–5	4–7	4–8	7–12	11–14
Mg#	55–72	33–60	19–45	17–35	14–30	39–62	40–65	50–69	65–75	45–60	20–45	19–35	35–38
Zr/Y	1–4	1–4	1–5	2–6	2–3	8–12	8–10	2–13	8–10	10–13	9–15	15–19	16–20
Nb/Zr	0.01–0.07	0.02–0.06	0.04–0.08	0.03–0.07	–	0.15–0.28	0.15–0.26	0.06–0.15	0.1–0.15	0.06–0.15	0.1–0.2	0.1–0.2	0.09–0.28
$(Ce/Yb)_N$	0.1–4	0.7–3	1.5–3	1.8–3	1.37	10–30	6–15	5–15	6–13	7–17	7–17	10–20	14–17
$(La/Sm)_N$	0.2–1.5	0.6–13	0.75–1.5	0.7–2	0.9–1.4	2–6	2–5	1–3	1–3	2–3	2.3–4.5	3–6	2–4
Zr ppm	70–200	120–350	150–400	300–600	345–445	300–500	220–400	170–300	140–290	300–500	450–721	700–1000	800–1200
Nb	<7	3–20	20–31	30–41	–	60–115	50–80	19–40	20–52	40–70	50–81	80–120	77–130
Ce	3–25	10–45	60–70	90–100	68	70–180	70–90	50–90	45–72	80–170	120–201	150–220	140–270
Plagioclase	an_{71-85} $or_{0.01-0.06}$	ab_{30-40} $or_{0.40-0.1}$	an_{70-50} $or_{0.5-1}$	an_{35-50} $or_{0.8-1.2}$	–	–	ab_{20-35} $or_{6.3-1.0}$	ab_{19-45} $or_{0.4-2.5}$	ab_{15-30} $or_{0.5-1.9}$	ab_{90-70} or_{4-16}	ab_{30-45} or_{2-4}	ab_{55-75} or_{6-17}	–
Olivine	Fo_{76-88}	Fo_{65-80}	Fo_{40-70}	Fo_{30-40}	–	Fo_{77}	Fo_{75-85}	Fo_{75-83}	Fo_{80-89}	Fo_{55-73}	Fo_{70-80}	Fo_{30-40}	–

East Pacific Rise data are from Hekinian et al. (1989, 1999), Mahoney et al. (1994), Bach et al. (1994), Sinton et al. (1994), Niu et al. (1999) and unpublished data. Hotspot data are from Binard (1991), Hekinian et al. (1991, 2003), Garcia et al. (1993), Devey et al. (1990), Hémond et al. (1994), Fretzdorff et al. (1996) and Haase and Devey (1996). GSC = Galapagos Spreading Center near 95.5° W (Byerly 1980; Clague et al. 1981). Easter Seamounts are located west of Easter Island. Tephrite contains modal nepheline. N = normalized to chondrite C1. Mg# = atomic proportion $(Mg^{2+}/Mg^{2+} + Fe^{2+})$. MORB = normal (N–), transitional (T–) and enriched (E–) Mid-Ocean Ridge Basalt.

14.4.2
Volcanic Stratigraphy

Volcanic stratigraphic sequences are due to cyclical lava eruptions, and these are important phenomena giving rise to the construction of oceanic crust at spreading ridges and on intraplate volcanoes. Cyclic volcanic events are observed on subaerial volcanoes, and it has been inferred that they took place during a short period (over a month or a year) as well as during a relatively longer period (hundreds or thousands of years). Unfortunately, visual observations are often difficult to make and interpret in submarine environments. The most obvious indication of sequential magmatic variability is when there is a drastic change in the mode of extrusion and/or some major tectonic disruption such as explosive events and other types of forceful injections giving rise to contrasting lava forms (Sect. 5.6.1). To my knowledge, there are very few detailed stratigraphic observations in deep sea environments that have reported sequential eruptions related to volcanic construction. One rare example is the detailed video camera observations and dredging along segments of the SEPR-PAR axis near 37°40'-38°16' S. The stratigraphic profile follows along three domed-shaped structures about 120 m high that are capped with aphyric andesitic flows (Hekinian et al. 1999; Stoffers et al. 2002). The silica-enriched domes lie on top of tubular lava flows having a basaltic MORB composition.

The volcanic stratigraphy of submarine edifices in intraplate regions of hotspot origin is often complicated by the presence of numerous cinder-spatter cones and parasitic vents (adventive cones) on their flanks. Also, the migration of eruptive centers along rift zones and the variable depths of accidental debris (hyaloclastites and pyroclasts) lying on top of the lava flows make it difficult to pinpoint the volcanoes' extrusive vents. Nevertheless, continuous field observation on the morphology with relation to compositional variation of the flows will help to identify the sequential distribution of the volcanic events taking place on some of these edifices. The individual flows are characterized by their different morphology. The pillow lavas are bulbous or tubular with radial jointing, abundant surface corrugations and a thin glassy crust (<1 cm thick). The flattened tubular and short oval-shaped flows with smoother surfaces represent an intermediate morphology, and often their composition (trachybasalt) is different from the bulbous pillows or tubes (Sect. 5.2.3.5, Fig. 5.8i). The most viscous lavas such as the trachy-andesite, andesite and trachyte have given rise to rocks with a blocky and tabular appearance and are often coated with a "scoriaceous" (agglutinized water-quenched) glassy crust (up to 10–25 cm thick) (see Sect. 5.2.3.5, Fig. 5.8h,j). These rocks show a fluidal arrangement of their cavities, vesicles and plagioclase lath orientation. Pyroclastic flows composed of fragmented glass and angular blocks are often accompanied by hyaloclastites and found lying on top of silicic (evolved) volcanics (Sect. 5.2.3.1 and 5.3.1.1; Fig. 5.8h,n).

Since the most detailed observations and sampling were carried out on the Bounty Seamount, a summary of the compositional and morphological sequences is able to be shown in Fig. 14.6, which was constructed from the geological profile shown in Binard et al. (Sect. 5.4.1.3 and Fig. 5.14a; Hekinian et al. 2003, Table 5). Six submersible profiles conducted along the Bounty Seamount revealed at least nine sequential volcanic eruptions between 3 000 to 500 m depth (Hekinian et al. 2003 and Sect. 5.4.1.3, Fig. 5.14a). Each of these extrusive sequences varies between 250–300 m in thickness and includes different types of lava flows. A schematic representation and a compositional variation diagram using some characteristic oxides are shown (Fig. 14.6). The variations of MgO (1–12%), total alkalis (3–10%) and SiO_2 (45–63%) show a spider type of distribution correlating to the different flow morphologies encountered. In addition, another dive profile (dive PN13) made on a small adventive cone (<500 m high) located on the southern slope of the Bounty Seamount also revealed sequential volcanism of essentially silica-enriched flows (SiO_2 > 52%) consisting of trachy-andesite, trachyte and hyaloclast/pyroclast associations (Sect. 5.4.1.3, Fig. 5.14b). In this example, the volcanic stratigraphy is thinner and the thickness of the flows varies from a few tens of meters (10–30 m) up to 150 m thick (Hekinian et al. 2003). Other volcanic edifices showing compositional diversities in their volcanic sequences are the Adams and Rocard Seamounts in the Pitcairn and Society hotspots, respectively (see Sect. 5.2.3.5, 5.4.1.3 and Fig. 5.7ad, 5.14b,c). Although these edifices were not sampled as extensively as the Bounty Seamount, they also indicate sequential eruptive events. Trachytic flows were encountered in the lower (2 400 m depth) and middle part (1 800 m depth) of Adams. The Rocard Volcano in the Society hotspot is an intermediate-sized edifice made up of multiple composite cones of silicic flows (trachy-andesite and trachyte) (Sect. 5.2.2.5 and Fig. 5.7a–d).

Volcanic eruptions have taken place during several magmatic pulses. During an eruption, pyroclasts and hyaloclasts might indicate the end of volcanic events on the individual edifices (Batiza 1989; Hekinian et al. 2003). The sequential eruptive events are inferred on the basis of the alternance of hyaloclastites and pyroclastic deposits and due to morpho-structural breaks on the slope of the edifice. These "breaks" are marked by the presence of volcanic cones and morphological discontinuities such as the transition from giant pillow lava to more blocky/tabular flows, which also indicates the lava's compositional variability.

14.5
Summary and Conclusions

The volcanics from intraplate hotspots differ from those from the North and South Pacific spreading ridges because of the common occurrence of K-enriched plagioclase, Na-enriched clinopyroxene, and K-feldspars. The most evolved rocks from the hotspots often contain biotite and amphibole. The plagioclase, olivine and clinopyroxene are the most sensitive minerals to the bulk rock's chemical variation for the various rock types. Cr-spinel is one of the earliest formed minerals during melt solidification, and often its composition reflects the magmatic source rather than melt evolution during crystal-liquid fractionation.

The compositional variability of the plagioclase (for comparison between hotspot and spreading-ridge volcanics) is best seen in the evolved (silicic) lavas. The trachy-basalts/trachy-andesites/trachytes are more enriched in K-feldspar, and their plagioclase is enriched in normative orthoclase with respect to that from the spreading ridge basaltic-andesite/andesite/dacite suites.

The petrological variation of hotspot volcanics includes some less evolved rock types such as tephrites, basanites and picritic-alkali basalts, which often accompany the alkali basalt-trachybasalts-trachy-andesites-trachytes suite. The hotspot volcanics are more enriched in alkalis ($Na_2O + K_2O$ > 3.4% up to 9%) and other incompatible element ratios

{(K/Ti > 0.45, Zr/Y > 4, (La/Sm)$_N$ > 2 and (Ce/Yb)$_N$ > 3, Nb/Zr > 0.07} than the spreading-ridge lava. The hotspot rocks also have more extensive Sr (^{87}Sr/^{86}Sr = 0.7030–0.7060) and Nd (^{143}Nd/^{144}Nd = 0.5124–0.5130) isotope ratio variations (Sect. 8.2.6) than do samples from spreading-ridge segments (^{87}Sr/^{86}Sr < 0.70320). This suggests that hotspot volcanoes have tapped a more variable heterogeneous mantle source than the mantle source for the spreading ridges.

The hotspot regions show a higher proportion of evolved silica-enriched lava than the spreading ridges. However, the distribution and relative abundance of silicic lava could be due to a more biased sampling along the spreading ridge. Nevertheless, silica-enriched volcanics on the ocean floor have occasionally been found in a limited area along various accreting ridge segments of the southern (SEPR) and northern (NEPR) East Pacific Rise, and on the Galapagos Spreading Center (GSC), although they are much more commonly found on intraplate volcanoes of hotspot origin. Along accreting ridge segments, the silicic lava of basaltic-andesite, andesite and dacite composition are commonly found on ridge segments influenced by ridge-hotspot interaction. For example, silicic lava of MORB-andesite-dacite suites is found along the South East Pacific Rise-Pacific Antarctic Ridge (SEPR-PAR) segments near 37°30' S and on the Galapagos Spreading Center (GSC) near 95°5' W. The silicic-enriched (SiO$_2$ = 51–65%) lava from the spreading ridges contains relatively low incompatible element ratios [Zr/Y < 6, (Ce/Yb)$_N$ < 5, Nb/Zr < 0.1] when compared to the intraplate trachybasalt-trachy-andesite-trachyte suites [Zr/Y = 8–19, (Ce/Yb)$_N$ = 6–25, Nb/Zr = 0.07–0.30] found on large (>1 000 m high) and small (<500 m high) volcanic edifices. One of the most attractive hypotheses for the origin of the silicic lava is related to the process of crystal-liquid fractionation inside or close to a magma reservoir.

Volcanic stratigraphy based on detailed (submersible) field observations taking into consideration both lava morphology and the rocks' composition suggests that cyclic eruptions of mafic and evolved lava have taken place on small as well as large edifices. Sampling and observations during several dives (Sect. 5.4.1.3 and Fig. 5.14a) on the Bounty Seamount in the Pitcairn hotspot between 3 000 to 400 m deep indicate that there were at least nine eruptive phases giving rise to a series of layers less than 350 m thick (Hekinian et al. 2003). For each of the individual sequences, the initial eruptive event probably started with the least evolved rock type and moved to more evolved phases (i.e., picritic basalt to alkali basalt to trachybasalt) forming pillows and giant tubular flows. Successive events gave rise to more viscous silica-enriched flows (>53%, SiO$_2$).The silicic rocks are flatter and more tabular with a blocky appearance and large cavities (up to 5 cm long) and have thicker (>3 cm up to 25 cm) glassy margins. The observed stratigraphic sequences suggest that the silicic lava is the final product of the Bounty Seamount's cyclic eruptive events.

The geochemical and structural relationship between the various edifices and mostly between the small and the large volcanoes of the same hotspot area were previously pointed out and discussed elsewhere (Hekinian et al. 2003; Devey et al. 2003; Sect. 5.2.3). It is observed that both the silicic lavas and the less evolved alkali basalts from the small and large edifices at the same hotspot (i.e., Pitcairn) lie in the same field of extremely high Sr isotopic ratios (0.07050–0.7060) and low Nd ratios (0.5124–0.5125) (Devey et al. 2003). This suggests that the volcanics forming the small and large edifices could be genetically related. The relationship between the various edifices on the same hotspot could be the result of two different modes of magma emplacement: (1) either each

Chapter 14 · Petrology of Young Submarine Hotspot Lava: Composition and Classification

edifice is fed directly from its individual mantle source and/or (2) the larger edifices are the site where most of the magma from the mantle source rises and then it could subsequently be channeled laterally to the smaller edifices through sub-crustal pathways (i.e., rift zones, fractures, etc.). However, the significant variation of the isotope ratios for the volcanics found on the large edifices suggests that several melting sources are involved. These various melting sources could replenish the magmatic reservoirs of the edifices during their building stage.

At the present time, it is unclear if the magma that fed the small volcanic edifices was directly connected to the mantle's partial melting source region or if this magma was headed for the larger edifices but was sidetracked and channeled laterally to the smaller edifices. It is likely that both hypotheses are valid and that hotspot volcanism is generated from multiple parental melts during variable degrees of melting of a heterogeneous mantle plume source located underneath the various hotspots.

Acknowledgements

Sincere thanks to Daniel Bideau for his advice and assistance with the manuscript and to my colleagues at the Geosciences Department of IFREMER and the Geology Department of the University of Kiel for their support during the preparation of this paper.

References

Ackermand D, Hekinian R, Stoffers P (1998) Magmatic sulfides and oxides in volcanic rocks from the Pitcairn hotspot (South Pacific). Mineral Petrology 64:149–152

Allan JF, Batiza R, Perfit MR, Fornari DJ, Sack RO (1989) Petrology of lavas from the Lamont Seamount Chain and adjacent East Pacific Rise, 10° N. J Petrol 30:1245–1998

Aumento F (1969) Diorites from the Mid-Atlantic Ridge. Nature 165:112–113

Bach W, Hegner E, Erzinger E, Satir M (1994) Chemical and isotopic variations along the superfast spreading East Pacific Rise from 6 to 30° S. Contr Miner Petrol 116:365–380

Batiza R (1989) Seamount and seamount chains of the Eastern Pacific. In: Winterer EL, Hussong DM, Decker RW (eds) The Eastern Pacific Ocean and Hawaii, the geology of North America. Geol Soc Amer N:145–159

Batiza R, Niu Y (1992) Petrology and magma chamber processes at the East Pacific Rise. J geophys Res 97:6779–6797

Batiza R, Niu Y, Karsten JL, Boger W, Potts E, Norby L, Butler R (1996) Steady and non-steady state magma chambers below the East Pacific Rise. Geophys Res Lett 23:221–224

Bideau D, Hekinian R (1995) A dynamic model for the generating small-scale heterogeneities in ocean floor basalts. J Geophys Res 100:10141–10162

Binard N (1991) Les points chaud de la Sciété et des Australes et de Pitcairn (Pacifique Sud): Approche volcanologique et petrologique. PhD theses, Université de bretagne Occidentale, Avenue Le Gorgeu, Brest, France, pp 372

Binard N, Hekinian R, Stoffers P (1992) Morphostructural study and type of volcanism of submarine volcanoes over the Pitcairn hot spot in the South Pacific. Tectonophysics 206:245–264

Byerly G (1980) The nature of differentiation trends in some volcanic rocks from the Galapagos Spreading Center. J Geophys Res 85:3797–3810

Carmichael ISE (1964) Petrology of the Thingmuli, a tertiary volcano in easter Iceland. J Petrol 5: 535–460

Clague DA (1988) Petrology of ultramafic xenoliths from Loihi Seamount, Hawaii. J Petrology 29: 1161–1186

Clague DA, Frey FA, Thompson G, Rindge S (1981) Minor and trace geochemistry of volcanic rocks dredged from the Galapagos spreading center: Role of crystal fractionation and mantle heterogeneity. J Geophys Res 86:9469–9482

Coogan LA, KM Gillis, CJ MacLoad, JM Thompson, Hekinian R (2002) Petrology and geochemistry of lower crust formed at the East Pacific Rise and exposed at Hess Deep: A synthesis and new results. Electr J Earth Sci Geochemistry, Geophys, Geosystem G3, Am Geophys Union 3 (11):1–30

Devey CW, Albarède F, Cheminée J-L, Michard A, Mühe R, Stoffers P (1990) Active submarine volcanism on the Society hotspot swell (west Pacific): A geochemical study. J Geophys Res 95:5049–5066

Devey CW, Hekinian R, Ackermand D, Binard N, Francke B, Hémond C, Kapsimalis V, Lorenc S, Maia M, Möller H, Perrot K, Pracht J, Rogers T, Stattegger K, Steinke S, Victor P (1997) The Foundation Seamount Chain: A first survey and sampling. Mar Geol 137:191–200

Devey CW, Lackschewitz KS, Mertz DF, Bourdon B, Cheminée J-L, Dubois J, Guivel C, Hekinian R, Stoffers P (2003) Giving birth to hotspot volcanoes: Distribution and composition of young seamounts from the seafloor near Tahiti and Pitcairn Islands. Geology 31(5):395–398

Fretzdorff S, Haase KM (2002) Geochemistry and petrology of lavas from the submarine flanks of Reunion Island (western Indian Ocean): Implications for magma genesis and the mantle source. Mineralogy Petrology 75:153–184

Fretzdorff S, Haase KM, Garbe-Schonberg C-D (1996) Petrogenesis of lavas from the Umu volcanic field in young hotspot region of Easter Island, Southeastern Pacific. Lithos 38:23–40

Frey FA (1980) The origin of pyroxenite and garnet pyroxenites from Salt Lake Crater, Ohau, Hawaii: Trace element evidence. Am J Sci 280A:427–449

Frey FA, Suen CJ, Stockman HW (1985) The Ronda peridotite: Geochemistry and petrogenesis. Geochm Acta 49:2468–2491

Garcia MO, Jorgenson BA, Mahoney JJ (1993) An evaluation of temporal geochemical evolution of Loihi Summit lavas: Results from *Alvin* submersible dives. J Geophys Res 98:537–550

Graham DW, Zinlder A, Kurz, MD, Jenkins WJ, Batiza R, Staudigel H (1988) He, Pb, Sr and Nd isotope constraints on magma genesis and mantle heterogeneity beneath young Pacific seamounts Contrib to Mineral Petrol 99:446–463

Haase KM (2002) Geochemical constrats on magmasources and mixing processes in Easter Microplate MORB (SE Pacific): A case study of plume-ridge interaction. Chem Geol 182:335–355

Haase KM, Devey C (1996) Geochemistry of lava from the Ahu and Tupa volcanic fields, Easter hotspot, southeast Pacific: Implications for intraplate magma genesis near spreading axis. Earth Planet Sci Lett 137:129–143

Hawkins J, Melchior J (1983) Petrclogy of basalts from Loihi Seamount, Hawaii. Earth Planet Scie Let 66:356–369

Hekinian R, Walker D (1987) Diversity and spatial zonation of volcanic rocks from the East Pacific Rise near 21°N. Contrib Mineral Petrol 96:265–280

Hekinian R, Thompson G, Bideau D (1989) Axial and off-axial heterogeneity of basaltic rocks from the East Pacific Rise at 12°38'N–12°51'N and 11°26'N–11°30'N. J Geophys Res 94:17437–17463

Hekinian R, Bideau D, Stoffers P, Cheminée JL, Muhe R, Puteanus D, Binard N (1991) Submarine intraplate volcanism in the South Pacific: Geological setting and petrology. J Geophys Res 96:2109–2138

Hekinian R, Bideau D, Hébert R, Niu Y (1995a) Magmatism in the Garrett Transform fault. J Geophys Res 100:10163–10185

Hekinian R, Stoffers P, Ackermand D, Binard N, Francheteau J, Devey CW, Garbe-Shonberg D (1995b) Magmatic evolution of the Easter Microplate-Crough Seamount region (South East Pacific). Marine Geophys Res 17:375–397

Hekinian R, Francheteau J, Armijo R, Cogné JP, Constantin M, Girardeau J, Hey R, Naar DF Searle R (1996) Petrology of the Easter Microplate region in the South Pacific. J Volc Geotherm Res 72:259–289

Hekinian R, Stoffers P, Ackermand D, Révillon S, Maia M, Bohn M (1999) Ridge-hotspot interaction: The Pacific-Antarctic Ridge and the Foundation seamounts. Marine Geol 160:199–223

Hekinian R, Cheminée J-L, Stoffers P, Dubois J, Scott S, Guivel C, Garbe-Schoberg D, Devey CW, Bourdon B, Lacschewitz K, McMurtry G, Le Drezen E (2003) Pitcairn hotspot in the South Pacific: Distribution and composition of submarine volcanic sequences. J Volc Getherm Res 121:219–245

Hémond C, Devey CW, Chauvel C (1994) Source compositions and melting processes in the Society and Austral plumes (South Pacific Ocean): Element and isotope (Sr, Nd, Pb, Th) geochemistry. Chemical Geology 115:7–45

Hofmann AW, White WM (1982) Mantle plumes from ancient oceanic crust. Earth Planet Sci Lett 57:421–436

Irving AJ (1980) Petrology and geochemistry of composite ultramafic xenoliths in alkali basalts and implication for magmatic processes within the mantle. Am J Sci 280:389–426

Juster TC, GroveTL (1989) Experimental constraints on the genesis of FeTi basalts, andesites, and rhyodacites at the Galapagos Spreading Center, 85°W and 95°W. J Geophys Res 94:9251–9274

Juteau T, Maury R (1997) Geologhie de la croute oceanique. Masson, Paris, pp 367

Langmuir CH, Bender JF, Bence AE, Batiza R (1986) Petrological and tectonic segmentation of the East Pacific Rise, 5°30'–14°30'N. Nature 332:422–429

Le Bas MJ, Streickeisen AL (1986) The IUGS systematics of igneous rocks. J Geol Soc London 148: 825–833

Mahoney JJ, Sinton JM, Kurz MD, Macdougall JD, Spencer KJ, Langmuir CW (1994) Isotope and trace elements characteristics of a super-fast spreading ridge: East Pacific Rise, 13–23° S. Earth Plant Sci Lett 121:173–193

Maia M, Hémond C, Gente P (2001) Contrasted interactions between plume and lithosphere: The Foundation chain case. Geochem Geophys Geosyst 2 (article), 2000GC000117

Middlemost EAK (1980) A contribution to the nomenclature and classification of volcanic rocks. Geol Mag 117:51–57

Nielsen RL, Delong SE (1992) A numerical approach to boundary layer fractionation: Application to differentiation in natural magma systems. Contrib Mineral Petrol 110:355–369

Pan Y, Batiza R (1998) Major element chemistry of volcanic glasses from the Easter Seamount Chain: Constraints on melting conditions in the plume channel. J Geophys Res 103:5287–5304

Sacks RO, Carmichael ISE, Rivers M, Ghiorso MS (1980) Ferric-ferrous equilibria in naturalsilicate liquids at 1 Bar. Contr Mineral Petrol 75:369–376

Schilling J-G (1971) Sea-floor evolution: Rare earth evidence. Philos Trans R Soc London Ser A 268: 663–706

Schilling J-G (1973) Iceland mantle plume: Geochemical study of Reykjanes Ridge. Nature 242: 565–571

Schilling J-G (1985) Upper mantle heterogeneities and dynamics. Nature 314:62–67

Sigurdsson H, Sparks RSJ (1981) Petrology of rhyolite and mixed magma ejecta from the 1875 eruption of Askja, Iceland. J Petrol 22:41–84

Sinton JM, Smaglik SM, Mahoney JJ, Macdonald KC (1991) Magmatic processes at superfast spreading oceanic ridges: Glass variations along the East Pacific Rise, 13° S–23° S. J Geophys Res 96: 6133–6155

Stoffers P, Botz R, Cheminée J-L, Devey CW, Froger V, Glasby GP, Hartmann M, Hekinian R, Kögler F, Laschek D, Larqué P, Michaelis W, Mühe RK, Puteanus D, Richnow HH (1989) Geology of Macdonald Seamount region, Austral Islands: Recent hotspot volcanism in the south Pacific. Mar Geophys Res 11:101–112

Stoffers P, Worthington T, Hekinian R, Petersen S, Hannington M, Turkey M, et al. (2002) Silicic volcanism and hydrothermal activity documented at Pacific-Antarctic Ridge. EOS 83 (28):301–304

Sun S-S, McDonough WF (1989) Chemical and isotopic systematics of oceanic basalts: Implications for mantle compositions and processes. In: Saunders AD, Norry MC (eds) Magmatism in the ocean basins. Geol Soc Spec Publication 42:313–345

Tatsamuto M, Hedge CE, Engel AEJ (1965) Potassium, thorium, uranium and Sr/86Sr in oceanic tholeiitic basalt. Science 150:886–888

Thompson G, Bryan WB, and Humpris SE (1989) Axial volcanism on the East Pacific Rise, 10–12° N. In: Saunders AD, Norry MJ (eds) Magmatism in the ocean basins. Geol Soc Spec Pub 42:181–200

Turner S, George R. Jerram DA, Carpenter N, Hawkesworth C (2003) Case studies of plagioclase growth and residence times in island arc lavas from Tonga and the Lesser Antilles, and a model to reconcile discordance age information. Earth Planet Sci Lett 214:279–294

White WM (1985) Sources for oceanic basalts: Radiogenic isotopic evidence. Geology 13:115–118

White WM, Duncan RA (1996) Geochemistry and geochronology of the Society Islands: New evidence for deep mantle recycling. In: Basu A, Hart SR (eds) Earth processes: Reading the isotopic code. Amer Geophys Union Washington, Geophysical Monograph 95:183–206

White WM, Hofmann AW (1982) Sr and Nd isotope geochemistry of oceanic basalts and mantle evolution. Nature 290:821–825

Weaver JS, Langmuir CH (1990) calculation of phase equilibrium in mineral-melt sestems. Comput Geosci 16:1–19

Woodhead JD, Devey CW (1993) Geochemistry of the Pitcairn Seamounts: I. Source character and temporal trends. Earth Planet Sci Lett 116:81–99

Woodhead JD, McCulloch MT (1989) Ancient seafloor signals in Pitcairn Island lavas and evidence for large amplitude, small length-scale mantle heterogeneities. Earth Planet Sci Lett 94:257–273

Woodhead JD, Greenwood P, Harmon RS, Stoffers P (1993) Oxygen isotope evidence for recycled crust in the source of EM-type ocean island basalts. Nature 362:809–813

Zindler A, Hart S (1986) Chemical geodynamics. An Rev Earth Planet Sci 14:493–571

Index

A

abiogenic
- -, methane 424–426
- -, reactions in the Earth's crust 407, 423

absorption 73

abyssal hill 19, 158, 161, 163–165, 180, 186, 191, 195
- -, fabric 19
- -, region 158, 161, 164, 180, 186, 195
- -, structure 164

accumulation 57, 92, 134, 197, 203, 230, 331

actinolite 320, 323, 326, 328, 330, 331, 337, 338, 343, 344

activity
- -, explosive type 197
- -, hydromagmatic 183
- -, hydrothermal 6, 55, 58, 161, 311, 376, 395, 398, 407, 409–411
- -, magmatic 1, 33, 39, 58, 69, 83, 312

Adams
- -, Mn crusts 376
- -, Seamount 180, 183–187, 190, 200, 201, 376–378, 380, 383, 385, 388, 398, 409, 412–418, 451, 455
  - -, hydrothermal fluid 412–417
  - -, stratigraphy 455
- -, Volcano 183–185, 190, 201, 383

Adamstown Volcanics 182

adsorption 375

Afareaitu 30, 51, 60

Agassiz Fracture Zone 349

age
- -, $^{40}Ar/^{39}Ar$ 350, 352, 353, 363
- -, crustal 92, 97, 125, 131
- -, dating 219, 381
  - -, Bora Bora 219
- -, fission track grain (FTGA) 151
- -, Foundation Chain 350, 355, 367
  - -, $^{40}Ar/^{39}Ar$ 352
- -, lithosphere 364
- -, Marquesas volcanoes 255
- -, progression 9, 25, 143, 146, 148, 175, 272, 301
- -, radiometric 116
- -, Tahiti 255
- -, VER (see also volcanic elongated ridge) 367
- -, volcanic 143

aggregation 296

aging 90

Airy 4, 75–79, 83, 101, 135, 136
- -, compensation 77, 78
  - -, isostatic 75
  - -, mechanism 4

- -, scheme 79, 136
- -, type 76, 78, 135, 136

albite (see also sodic plagioclase) 320, 323, 330, 331, 337, 338, 433

Alexander Selkirk (Mas Afuera) 268, 269

algae 65

alignment 157, 161, 165, 212, 312, 368

alkali
- -, basalt 168, 177, 269, 297, 341, 435
  - -, Canary Islands 341
  - -, composition 444
- -, olivine basalt 183

alkali-feldspar 433

alkalinity 262

altimetry 3, 14, 15, 20, 157, 259, 372

aluminum 17, 326, 328, 337, 353

American lithospheric plates 120

amphibole 320, 326, 328, 330, 337, 343

amphibolite 337

Amsterdam hotspot 353

analysis
- -, isotopic 431
- -, trace element 431

Andes 75

andesine 320, 433

andesite 165, 168–170, 173, 201, 435, 436, 444
- -, composition 445

ankaramite 168, 196, 315, 339

anorthite 337

Antarctic 272, 297, 301, 349, 351, 352, 359, 363, 365–369, 371

antimony 30, 389

apatite 320, 330, 433

Arago Seamount 2, 175, 176, 212, 226, 227, 229, 258
- -, bathymetric map 226

arc
- -, Bonin 396
- -, lava 240
- -, volcanism, Chukotka 153

archaebacteria 426
- -, hyperthermophilic 426

Argentine Basin 90

argon 175, 183, 353
- -, age 145, 350, 352, 353, 363
- -, dating 6, 145
- -, geochronology 351

Ascension
- -, Fracture Zone 112
- -, Island 112–115
  - -, regional map 113

462 Index

ash 59, 174, 177, 183, 187, 190, 312, 340
-, flow 312
-, formation 183
-, volcanic 59, 174, 177, 183
Asia 149
assessment, geologic hazard 9
asthenosphere 4, 73, 78, 80, 81, 88–96, 99, 109, 114,
116, 120, 124, 127, 132, 136, 267, 286, 288–290, 292,
294, 300, 302
-, mantle 288, 289, 292
Atlantic
-, hotspots 275
-, Ocean 3, 5, 90, 92, 93, 112, 113, 115, 116, 120, 128,
131, 132, 275, 286, 288, 292, 295, 296, 327, 333,
336, 338, 343, 370
Atlantis Fracture Zone 333, 335, 336
Atlantis-Plato-Cruiser-Great Meteor group 116
atmosphere 6, 137, 355, 411
-, methane 411
atoll 38, 67, 210, 212, 213, 220, 235, 264
-, Fangataufa 264
-, Maria 211
-, Mururoa 264
-, Rangiroa 38, 43, 55, 68
-, Tetiaroa 67
-, Tupai 220, 221
attenuation 33, 34
augite 326, 330
Austral
-, archipelago 157, 175, 210, 231, 235
-, chain 4, 191, 211, 259, 262, 263, 272, 407
-, Fracture Zone 20, 210, 224
-, geochemistry 256
-, hotspot 6, 29, 34, 59, 65, 68, 69, 160, 169, 173,
175, 178, 180, 194, 198, 199, 202, 309, 312, 343, 421
-, seismicity 29
-, submarine edifices 175
-, volcanics 450
-, Island chain 96, 175
-, Islands 14, 23–25, 59, 96, 97, 175, 210–213, 253,
255–258, 260, 264, 309, 312
-, bathymetric map 211, 213
-, South Alignment 59
-, trace element patterns 257
-, landslide, classification 232
-, magma 257, 262
-, petrogenesis 257
-, Seamounts 407, 451
-, volcanics 451
Australia 286
Australian Plate 151
avalanche 5, 209, 213–215, 217, 220, 222, 224–228, 231,
233, 235, 236
-, debris 5, 209, 213–215, 217, 220, 222, 225–228,
231, 235, 236
Axial Volcano 383
azimuth 43
Azores 119, 246, 286, 296, 327
-, plateau 327

B

bacteria 6, 383, 384, 411, 424, 425
-, CH_4

-, formation 425
-, production 424
-, communities, development 6
-, iron oxidizing 384
barite 390, 392
barium 178, 239, 241, 242, 249, 273, 296, 331, 341, 387,
388, 390
basal diameter 112, 165, 168, 170, 194
basalt
-, alkali 168, 177, 269, 297, 341, 435, 436
-, composition 444
-, crust 151, 246
-, flow 396
-, HIMU 259
-, magmas 263, 341
-, melt 240, 248
-, ocean floor 431
-, oceanic 246–249
-, picritic 168, 436, 437
-, spreading-ridge axis basalts (MORBs) 437
-, volcanism 239
basanite 168, 261, 269, 309, 311, 320, 329, 340, 433, 435
-, composition 444
-, lapilli 309, 311
basin
-, Argentine 90
-, oceanic 1, 3, 29, 73, 76, 81, 84, 90, 95
bathymetric
-, anomaly 120
-, Superswell 280
-, profile 24, 228, 230, 233, 311
bathymetry 3, 9, 10, 12, 13, 15, 17, 19–25, 29, 35, 37, 64,
90, 102, 117, 121, 158, 174, 180, 190, 202, 210, 213, 215,
228, 233, 273, 298, 359, 418–420
belt 149, 311
bend 102, 148
-, 43 Ma 143, 144, 146–148, 151, 153
-, origin 144, 148
bifurcation 366, 367
-, Foundation Chain 366, 367
biogenic 383, 407, 408, 423–426
-, gas formation 423
-, methane formation 407
biological community 6
biomineralization 382
biotite 320, 323, 329–331, 341, 343
-, Macdonald 329
block 65, 77, 80, 131, 167, 168, 171–173, 183, 186–188, 213,
214, 219, 222, 227, 229, 235, 314, 315, 320, 344, 355
Bode Verde Fracture Zone 112
boiling 61, 312, 315, 396, 417
-, seawater 315
bomb 184, 199, 315, 340, 353
-, volcanic 199, 315
Bonin
-, arc 396
-, Island 35
Bora Bora 219
borehole 266
bottom reflectivity 190, 202
Bouguer anomalies 75
Bounty 157, 180, 183–187, 190, 200, 201, 376, 378, 383,
388, 390, 392–396, 398, 408, 409, 411, 418–423, 425
-, hydrothermal fluids 394, 395

Index 463

-, Seamount 180, 183, 376, 377, 385, 388, 407–409, 411–417, 419–425, 452, 454–456
 -, bathymetry map 419, 420
 -, gas sample 425
 -, hydrothermal fluid 412–417
 -, stratigraphy 454, 455
 -, vent waters, rare earth elements 393
 -, Volcano 157, 183–185, 190, 201, 383
breccia 184, 186, 320
brecciation 168, 186
bubble 59, 197, 312, 314, 417
bulge 35–37, 59, 95, 101, 163–165, 180, 192
bulging 37, 134, 178, 201
bulk
 -, composition 249, 288
 -, distribution coefficient 243, 244
buoyancy 17, 18, 75, 149–151, 153, 247–249, 293, 375
 -, thermal 17, 149
bytownite 320, 433

C

Cabo Verde Islands 210
cadmium 353
calcite 379
caldera 22, 186, 196, 202, 209, 217, 220, 222, 340, 342, 343, 383
 -, morphology 196
California 29, 338
 -, coastline 29
Cameroon 286
 -, volcanic line 286
Canada 3
Canary Islands 235, 278, 332–335, 340, 341
 -, alkali basalt 341
 -, gabbroic xenoliths 340, 341
 -, hotspot 278
carbon 6, 383, 423, 425
 -, dioxide 197, 311, 320, 396, 408, 410, 417, 422–425
  -, exsolution 197
  -, reduction 425
carbonate 259, 270, 320, 329, 331, 338
cavity 65
cell 1, 375, 396
Cenozoic 286
 -, volcanic activities 286
cerium 241, 255, 393–395
cesium 239, 241, 249
Chain transform fault 333
chalcopyrite 331
channel 30, 34, 68, 187, 201, 202, 219, 223, 231, 311, 313
 -, landslide 220, 223
 -, volcanic 201
channelization 102, 191, 200, 202, 203, 272, 300, 369
chemistry, magmatic 254
Chile 349
 -, Fracture Zone 349
chimney 167, 376, 411, 417
 -, hydrothermal 167
China 286
chlorite 320, 323, 329–331, 338
chondrite 243, 334, 335, 393–395
Christian Seamount 187, 190, 196, 411–417
 -, hydrothermal fluid 412–417

Christians Cave Formation 182
chromatography 410
chrome-spinel, variability 434
chromite 320, 329
Chukotka 144, 149, 151, 153
 -, continental arc volcanism 153
Circe Seamount 112
classification
 -, intraplate volcanoes 194–196
 -, landslides 232–234
 -, volcanic rocks 431, 432
clast 309, 311, 315, 331, 336–344
 -, gabbroic 309, 311, 331, 336, 337, 341, 342, 344
clay 321, 330, 331, 338
 -, mineral 321, 330, 331
cliff 171
clinopyroxene 320, 321, 323, 325, 328–331, 339, 341, 343, 433, 437
 -, composition 435
 -, Macdonald 325
 -, variability 434
 -, xenocrysts 435
Clipperton Fracture Zone 194
coast 30, 128
coating 167, 177, 186, 196, 363
cobalt 375, 385, 387–390
colorimetry 320
column, magmatic 197
combustion 410
community
 -, bacterial, development 6
 -, biological 6
 -, coral 314
compensation depth 80
conductive cooling 289, 290, 300
conductivity 410
cone 20, 23, 25, 64, 65, 69, 161, 164, 165, 168–172, 180, 183–188, 190, 195, 196, 198, 200–202, 312–315, 339
 -, adventive 171, 172
contamination 239, 265, 311, 342, 344, 388
continent 73, 75, 88, 131, 151
continental 7, 73, 80, 82, 83, 86, 87, 89, 144, 149, 151, 153, 239, 245, 246, 262, 311, 336
 -, drift 73
 -, margin 86, 144, 149, 151, 153
convection 1, 73, 240, 246, 247, 285, 343, 350, 370, 375
 -, mantle 240, 246, 247, 285
 -, thermal 285
 -, Earth 285
convergence 84
Cook 59, 175, 212, 255, 256, 258
 -, Islands 175, 212, 255, 256, 258
cooling 81, 83, 90, 94, 95, 113, 119, 133, 134, 136, 199, 202, 249, 290, 294, 300, 338, 341, 343, 344, 417
 -, conductive 289, 290, 300
copper 353, 375, 385, 387, 388, 390, 391, 398, 410
coral 65, 69, 171, 183, 220, 222, 231, 313, 314, 409
 -, communities 314
 -, fossil debris 313
 -, reef 171, 220, 222, 231
 -, sand 183, 409
core 116, 149, 164, 239, 246–249, 267, 285, 315, 321, 329, 337, 341, 375, 389
 -, -mantle boundary 149, 239, 246–249, 375

Index

coring 178, 378
crater 6, 45, 48, 57, 63, 167, 168, 170, 171, 177, 186, 196, 199, 311, 315, 340, 426
-, circular 168, 199
crest 73, 81, 84, 89, 95, 96
Cretaceous 15, 19, 144, 148, 149, 339
-, oceanic crust 339
-, superchron 19
crome, solubility 436
Crough Seamount 434, 435
Cr-spinel 433, 436
-, composition 436
crust
-, ancient oceanic 170, 190, 200, 239, 243, 245, 246, 248, 286, 333
-, basaltic 151, 246
-, density 86
-, Earth 73, 75, 80, 89, 407, 423
-, abiogenic reactions 407, 423
-, formation 289, 337, 376
-, ridge 289
-, hydrogenetic 375
-, iron 6, 376, 378, 381, 385, 387-395, 398, 401, 404, 405
-, Moua Pihaa 388
-, Pitcairn 375, 395, 398
-, manganese 6, 167, 168, 170, 186, 196, 375, 376, 378, 380, 381, 385, 387, 390-398, 401, 404, 405
-, Pitcairn 375, 395, 398
-, oceanic 1, 5, 19, 35, 84, 87, 89, 90, 100, 105, 107, 112, 116, 122, 124, 125, 170, 171, 178, 190, 200, 210, 234, 235, 240-243, 245-249, 257, 275, 278, 286, 290, 302, 304, 333, 339, 340, 342, 344
-, recycled 240, 241, 245, 246, 249
-, Cretaceous 339
-, formation 431
-, melting 240
-, subducted 247
crustal
-, accretion 92-95, 109, 116, 118, 120, 190
-, age 92, 97, 125, 131
-, alteration 1
-, density 75, 79, 131, 135
-, elevation 76, 134, 135
-, load 79, 83, 85, 94, 136
-, renewal 3
-, root 134, 135
-, seismic velocity 88
-, thickness 43, 79, 86, 87, 90-94, 97, 98, 113, 115-117, 121, 125, 127, 129, 131-133, 135, 136
crystal 198, 241, 278, 315, 320, 322, 323, 328-339, 342, 353, 432
-, growth 432
-, subhedral 323
crystallization 201, 296, 328, 329, 331, 333, 336-341, 343
-, fractional 296, 331
cumulus 320, 329, 338, 341, 343
Cyana (submersible) 3, 158, 163, 166-173, 203, 312
Cyana
-, Seamount 158, 167, 168
-, Volcano 163, 171, 172
$\delta^{13}C$ value 410, 423, 424

D

dacite 435, 444
-, composition 445
Darwin Rise 15
dating 219, 381
-, $^{40}Ar/^{39}Ar$ 6, 145
-, $^{230}Th/^{234}U$ 381
-, age 381
-, radiometric 112, 227
debris 5, 65, 164, 175, 187, 188, 197, 199, 203, 209, 213-215, 217, 220, 222, 224-228, 231, 233, 235, 236, 280, 309, 311, 312, 314, 320, 343
-, avalanche 5, 209, 213-215, 217, 220, 222, 225-228, 231, 235, 236
-, flow 217
-, fossil coral 313
-, gabbroic rock 309
-, lithic 175
decay 50, 51, 57, 62, 275, 278, 381
decompression 25, 150, 151, 239, 288, 289, 292-302, 370
-, melting 25, 150, 239, 289, 293, 294, 298, 299, 370
decoupling 197, 290, 294, 302
deflation 33
deflection 75
deformation 200, 322, 323, 329, 330, 336, 340, 341, 343
dehydration 239, 241-243, 245, 246, 248, 249, 257
-, subduction zones 248
Del Cano Lineament 370
density
-, crustal 75, 79, 86, 109, 131, 135
-, Earth 82
-, lithosphere 4, 78, 90, 133, 136
-, changes 73, 81, 89, 95
-, lithospheric evolution 90
-, seawater 86
depleted MORB mantle (DMM) 245, 253, 268, 274, 275, 431
deposit
-, ferrihydrite 382
-, hyaloclastite 177, 198
-, hydrogenetic 375
-, hydromagmatic 183
-, hydrothermal 6, 65, 168, 174, 184, 199, 376, 385, 388
-, iron, chemical composition 385
-, landslide 212-217, 220, 222, 227, 232, 233
-, manganese, chemical composition 385
-, phreatomagmatic 171
-, pyroclastic 198, 199
deposition 396
-, lapilli 396
depression 76, 90, 108, 198, 217
depth anomaly 3, 15-17
detection of earthquakes 4
detrital 241, 393
Detroit Seamount 143, 145, 146
-, drill holes 146
-, lava 146
-, lavas 146
diatoms 388
dike 201
diopside 320, 329, 337, 343
DMM (see *depleted MORB mantle*)

dolerite 165, 315, 320–322, 329, 330, 338, 341, 343
doming, vertical 92
dredge 65, 69, 116, 178, 196, 227, 313, 363
–, sample 351
dredging 65, 69, 378
drift 6, 73, 112, 143, 146–148, 152, 195, 196, 222, 255, 364, 365, 369–372
–, continental 73
–, Hawaiian hotspot 146–148, 152
–, Pacific Plate 370, 371
–, plate 255
Duke of Gloucester Island 376
dunite 339, 436
dyke 173, 196, 200, 202, 249, 287, 288, 298–301, 330, 338
–, propagation 200
–, swarms 202

E

E.V. HENRY 59, 63, 64, 313
Earth
–, crust 73, 75, 80, 89, 407, 423
–, abiogenic reactions 407, 423
–, density 82
–, gravity potential field 80
–, mantle 1, 253, 254, 285
–, evolution 1
–, origin 1
–, outer shell 80
–, surface 1, 4, 89, 241
–, thermal
–, convection 285
–, evolution 285
–, volcanic structures 1
earthquake 4, 30, 32–34, 36, 37, 39–41, 44, 45, 48–58, 60, 67, 68, 70, 76, 174
–, activity 55
–, body wave analyses 81
–, detection 4
–, distribution 39, 40, 50
–, South Pacific
–, detection 4
East Indian Ocean 83
East Pacific Rise (EPR) 68, 89, 166, 178, 184, 186, 187, 195, 196, 268, 292–294, 296–298, 300, 302, 349, 390, 423, 425–425
–, MORB 296, 297, 302
–, venting fluids 423
–, volcanics 442
East Rift (Easter Microplate) 273, 298–300
–, spreading center 273
–, zone 298
Easter
–, hotspot 265, 273, 275, 296, 300
–, Island 2, 255, 265, 273, 300
–, mantle plume 298
–, Microplate 157, 273, 298–300, 333, 435
–, plume 265, 300
–, Seamount chain (ESC) 273, 274, 298–301, 302, 434, 435, 442, 450, 451
–, lava 273, 450, 451
Easter-Microplate-Crough-Seamount (EMC) 442
Eastern Volcanic Zone 120
echo sounder 9, 210, 212, 218

echogram 65
eclogites 240, 246
edification 65
edifice
–, conical 194
–, intraplate
–, large submarine 165
–, stratigraphy 454
–, truncated 194–196
EEZ (see Exclusive Economic Zone)
effusion
–, lava 37
–, rate 197
ejecta 171, 173, 177, 182, 183, 188, 198, 199, 203, 315–317, 331, 334–336, 339
–, gabbroic 316, 317, 331, 334, 335, 338
–, strombolian 171, 182
–, volcanic 177
ejection 56, 57, 65, 175, 309, 315, 320, 336, 338, 341–344
elevation 15, 17, 35, 73, 75–78, 81, 82, 84, 87, 89, 92–96, 107, 112, 115, 134, 135, 163, 213, 241, 293, 297, 354
EM (see enriched mantle)
EM1 (see enriched mantle 1)
EM2 (see enriched mantle 2)
EMC (see Easter-Microplate-Crough-Seamount)
Emperor Seamount 145
–, chain (E-SMC) 143–145, 147, 152
–, formation 147
–, paradox 145
–, volcanism 145
endiopside 320, 329
energy, seismic 38–40, 49, 50, 52
enriched
–, mantle (EM) 259, 274
–, 1 (EM1) 253, 273, 278
–, 2 (EM2) 243, 246, 253, 262, 268, 274, 278
–, MORB 286, 292
epicenter 34, 36, 43–45, 48, 49, 57, 59
–, relocation 43
epidote 320, 323, 330, 331, 338
EPR (see East Pacific Rise)
equation, isostatic load 76, 79, 84
erosion 5, 65, 171, 209, 230, 232, 235, 254, 261, 300, 312
eruption
–, hotspot edifices 197
–, hydromagmatic 175, 182, 183, 203
–, Macdonald Seamount 411
–, magma 32
–, quiet 197, 199, 203
–, submarine 33, 34, 57, 58, 61, 68, 312
–, volcanic 30
–, Surtsean type 174
–, volcanic 9, 199, 396
ESC (see Easter Seamount chain)
E-SMC (see Emperor Seamount chain)
ethane 423
Eurasia 120, 131, 143
Eurasian lithospheric plates 120
europium 393–395, 398
–, anomaly 393–395, 398
evolution
–, hotspot volcanoes 254
–, hydrothermal fluids 396
–, lithospheric densities 90

466 Index

-, mantle flow 5
-, swarm 45
-, thermal, of Earth 285
Exclusive Economic Zone (EEZ) 9
explosion, hydromagmatic 184, 199, 336
exsolution 197, 315, 336
-, CO_2 197
extrusion 63, 186, 199, 344
extrusive 61, 94, 99, 109, 118, 312, 313, 344
-, alkali-enriched 431
-, silica-enriched 431

F

F.S. SONNE 63, 158, 166, 169, 177, 203, 309, 310, 316, 317, 344, 350, 376, 378, 387, 410, 418, 420, 421, 426
Faanui Valley 220
fabric 15, 19, 20, 161, 163–165, 178, 191, 194, 200, 265, 311, 313, 350
-, abyssal hill 19
Faeroe 120, 131
-, Iceland Ridge Project (FIR) 131
-, Island 120, 131
-, Ridge 131
Failed Rift 363–365, 369
fallout 315
Fangataufa Atoll 264
Farallon
-, Plate 180, 187, 189–191, 202
-, ridge axis 161
Fareura Peak 171
Fatu 267
fault 20, 158, 161, 191, 194, 201, 209, 215, 217, 310, 313, 328, 333, 340, 368
faulting 161, 163, 164, 340
-, abyssal hill 164
fayalite 435
feldspar 379
-, composition 433
ferrihydrite 379, 381, 383, 384, 388
-, deposits 382
-, precipitation 383
FID (see *flame ionization detector*)
FIR (sse *Faeroe Iceland Ridge Project*)
fishing 9
fission track grain age (FTGA) 151
fissure 65, 158, 161, 166, 171, 172, 174, 201, 314, 340, 417
fjord 131
flame ionization detector (FID) 410
flatness
-, degree of 158, 196
-, factor 194
flood 239
flow
-, ash 312
-, basalt 396
-, debris 217
-, differentiation 294, 300, 302
-, mantle, evolution 5
fluid 1, 6, 33, 75, 80, 168, 184, 198, 199, 320, 338, 375, 376, 381, 383, 384, 388, 390, 392–396, 398, 401, 404, 405, 407, 410, 411, 417, 419, 421,423–425
-, discharges, magmatic 1
-, hydrothermal 6, 320, 375, 381, 388, 392–396,

398, 401, 404, 405, 410, 417, 421, 424, 425
-, Bounty's summit 394, 395
-, Macdonald Seamount 425
fluidity 198
fluorescence 315
focal zone (FOZO) 273, 275
-, mantle types 273
foliation 336
fore arc 144
formation
-, crust 289, 337, 376
-, Fe 395
-, Mn 395, 396
-, Emperor Seamount chain 148
-, gaseous hydrocarbons 408
-, hotspot edifices 197
-, hotspot seamounts 201
-, Mn oxides 396
-, submarine hotspot volcanoes 158
Forsterite 435
fossil 65, 69, 189, 313
-, coral debris 313
Foundation
-, Chain 6, 272, 301, 349, 350, 363–365, 368, 369, 371, 432–435, 450
-, $^{40}Ar/^{39}Ar$ age 352
-, age 350, 355, 367
-, bifurcation 366, 367
-, creation 364
-, development 371, 372
-, migration of volcanism 363
-, seamounts 363, 371
-, VER (see also *volcanic elongated ridge*) 350, 359, 366
-, volcanism 350, 366, 370
-, hotline 298, 301
-, hotspot 301, 351, 363, 364, 366, 368, 371
-, magmatism 364, 366
-, intraplate volcanism 367
-, magmatism 371
-, mantle plume 350, 364, 371
-, upwelling 350
-, plume 272, 301, 350, 351, 363–371
-, Seamount 365, 368, 435
-, lava 450
-, plagioclase 433
-, volcanics 451
FOZO (see *focal zone*)
fractionation 432
-, crystal-liquid 431
fracture zone (FZ) 3, 18–20, 112, 157, 158, 194, 200, 201, 210, 211, 310, 333, 338, 349, 364
fracturing 32, 52, 58, 336, 341, 343, 344
-, hydraulic 341, 343
France 12, 203, 315, 344
Franklin
-, Seamount 389,390
-, Volcano 383
French Polynesia 9, 12–16, 18, 20, 22, 25, 26, 30, 37, 38, 96, 99, 209, 210, 213, 222, 236, 257, 309, 313
-, bathymetric map 18
-, depth/age relationship 17
-, sea floor 9, 25
-, submarine landslides 209

Index 467

French Polynesian
- -, chains 256, 260, 265
- -, hotspots 96, 99, 100, 267
- -, Islands 9
- -, network 61
- -, region 4
- -, seismic network ("Réseau Sismique Polynésien", *RSP*) 4, 29, 30, 35, 36, 60, 66, 174, 309, 312
- -, swarms 54
- -, volcanoes 20, 35
  - -, monitoring 35

Friday Volcano 269
FTGA (see *fission track grain age*)
FZ (see *fracture zone*)

G

gabbro 6, 311, 315, 320–323, 326, 328–341, 343, 344
- -, isotropic 6, 330, 344

gabbroic 6, 309, 311, 316–327, 329, 331–337, 339, 341–344
- -, clasts 309, 311, 331, 336, 337, 341, 342, 344
- -, cumulates 6, 341
- -, ejecta 316, 317, 331, 334, 335, 338
- -, rock 318, 320, 324–327, 329–333, 341, 342, 344
  - -, debris 309
- -, xenoliths
  - -, Canary Islands 340, 341
  - -, Hawaii 339

gabbronorite 339
gadolinium 393
Galapagos 35, 194, 273, 275, 336, 398, 423
- -, Fracture Zone 194
- -, hotspot 273, 275
- -, Islands 35
- -, spreading center (GCS) 398, 445, 456
- -, volcanics 446
- -, venting fluids 423

Gallionella ferruginea 383
gallium 243, 245, 249
Gambier Islands 60, 96, 97, 178, 179, 264, 265, 267, 376
Garrett transform 328, 333, 337
- -, fault 328

gas 1, 59, 61, 197, 199, 201, 203, 242, 312, 314, 353, 354, 407, 408, 410, 417, 423, 424
- -, biogenic formation 423
- -, discharges, magmatic 1
- -, hydrothermal 423
- -, pocket 197, 201
- -, vesicles 197

GCS (see *Galapagos spreading center*)
Geisha Seamounts 314
geochemistry, Society Islands 210
geochronology
- -, $^{40}Ar/^{39}Ar$ 351
- -, seamount 143
- -, Society Islands 210

geoid 15, 17–20
- -, anomaly 17, 18

geology 73, 210, 315
- -, Society Islands 210

geomorphology 25, 210, 212, 220
geophysics 1, 3, 4, 19, 25, 59, 151, 157, 201, 254, 260, 272, 292, 309, 313, 349

Georoc
- -, compilation 257, 261, 262
- -, database 263, 264, 268

geothermal 338
Germany 3, 12, 203, 253, 273, 280, 304, 315, 344, 389, 410
giant tubes 444
Gibbs transform 295
gjà (see *fissure*)
glaciation 313
goethite 379, 383
gold 390
grab 378
graben 161, 170, 191, 201
- -, structures 161

granulation 320, 336
granulite 337
Great Meteor Seamount 116–118
- -, crustal structure 118
- -, regional map 117

Greenland-Iceland ridge 120
greenschist 337, 338, 344
ground velocity 60
Gutemberg and Richter's relation 38

H

H.M. Bounty 180
hafnium 239, 241–243, 245, 249
- -, isotopes 241

Hamakua Volcanics 339
harzburgite 339, 436
hastinsite 328
Hawaii 34–38, 45, 55, 56, 69, 102, 109, 112, 131, 136, 143, 145, 147, 157, 168, 200, 209, 262, 286, 332, 333, 339, 363
- -, gabbroic xenoliths 339
- -, regional map 102
- -, Volcano Observatory (HVO) 38, 43

Hawaiian
- -, -Emperor Seamount chain (H-E SMC) 143–145, 147, 152
- -, hotspot 54, 109, 143–148, 152, 292, 390
  - -, hydrothermal Mn crusts 390
  - -, drift 146–148, 152
- -, Institute of Geophysics Network 59, 309
- -, Islands 35, 69, 331
- -, mantle plume 144, 151, 153, 293, 294, 371
  - -, head 144, 151, 153
- -, Seamounts 144–147, 152
- -, shield volcano 164, 340
- -, swell 103, 133
- -, volcanoes 56, 67, 68, 261, 339
- -, xenoliths 5

hawaiite 339
hazard, geologic 9
H-E SMC (see *Hawaiian-Emperor Seamount chain*)
helium 80, 101, 134, 135, 269, 270, 272
Hellenic colcanic arc 389
hematite 383
Hess Deep 326, 333, 336, 337
- -, rift valley 336

Hilina Fault Zone 340
Himalayan Mountains 75
HIMU (see also *mantle, enriched, high μ*) 243, 246, 253, 259, 268, 273, 274, 278

468 Index

-, basalts 259
-, mantle 273
-, source 259, 274
-, trace element characteristics 259
-, volcanoes 259
Hollister Ridge 35
hornblende 320, 326, 328, 330, 331, 337, 341, 343
-, kaersutitic 330
horst structures 161
hotspot
-, activity 1, 3-5, 36, 74, 109, 158, 187, 189, 190, 200, 202, 220, 312
-, Amsterdam 353
-, Austral 6, 29, 34, 59, 65, 68, 69, 160, 169, 173, 175, 178, 180, 194, 198, 199, 202, 309, 312, 343, 421
-, Easter Islands 265, 275, 296, 300
-, edifices 170, 194, 200, 451
-, eruption 197
-, formation 197
-, Foundation 301, 351, 363, 364, 366, 368, 371
-, Galapagos 273, 275
-, Hawaiian 54, 109, 143-148, 152, 292, 390
-, intraplate, rock types 432
-, Iceland 5, 83, 131
-, lava 431, 447
-, location 157
-, magmatism 4, 158, 164, 196, 202, 343, 364, 366
-, oceanic 2
-, rejuvenation 189
-, structures, subsidence 133
-, swells 18, 84, 92, 94, 95, 100, 105, 107, 136
-, lithospheric density structure 73, 95
-, theory 9, 370, 372
-, tracks 94
-, Pacific Plate 144
-, volcanism 3, 6, 190, 196, 200, 203, 210, 240, 246, 272, 351, 365-369, 371, 372, 375, 426
-, sources 253
-, volcanics 432, 455
-, volcanoes 5, 158, 161, 200, 202, 254, 255, 269, 301, 319, 376, 395
-, evolution 254
Hoyo Negro Crater 340
HREE (see rare earth elements, heavy)
Huahine 99, 217
-, Iti 217
-, Nui 217
hummock 213, 217, 220, 222, 224, 227, 228, 231, 233, 235
Hurai Peak 171
HVO (see Hawaii Volcano Observatory)
hyaloclast 198
hyaloclastite 163, 167, 171, 172, 174, 177, 183, 187, 188, 197, 198, 199, 203, 454
hydration 336-338, 341
hydraulic fracturing 341, 343
hydrobrecciation 320
hydrocarbon 407
-, formation 407, 408
-, oxidation 407, 408
hydrogen 59
-, sulfide 59
hydrogenetic
-, crusts 375
-, deposits 375

hydromagmatic
-, activity 183
-, deposits 183
-, eruption 175, 182, 183, 203
-, explosion 184, 199, 336
hydrothermal
-, activity 6, 55, 58, 161, 311, 376, 395, 398, 407-411
-, chimneys 167
-, deposits 6, 65, 168, 174, 184, 199, 376, 385, 388
-, fluids 6, 320, 375, 381, 388, 392-396, 398, 401, 404, 405, 410, 417, 421, 424, 425
-, Bounty's summit 394, 395
-, Macdonald Seamount 425
-, methane 423, 425
-, origin 423
-, Teahitia Seamount 426
-, venting fluids 423
-, mineralization 376, 407
-, Mn crusts 392
-, Hawaiian hotspot 390
-, morphology 66
-, plumes 389
-, precipitates 381
-, vent 6, 382, 407
-, sources 408
-, systems 407, 408
hydrothermalism 1, 6, 35, 407
hydroxide 375
hyloclastite 197
hypocenter 43, 44, 49
-, determination 43
hypocentral 43, 49, 55
-, depth 43

I

IAB (see island arc basalts)
Iceland 83, 120, 121, 124-128, 130-132, 136, 200, 292-296, 301, 327, 370, 446
-, crust 120, 132
-, hotspot 5, 83, 131
-, plateau 327
-, plume 292-295, 370
-, center 131
-, material 295
-, regional map 121
-, rift system 120
-, shore 126
ICEMELT 120, 128, 131
-, profile 131
IFREMER (see Institut Français de Recherche pour l'Exploitation de la Mer)
ilmenite 320, 323, 328, 330, 336, 339
India 143
Indian Ocean 83, 90, 134
injection 173, 200, 201, 340, 417
Institut Français de Recherche pour l'Exploitation de la Mer (IFREMER) 2, 203, 236, 320, 344, 426
Institut pour la Recherche et Developpement (IRD) 66
intraplate volcanism, oceanic 6
intraplate
-, activity 190
-, hotline 310

Index 469

-, magnetism 5
-, plume 3
-, region 1, 3-6, 29, 190, 311, 331, 341, 431
-, volcanic activity 3, 59, 157, 196
-, volcanic provinces 1
-, volcanism 1
    -, Foundation 367
-, volcano 4, 6, 157, 196, 202, 254, 338, 343, 388
    -, active 4
    -, morphological classification 194
intrusion 14, 50, 58, 69, 182, 215, 344, 396
-, magma 344
intrusive flow 173, 198, 203
ionization 410
IRD (see *Institut pour la Recherche et Developpement*)
iron 6, 167, 168, 171, 172, 187, 326, 328, 337, 343, 375, 376, 378, 379, 381, 383-396, 398, 401, 404, 405, 411, 417
-, crust 6, 168, 376, 378, 381, 385, 387-395, 398, 401, 404, 405
    -, formation 395
    -, Moua Pihaa 388
    -, PCA 389
    -, Pitcairn 375, 395, 398
-, deposits 375
    -, chemical composition 385
-, oxide 384
    -, precipitation 383, 384
    -, sediments 411
-, oxidizing bacteria 384
-, oxyhydroxide 375, 376, 379, 383, 384
    -, Pitcairn 379
irradiation 353
island
    -, arc basalts (IAB) 239, 241
        -, genesis 241
        -, geochemistry 241
    -, oceanic 209, 239
        -, volcanism 1
isobath 163, 212, 213, 233
isochron 290, 355
isostasy 73, 75, 76, 79, 84, 90, 91
    -, lithospheric plate 74
    -, principles 75
isostatic
    -, equilibrium 4, 80, 86, 101, 135, 136
    -, load model 4, 73, 74, 82, 84, 86-88, 90, 92-96, 98, 100, 101, 103, 111, 114-119, 121-123, 133, 136
Izu Island 35

J

Jan Mayen plume 295
Japan 29, 35
Josephine Seamount 116-119
    -, crustal structure 119
    -, regional map 117
Juan Fernandez
    -, Chain 268
    -, Island 2, 269
    -, magma 269
    -, Microplate 157
    -, Seamounts 271
    -, volcanoes 270

K

kaersutite 437
Kahoolawe Island 331
Kamchatka 144, 149, 151, 153
    -, arc 151, 153
    -, lithosphere 151, 153
    -, forearc 144
Kane
    -, Fracture Zone 93
    -, transform 336
Kauai 109, 331
Kealakekua Bay 177
Kerguelen 286
Kermadec 35, 392
K-feldspathoids 433
Kilauea 32, 33, 38, 45, 49, 50, 54, 56, 57, 63, 67, 331, 340
    -, Volcano 32
Kolbeinsey Ridge 120, 295
    -, spreading center 120
Kula Plate 151

L

La Palma (Canary Islands) 340
labradorite 320, 330, 433
lagoon 217, 224
laminae 378
lamination 379
landslide 5, 9, 168, 209, 210, 212-217, 220, 222-236
    -, Austral Islands
        -, classification 232
    -, channel 220, 223
    -, deposits 212, 213, 215-217, 220, 222, 227, 232, 233
    -, Society Islands 214
        -, classification 232
    -, submarine 209, 210, 217, 224
    -, superficial 209, 214, 224, 225, 227, 228, 230-232
    -, Young Island 235
lanthanum 267, 393
lapilli 175, 184, 309, 311, 314, 315, 342, 396
    -, basanite 309, 311
    -, deposition 396
laser ablation 395
lattice 337
Laupahoehoe Volcanics 339
lava
    -, arc 240
    -, classification 431
    -, composition 5, 431, 452
    -, cyclical eruptions 454
    -, Detroit Seamount 146
    -, Easter Seamount Chain 273
    -, effusion 37
    -, flow 112, 163, 167, 170-172, 180, 182, 183, 186, 196, 199, 202, 203, 215, 313, 314, 340
        -, frozen 171, 172
    -, ocean-island basalts (OIB) 249
    -, petrology 431
    -, seamount 145, 298
    -, silica-enriched 432
    -, silicic 445, 446
        -, hotspot 447
        -, plagioclase composition 433

Index

-, tube 166, 168, 186, 187, 199, 444
-, tubular flow 171, 172
-, tunnel 167, 183, 184, 199
leaching 6, 392
lead 241
Leptothrix ochracea 383, 384
leucodiorite 330, 331
lherzolite 339, 340
limestone 116
-, Foraminiferical 116
Line Islands 157, 264
-, basalt 265
lineament 6, 310, 370
lithology 169, 184, 185, 199, 288, 294, 302
lithosphere
-, age 364
-, cold 92, 93, 98, 100, 101, 111, 114, 115, 118, 119, 122, 123
-, composition 5
-, cooling: 113
-, density 4, 73, 78, 79, 82, 84–102, 105, 109, 111–119, 121–127, 130, 131, 133–136
-, changes 73, 81, 89, 95
-, evolution 90
-, hotspot swells 73, 95
-, development, conductive cooling 300
-, hot 92, 93, 98, 100, 101, 111, 114, 115, 118, 119, 122, 123
-, loading, flexural modeling 24
-, oceanic 18, 73, 80, 81, 94, 124, 133, 150, 158, 191, 246, 249, 290
-, subsidence 94
-, plate 4, 15, 19, 73, 74, 76–80, 82, 84, 87, 89, 94, 120, 136, 158, 255, 285, 289, 290, 302, 367
-, migration 367
-, sub-crustal 4
lithostatic
-, load 74–76, 79, 82, 83, 87, 90, 108, 109, 136
-, pressure 74, 80, 197
Llano del Banco 340
Loihi 32, 34–36, 43, 48, 50, 54, 55, 63, 67, 191, 269, 376, 382, 388, 389, 450
-, alkali basalts 450
-, Seamount 36, 177, 191, 197, 389, 434, 436, 442
-, tholeiites 450
-, volcanics 450
-, Volcano 32, 35
Lookout Point 183
Louisville
-, hotspot 2, 144, 297
-, Seamount chain 148, 151, 154, 157
low velocity zone (LVZ) 249, 289, 290, 294, 302
LREE (see *rare earth elements, light*)
lutetium 245
LVZ (see *low velocity zone*)

M

Macdocald Seamount 175, 178, 195, 196
Macdonald
-, basalt 262
-, biotite 329
-, Ca-amphibole 326
-, clinopyroxene 325

-, explosive activity 61
-, gabbro 337, 338, 341
-, gas 424
-, hotspot 191, 211, 222
-, hydrothermal
-, CH_4 425
-, fluids 417
-, magmatic activity 69
-, oxides 327
-, plagioclase 324
-, Seamount 2, 4, 21, 24, 25, 29, 35, 59, 61–64, 66, 68, 69, 157, 173, 175–178, 199, 222, 223, 257, 259, 309–318, 321–329, 331–337, 340–344, 376, 393, 407–409, 411–417, 421, 423–425, 450
-, activity 342
-, archaebacteria 426
-, bathymetry 21, 63
-, chain 151
-, eruption 411
-, gabbroic clasts 337, 344
-, gas sample 423–425
-, history 62, 312
-, hydrothermal fluid 412–417, 425
-, hyperthermophilic archaebacteria 426
-, magma chamber 309
-, plate thickness 25
-, region 4
-, swarm duration 62
-, trace element patterns 257
-, volcanics 450
-, seismic swarms 60
-, summit area 6
-, bathymetry 421
-, volcano 6, 24, 35, 175, 190, 212, 311
Madeira-Tore Rise 116
magma
-, alkali basaltic 341
-, basaltic 263, 341
-, chamber 200, 220, 267, 296, 309, 311, 330, 338, 339, 341–343
-, convection cells 1
-, degassing 57, 58, 423
-, eruption 32
-, intrusion 344
-, pockets 67
-, pressure 32
-, reservoir 6, 199–201, 340–343
-, source 5, 253, 254, 265, 267, 275, 339
-, nature 254
-, possible origin 254
magmatic
-, activity 1, 33, 39, 58, 69, 83, 312
-, chemistry 254
-, column 197
-, fluid discharges 1
-, gas discharges 1
-, petrogenesis 272
-, pressure 33
-, reservoir 196
-, segregation 196
-, solidification 197
magmatism 4, 6, 95, 158, 164, 196, 202, 262, 269, 341, 343, 363, 364, 366, 369–371
-, Foundation hotspot 364, 366, 371

-, hotspot 4, 158, 164, 196, 202, 343, 364, 366, 449
-, plume 343
-, spreading ridge 158, 449
magnesiowuestite 248
magnesium 328
-, melt 240
magnetic 1, 124, 148, 153, 178, 191, 309, 311, 313, 340, 351
-, anomaly 1, 124, 148, 153, 178, 191, 309, 311, 313
magnetism 5
magnetite 320, 323, 329–331, 336, 338, 339
manganese 6, 167, 168, 170–172, 186, 187, 195, 196, 375, 376, 378, 380, 381, 385, 387, 390–392, 394–396, 398, 401, 404, 405, 408
-, crust 6, 167, 168, 170, 186, 196, 375, 376, 378, 380, 381, 385, 387, 390–396, 398, 401, 404, 405
   -, formation 395, 396
   -, hydrothermal 392
   -, PCA 392
   -, Pitcairn 375, 395, 398
-, deposits 375
   -, chemical composition 385
-, oxide 379, 392, 396
   -, formation 396
   -, precipitate 396, 398
-, oxyhydroxide 376
manganite 380
mantle
-, asthenospheric 288, 289, 292
-, convection 240, 246, 247, 285
-, Earth 1, 253, 254, 285
   -, evolution 1
   -, origin 1
-, enriched 431
   -, high μ (HIMU) 431, 432
   -, type I (EM1) 273, 431
   -, type II (EM2) 431
-, flow
   -, evolution 5
   -, models 143
-, FOZO/C 273
-, HIMU 273
-, hotspot 149
-, lower 246–249, 292
-, peridotites 240, 248
-, plume 3–5, 90, 91, 120, 143, 144, 149, 150, 153, 158, 245, 246, 248, 249, 255, 259, 273–275, 285, 286, 290, 292–294, 296, 298, 301, 304, 311, 350, 364, 368, 370–372, 431
   -, Easter 298
   -, Foundation 350, 364, 371
   -, Hawaiian 144, 153, 294, 371
   -, head 144, 150, 153
   -, initiation 150
   -, midplate 311
   -, rise 150
   -, sources 245, 249
-, prevalent (PREMA) 270, 275
-, primitive helium (PHEM) 275
-, temperature 247, 278
-, upper 5, 16, 25, 83, 120, 200, 240, 246–249, 254, 275, 370
-, upwelling 239, 254, 288, 293, 294, 304
-, viscosity 289, 290
-, viscous 67

MAR (see *Mid-Atlantic Ridge*)
margin, continental 86, 144, 149, 151, 153
Mariana
-, Archipelago 35
-, Islands 34
-, Volcano 35
MARK area 333
Marotiri 309
Marquesas 9, 18–20, 68, 96, 97, 100–102, 157, 210, 253–256, 260, 265–269, 275
-, Chain 100, 266
   -, islands 266
-, crustal structure 101
-, Fracture Zone 20, 96, 97, 100, 102, 210, 265
-, hotspot 2, 68
-, Island 18, 96, 97, 100, 102, 157, 265, 267, 268
-, plateau 102
-, plume 266, 267
-, swell 18, 19
-, volcanoes 255, 267
   -, age 255
Mascarene 109, 133
-, plateau 133
-, ridge 109
mass
-, flow, ridgeward 289, 304
-, fractionation 353
-, movement 209
Maui 37, 109
Mauna 38, 45, 56, 331–333, 339, 344
-, Kea 331–333, 339, 344
   -, Volcano 339
-, Loa 38, 45, 56, 331, 339, 340
   -, Southwest Rift Zone 339
Maupiti 99, 158, 210
-, Volcano 158
Mauritius 109
-, Island 109
Mehetia 33, 35–39, 41, 43–46, 48–50, 52–58, 61, 66–68, 99, 157, 158, 161, 164, 165, 170, 171, 174, 183, 192, 203, 210, 214, 235, 261, 312, 409, 450
-, edifices 33
-, Island 46, 157, 171, 174, 214, 409
-, recent activity 67
-, Seamounts 29, 163, 261
-, seismic swarm 39
-, swarm 39, 41, 43, 45, 48, 49, 53, 55
-, Volcano 33, 158
Meiji Seamount 146, 152
melt
-, alkali-enriched 433
-, alkalinity 433
-, basaltic 240, 248
-, crystal growth 432
-, generation, source regions 287
-, high magnesian 240
-, picritic 240
-, solidification 433
melting 5, 25, 150, 158, 200, 239–241, 243, 248, 249, 255, 260, 265, 268, 275, 286–289, 293, 294, 298–302, 331, 354, 368, 370
-, anomaly 260
-, decompression 25, 150, 239, 289, 293, 294, 298, 299, 370

472 Index

-, degree of 268
-, oceanic crust 240
-, partial 431
metabasalt 311, 320, 323
metabasite 336, 338
metadolerite 311, 315
metagabbro 311, 315, 323, 330, 336
metal 168, 375, 385, 388, 390, 396
metamorphism 246, 257, 311, 320, 337, 338, 341, 343
metasomatism 249, 259, 270, 273, 286, 287, 301, 320
methane 6, 311, 314, 407–411, 417, 419, 421, 423–425
-, abiogenic 424–426
-, anomaly 417
-, atmospheric 411
-, biogenic 407
-, concentration 6, 314, 409–411, 418–422, 424, 425
    -, profile 411
    -, sea surface water 411
-, distribution in the water column 408
-, enrichment 411, 421
-, hydrothermal 423, 425
    -, origin 423
    -, Teahitia Seamount 426
    -, venting fluids 423
-, oxidation 426
-, production, bacterial 424, 425
-, sea surface equilibrium concentration (MSE) 411
-, venting 407
microlites 435
microphenocrysts 433, 435, 443
microplate 157, 273, 298, 299, 349, 350, 363–366, 368, 369, 371
    -, boundaries 310
    -, creation 364
microthermometry 338
Mid-Atlantic Ridge (MAR) 18, 115, 292, 296, 297, 302
    -, bathymetric profile 115
    -, MORB 296, 302
Mid-Ocean Ridge
    -, basalt (MORB) 165, 178, 196, 239, 243, 245, 253, 264, 269, 273, 286, 288, 292, 294–297, 302, 341, 431, 436, 437
        -, clinopyroxene 435
        -, database 296, 297
        -, enriched 286, 292, 431, 437
        -, global database 296, 297
        -, Hawaiian-like 294
        -, mantle 239, 243, 245, 253, 302
        -, MAR 296, 302
        -, normal 431, 437
        -, sources 288
        -, transitional 431, 437
    -, systems 343
    -, hydrothermal fluids 393
midplate
    -, mantle plume 311
    -, swell 17, 18
migration 37, 48, 49, 52, 58, 67, 195, 340, 351, 352, 363–369, 371
    -, rate, volcanism 352, 365
    -, seismic activity 50, 67
    -, volcanism 363

mineral 9, 17, 240, 247–249, 311, 320, 321, 329–331, 336, 337, 341, 379, 383
    -, crystallization order 433
    -, submarine 9
mineralization 315, 376, 396, 407
    -, hydrothermal 376, 407
mineralogy 1, 330, 331, 336, 343, 375, 393
Miocene 116
Mn crust, hydrothermal 392
    -, Hawaiian hotspot 390
Mn-Seamount 195
modeling, lithospheric loading 24
Mohorovicic (Moho) 32, 43–45, 48, 55, 57, 58, 82, 83, 85, 90, 101, 108, 124
    -, depth 83, 85, 90
    -, discontinuity 32, 43–45, 48, 55, 57, 58, 82, 83, 85, 90, 101, 108, 124
    -, interface 82
molybdenum 38, 256, 387, 388, 392
Monowaii Seamount 35
Moorea Island 217
MORB (see Mid-Ocean Ridge basalt)
morphology 15, 20, 157, 181, 187, 196, 197, 202, 313
    -, caldera 196
    -, hydrothermal 66
    -, starfish 168
    -, volcanic 22, 23, 66, 175
Moua Pihaa 36, 39, 52, 67, 158, 161, 163, 168, 170, 180, 192, 200, 214, 215, 228, 388, 409, 450
    -, Fe crust 388
    -, main volcanic axis 170
    -, Seamount 67, 161, 163, 180, 199, 214, 215, 228, 409
    -, Si concentrations 388
MREE (see rare earth elements, medium)
MSE (see methane sea surface equilibrium concentration)
mugearite 339
multibeam
    -, bathymetry 10, 12, 13, 15, 20, 21, 23, 359
    -, sonar systems 9
multichannel, bathymetry 190, 202
Mururoa 178, 179, 264, 376
    -, Atoll 264
Myojin Volcano 29

N

N.O. JEAN CHARCOT 309
N.O. L'ATALANTE 158, 178, 179, 188, 203, 218, 236, 309, 410, 426
N.O. LE SUROIT 63, 312
N.V. ESTAFETTE 311
N.V. MELVILLE 59
Na-feldspathoids 433
NAS (see North American Shale)
NASP (see North Atlantic Seismic Project)
Nautile (submersible) 3, 169, 171–173, 178, 179, 203, 376, 378, 410, 411, 419, 426
Nazca Plate 268, 296, 349, 364, 369
neodymium 242, 243, 245, 262, 263, 296, 297, 302, 341, 393
nepheline 177, 433, 443
nephelinite 443

NEPR (see *North-East Pacific Rise*)
New Zealand 35, 338
Ngatemato 23, 368, 369, 371
  –, Chain 23, 369, 371
  –, Seamounts 23, 368, 369
nickel 331, 375, 385, 387–391, 398
niobium 239, 241, 242, 245, 249, 262, 273, 296, 331
N-MORBs, plagioclase composition 433
nontronite 388
norite 339
North America 90, 393
North American Shale (NAS) 393
North Atlantic Seismic Project (NASP) 131
North Pacific 5
North-East Pacific Rise (NEPR) 445, 456
  –, volcanics 446
Northern Volcanic Zone (NVZ) 120, 131

O

Oa 267
Oahu Island 106, 234
ocean
  –, basin 29, 32, 68, 302
  –, Floor Observation System (OFOS) 186, 410, 417, 421
  –, -island basalts (OIB) 5, 239–243, 245, 246, 249, 253, 286, 288, 292, 293, 431
    –, lavas 249
    –, sources 245, 288
    –, suites 240, 243, 245, 246, 249
  –, ridge 5, 253, 285, 286, 288–290, 294, 302, 311, 333
    –, actual role 5
oceanic
  –, basalt 246–249
  –, basin 1, 3, 29, 73, 76, 81, 84, 90, 95
  –, crust 1, 5, 19, 35, 84, 87, 89, 90, 100, 105, 107, 112, 116, 122, 124, 125, 170, 171, 178, 190, 200, 210, 234, 235, 240–243, 245–249, 257, 275, 278, 286, 290, 302, 304, 333, 339, 340, 342, 344
    –, recycled 240, 241, 245, 246, 249
    –, subducted 247
  –, hotspots 2
  –, intraplate volcanism 6
  –, island 209, 239
    –, volcanism 1
  –, lithosphere 18, 73, 80, 81, 94, 124, 133, 150, 158, 191, 246, 249, 290
  –, plateau 144, 150, 151, 153
OFOS (see *Ocean Floor Observation System*)
OIB (see *ocean-island basalts*)
Okhotsk 144, 149, 151, 153
  –, Sea 151, 153
Oligocene 116, 180
olivine 183, 320, 321, 323, 329, 330, 339, 433, 436, 437
  –, composition 435
  –, minerals 435
  –, variability 434
Ontong Java Plateau 151
opaque 323, 330, 336, 339
ophiolite 338
organism 383, 407
  –, chemosynthetic 407

orthoclase 433, 443
  –, normative 433
orthopyroxene 320, 330, 341
osmium 242
oxidation 59, 393, 407, 408, 411, 424–426
  –, hydrocarbon 407, 408
  –, methane 426
oxide 320–323, 326, 327, 330, 336, 339, 375, 379, 381, 383, 384, 389, 390, 392, 396, 398, 410, 411, 417
oxycline 396
oxygen 375, 410
oxyhydroxide 6, 168, 187, 375, 376, 379, 383, 384
  –, Fe- 167
  –, Fe-Mn, deposits 6
  –, Fe-Mn-Si, deposits 171, 172

P

Pacific
  –, -Antarctic Ridge (PAR) 301, 432
    –, volcanics 446
  –, crust 107, 165
  –, -Farallon Ridge 210
  –, guyots 236
  –, hotspot 3, 254, 265, 278
    –, activity 3
  –, lithosphere 364
  –, midplate volcanism 372
  –, Ocean 3, 34, 35, 90, 157, 175, 253, 259, 310, 407, 411
  –, Plate 20, 67, 96, 99, 102, 106, 143, 144, 146–149, 151, 152, 158, 175, 194, 210, 265, 272, 292, 294, 298, 301, 309, 349, 350, 363, 369–372
    –, consumption 148
    –, drift 370, 371
    –, motion 67, 99, 102, 143, 146–149, 153, 175, 210, 294, 349, 350, 369
    –, production 148
    –, reorientation 143, 148, 149, 153
    –, subduction 151
    –, velocity 106, 158
  –, sea floor 175, 349, 363
  –, seamounts, shape 194
paleolatitude 145, 148
Palva Valley Point 183
PAR (see *Pacific-Antarctic Ridge*)
parasite 164, 165
pargasite 328, 337
PCA (see *principal component analysis*)
peridotite 239, 240, 247–249, 287, 288, 294, 301
  –, mantle 240, 247, 248
perovskite 248
petrogenesis 242, 257, 272
  –, Austral 257
  –, magmatic 272
  –, South East Pacific hotspots 274
petrography 331, 341, 343
petrology 5, 201, 240, 249, 253, 309, 311, 336
Pg wave 43, 44
pH anomaly 411
PHEM (see *primitive helium mantle*)
phenocryst 320, 331, 433, 443
phonolite 261, 433, 447
  –, composition 449

Index

phyllosilicate 320
picrite 168, 240, 315, 320, 339
pillow lava 437, 444
piston coring 378
Pitcairn 4, 6, 18, 69, 96, 157, 158, 171–175, 178–185, 188, 190, 191, 195–197, 199–203, 253, 260, 264, 265, 267, 268, 375, 376, 379–381, 383–385, 388, 390, 392–396, 398, 407, 409, 411, 418–421
  –, Fe
    –, crusts 375, 388, 393, 395, 398
    –, oxyhydroxide 379
  –, hotspot 4, 157, 171–173, 175, 178, 180, 181, 189, 191, 195–197, 200, 201, 253, 268, 376, 384, 409, 418–420, 432
    –, bathymetry map 418
    –, recent volcanic activity 180
    –, region 178, 180, 181, 196
    –, volcanic edifices 180
    –, volcanics 450
  –, Island 69, 178–183, 190, 203, 264, 376, 407, 451
    –, chain 376
  –, Mn crusts 375, 380, 381, 390, 392, 395, 396, 398
  –, region 178–180, 183–185, 188, 190, 202
  –, Seamounts 6, 264, 376, 377, 381, 382, 385, 388, 390, 393–396, 398, 401–403
    –, formation of Fe crust 395
    –, formation of Mn crust 395, 396
    –, Mn crusts 390, 395, 396, 398
Piton de la Fournaise Volcano 338
plagioclase 320, 323, 324, 329–331, 337–339, 343, 351, 353, 432, 433, 437
  –, composition 433
  –, sodic 433
  –, variability 434, 455
plagiogranite 447
plate
  –, Australian 151
  –, boundary 1, 5, 18, 239, 285, 286, 310
    –, features 18
    –, history in South Pacific 18
  –, divergent 3
  –, drifting 255
  –, dynamics 73
  –, Farallon 180, 187, 189–191, 202
  –, lithospheric 4, 15, 19, 73, 74, 76–80, 82, 84, 87, 89, 94, 120, 136, 158, 255, 285, 289, 290, 302, 367
    –, American 120
    –, Eurasian 120
    –, isostasy 74
    –, migration 367
  –, motion 4, 9, 19, 143, 146, 151, 152, 265, 290
    –, relative 9, 19
    –, statistical analysis 143
  –, movement
    –, rate 260
  –, reconstruction 15, 143
  –, rotation vector 260
  –, separation 5, 286, 288, 289
    –, rate 286
  –, tectonic 4, 73, 76, 79, 80, 136, 239, 285, 286, 290, 301, 302
    –, models 81
  –, thickness 25, 79–81
    –, Macdonald Seamount 25

  –, velocity 109, 266
plateau, oceanic 144, 150, 151, 153
Pleistocene 65
plume
  –, dispersion model 288, 292, 294, 302
  –, Easter 265, 300
  –, flux 246
  –, Foundation 272, 301, 350, 351, 363–365, 367–371
  –, head 149
  –, hydrothermal 389
  –, intraplate 3
  –, magmatism 343
  –, material, nature 286
  –, sources 5, 245, 249
    –, composition 5
    –, nature 5
  –, tectonics 285
Pn wave 30, 32, 43, 44, 48–50, 52
  –, arrival 43, 50, 52
  –, seismogram 50
pole 20
Polynesian Seismic Network (Réseau Sismique Polynésien, RSP) 29, 30, 35, 60, 66, 309, 312
potassium 145, 158, 175, 178, 179, 183, 273, 300, 341
praseodymium 393
Pratt 4, 75, 76, 78, 79, 116, 135, 136
  –, compensation
    –, mechanism 4
    –, scheme 79, 136
  –, equation 78, 79
  –, type of isostatic compensation 76
precipitate, hydrothermal 381
precipitation 6, 375, 383, 384, 388, 398
  –, ferrihydrite 383
  –, iron oxide 383, 384
  –, manganese oxide 398
PREMA (see *prevalent mantle*)
pressure
  –, hydrostatic 199
  –, magma 32
  –, magmatic 33
prevalent mantle (PREMA) 270, 275
primitive helium mantle (PHEM) 275
principal component analysis (PCA) 388
  –, Fe crusts 389
  –, Mn crusts 392
profile, net bathymetric 24
propane 423
protolite 321, 329
pseudofault 202
pseudomorphs 329
Puka Puka 22, 23, 254
  –, Chain 22
  –, lavas 25
  –, Ridge 254
Pulawana Volcanics 182
pumice 167, 171, 172, 184, 199
  –, flow 199
pyrite 320, 331
pyroclast 167, 182, 183, 187, 197–199, 454
pyroclastic 183, 188, 190, 198, 199, 312
  –, deposits 198, 199
pyroclastite 182, 203
pyroxene 330, 338, 341, 380

Index   475

Q

quartz 320, 323, 330, 338, 353, 380

R

R.V. HAVAIKI 63
R.V. J. CHARCOT 35
R.V. KAWAMEE 64
R.V. L'ATALANTE 210, 212
R.V. LA PAIMPOLAISE 45, 63, 64
R.V. MARARA 63-65
R.V. MELVILLE 45, 69, 312
Rà
  -, Seamount 176, 177, 409, 450
    -, volcanics 450
  -, Volcano 175
radiometric
  -, age 116
  -, dating 112, 227
Raiatea 99, 217, 235
Raivavae 211, 224, 225, 235
  -, bathymetric map 224
  -, Island 224, 225
Rangiroa Atoll 38, 68
Raoul Island 35
Rapa 211, 222, 223, 235
  -, bathymetric map 223
  -, topography 222
rare earth elements (REE) 241, 334, 335, 339, 341,
    393-395, 398
  -, Fe crusts 393, 394
  -, heavy (HREE) 341, 394
  -, hydrothermal fluids 393
  -, light (LREE) 331, 393, 394
  -, medium (MREE) 394
  -, Mn crusts 395
  -, seawater 393
  -, vent waters 393
rate
  -, effusion 197
  -, migration, volcanism 352, 365
  -, plate
    -, movement 260
    -, separation 286
  -, spreading 121, 145, 290, 291, 294, 295, 298,
    300, 302, 309
Rayleigh 32, 33, 45, 424
  -, fractionation 424
reactions in the Earth's crust, abiogenic 407, 423
recrystallization 329, 330, 336-338, 343
  -, syn-tectonic 336
REE (see *rare earth elements*)
reef 171, 220, 222, 231
  -, coral 171, 220, 222, 231
refertilisation 242
reflection, seismic 18, 19, 161
refraction, seismic 19, 45, 48, 116, 128
rejuvenation 73, 94, 103, 115, 116, 133, 189, 260, 339, 340
  -, thermal 94, 103, 116, 133
reorientation, Pacific Plate 143, 148, 149, 153
Réseau Sismique Polynésien (RSP, Polynesian
    Seismic Network) 4, 29, 30, 35, 36, 60, 66, 174,
    309, 312

reservoir, magmatic 196
Resolution/Del Cano Fracture Zone 349
Réunion 69, 109, 111, 112, 136, 210, 311, 338, 341
  -, clast 339
  -, crustal structure 111
  -, Island 69, 109, 111, 112, 136, 210, 311, 338, 341
    -, gabbros 338
    -, volcanic edifice 112
  -, volcanics 435
Reykjanes Ridge 93, 120-125, 127, 292, 293, 295, 298,
    300, 301
  -, axial lavas 292
  -, axis 127
  -, crustal structure 122, 123
  -, MORB 292
  -, spreading center 120, 121
rheology 80
rhyolite 201, 273, 444
  -, composition 445
ridge
  -, aseismic 133
  -, -centered hotspots 295
  -, spreading 68, 89, 158, 166, 187, 189-191, 197,
    202, 290, 298, 302, 327, 336, 341
  -, suction 288, 290
  -, volcanic 4, 22, 161, 301
Riedel shear 191
rift 5, 19, 20, 50, 54, 65, 120, 165, 168, 170-172, 174, 183,
    190-192, 194, 195, 199-202, 213, 215, 224, 227, 235,
    236, 300, 314, 336, 340, 350, 363, 364, 369
  -, zone 5, 50, 54, 165, 168, 170-172, 174, 183,
    190-192, 194, 195, 199-202, 213, 215, 224, 227,
    235, 236, 314, 340
  -, formation 200
  -, starfish type 192
Rikitea 60
Rimatara Island 175, 231, 232, 235
  -, bathymetric map 231
Rio Grande Rise 133
Robinson Crusoe (Mas a Tierra) 268, 270
  -, volcanism 269
Rocald Volcano 169, 173, 195, 196
Rocard
  -, Seamount 67, 451, 455
    -, stratigraphy 455
  -, Volcano 163, 173
rock
  -, basaltic 437
  -, falls 209
  -, gabbroic 318, 320, 324-327, 329, 331, 333, 341,
    342, 344
  -, mineral composition 431, 432
  -, samples 6, 320, 351
  -, silica-enriched 432
  -, silicic 435, 444
  -, types 5, 261, 320, 436
  -, vesicularity 197
  -, volcanic
    -, classification 432
    -, composition 448
    -, description 432
Rockne Seamount 157
rodingite 338
Romanche transform fault 333

root 18, 39, 40, 75–78, 92, 95, 99, 101, 134, 135
 –, crustal 134, 135
RSP (see *Polynesian Seismic Network* or *Réseau Sismique Polynésien*)
rubidium 178, 239, 241, 242, 245, 249, 273, 296, 341, 390
Rumble Seamount 35
Rurutu Island 175, 211, 213, 224, 227–230, 235, 255, 258
 –, acoustic imagery 229
 –, bathymetry 228
 –, volcanics 259

S

Sala y Gómez
 –, Island 2
 –, Seamounts 273
salinity 410, 411
samarium 2, 245, 300, 393, 394
Samoa Islands 2, 157, 246, 280
sand 183, 315, 396, 409
 –, coral 183, 409
sandstone 144
sanidine 353, 433
 –, Taylor Creek Rhyolite (TCR) 352, 353, 355
saponite 329
satellite altimetry 3, 14, 15, 20, 372
 –, South Pacific Ocean 259
schistosity 336
scuba divers 63–66, 313
sea floor
 –, central Pacific 9
 –, fabric 191
 –, French Polynesia 9, 15, 25
 –, depth 15
 –, morphology 15
 –, reflectivity 188
sea level 35, 37, 49, 55, 65, 67, 76, 79, 86, 87, 89, 94, 112, 165, 171, 180, 227, 233, 297, 312–314, 409
Sea of Okhotsk 149
seamount
 –, Adams 180, 183, 376, 377, 385, 388, 409, 412–417, 451, 455
 –, Arago 2, 175, 176, 212, 226, 227, 229, 258
 –, Bounty 180, 183, 376, 377, 385, 388, 407–409, 411–417, 419–425, 452, 454–456
 –, Christian 196, 411–417
 –, Circe 112
 –, Crough 434, 435
 –, Cyana 167, 168
 –, Detroit 143, 145, 146
 –, Easter-Microplate-Crough-(EMC) 442
 –, Franklin 389, 390
 –, geochronology 143
 –, Great Meteor 116, 118
 –, Hawaiian chain 146, 147
 –, Josephine 116–119
 –, lava 145, 298
 –, Loihi 36, 177, 191, 197, 389, 434, 436, 442
 –, Macdocald 195
 –, Macdonald 2, 4, 21, 24, 25, 29, 35, 59, 61–64, 66, 68, 69, 157, 173, 175–178, 199, 222, 223, 257, 259, 309–318, 321–329, 331–337, 340–344, 376, 393, 407–409, 411–417, 421, 423–425, 450

 –, Mehetia 29, 163, 261
 –, Meiji 146, 152
 –, Mn- 195
 –, Monowaii 35
 –, Moua Pihaa 67, 161, 163, 180, 199, 214, 215, 228, 409
 –, Rà 176, 177, 409, 450
 –, Rocald 169, 173, 195, 196
 –, Rocard 67, 451, 455
 –, Rockne 157
 –, structure 202
 –, Suiko 143, 145
 –, Tahiti 261
 –, Taukina 20, 369
 –, Teahitia 29, 37, 49, 58, 63, 157, 165, 166, 170, 264, 408, 409, 411–417, 422–426
 –, Turoi 165, 166, 199
 –, Young 186
seamounts (chain)
 –, Austral 407, 451
 –, Easter 273, 274, 298, 299, 301, 302, 434, 435, 442, 450, 451
 –, Emperor 143–145, 147, 152
 –, Foundation 272, 301, 350, 363–365, 368, 369, 371, 432–435, 450
 –, Geisha 314
 –, Hawaiian 144–147, 152
 –, Juan Fernandez 271
 –, Louisville 148, 151, 154
 –, Ngatemato 23, 368, 369
 –, Pitcairn 6, 264, 376, 377, 381, 382, 385, 388, 390, 393–396, 398, 401–403
 –, Rumble 35
 –, Sala y Gómez 273
 –, Society 261
 –, Tarava 21, 22
 –, Tuamotu 151
seawater 1, 6, 34, 61, 86, 199, 312, 314, 315, 320, 375, 381, 384, 390, 392–396, 401, 404, 405, 410, 411, 417, 422, 424
 –, boiling 315
 –, density 86
sediment
 –, cover 164, 187, 188, 222
 –, iron oxide 411
 –, pelagic 161, 167, 234
 –, terrigenous 241, 246
 –, thickness 14, 35
 –, waves 222, 227, 228, 230
sedimentation 15, 168, 190, 199
segregation 196, 199, 246, 311
 –, magmatic 196
seismic 4, 29, 30, 32, 35, 36, 38, 45, 55, 57, 59, 60, 65, 120, 131, 174, 186, 195, 196, 233, 234, 313
 –, activity, monitoring 29
 –, detection 56, 63, 68
 –, energy 38–40, 49, 50, 52
 –, map 32
 –, migration 50, 67
 –, profile 99, 163, 233
 –, reflection, profile 161
 –, sensing stations 33
 –, swarm
 –, Macdonald 60

-, Mehetia 39
-, tremor 30, 32, 41, 49, 50, 56, 60
-, velocity 5, 15, 16, 74, 82, 83, 86, 88, 116, 124, 233
   -, anomaly 16
   -, crustal 88
-, waves 30
   -, conventional 30, 32, 60
seismicity 5, 18, 32, 33, 36, 37, 39, 48, 49, 52, 54, 55, 57, 67, 68, 70, 149, 157, 290, 340
   -, Austral hotspot 29
   -, Society hotspot 29, 36, 67
   -, volcanic 32, 52, 55
seismogram 45, 48, 50
seismograph 29
seismology 80
Selkirk Microplate 349, 350, 363–366, 368, 369, 371
   -, lithosphere 371
   -, migration 368
SEPR (see *South East Pacific Rise*)
SEPR-PAR (see *South East Pacific Rise-Pacific-Antarctic Ridge*)
serpentine 329, 330
serpentinization 248, 338
Service Hydrographique et Océanographique de la Marine (SHOM) 13
sheet flow 166, 171, 172, 196–199, 203
shield 29, 56, 164, 196, 197, 202, 203, 215, 235, 255, 261, 266, 267, 269, 270, 275, 339, 340, 343
   -, volcano 164, 196, 197, 202, 215, 340
shielding 34, 353
   -, cadmium 353
ship sounding 3, 15, 16
SHOM (see *Service Hydrographique et Océanographique de la Marine*)
silica 165, 168, 169, 187, 261, 332, 333, 383, 388
silicate 300, 320
silicic rocks 444
silicon 168, 170–172, 187, 255, 326, 328, 337, 376, 388, 389
   -, oxyhydroxide 376
slab 149, 151, 184, 186, 239, 248, 378
   -, pull 151
smectite 320, 323, 329, 330
Sn wave 43
Society
   -, Archipelago 157, 158, 213
      -, map 158
   -, basic magmas 262
   -, hotspot 4, 432
      -, activity 220
      -, bathymetric map 166
      -, lavas 450
      -, magmatic activity 55
      -, regions 160, 169, 196
      -, sea floor fabric 191
      -, seismic activity 29, 36, 55, 67
      -, volcanic edifices 165
      -, volcanics 450
   -, Islands 14, 21, 22, 35, 67, 96, 97, 99, 101, 157, 210–212, 214, 218, 260, 261, 263, 376
      -, bathymetric map 211, 212
      -, chain 67, 96, 99, 158, 164, 192, 210, 215, 235, 260, 262
      -, geochemistry 210
      -, geochronology 210

-, geology 210
-, landslides 214
-, volcanism 210, 260
-, landslide classification 232
-, magmas 280
-, regions 180
-, Seamounts 261
-, swells 18
-, volcanoes 262
SOFAR (see *sound fixing and ranging*)
solidification 432
   -, magmatic 197
sonar systems 3, 9
sound fixing and ranging (SOFAR) 30, 34, 68
South Alignment (Austral Islands) 59
South East Pacific 6, 96, 97, 253–255, 274, 298, 349, 370, 456
   -, hotspot
      -, chains 254
      -, petrogenesis 274
   -, intraplate volcanoes 254
   -, mantle plumes 370
South East Pacific Rise (SEPR) 96, 97
South East Pacific Rise-Pacific-Antarctic Ridge (SEPR-PAR) 445, 456
   -, volcanics 442, 446
South Indian Ocean 84
South Kolbeinsey Ridge 295
South Pacific Ocean 4, 5, 15, 17, 18, 29, 35, 96, 100, 157, 178, 179, 253, 254, 259, 275, 278, 279, 309, 310, 318, 319, 407, 411, 426
   -, earthquake detection 4
   -, floor 310
   -, plate boundaries 18
   -, satellite altimetry 259
   -, Superswell 15, 96, 100, 254, 275, 278
Southern Cooks 259
Southwest Indian Ridge 333, 335, 336
Southwest Rift Zone, Mauna Loa 339
spatter cone 64, 65, 69, 314, 315
specimen 320, 329, 330, 337
spectrometer 315, 353, 410
spinel 436, 437
spreading
   -, activity 191
   -, center 1, 9, 83, 89, 90, 94–96, 109, 111, 116, 120, 121, 125, 127, 128, 131, 136, 190, 191, 196, 200, 239, 273, 310, 327, 350–352, 359, 363–367, 369, 371, 375
      -, East Indian Ocean 83
      -, East Rift (Easter Microplate) 273
      -, Galapagos 398
      -, reorganizations 9
   -, rate 121, 145, 290, 291, 294, 295, 298, 300, 302, 309
      -, half- 121
   -, ridge 68, 89, 158, 166, 187, 189–191, 197, 202, 290, 298, 302, 327, 336, 341, 431
      -, axis basalt (MORB) 437
      -, lava 445, 446
      -, magmatism 158, 449
      -, provinces 432
      -, volcanics 447, 455
St. Paul transform fault 333

478 Index

stalactite 199
starfish
-, morphology 168
-, shape 195
-, type 192
stratification 370
stratigraphy 254, 257, 264
-, volcanic 454, 456
strombolian ejecta 171, 182
strontium 178, 239, 241–243, 245, 249, 256, 262, 273, 296, 297, 302, 331, 341, 390
subduction 5, 73, 90, 144, 149–151, 153, 239–243, 245–249, 256, 262, 278, 280, 285, 286
-, Pacific Plate 151
-, zone 149–151, 153, 239, 241, 248, 285, 286
-, dehydration 241–243, 245, 246, 248, 249
submarine
-, eruption 33, 34, 57, 58, 61, 68, 312
-, volcanic 30
-, intraplate volcanoes 157
-, iron deposits 375
-, landslides 209, 210, 217, 224
-, French Polynesia 209
-, manganese deposits 375
-, minerals 9
-, noise 32, 34
-, photographs 65
-, topography 9
-, volcanic
-, activity 29, 30
-, edifices 35, 175, 181
-, volcanoes 6, 165, 202, 235, 269, 309, 310, 383
-, construction 6
submersible
-, Cyana 3, 158, 163, 166–168, 171–173, 203, 312
-, Nautile 3, 169, 171–173, 178, 179, 203, 376, 378, 410, 411, 419, 426
Suiko Seamount 143, 145
sulfide 59, 315, 320, 331, 384, 396
summit
-, collapse 195
-, flow 171
superchron, Cretaceous 19
supersaturation 383
Superswell 15–17, 96, 97, 100, 253, 254, 260, 275, 278, 280
-, bathymetric
-, anomaly 280
-, expression 15
-, hotspots 275, 278
-, mantle 260, 278, 280
swarm 30, 33, 36–41, 43–45, 48–63, 67, 68, 174, 202, 309, 312, 344
-, duration
-, Macdonald Seamount 62
-, dyke 202
-, evolution 45
-, Mehetia 45
-, seismic 60
-, Teahitia 41
swell 3, 17–19, 35, 69, 84, 89, 91, 92, 94–96, 100, 103, 106, 112, 120, 133, 136, 161, 180, 247
-, hotspot 18, 84, 92, 94, 95, 100, 105, 107, 136
-, midplate 17, 18

T

Tahaa 217, 235
-, Islands 217
Tahiti
-, age 255
-, crustal structure 100
-, Island 67, 164, 215, 216, 235, 312
-, Iti 215
-, landslide deposits 216
-, Nui 215, 261
-, Seamounts 261
-, seismic stations 29
-, southern landslide 233, 234
-, station 33
-, Volcano 67, 261
Tahuata 267
Taiarapu Peninsula (Tahiti) 161, 164
talc 329, 330, 337
tantalum 239, 241, 242, 249, 262, 273, 296
Tarava Seamounts 21, 22
Taukina Seamounts 20, 22, 369
Taylor Creek Rhyolite sanidine (TCR) 352, 353, 355
-, monitor age 352, 355
TCR (see Taylor Creek Rhyolite sanidine)
Teahitia
-, earthquake activity 55
-, edifices 33
-, recent activity 67
-, Seamount 29, 37, 49, 58, 63, 157, 165, 166, 170, 264, 408, 409, 411–417, 422–426
-, gas sample 425
-, hydrothermal fluid 412–417
-, hydrothermal methane 426
-, vent fluids 424
-, summit area
-, bathymetry map 422
-, swarm 30, 40, 41, 44, 49, 50, 52, 54, 55, 57
-, vent waters
-, rare earth elements 393
-, Volcano 30, 32, 66, 163, 168
tectonic
-, lineation 9, 202
-, plume 285
-, readjustments 58, 174
tectonism 45
-, extensional 45
Tedside Volcanics 183, 203
temperature
-, anomaly 411
-, mantle 247, 278
-, profile 418, 419, 422
tephrite 168, 177, 261, 340, 433
-, composition 437
terbium 394
Terevaka Fracture Zone 333
Tertiary 144
thermal
-, anomaly 4
-, boundary layer 149, 150, 289
-, buoyancy 17, 149
-, convection 285
-, rejuvenation 94, 103, 116, 133
Thingmuli Volcano 446

tholeiite 255, 257, 263, 266, 269, 270, 273, 327, 331–333, 339, 340
thorium 239, 241, 242, 249, 296, 388, 390
–, dating 381
–, -/uranium dating 381
Ti-magnetite 433
time $t^{1/2}$ law 133
titanite 330
titanium 178, 239, 241, 242, 249, 273, 300, 320, 326, 328, 331, 336, 337, 341, 343, 387, 390
titano-magnetite 447
tomography 18, 246, 339
Tonga 35, 392, 396
–, -Kermadec Ridge 392
–, Ridge 396
topography 15, 76, 149, 163, 180, 201, 213, 290, 292, 295, 296, 327
–, sea floor 349
–, submarine 9
Tori-Shima Island 35
total alkali 332, 333
TPW (see *true polar wander*)
trace element 240, 242, 249, 253, 257, 259, 262, 263, 267–270, 278, 296, 315, 341, 375, 387, 388, 390, 391, 398
–, concentration 387, 388
–, distribution 391
–, enrichment 278
–, patterns 257, 262, 263, 267
–, variations 269
tracer 408
trachy-andesite 433, 435, 444, 447
–, composition 447
trachybasalt 433, 435, 444, 447
–, composition 447
trachyte 165, 168–170, 182, 201, 269, 273, 433, 435, 444, 447
–, composition 449
transform fault 194, 310, 333, 368
tremolite 328
tremor 30, 32, 33, 41, 49–51, 56–58, 60, 70
–, seismic 30, 32, 41, 49, 50, 56, 60
trench 149–151, 153
–, jam 149–151, 153
triple junction 19
Tristan hotspot 278
troctolite 315, 320, 321, 323, 329, 331, 336, 338, 339, 341, 343
true polar wander (TPW) 146, 148
truncated 170, 183, 186, 194–196, 202
–, edifice 195
–, Rocald 196
truncation 195
–, intraplate edifice 195
tschermakite 328
tsunami 9
Tuamotu 9, 23, 60, 96, 97, 151, 273
–, hotspot 144
–, Islands 273
–, plateau 96, 97
–, Seamount chain 151
tube
–, giant 444
–, lava 444

Tubuai 18, 60, 175, 211, 224–226, 235, 259
–, bathymetric map 225
–, Island 18, 60, 175, 211, 224–226, 235, 259
Tupai Atoll 220, 221, 231, 235
–, landslide deposits 220
turbidites 164
turbidity 222
Turoi Seamount 163, 165–167, 171, 172, 199
Tyrrhenian Sea 396
–, Mn crusts 396

U

Ua Pou 266
Ukelayat flysch sandstones 144
undersaturation 255, 261, 266
upwelling 5, 37, 57, 69, 78, 88–90, 94, 96, 99, 124, 161, 191, 239, 254, 288, 289, 293, 294, 298, 299, 302, 304, 350, 371
–, Foundation mantle plume 350
–, mantle 239, 254, 288, 293, 294, 304
–, passive 289
uranium 3, 82, 239, 241, 242, 249, 273, 296, 313, 381, 389, 390
–, dating 381

V

Va'a Tau Piti Ridge 23
vacuole 320, 323, 330
Vaihoa 60
Valu Fa Ridge 392
vapor 59
–, hydrogen sulfide 59
vaporization 34, 68
Vatnajökull highlands 120
vein 249, 287, 288, 298–301, 320, 323, 328, 331, 337, 344, 411
veinlet 315, 320, 321, 329, 331
velocity
–, ground 60
–, plate 109, 266
–, seismic 5, 15, 16, 74, 82, 83, 86, 88, 116, 124, 233
–, crustal 88
vent 6, 57, 65, 166, 170–172, 186, 195, 199, 339, 340, 376, 382–384, 393, 396, 398, 407, 408, 410, 411, 417, 421, 423, 424, 426
–, fluids 383, 384, 393, 398, 410, 422, 424
–, hydrothermal 6, 382, 407
–, sources 408
–, systems 407, 408
venting, methane 407
vernadite 375
VERs (see *volcanic elongate ridges*)
vesicle 197, 314
–, gas 197
vesicularity, rock 197
viscosity, mantle 289, 290
volatile 6, 197, 201, 249, 286–288, 290, 298, 300, 301, 315, 396
volcanic
–, activity 3, 15, 29, 30, 33, 56, 57, 59, 68, 94, 99, 102, 109, 157, 170–175, 180, 183, 191, 196, 214, 216, 217, 229, 236, 254, 255, 266, 312, 315, 342, 368, 376

480 Index

-, sea floor 3, 376
-, age 143
-, alignments 157
-, aprons 313
-, ash 59, 174, 177, 183
-, bombs 199, 315
-, channels 201
-, cones 25, 161, 164, 165, 168, 180, 190, 201, 202
-, construction 3, 4, 189–191, 196, 197, 203, 298
-, edifice 4, 35–38, 45, 55, 57, 59, 66, 67, 89, 101, 107, 112, 116, 117, 158, 160, 161, 163, 164, 169, 174, 175, 178–181, 192, 194, 196, 199–201, 210, 272, 315, 340, 376
-, ejecta 173, 177, 183, 188, 199, 203, 315
-, elongate ridges (VERs) 350–352, 359, 363, 366–372
  -, Foundation Chain 350, 359, 366
  -, volcanism 352, 359, 366, 367
-, eruptions 9, 199, 396
-, events 6, 175, 342
-, fountaining 33
-, glass 164
-, island chains 96, 194
-, islands 1, 158, 178, 179, 311, 331
-, morphology 22, 23, 66, 175
-, ridges 4, 22, 161, 301
-, seismicity 32, 52, 55
volcaniclastic 309, 314, 379, 380, 388–390, 393, 394, 398
volcanics 165, 168, 170, 177, 183, 187, 190, 196, 197, 200, 202, 246, 253, 254, 259, 267, 268, 292, 301, 309, 311, 315–317, 320, 331–333, 342, 343
-, classification 432
-, composition 442
-, intraplate hotspots 455
-, mineral assemblages 435
-, spreading-ridge 447
-, volume 202
volcanism
-, age-progressive 264, 265
-, arc, Chukotka 153
-, basaltic 239
-, Emperor Seamount 145
-, explosive 4
-, Foundation Chain 350, 366, 370
-, hotspot 3, 6, 190, 196, 200, 203, 210, 240, 246, 272, 351, 365, 366, 368, 369, 371, 372, 375, 426
-, intraplate 1
  -, Foundation 367
-, migration 363
  -, rate 352, 365
-, quiet 4
-, underwater 32, 34, 35, 68, 431
  -, interpretation 29
Volcano Island 35
volcano
-, abyssal hill 165
-, axial 383
-, Hawaiian-type 63
-, HIMU 259

-, hotspot 5, 158, 161, 200, 202, 254, 255, 269, 301, 319, 376, 395
  -, evolution 254
-, intraplate 4, 6, 157, 196, 202, 254, 338, 343, 388
  -, active 4
  -, hotspot 449
-, submarine
  -, construction 6
-, types of activities 4
volcanology 210
-, Society Islands 210

W

Walvis Ridge 133
warming 260
wave 15, 29, 30, 32–35, 43–45, 48–50, 55–61, 63, 66, 68, 74, 76, 81, 125, 222, 227, 228, 230, 309, 312
-, sediment 222, 227, 228, 230
-, seismic 30
  -, conventional 30, 32, 60
waveform 49, 55, 56
wavelength 14, 187, 227, 230
weathering 186, 337
wehrlite 311, 339
Western Fjords 131
Western Volcanic Zone 120
wet spots 286
Windward group 67
Woodlark Basin 389

X

xenocryst 311
xenolith 5, 311, 320, 329, 331–335, 339–341, 343
-, felsic 332, 333
-, gabbroic
  -, Canary Islands 340, 341
  -, Hawaii 339
-, hornblendite 334, 335, 341
X-ray diffraction patterns
-, iron oxide crust 381

Y

Young
-, Seamount 186
-, Volcano 187, 189, 190
  -, MORB 196

Z

zeolite 323, 330, 337, 338
zinc 375, 387, 388, 391, 398
zircon 241
zirconium 239, 241, 242, 249, 331
zone
-, hotspot volcanism 351, 365, 366, 368
-, subduction 149–151, 153, 239, 241, 248, 285, 286

Printing: Mercedes-Druck, Berlin
Binding: Stein + Lehmann, Berlin